D1266738

The chemistry of
the nitro and nitroso groups

THE CHEMISTRY OF FUNCTIONAL GROUPS

A series of advanced treatises under the general editorship of
Professor Saul Patai

The chemistry of alkenes (published)
The chemistry of the carbonyl group (published)
The chemistry of the ether linkage (published)
The chemistry of the amino group (published)
The chemistry of the nitro and nitroso groups (published in 2 parts)
The chemistry of carboxylic acids and esters (published)
The chemistry of the carbon–nitrogen double bond (published)

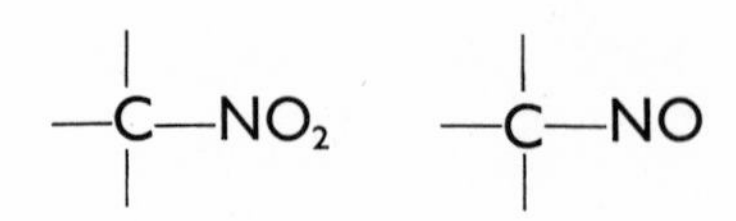

The chemistry of
the nitro and nitroso groups

Edited by

HENRY FEUER

Purdue University
Lafayette, Indiana

Part 2

1970

INTERSCIENCE PUBLISHERS

a division of John Wiley & Sons

NEW YORK—LONDON—SYDNEY—TORONTO

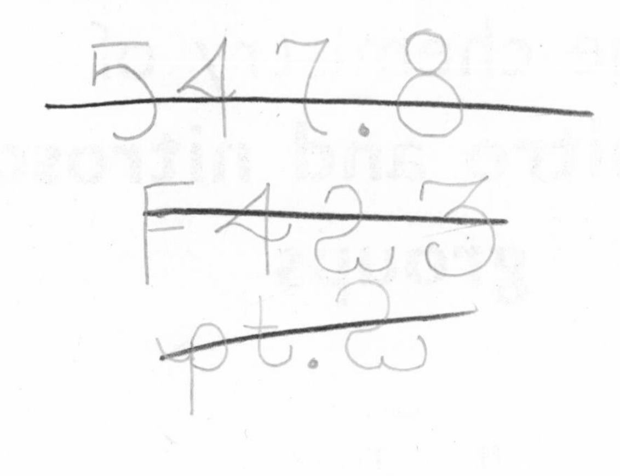

Library of Congress Catalog Card Number: 68-29395

ISBN 0-471-25791-5

Printed in the United States of America

10 9 8 7 6 5 4 3 2 1

The Chemistry of Functional Groups

Preface to the Series

The series 'The Chemistry of Functional Groups' is planned to cover in each volume all aspects of the chemistry of one of the important functional groups in organic chemistry. The emphasis is laid on the functional group treated and on the effects which it exerts on the chemical and physical properties, primarily in the immediate vicinity of the group in question, and secondarily on the behavior of the whole molecule. For instance, the volume *The Chemistry of the Ether Linkage* deals with reactions in which the C–O–C group is involved, as well as with the effects of the C–O–C group on the reactions of alkyl or aryl groups connected to the ether oxygen. It is the purpose of the volume to give a complete coverage of all properties and reactions of ethers in as far as these depend on the presence of the ether group, but the primary subject matter is not the whole molecule, but the C–O–C functional group.

A further restriction in the treatment of the various functional groups in these volumes is that material included in easily and generally available secondary or tertiary sources, such as *Chemical Reviews*, *Quarterly Reviews*, *Organic Reactions*, various 'Advances' and 'Progress' series as well as textbooks (i.e. in books which are usually found in the chemical libraries of universities and research institutes (should not, as a rule, be repeated in detail, unless it is necessary for the balanced treatment of the subject. Therefore each of the authors is asked *not* to give an encyclopedic coverage of his subject, but to concentrate on the most important recent developments and mainly on material that has not been adequately covered by reviews or other secondary sources by the time of writing of the chapter, and to address himself to a reader who is assumed to be at a fairly advanced post-graduate level.

With these restrictions, it is realized that no plan can be devised for a volume that would give a *complete* coverage of the subject with *no* overlap between the chapters, while at the same time preserving the readability of the text. The Editor set himself the goal of attaining *reasonable* coverage with *moderate* overlap, with a minimum of cross-references between the chapters of each volume. In this manner, sufficient freedom is given to each author to produce readable quasimonographic chapters.

The general plan of each volume includes the following main sections:

(a) An introductory chapter dealing with the general and theoretical aspects of the group.

(b) One or more chapters dealing with the formation of the functional group in question, either from groups present in the molecule, or by introducing the new group directly or indirectly.

(c) Chapters describing the characterization and characteristics of the functional groups, i.e. a chapter dealing with qualitative and quantitative methods of determination including chemical and physical methods, ultraviolet, infrared, nuclear magnetic resonance, and mass spectra; a chapter dealing with activating and directive effects exerted by the group and/or a chapter on the basicity, acidity or complex-forming ability of the group (if applicable).

(d) Chapters on the reactions, transformations and rearrangements which the functional group can undergo, either alone or in conjunction with other reagents.

(c) Special topics which do not fit any of the above sections, such as photochemistry, radiation chemistry, biochemical formations and reactions. Depending on the nature of each functional group treated, these special topics may include short monographs on related functional groups on which no separate volume is planned (e.g. a chapter on 'Thioketones' is included in the volume *The Chemistry of the Carbonyl Group*, and a chapter on 'Ketenes' is included in the volume *The Chemistry of Alkenes*). In other cases, certain compounds, though containing only the functional group of the title, may have special features so as to be best treated in a separate chapter as e.g. 'Polyethers' in *The Chemistry of the Ether Linkage*, or 'Tetraaminoethylenes' in *The Chemistry of the Amino Group*.

This plan entails that the breadth, depth, and thought-provoking nature of each chapter will differ with the views and inclination of the author and the presentation will necessarily be somewhat uneven. Moreover, a serious problem is caused by authors who deliver their manuscript late or not at all. In order to overcome this problem at least to some extent, it was decided to publish certain volumes in several parts, without giving consideration to the originally planned logical order of the chapters. If after the appearance of the originally planned parts of a volume, it is found that either owing to non-delivery of chapters, or to new developments in the subject, sufficient material has accumulated for publication of an additional part, this will be done as soon as possible.

It is hoped that the series 'The Chemistry of Functional Groups' will include the titles listed below:

The Chemistry of the Alkenes (published)
The Chemistry of the Carbonyl Group (published)
The Chemistry of the Ether Linkage (published)
The Chemistry of the Amino Group (published)
The Chemistry of the Nitro and Nitroso Groups parts 1 *and* 2 *(published)*
The Chemistry of Carboxylic Acids and Esters (published)
The Chemistry of the Carbon–Nitrogen Double Bond (published)
The Chemistry of the Cyano Group (in press)
The Chemistry of the Carboxamido Group (in press)
The Chemistry of the Carbon–Halogen Bond
The Chemistry of the Hydroxyl Group (in preparation)
The Chemistry of the Carbon–Carbon Triple Bond
The Chemistry of the Azido Group (in preparation)
The Chemistry of Imidoates and Amidines
The Chemistry of the Thiol Group
The Chemistry of the Hydrazo, Azo, and Azoxy Groups
The Chemistry of Carbonyl Halides
The Chemistry of the SO, SO_2, $-SO_2H$, *and* $-SO_3H$ *Groups*
The Chemistry of the –OCN, –NCO, and –SCN Groups
The Chemistry of the $-PO_3H_2$ *and Related Groups*

Advice or criticism regarding the plan and execution of this series will be welcomed by the Editor.

The publication of this series would never have started, let alone continued, without the support of many persons. First and foremost among these is Dr. Arnold Weissberger, whose reassurance and trust encouraged me to tackle this task, and who continues to help and advise me. The efficient and patient cooperation of several staff-members of the Publisher also rendered me invaluable aid (but unfortunately their code of ethics does not allow me to thank them by name). Many of my friends and colleagues in Jerusalem helped me in the solution of various major and minor matters and my thanks are due especially to Prof. Y. Liwschitz, Dr. Z. Rappoport, and Dr. J. Zabicky. Carrying out such a long-range project would be quite impossible without the non-professional but none the less essential participation and partnership of my wife.

The Hebrew University
Jerusalem, ISRAEL

SAUL PATAI

Foreword

The concept of this book arose from several discussions with Professor Saul Patai during my sabbatical leave at the Hebrew University of Jerusalem in the Spring of 1964. While disclosing his plans to edit a series of treatises concerned with the chemistry of functional groups, inclusion of the nitro and nitroso groups as the subject-matter for one of the volumes was brought forward. I accepted the editorship of such a treatise as part of the series because I considered it very worthwhile that up-to-date discussions on the theoretical, physical, and mechanistic aspects of these groups be unified in one publication by active workers in the field. For although several review articles, proceedings of various symposia, and isolated chapters in various books have concerned themselves with certain aspects of the chemistry of the nitro and nitroso groups—an active and exciting field of research—no self-contained book on the subject has been available.

As in the already-published books of this series, the subject-matter in this treatise has been considered from the viewpoint of the functional group. Instead of an encyclopedic coverage of all known reactions and compounds, the emphasis has been placed on basic principles, mechanisms, and recent advances in both theory and practice. It is hoped that by choosing this approach, a broad and concise picture of the importance of the nitro and nitroso groups has been attained.

The editing and publishing of a book which is made up of contributions from several authors are usually delayed by the fact that the deadline agreed upon is exceeded by some of the contributors. Such delay is unfortunate because it can sometimes result in obsolescence on some parts of a manuscript. To minimize such possibilities, which invariably occur when discussions in active fields of research are involved, and to keep the format of the book to a manageable size, it was decided to publish the treatise in two volumes.

It is with great pleasure that I acknowledge the cooperation of Professor Saul Patai, and the advice and suggestions in editorial matters of the Publishers.

I also express my gratitude to Dr. M. Auerbach, who did most of the painstaking work involved in preparing both the Author and Subject Index.

Lafayette, January 1970 HENRY FEUER

Authors of Part 2

Hans. H. Baer	Department of Chemistry, University of Ottawa, Ottawa, Ontario, Canada
Thomas N. Hall	U.S. Naval Ordnance Laboratory, White Oak, Silver Spring, Maryland
Lloyd A. Kaplan	Chemistry Research Department, U.S. Naval Ordnance Laboratory, Silver Spring, Maryland
Chester F. Porański, Jr.	U.S. Naval Research Laboratory, Washington, D.C.
Tadeusz Urbański	Warsaw Institute of Technology (Politechnika), Warszawa, Poland
Ljerka Urbas	Department of Chemistry, University of Ottawa, Ottawa, Ontario, Canada
Robert L. VanEtten	Purdue University, Lafayette, Indiana
Jan Venulet	Drug Research Institute, Warszawa, Poland
William M. Weaver	Department of Chemistry, John Carroll University, Cleveland, Ohio

Contents—Part 2

Contents—Part 1

The chemistry of

the nitro and nitroso groups

CHAPTER I

Introduction of the nitro group into aromatic systems

William M. Weaver

Department of Chemistry,
John Carroll University,
Cleveland, Ohio

The preparation of aromatic nitro compounds is most often achieved with reagents capable of forming the nitronium ion, NO_2^+. The reagents capable of producing this ion are numerous, and the conditions employed are as varied as the aromatics being nitrated. Besides the common sulfuric acid–nitric acid combination for the production of NO_2^+, nitronium fluoroborate and other nitronium salts, nitrate esters, N_2O_4, N_2O_5, and metal nitrates plus sulfuric or Lewis acids are reagents which are believed to involve nitronium ion as the nitrating species.

Nitric acid in acetic anhydride might at first appear to behave similarly. Yet, the resulting acetyl nitrate is somewhat anomalous in its action. Moreover, the active nitrating agent from nitric acid in acetic anhydride is uncertain, although it is probably protonated acetyl nitrate and not nitronium ion.

The atypical procedures for the introduction of a nitro group into aromatic systems consist of oxidation of nitroso and amino compounds, replacement of diazonium ion, rearrangement of nitramines, nucleophilic displacement by aryl anions on nitrate esters and other suitable reagents such as N_2O_4 and tetranitromethane, and free-radical processes involving $\cdot NO_2$. These procedures are employed most often to overcome problems of orientation, but the sensitivity of some aromatic systems to oxidation by the usual nitrating media necessitate other methods of preparing the nitroarene.

I. ELECTROPHILIC NITRATION

Preparative electrophilic nitration can be done in a variety of media, but those most often employed are mixed acid (nitric plus sulfuric), aqueous nitric acid, nitric acid in acetic acid, and nitric acid in acetic anhydride. However, in theoretical and mechanistic studies, the number of electrophilic nitrating agents and the variety of solvents employed are numerous. Nitronium tetrafluoroborate (and similar salts of P, As, and Sb) and dinitrogen pentoxide are excellent nitrating agents whose preparative value would be greater if they were more readily available.

A. Theory

As most commonly effected, nitration of aromatics is a typical electrophilic substitution by the nitronium ion, NO_2^+ (equation 1).

$$C_6H_6 + NO_2^+ \rightleftharpoons \underset{(\mathbf{1})}{[C_6H_6NO_2]^+} \xrightarrow{B} C_6H_5NO_2 + HB \quad (1)$$

The formation of the σ-complex (**1**) is an ionic bimolecular process sensitive to the individual reactivity of a particular aromatic and to solvation effects. The exact products produced are governed by the typical rules of orientation in electrophilic substitution, and strong solvation of the nitronium ion retards the rate of nitration. The velocity of the formation of the σ-complex is very rapid; where the nitronium ion is involved in complex equilibrium with the nitration medium, this formation is too fast to be rate determining. Because the loss of the proton from the σ-complex is also too fast to be even partly rate determining, a primary deuterium isotope effect is not observed in nitration. An exception has been found by Myhre[1] in the nitration of *sym*-nitrotri-*t*-butylbenzene wherein steric strain causes sufficient reversibility to the formation of the σ-complex to make the isotope effect detectable.

The major body of evidence in favor of nitration via the nitronium ion was first amassed by C. K. Ingold and coworkers in 1950 and has been summarized in numerous places[2–5]. Ingold himself was fully aware that dinitrogen pentoxide in carbon tetrachloride and acyl nitrates behave anomalously, and he maintained an open mind toward the possibility that species other than the nitronium ion might be responsible for nitration. Some writers[5–7], however, have tried to retain a simple picture with nitration always occurring via the nitronium ion. Such a simple concept seems unrealistic. Not only must one consider the nitrating agent to vary with the nitrating medium, but the transition to the σ-complex must be considered in greater detail.

The σ-complex or Wheland intermediate[8] is representable as a minimum in an energy–reaction coordinate profile and can be isolated under proper conditions; a σ-complex has been isolated from the reaction of trifluoromethylbenzene, nitryl fluoride, and boron trifluoride at −50° [9]. Even prior to its formation is the possibility of a less stable intermediate—a π-complex[10–11] —a multicentered, less directed interaction[12] of the nitronium ion with the electrons of the aromatics. Olah[13] has given evidence that formation of π-complexes may be rate determining in nitration by nitronium fluoroborate in sulfolane.

But more correctly, rates are controlled by activation energies which are determined by energy maxima, transition states, and not by intermediates, which are energy minima. Although the presence of an intermediate is useful in elucidating a reaction mechanism, an intermediate only restricts and hints at the dynamic process that leads to it[14]. In nitration this process is electrophilic yet it must also be partly nucleophilic because the electrophile receives its electron-pair from a donor, a nucleophile. More importantly, an electrophile sufficiently electrophilic to disrupt the resonance stabilization energy of benzene is going to be associated with some electron-rich species which must be displaced during formation of the σ-complex. Thus, in aromatic bromination there is a clear distinction between bromination by protonated hypobromous acid and by molecular bromine wherein displacement of the bromide ion by the electrons of the aromatic is a significant portion of the transition state[15] (equation 2). From another viewpoint, electro-

$$C_6H_6 + Br_2 \longrightarrow [C_6H_6^{\delta+}\cdots Br\text{---}Br^{\delta-}] \longrightarrow C_6H_6Br^{+} + Br^{\ominus} \quad (2)$$

philes, depending on the medium, can be in various ground states. The two extremes are: (1) the electrophile is coordinated to a base by a directed covalent bond and is essentially a molecular entity and (2) the electrophile is a 'free' cation surrounded by several basic species through non-directed electrostatic interaction in a manner analogous to a cation in a crystal or an alkali metal cation in solution in water. Intermediate between these extremes are ion-pairs and a whole continuum of weak to strong, non-directed to directed, interactions of an electrophile with electron-donating substances.

At this time, three nitration systems have been sufficiently studied to warrant the conclusion that at least three distinctive electrophilic nitrating agents exist:

1. The complex fluoranion nitronium salts, particularly nitronium fluoroborate in sulfolane which acts as a solvated ion pair.
2. Nitric acid in acetic anhydride which reacts to produce acetyl nitrate which nitrates via its protonated form.
3. Nitric acid in concentrated sulfuric acid—mixed acid—which gives the solvated nitronium ion in a protic medium of high dielectric.

That these three systems contain distinctive nitrating entities is shown by their differences in both substrate selectivity and positional selectivity.

Competitive nitration with benzene shows that both nitronium fluoroborate and mixed acid have low substrate selectivity. The rate ratios are close to one (Table 1)[16]. Yet nitration in acetic anhydride gives much greater substrate selectivity, toluene being 27 times more reactive than benzene and biphenyl 16 times more reactive. The low substrate selectivity by nitronium fluoroborate is interpreted by Olah as evidence of the transition to the π-complex being rate determining. This interpretation is quite reasonable if one assumes that the nitrating entity is in a high ground state so that the rate-controlling transition state is closer to the starting materials. In acetic anhydride the actual nitrating agent is in a much lower

TABLE 1. Isomer distribution and relative rate for nitration with nitronium fluoroborate, mixed acid, and protonated acetyl nitrate[16].

		NO_2BF_4(25°)/ Sulfolane	HNO_3/ H_2SO_4(25°)	HNO_3/Ac_2O(0°)
o-Xylene	% 3 (*o* and *m*)	79.7[16a]	55[16c]	33[16d]
	% 4 (*p* and *m*)	20.3	45	67
	k_{Ar}/k_B	1.75[16b]	1.02[16c]	—
Biphenyl	% 2	75[16b]	37[16e] (35–40°)	68[16f] (58[16g], 53°)
	% 4	23.8	63	32 (42, 53°)
	k_{Ar}/k_B	2.08[16b]	—	16[16f]
Toluene	% 2	65.4[16a]	56.4[16c]	61.4[16h]
	% 4	31.8	38.8	37.0
	k_{Ar}/k_B	1.67[16b]	1.24[16c]	27
Chlorobenzene	% 2	22.7[16a]	30[16i]	10[16j]
	% 4	76.6	70	90
	k_{Ar}/k_B	0.14[16a]	—	0.033[16k]
Acetanilide	% 2	—	19[16l]	68[16l] (20°)
	% 4	—	79(20°)	30(20°)
Anisole	% 2	69[16m]	31[16n]	71[16o]
	% 4	31	67	28

energy state and the rate-controlling transition resembles the σ-complex. Figure 1 gives a pictorial representation of this concept.

Isomer distribution in the products is also quite different depending on the nitrating medium. This positional selectivity of various nitrating media is quite obvious in the nitration of *o*-xylene but only slightly evident in the nitration of toluene. The relatively invariant isomer distribution in the nitration of toluene is often quoted[17] as evidence for a single active nitrating entity, the nitronium ion. However, this insensitivity of toluene is general for all electrophilic substitutions and is characteristic of all monoalkylated benzenes. Knowles, Norman, and Radda[18] ascribed this insensitivity of toluene to the low polarizability of an alkyl group to the electron demands of an electrophilic reagent. Therefore, the fact that toluene always gives *ca.* 60 % *o*-nitrotoluene with varied nitrating agents is irrelevant.

Kinetic evidence for the nitronium ion as the active nitrating agent comes from the nitration of alkylbenzenes with nitric acid in either acetic acid or nitromethane. In an excess of nitric acid the rate is zero order, catalyzed by strong mineral acid and retarded by added nitrate ion without altering the zero order of the reaction. This finding was interpreted[19] as showing that the slow step in the reaction was formation of the nitronium ion (equation 3).

$$\mathrm{HONO_2 + HA \longrightarrow HO\overset{H}{\overset{|}{\oplus}}NO_2 + A^{\ominus}} \quad \text{(fast)}$$

$$\mathrm{\underset{\underset{H}{|}}{HO}{\oplus}NO_2 \longrightarrow H_2O + NO_2^{\oplus}} \quad \text{(slow)} \tag{3}$$

$$\mathrm{NO_2^{\oplus} + C_6H_6 \longrightarrow C_6H_5NO_2} \quad \text{(fast)}$$

Identical kinetic behavior would be observed also if the nitracidium ion, $H_2O^{\oplus}NO_2$, reacted with acetic acid to produce protonated acetyl nitrate (equation 4).

$$\mathrm{H_2ONO_2^{\oplus} + CH_3{-}C({=}O){-}OH \longrightarrow CH_3{-}C(O{-}NO_2){=}\overset{\oplus}{O}{-}H + H_2O} \quad \text{(slow)} \tag{4}$$

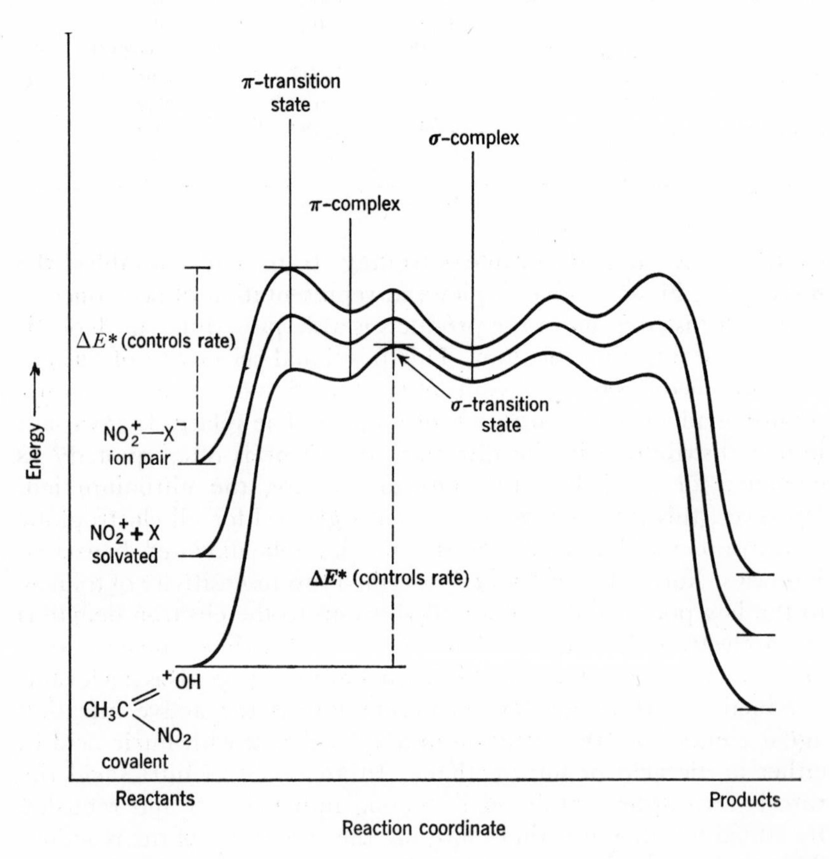

FIGURE 1. Effect of changing nitrating species on the rate-controlling transition state.

Nitronium fluoroborate reacts with explosive violence with acetic acid[20]. It seems unlikely that the nitronium ion can exist as such in acetic acid, and it is much more reasonable that protonated acetyl nitrate is also the active nitrating agent from nitric acid in acetic acid. Substrate selectivity for nitric acid in acetic acid is similar to that for acetyl nitrate, but comparative isomer distribution for a suitably sensitive substrate, other than toluene which is not subject to solvation itself, is unavailable. More work with *o*-xylene is needed.

Fisher, Packer, and Vaughan[16d], in reporting their evidence for protonated acetyl nitrate as the active nitrating agent in the nitration of *o*-xylene with nitric acid in acetic anhydride, point out that

TABLE 2. Relative rates and isomer distributions for the bromination and nitration of toluene[24].

	k_T/k_B	% *ortho*	% *meta*	% *para*
Bromination:				
85% HOAc, 25° [24a]	605	32.9	0.3	66.8
CF_3CO_2H, 25° [24b]	2580	17.6	0	82.4
Nitration:				
90% HOAc, 45° [24c]	24	56.5	3.5	40.0
CH_3NO_2, 30° [24d]	21	58.5	4.4	37.1
Ac_2O, 30° [24d]	23	58.4	4.4	37.2
Ac_2O, 0° [24d]	27	58.1	3.7	38.2
CF_3CO_2H, 25° [24b]	28	61.6	2.6	35.8

a second less active nitrating agent is probably present in the medium. This less active nitrating agent becomes apparent in the nitration of the more reactive substrate *m*-xylene for which the kinetics require a second term.

Azulene, a highly active aromatic compound, is successfully nitrated by cupric nitrate in acetic anhydride[22]. This system initially contains no protic hydrogens, so protonated acetyl nitrate as a nitrating agent is unlikely. However, acetyl nitrate itself is a reasonable nitrating species but one which is only active with very reactive arenes such as azulene and *m*-xylene.

Brown[23] has shown recently that trifluoroacetic acid is an effective solvent for electrophilic substitution. Typically, toluene is the substrate nitrated, but even though toluene is mostly insensitive to changes in nitrating medium, there is an unmistakable, albeit small, change in both substrate and positional selectivity with change in solvent. Table 2[24] includes bromination data along with the nitration data. The solvent change from acetic acid to trifluoroacetic acid produces a large effect in the bromination but only a slight effect with nitration. This lowered sensitivity in the nitration might be ascribed to a peculiarity of nitrogen in the plus-five oxidation state: regardless of whether the nitronium ion is complexed or free, the nitrogen always carries a formal positive charge. This is not the case with the bromonium ion. Where solvent complexed molecular bromine is the active brominating agent, a formal positive charge does not even reside on the halogen atom entering the aromatic compound (equations 5 and 6). The consequence is that

$$\oplus NO_2 + :A \longrightarrow \ominus \begin{matrix} O \\ \rangle N^{\oplus}\text{—}A^{\oplus} \\ O \end{matrix} \quad (5)$$

$$\begin{aligned} Br^{\oplus} + :A &\longrightarrow Br\text{—}A^{\oplus} \\ Br\text{–}Br + :A &\longrightarrow Br\text{–}Br^{\ominus}\text{—}A^{\oplus} \end{aligned} \quad (6)$$

nitration is 10^5–10^6 times faster than bromination, the specific rate for bromination in acetic acid being of the order of 10^{-10} mole l.$^{-1}$ sec^{-1}* and that of nitration by protonated acetyl nitrate in acetic acid[25] of the order of 10^{-4} mole l.$^{-1}$ sec^{-1}.

The low substrate sensitivity reported by Olah for nitronium fluoroborate as determined by competitive rate studies has been criticized[5,23,26]. The rate of mixing, in the rapid reaction with alkylbenzenes, seems to affect the competitive rate ratios. Nevertheless, this concern over the reliability of Olah's numbers should not obscure the fact that substrate selectivity does indeed vary with the nitrating medium. Ingold found small differences for the relative rates of nitration of toluene and benzene for nitric acid in acetic anhydride ($k_T/k_B = 23$) and nitric acid in nitromethane ($k_T/k_B = 21$)[24d]. This 10% difference can be attributed to experimental uncertainty, but when one compares toluene with *t*-butylbenzene in the same media, whose reactivities are more nearly alike, the percentage difference is greater and in the same direction for the two solvents: the ratio, in acetic anhydride ($k_T/k_{t\text{-BuB}} = 2.0$) being larger than in nitromethane ($k_T/k_{t\text{-BuB}} = 1.4$)[27]†.

* Calculated from the rate for bromination of benzene in trifluoroacetic acid, 7.6×10^{-7} l. mole^{-1} sec^{-1} and the 2500-fold rate difference between trifluoroacetic and acetic acid[23].

† Recently, C. A. Cupas and R. L. Pearson, reported that *N*-nitropyridinium tetrafluororoborates are effective nitrating agents of aromatic substrates in acetonitrile at 25° and show high substrate selectivity. For the 2,6-lutidine salt, k_T/k_B is 39, and the yield of the *ortho* isomer is 63.9%[166].

Large variation in isomer distribution with nitrating medium is observed with strongly activated substituted aromatics which have polarizable non-bonding electrons on the substituent such as anisole and other ethers, amines, and *N*-arylamides. Mixed acid nitration of this type of aromatic gives a higher proportion of *para*- than of *ortho*-nitro compound while nitration in acetic anhydride gives a very high proportion of *ortho* product, sometimes in excess of the 67 % statistically predicted for *ortho*/*para* activation.

To explain this '*ortho* effect' associated with nitrations in acetic anhydride both cyclic[28] (**2, 3**) and linear[29] coordination (**4** → **8**)

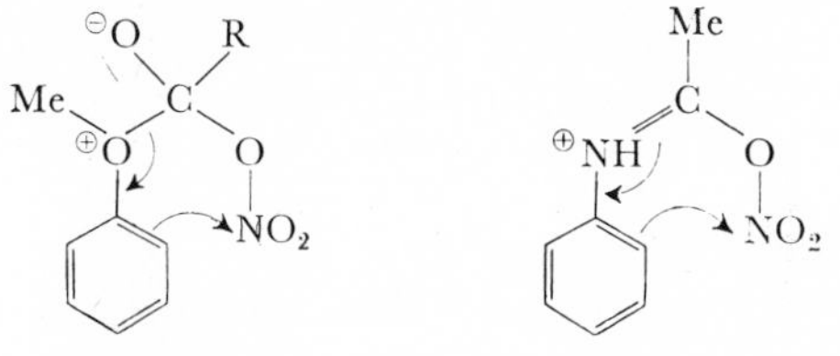

(**2**) (**3**)

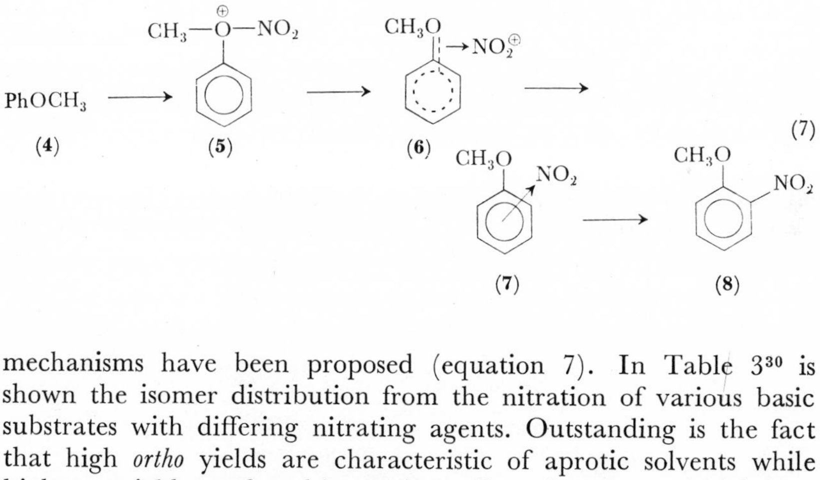

$PhOCH_3$ (**4**) → $CH_3—\overset{\oplus}{O}—NO_2$ (**5**) → $CH_3O \rightarrow NO_2^{\oplus}$ (**6**) → CH_3O, NO_2 (**7**) → CH_3O, NO_2 (**8**) (7)

mechanisms have been proposed (equation 7). In Table 3[30] is shown the isomer distribution from the nitration of various basic substrates with differing nitrating agents. Outstanding is the fact that high *ortho* yields are characteristic of aprotic solvents while high *para* yields are found in protic medium. It seems quite obvious that a protic medium is inhibiting *ortho* attack: solvation of the electron-rich atom by hydrogen bonding increasing the bulk in the vicinity of the *ortho* positions thereby causing a steric inhibition to *ortho* attack. The influence of steric factors is readily discernable in comparing the isomer distribution in the nitration of toluene, cumene, and *t*-butylbenzene (Table 4)[31].

TABLE 3. Isomer distribution for nitration of aromatics containing basic substituents.

Compound	Conditions	% *ortho*	% *para*	Ref
Anisole	HNO_3–H_2SO_4	31	67	30a
	HNO_3	40	58	30a
	HNO_3 in HOAc	44	55	30a
	NO_2BF_4 in sulfolane	69	31	30b
	HNO_3 in Ac_2O	71	28	30c
	$BzONO_2$ in MeCN	75	25	30c
Acetanilide	HNO_3–H_2SO_4	19	79	30d
	90% HNO_3	24	77	30e
	HNO_3 in Ac_2O	68	30	30d
Methyl phenethyl ether	HNO_3–H_2SO_4	32	59	30f
	HNO_3	40	53	30f
	HNO_3 in $MeNO_2$	41	56	30f
	HNO_3 in Ac_2O	62	34	30f
	$AcONO_2$ in MeCN	66	30	30f
	N_2O_5 in MeCN	69	28	30f
Benzeneboronic acid	HNO_3–H_2SO_4	22	5	30g
$PhB(OH)_2$	HNO_3 in Ac_2O	63	14	30g
(*meta* directed)				
Biphenyl	HNO_3–H_2SO_4	37	63	30h
	HNO_3 in HOAc	36	64	30i
	HNO_3 in Ac_2O	69	31	30j

This behavior of the more basic aromatics is quite similar to ambident anion alkylations and is indicative of a large nucleophilic contribution in the nitration process. Protic solvents inhibit alkylation of an ambident anion at the center of highest electron density[32]. Since electron density in a basic aromatic is concentrated near the

TABLE 4. Isomer distribution in the mononitration of monoalkylbenzenes.

	$PhCH_3$[31a]	$PhCH(CH_3)_2$[31b]	$PhC(CH_3)_3$[31a]
% *ortho*	63	28	10
% *para*	34	68	80

substituent, nitration would be anticipated to be on the substituent, then on the *ortho* position and finally at the *para* position when hydrogen bonding to the solvent is absent. This reaches its extreme in amines. Pyridine, when nitrated with nitronium fluoroborate, gives only the *N*-nitropyridinium fluoroborate[33].

Anilines, also, are *N*-nitrated in aprotic medium[34], the *N*-nitroamines being obtained from the action of dinitrogen pentoxide in carbon tetrachloride or by the addition of solid anilinium nitrate to acetic anhydride at $-10°$. If the ring is deactivated, as in 2,4-dinitromethylaniline, *N*-nitration even occurs in protic medium[35]. The *N*-nitramines will rearrange to the ring-nitrated anilines on treatment with sulfuric or hydrochloric acid. These rearrangements go essentially *ortho*; treating phenylnitramine with 74% sulphuric acid at $-20°$ gives 95% *o*-nitroaniline[36]. Ingold and Jones[35] have shown that nitration with nitric acid, and 85% sulfuric acid does not occur via the *N*-nitroamine but is a direct ring nitration, the *para* and *meta* positions being nitrated 59 and 34%, respectively. Here nitration at the *ortho* position (6%) has been inhibited, again in protic medium. The linear coordination model of Kovacic and Hiller[29] associated with protic solvation of electron-rich centers nicely explains the difference in orientation by mixed acid and nitric acid in acetic anhydride.

Phenols also show considerable variation in isomer distribution with nitrating medium. Nitration in aqueous medium, 0.5 *M* in nitric acid and 1.75 *M* in sulfuric acid, gives 73% *o*-nitrophenol, whereas nitration in acetic acid, 3.2 *M* in nitric acid, gives only 44% *o*-nitrophenol. The difference is *p*-nitrophenol since *m*-nitrophenol has not been detected. Ingold[37] was of the opinion that in aqueous medium the nitracidium ion, $H_2NO_3^{\oplus}$, was responsible for phenolic nitration. In dilute aqueous nitric acid, nitration can occur via nitrosation followed by oxidation. Since phenols (and amines also) are easily oxidized, the nitrosating agents are readily available from the reduction of the nitric acid.

The yields for nitration of phenols quoted above are for reactions essentially free of nitrous acid or dinitrogen tetroxide. It is interesting to note that in the two solvents, water and acetic acid, nitrosation gives the opposite specificity for *ortho* or *para* positions. In aqueous medium nitrosation gives 91% *p*-nitrosophenol; oxidative nitration of phenol with 1.0 *M* nitrous acid and 0.5 *M* nitric acid gives, likewise, 91% *p*-nitrophenol. Direct nitration of phenol in water (0.5 *M* nitric acid) gives only 27% *p*-nitrophenol. In acetic acid, oxidative nitration (4.5 *M* N_2O_4; 3.2 *M* HNO_3) gives only 26% *p*-nitrophenol, but direct nitration gives 56% *p*-nitrophenol.

In summary, nitrosation of phenol in water is more *para* seeking than nitration; in acetic acid nitration is more *para* seeking than nitrosation. The *ortho*/*para* ratio in the nitration of phenol can then be controlled to a very high degree through choice of solvent or mechanism. Table 5[37] gives the amount of *ortho* product obtainable under varied nitrosating agent concentration.

A simple rationale of this medium difference is not obvious since both the active nitrating and nitrosating agents may be different in the two media. But, the results themselves indicate that in aqueous medium the nitrosating agent—generally a weak electrophile—is more deterred from the *ortho* position by the steric bulk of the solvent hydrogen bonded to the hydroxyl group than is the more active nitrating agent, most likely a nitracidium ion. That nitration of phenol in acetic acid is more *para* orientated than in water can be

TABLE 5. Concomitant nitration and oxidative nitrosation of phenol in water and acetic acid[37].

H_2O, 20°		HOAc, 0°	
[PhOH] = 0.45 *M*; [HNO_3] = 0.5 *M*; [H_2SO_4] = 1.75 *M*		[PhOH] = 0.6 *M*; [HNO_3] = 3.2 *M*	
[HNO_2], *M*	% *o*-nitrophenol	[N_2O_4], *M*	% *o*-nitrophenol
0.00	73	0.03	45
0.25	55	1.8	64
1.00	9	4.5	74

explained by proposing that the steric size of acetyl nitrate is greater than that of the nitracidium ion. To account for the *ortho* nitrosation in acetic acid it is necessary to suggest that a careful balance exists between the degree of solvation of the aromatic substrate and the coordination of the incipient nitrosonium ion with its leaving group. Whatever the nitrosating agent, the facts say that acetic acid is too weak in its solvation of the phenolic hydroxyl group to prevent coordination of the nitrosating agent with the hydroxyl substituent. An analogous effect is observed in the nitration of nitronaphthalenes and other arenes substituted with *meta* directors.

This different kind of *ortho* effect is observed in the nitration of nitro-*p*-xylene: the second nitro group more often enters adjacent to the first nitro group rather than going to the open side of the molecule[38]. The ratio of 2,3-dinitro-*p*-xylene to 2,5-dinitro is 1.5–2.3 to 1. A similar effect is observed in the dinitration of *p*-bromotoluene, 2,3-dinitro-4-bromotoluene being the only product reported. The material balance for this reaction is very poor, however[39].

Nevertheless, high *ortho*:*para* ratios are typical of compounds containing *meta* directors. Examination of Table 6[40] shows the high

TABLE 6. Isomer distribution for *meta*-directed arenes[40].

	% *ortho*	% *para*	% *meta*
$PhNO_2$[40a]	6.4	0.3	93.2
PhCN[40b]	17.1	2.2	80.7
$PhCO_2H$[40a]	18.5	1.3	80.2
PhCHO[40c]	19	9	72
$PhCONH_2$[40d]	27	<3	70
$PhCO_2Et$[40a]	28.3	3.3	68.4

yields of *ortho* nitro compounds associated with substituents directing predominately *meta*.

A similar, but not identical, phenomenon is observed in the nitration with mixed acid of 1-nitronaphthalene and, especially, of 1,5-dinitronaphthalene[41]. The 1-nitronaphthalene gives an excess of 1,8-dinitro over 1,5- in the ratio 67:33. The 1,5-dinitronaphthalene gives 94% of 1,4,5-trinitronaphthalene but only 6% of 1,3,5-trinitro compound (equation 8).

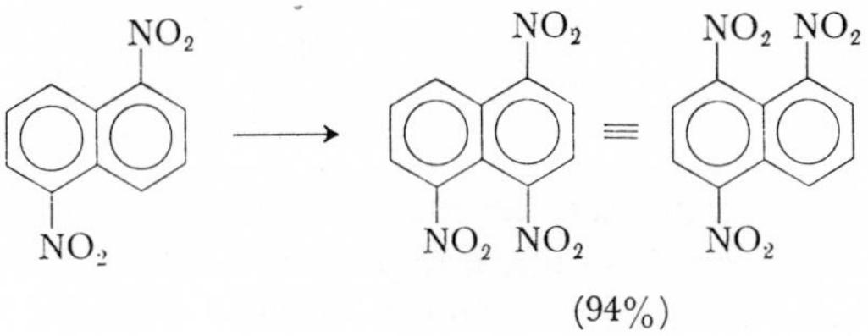

NO_2 NO_2 NO_2 NO_2 (8)

NO_2 NO_2 NO_2 NO_2

(94%)

Obviously, the nitro groups already on the ring are preferentially directing substitution in their vicinity. Coordination of nitronium ion with the electronegative atom of a *meta* director in a cyclic process (**9**) was proposed early by Hammond[42] for the high percentage of *ortho* product derived from *meta*-substituted benzenes. This explanation was felt by Hammond, himself, to be inadequate when

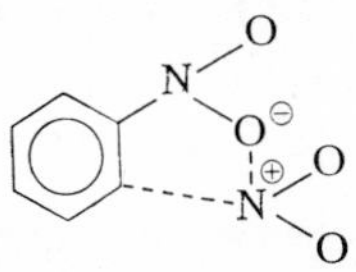

O N O⊖ ⊕ O N O

(**9**)

it was realized that benzonitrile[43] also gives a high *ortho*:*para* ratio in nitration. The objection is based on the linearity of the cyano group; but coordination of the nitrogen of the cyano group to the nitronium ion might alter the carbon atom's hybridization from that in a nitrile to that of an imine and thus permit a cyclic mechanism (equation 9). Furthermore, if one accepts that there is π-interaction

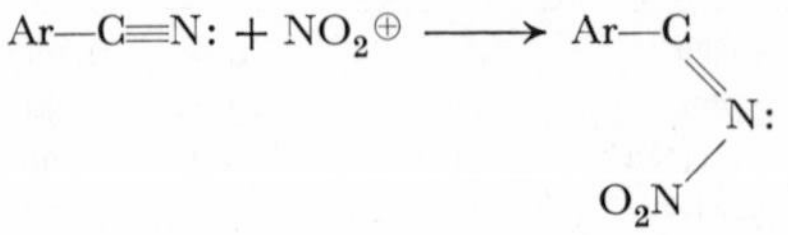

$$\text{Ar—C}\equiv\text{N:} + \text{NO}_2^{\oplus} \longrightarrow \text{Ar—}\overset{\oplus}{\text{C}}\text{=N:—NO}_2 \quad (9)$$

between a *meta*-directing group and an aromatic system then the linear coordination model would be applicable, the nitronium ion simply 'slithering' along the π-cloud (equation 10).

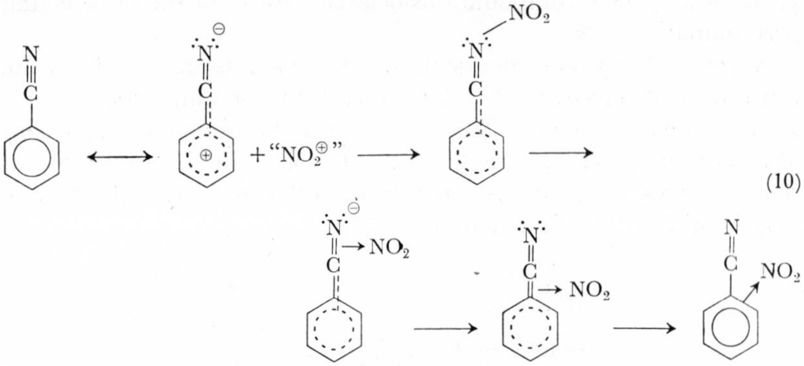

$$(10)$$

Apropos to substitution at deactivated positions, one should realize that rates of reaction are not entirely governed by activation energies of transition states leading to intermediates of lower energy, but that there is still the probability or entropy factor controlling rates of reaction.

The attempts at Hammett-type correlations[44] are full of compounds which fail to be correlated. These numerous failures[45] clearly point out that activation energies are not the only factors governing position of electrophilic substitution. Outstanding among the failures is the nitration of *p*-methoxyacetanilide in aqueous acetic acid. The σ-constants (CH_3O; $\sigma = -0.268$; CH_3CONH, $\sigma = -0.015$)[46] predict nitration *ortho* to the methoxy group. In fact, the acetamido group directs and the predominate product (79%) is 4-methoxy-2-nitroacetanilide[47].

The nitration of the nitronaphthalenes is subject to solvent effects[41]. In the nitration of 1-nitronaphthalene there is a small increase of 1,5-dinitronaphthalene from 33 to 41% in going from mixed acid to 70% (ordinary concentrated) nitric acid. With 1,5-dinitronaphthalene the ratio of 1,3,5-trinitronaphthalene to 1,4,5-trinitronaphthalene changes more drastically in going from mixed acid (6:94) to 70% nitric acid (58:42). The fact that 92.5% aqueous nitric acid gives the same yields as mixed acid suggests that fuming nitric acid and mixed acid contain the same active nitrating species, a 'free' nitronium ion, which is capable of coordinating with basic atoms in a substituent, but that the more aqueous concentrated nitric acid contains a nitrating species less capable of coordinating with the oxygen of the nitronaphthalene, the nitracidium ion, $H_2ONO_2^{\oplus}$.

In general it seems that *ortho* nitration tends to exceed *para* nitration as the statistical factor suggests. Factors which contradict this conclusion are steric and solvent. Fuson[48] seems to be of the opinion

TABLE 7. Isomer distribution for the nitration of halobenzenes and benzylic compounds with nitric acid in acetic anhydride[49].

	% *ortho*	% *para*	% *meta*	[*ortho*/*para*]
PhF[49a]	9	91	—	0.1
PhCl[49b]	10	90	—	0.11
PhBr[49b]	25	75	—	0.33
PhI[49c]	38.6	59.5	1.8	0.65
$PhCH_2CO_2Et$[49d]	54.4	32.6	12.9	1.62
$PhCH_2H$[49d]	56.1	41.4	2.5	1.36
$PhCH_2OMe$[49d]	51.3	41.9	6.7	1.22
$PhCH_2Me$[49a]	46.0	50.8	3.4	0.91
$PhCH_2NO_2$[49d]	22	23	55	0.96
$PhCH_2Cl$[49d]	33.6	42.9	13.9	0.78
$PhCH_2CN$[49d]	24.2	55.5	20.3	0.44

that greater *para* substitution in *ortho–para*-directed aromatics is general. It is the protic nature of the commonly employed nitrating agents or steric factors, not electronic factors, which however, cause the high proportion of *para* substitution. Only in the nitration of halobenzenes are solvent or steric factors inadequate for explaining the high proportion of *para* nitration. The electronic factors governing stability of the transition states as determined by the negative inductive effect of the halogens must govern orientation. Table 7[49] shows just how *para* directing the halobenzenes are. Some negatively α-substituted toluenes also show a predominance of *para* nitration and although there is a −I effect present, the similarity of the *ortho*:*para* ratios with ethylbenzene indicates, however, that steric effects are operative here.

Because of solvent and steric and electronic effects, identification of the active nitrating agent in nitration has been elusive. Only in concentrated sulfuric acid is nitronium ion conclusively the active nitrating agent. As carriers of the nitronium ion Ingold[37] contemplated the following series: $NO_2^{\oplus}$, $H_2NO_3^{\oplus}$, N_2O_5, and $BzNO_3$. The list is certainly longer than this. Acetyl nitrate, protonated acetyl nitrate, and nitronium fluoroborate clearly seem to belong to the list as distinctive nitrating agents. Orientation in nitration, however, is less dependent on the nitrating agent than on other factors. How simple nature is!

B. Choosing Experimental Conditions

Aside from the problem of orientation in nitration, experimental conditions of time, temperature, solvent, concentration, and reagents must be selected for the proper degree of nitration of a particular aromatic compound. How important these conditions are can be seen by an examination of Table 8 which summarizes the yields of

TABLE 8. Nitration of octaethylporphyrin under varied reaction conditions[50].

Conditions	Time, min	Method of product detn	Products, %[a]		
			Mononitro	Dinitro	Trinitro
HOAc–fuming HNO_3, 0° → room temp	1.5	Isolation	92	—	—
	12	Tlc	xxx	x	—
	30	Tlc	—	xxx	—
	160	Tlc	—	xxx	x
Concd H_2SO_4–concd HNO_3, 0° → room temp	0.5	Isolation	Trace	38	4
	1	Isolation	Trace	12	22
	1.5	Isolation	—	Trace	20
Fuming HNO_3, 20°	0.03	Isolation	26	Trace	—
	0.5	Isolation	—	4	—
Urea-treated fuming HNO_3, 22°	2	Isolation	72	—	—
	12	Isolation	2	46	—
	30	Isolation	—	5	8
Concd HNO_3, room temp	10	Tlc[b]	xx	—	—
NO_2BF_4–sulfolane, 100°	60	Tlc[b]	Trace?	—	—
NO_2BF_4–H_2SO_4, 18°	60	Tlc[b]	xx	—	—

[a] % yields refer to once-crystallized compounds; proportions estimated visually: xxx = major, xx = moderate, x = minor component; [b] Unchanged octaethylporphyrin was also detected.

mono-, di-, and trinitration of octaethylporphyrin (**10**) under

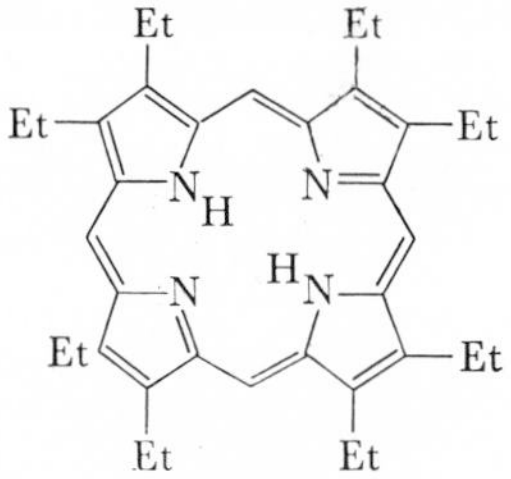

(**10**) 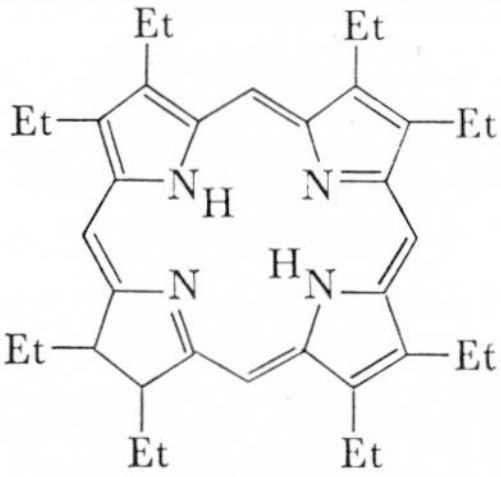 (**11**)

different conditions[50]. It will be observed that fuming nitric in acetic acid is a less active nitrating agent than concentrated nitric acid in concentrated sulfuric acid. Not only is the acetic acid solution of nitric acid the mildest nitrating agent of those listed with the possible exception of the concentrated nitric acid, but the oxidative properties of the fuming nitric acid are likewise lost in acetic acid. Even the urea-treated fuming nitric acid is not free of degradative properties; however, the higher temperatures employed may be responsible for the lowered yield of products. The behavior of nitronium fluoroborate toward the porphyrin is probably similar to its reaction with pyridine[33]; i.e., *N*-nitration has occurred with subsequent nucleophilic ring opening. It is interesting that octaethylchlorin (**11**) (from the reduction of one double bond in one isopyrrole ring of octaethylporphyrin), is successfully nitrated by nitronium fluoroborate, whereas fuming nitric acid in acetic acid, which nitrates the porphyrin in 92% yield, completely degrades the chlorin. A pronounced temperature effect is observed in the nitration of the chlorin with nitronium fluoroborate. At 24° for 2 hours, a 44% yield of mononitro compound is obtained along with 9% of the dinitro compound. An increase of 7° to 31° for the same period of time, causes dinitration exclusively (44%); the extent of degradation remains about the same at both temperatures (48–56%).

To show the applicability of various nitrating agents toward particular aromatic compounds Table 9[51] has been arranged with the more mild nitrating agents used with the more electron-rich aromatics and proceeds to the more vigorous reagents used for di- and trinitration and to nitrate deactivated aromatics.

The non-protonated acyl nitrates are mild nitrating agents and are quite often used to nitrate the electron-rich non-benzenoid aromatics. The so-called diacetylorthonitric acid, derived from

TABLE 9. Nitrating agent and conditions for various aromatic compounds.

Agent	Solvent	Compound	Temp, °C	Time	Ref
Mononitration:					
$BzONO_2$	CH_3CN	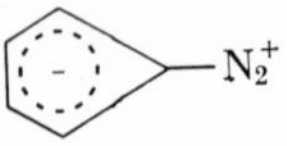 $-N_2^+$	0–5	2 hr	51a
$Cu(NO_3)_2$	Ac_2O	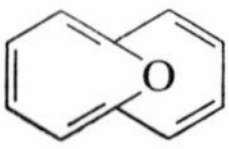 O	25	5 min	51b
HNO_3	AcOH	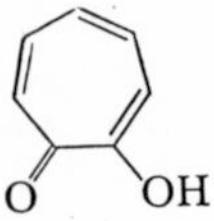 O OH	25	12 hr	51c
HNO_3	Ac_2O	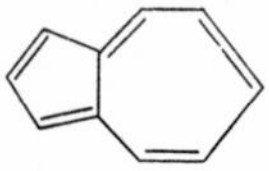	0–5	10 min	51d
HNO_3	AcOH	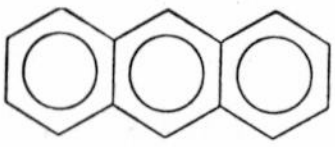	25	1.25 hr	51e
HNO_3	Ac_2O	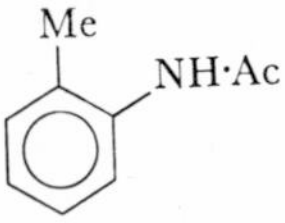 Me NH·Ac	10–12	2 hr	51f
HNO_3	Ac_2O	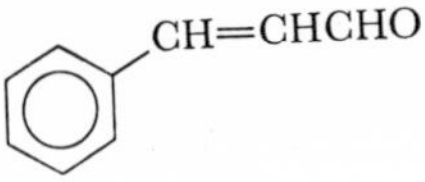 CH=CHCHO	5 25	4 hr 2 days	51g
HNO_3	aq AcOH	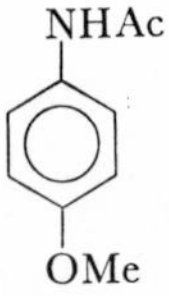 NHAc OMe	65	10 min	51h
HNO_3 (70%)	—	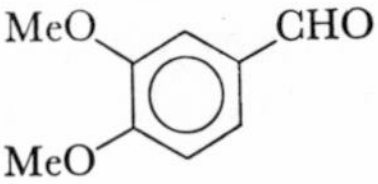 MeO CHO MeO	18–22	1 hr	51i
HNO_3 (70%) (107 mole %)	H_2SO_4 (1000 mole %)	 Me *i*-Pr	−10	2.2 hr	51j
HNO_3 (70%) (120 mole %)	H_2SO_4 (140 mole %)		50	1 hr	51k

TABLE 9—*Continued*

Agent and solvent		Compound	Temp, °C	Time	Ref
HNO_3 (70%) (105 mole %)	H_2SO_4 (800 mole %)	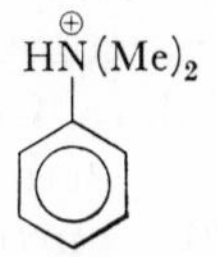 $H\overset{\oplus}{N}(Me)_2$	5–10	2.5 hr	51l
HNO_3 (90%) (200 mole %)	H_2SO_4 (1500 mole %)	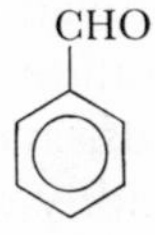 CHO	5–10 25	2–3 hr 12 hr	51m
HNO_3 (90%) (400 mole %)	H_2SO_4 (300 mole %)	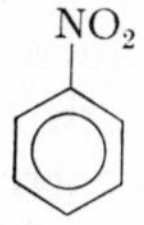 NO_2	95	30 min	51n
HNO_3 (90%) (600 mole %)	H_2SO_4 (700 mole %)	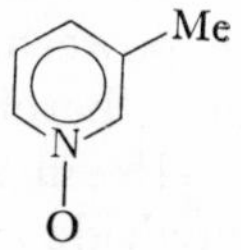 Me N O	10 105	Mixing 2.75 hr	51o
Dinitration:					
HNO_3 (90%) (600 mole %)	H_2SO_4 (1000 mole %)	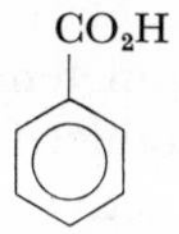 CO_2H	100 145	8 hr 3 hr	51p
HNO_3 (90%) (400 mole %)	H_2SO_4 (1000 mole %)	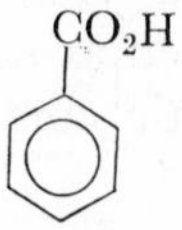 CO_2H	25	1000 hr (6 wk)	
KNO_3 (350 mole %)	H_2SO_4 (1200 mole %)	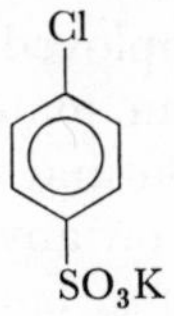 Cl SO_3K	40–60 145	Mixing 20 hr	51q
Trinitration:					
HNO_3 (red fuming) (2300 mole %)	H_2SO_4 (4800 mole %) HOAc (70 mole %)	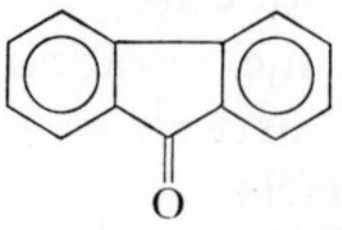 O	20–45 150	20 min 1 hr	51r

$Cu(NO_2)_2$ and acetic anhydride, most likely is acetyl nitrate in an acetate buffer. The acyl nitrates themselves are prepared in acetonitrile from silver nitrate and acetyl or benzoyl chloride. The acyl nitrates are somewhat thermally unstable and are prepared and used initially at low temperatures; subsequent gentle warming to 25–40° is acceptable procedure. Concentrated nitric (sp gr, 1.42) is 70% acid and is sufficiently active alone to nitrate anisole at 45° and veratraldehyde at 20°. Further dilution with acetic acid or water is common; cold dilute aqueous nitric acid is used for phenols and anilines. Because the dilute aqueous medium is a poor solvent for most organic compounds, acetic acid is very frequently used as a diluent. Anthracene is readily nitrated in 1 hour at 25° with nitric acid in acetic acid and *p*-methoxyacetanilide is nitrated in aqueous acetic acid at 65° in 10 minutes. Hot aqueous nitric acid is a good oxidizing agent and although probably active enough to nitrate alkylated benzenes, the occurrence of side-chain oxidation complicates its use. The addition of acetic anhydride or sulfuric acid to nitric acid gives a more powerful nitrating agent which can be used at lower temperatures thereby avoiding side-chain oxidation. *p*-Cymene is nitrated at −15° in 2 hours in sulfuric acid. Solutions of nitric acid in acetic anhydride are also non-oxidative, and xylenes and mesitylene are nitrated without oxidation in acetic anhydride. Cinnamaldehyde with its relatively sensitive side chain is nitrated in acetic anhydride at room temperature for 2 days. *o*-Nitrocinnamaldehyde is the only product formed in acetic anhydride; this is, of course, typical of the active reagent, protonated acetyl nitrate. As with acyl nitrates, the combination of acetic anhydride with nitric acid must be affected at low temperatures (0–5°), and initial reaction with the aromatic compound is generally carried out at temperatures less than 15°.

Benzene, itself, is nitrated with mixed acid, sulfuric and nitric. The amount of sulfuric employed is not great here, but temperatures are somewhat above room temperature (50–60°). Benzene reacts exothermically in nitration and this temperature is easily achieved without heating. In fact, on anything but small scale preparations cooling is necessary[52] to prevent significant dinitration which will occur if temperatures exceed 60°.

Both temperature and water content of the mixed acid have a pronounced effect on the nitration of deactivated aromatic compounds. Sulfuric acid, as well as selenic and perchloric acid, will cause complete conversion of nitric acid to the nitronium ion according to equation 11. Hydronium sulfate, however, is no good at effecting complete conversion. Inasmuch as ordinary concen-

$$HONO_2 + 2H_2SO_4 \longrightarrow NO_2^{\oplus} + H_3O^{\oplus} + 2HSO_4^{\ominus} \qquad (11)$$

trated nitric acid is 30% water, large excesses of sulfuric acid are employed to compensate for protonation of the water; or more often fuming nitric acid (*ca.* 90%, which is still only 70 mole % acid) is employed. Thus, the *N,N*-dimethylanilinium ion is nitrated with 70% nitric acid in a large excess of sulfuric acid (800 mole %) at 10° in 2½ hours, whereas, nitrobenzene has been nitrated at 95° for 30 minutes with fuming nitric acid in less sulfuric acid (300 mole %). Oleum is used to reduce the water content even further. For the production of *sym*-trinitrobenzene from *m*-dinitrobenzene[53] oleum and fuming (90%) nitric acid are employed at 110°.

The use of potassium nitrate, as the source of nitric acid, in sulfuric acid avoids the water content of nitric acid but the system may be less active (*vide infra*). Still, potassium nitrate (350 mole %) with sulfuric acid (1200 mole %) nicely dinitrates potassium *p*-chlorobenzenesulfonate in 20 hours at 145°.

C. Peculiarities of Mixed Acid

Some peculiarities are found in the mixed acid nitrating medium. Many nitrations are heterogeneous in spite of the generally good solubility of most aromatic compounds in concentrated sulfuric or nitric acid alone. Durene, which is readily soluble in sulfuric acid, fails to undergo mononitration in mixed acid but gives only dinitration[54]. It has been suggested that once mononitration has been effected, greater solubility of the mononitrated durene accounts for its subsequent dinitration[55]. This explanation is reasonable since the nitrodurene may be as much as 30 times more soluble in the mixed acid than the unsubstituted durene. The solubility of durene and nitrodurene should be somewhat analogous to that of hexafluoro-*m*-xylene and its nitro compound, whose solubilities are shown in Table 10[56]. The high solubility shown for fuming nitric acid suggests that the heterogenous character of the nitration could be avoided by using nitric acid as the solvent and adding only enough sulfuric acid to generate the necessary amount of nitronium ion. Cheronis[57] in his organic qualitative analysis books has long advocated the use of 100% nitric acid as an excellent nitrating agent in the preparation of hydrocarbon derivatives. The success of the method, to a large extent, probably results from the excellent solubility in the 100% nitric acid.

TABLE 10. Solubility of *m*-bistrifluoromethylbenzene and 5-nitro-1,3-*bis*(trifluoromethyl-benzene in mixed acids at 20° [56].

Acid composition		Solubility, gm/l.	
H_2SO_4, mole %	HNO_3, mole %	$C_6H_4(CF_3)_2$	$NO_2C_6H_3(CF_3)_2$
100	0	8.4	—
80	0	3.6	115
60	0	1.1	30.4
0	70	74.8	—
11	70	24.3	—
21	70	8.2	91.3
30	40	3.2	45.0
50	20	1.6	—
60	40	1.4	33.3
70	30	1.6	—

The high acidity of sulfuric acid through protonation of the aromatic substrate, can also cause deactivation of an aromatic compound making the nitration process much more difficult[58]. The degree of deactivation of aromatic substrates in sulfuric acid as a solvent can be appreciated from the fact that nitration of nitrobenzene in mixed acid is effected at 95° and occurs slowly only above 60°; yet, nitrobenzene can be nitrated at room temperature by adding stoichiometric amounts of sulfuric acid to the nitro compound dissolved in nitroglycerin[59].

Rates of nitration of nitrobenzene and other deactivated aromatics in sulfuric acid have been studied with respect to the water content of the sulfuric acid and its corresponding Hammett acidity function, H_0[58]. These studies show that nitration rates are at a maximum in 90–95 % sulfuric acid.

The slower rates observed below 90 % sulfuric acid are simply due to the incomplete conversion of nitric acid to the nitronium ion. The lowered activity for the medium containing less than 5 % water is not as straightforward, but is in part a consequence of protonation of the aromatic substrate: benzoic acid is about 20 times more active in 95 % than in 100 % sulfuric acid. Benzenesulfonic acid shows an eleven-fold factor; nitrobenzene and *p*-chloronitrobenzene show a four-fold factor[58]. Substituent protonation would seem an adequate explanation for the lowered rates in the more acidic 100 % sulfuric acid, but trimethylphenylammonium ion also exhibits rate retardation in going from 95 to 100 % sulfuric acid $k(95\%\ H_2SO_4)/k(100\%\ H_2SO_4) = 2.5$. Gillespie and Norton[58] felt that substrate protonation was impossible here and proposed hydrogen bonding or entropy effects as an explanation. Ring protonation was not considered.

Proton exchange[60] via ring protonation has been studied and protodesulfonation is a well-known synthetic process. The color produced by dissolution of anthracene in sulfuric acid suggests ring protonation[61], and it has recently been shown by NMR[62] that a methylene unit is present in the solution of anthracene in sulfuric acid. Ring protonation of durene might be an alternative explanation for its dinitration in mixed acid: ring protonation of the deactivated nitrodurene being less likely and thereby permitting dinitration. Proton exchange is very facile with durene, 1.7×10^6 more facile than with benzene[60].

Deactivation and the resulting change of orientation due to the protonation of amines is basic knowledge. This effect, however, can be complicated. 1,3-Dihydro-5-phenyl-2H-1,4-benzodiazepin-2-one (**12**), is nitrated in the 9 position with potassium nitrate in cold

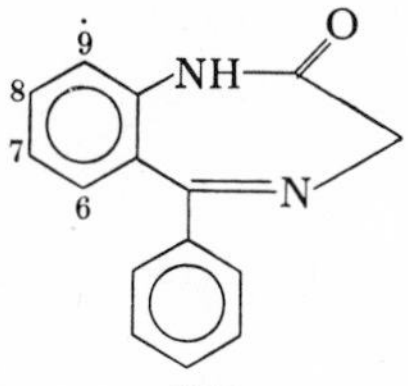

(**12**) 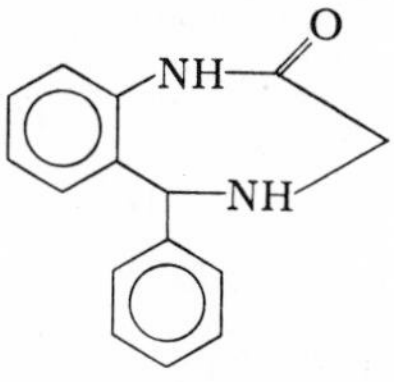 (**13**)

concentrated sulfuric acid[63]. The hydrogenated 1,3,4,5-tetrahydro-5-phenyl-2H-1,4-benzodiazepin-2-one (**13**) could not be nitrated with potassium nitrate in sulfuric acid but, 'a stronger reagent, fuming nitric acid in concentrated sulfuric acid, was required to effect nitration.'[64] Not only did the 'stronger reagent' effect nitration but it affected the position of nitration from the benzo ring to the *meta* (32%) and *para* (68%) positions of the 5-phenyl ring. No simple explanation seems obvious, but the acidity of the two nitrating media must have a bearing on the reactivity and orientation found with **12** and **13**.

'Forcing conditions' for polynitration call for the use of oleum yet the presence of sulfur trioxide causes a retardation of rate of nitration of nitrobenzene below that of 100% sulfuric acid. Inasmuch as maximum rates are observed in 95% sulfuric acid, it would seem that only enough oleum should be utilized to compensate for the water content of the nitric acid.

Temperatures in excess of 100° is the most necessary component of forcing conditions, especially if short reaction times are desired. The dinitration of benzoic acid is illustrative. Wherein the temperature is progressively raised from 100 to 145° the 3,5-dinitrobenzoic acid can be obtained in 11 hours; whereas 6 weeks are needed for comparable yields if the nitration is effected at room temperature.

Orientation in nitrogen heterocycles in general is complex and acidity dependent. A few examples for illustration: quinoline nitrates with mixed acid at the 5 and 8 positions in approximately equal amounts[65], but on nitration in acetic anhydride with lithium nitrate and cupric nitrate, 7-nitroquinoline is the product[66]. 2-Methylindole[67] is nitrated in cold concentrated sulfuric with sodium nitrate to give the 5-nitro-2-methylindole (**14**), but concentrated nitric acid has no effect until the temperature is raised sufficiently that oxidation begins, and then 3-nitration is effected with subsequent dinitration at the 6 position resulting in **15**.

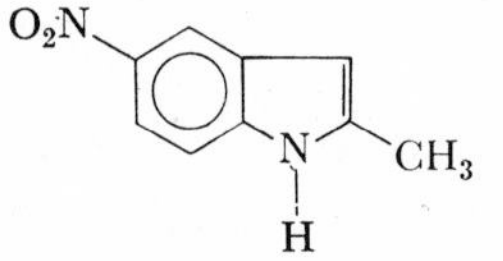

(**14**) 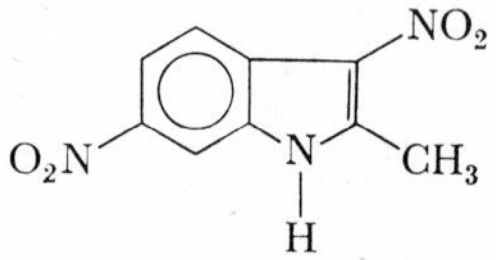 (**15**)

Schofield[68] in his review of nitration of heterocyclic nitrogen compounds could not offer much to correlate their diverse behavior. Recent work by Noland and coworkers[67,69] has provided a wealth of data on nitration of indoles. The further demonstration that onium ions are not always *meta* directing (Ridd has shown that a $NR_3^{\oplus}$ group may give as much as 38% *para* nitration[70]) may aid in future rationalization. Triphenyloxonium ion, $Ph_3O^{\oplus}$, nitrates almost 100% *para*, although the sulfur analog, $Ph_3S^{\oplus}$, nitrates *meta*[71]. An immonium ion, 1,2,3,3-tetramethylindoleninium (**16**) has been shown to nitrate exclusively in the 5, i.e., the *para* position. Under the same conditions of sodium nitrate in sulfuric acid at 0–10°, 2,3,3-trimethylindolenine gives the same 5-nitration[72]. These

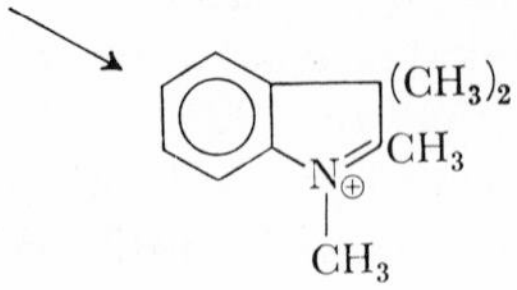

(**16**)

results support the hypothesis of Noland that 5-nitration of indoles in sulfuric acid is a consequence of prior protonation at the 3 position to give an indoleninium ion[69a].

D. Side Reactions

Side reactions are fairly common in nitration. Loss of substituents via replacement occurs with many polyalkylated benzenes and electrophilic replacement of halogen, sulfonyl, carboxyl, acyl, and aldehydo groups is known and sometimes useful. Oxidation can be extensive with phenols, phenolic ethers, and amines. An early summary of many of these side reactions has appeared, in *Chemical Reviews*[73]. The loss of substituents is a substitution process analogous to proton substitution. The replaced substituent, X, then resides in an activated position, i.e., *ortho* or *para* to an electron-donating

$$A\text{-}C_6H_4\text{-}X + NO_2^{\oplus} \longrightarrow [A^{\oplus}{=}C_6H_4(X)(NO_2)] \longrightarrow A\text{-}C_6H_4\text{-}NO_2 + X^{\oplus} \qquad (12)$$

group (equation 12). The ease of substituent replacement, as might be expected, is related to the stability of $X^{\oplus}$. It is found, therefore, that branched-chain alkyl groups are more readily nitrodealkylated than methyl or ethyl groups since secondary and tertiary carbonium ions are more stable than primary. Nitrodehalogenation occurs most readily with iodine and least readily with chlorine in accord with the ease of oxidizing halogens to the +1 oxidation state. Since positive fluorine is so unlikely only oxidative loss to quinones is observed[74] (equation 13).

$$2,6\text{-}Br_2\text{-}4\text{-}F\text{-}C_6H_2OCH_3 + HNO_3\ (90\%) \longrightarrow 2,6\text{-}Br_2\text{-}1,4\text{-}benzoquinone \qquad (13)$$

The *meta*-alkylated benzenes normally are not subject to nitrodealkylation. Thus, neither 1,3,5-triisopropyl nor tri-*t*-butylbenzene undergo nitrodealkylation (a report by Olah[75] that they do, has been shown to be incorrect[76]), whereas *p*-cymene gives about 8%

p-nitrotoluene[77], and *p*-diisopropylbenzene gives 56–83 % *p*-nitrocumene[75]. *sym*-Tri-*t*-butylbenzene, once mononitration deactivates

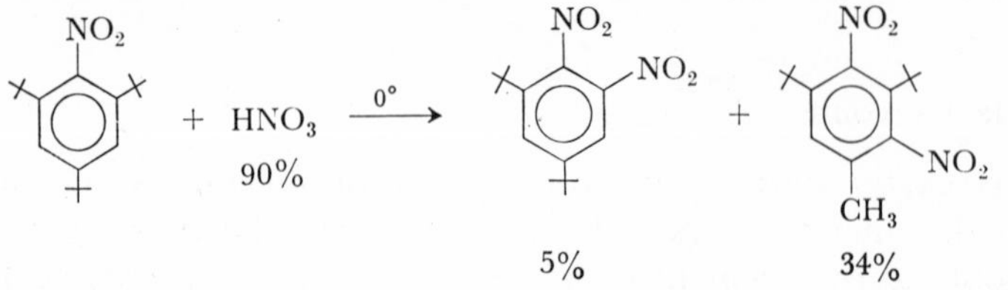

$$\text{2,4,6-tri-}t\text{-butylnitrobenzene} + \underset{90\%}{HNO_3} \xrightarrow{0^\circ} \underset{5\%}{\text{1,3-dinitro-2,5-di-}t\text{-butylbenzene}} + \underset{34\%}{\text{2,4-di-}t\text{-butyl-3,5-dinitrotoluene}} \quad (14)$$

the remaining open positions, on dinitration gives 5 % nitrodealkylation and 34 % 'rearrangement' involving loss of an isopropyl group (equation 14). The sequence in equation 15 has been proposed[76].

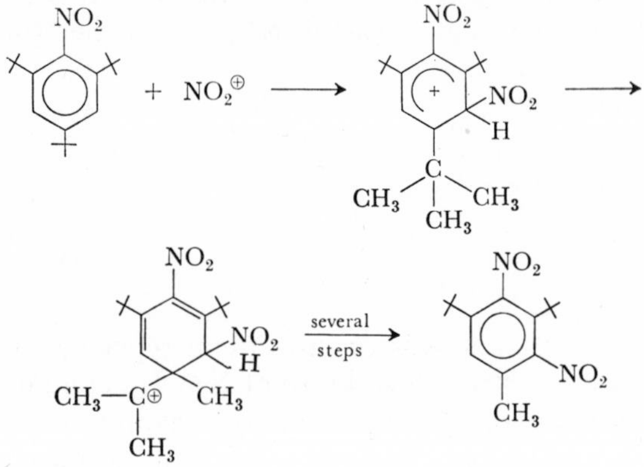

$$\text{2,4,6-tri-}t\text{-butylnitrobenzene} + NO_2^{\oplus} \longrightarrow \text{arenium ion (}NO_2\text{, H at ring C adjacent to } C(CH_3)_3\text{)} \longrightarrow$$

$$\text{cyclohexadiene with } NO_2\text{, H, } CH_3 \text{ and } CH_3{-}\overset{\oplus}{C}{-}CH_3 \xrightarrow{\text{several steps}} \text{2,4-di-}t\text{-butyl-3,5-dinitrotoluene} \quad (15)$$

Desilylation is a very facile process as shown by the fact that *p*-bis(trimethylsilyl)benzene gives an 80 % yield of trimethyl-*p*-nitrophenylsilane when nitrated with nitric acid in acetic anhydride[78]. In strong acid, protodesilylation can also occur. To avoid this during nitration, copper nitrate in acetic anhydride has been found useful[79].

In general, mixed acid gives more side reactions than does nitration with nitric acid in acetic anhydride, although it is not excluded in the latter. For example, *p*-diisopropylbenzene gives 83 % nitrodealkylation on mononitration in mixed acid, but gives only 59 % *p*-nitrocumene with nitric acid in acetic anhydride[80]; with nitronium fluoroborate in sulfolane the yield (56 %) of *p*-nitrocumene is quite comparable[75]. Nitric acid in acetic anhydride, however, can be an acetoxylating agent; *o*-xylene gives 43 % dimethylphenyl acetate, although *m*-xylene gives only 4 % of the ester.

Lower temperatures, also, favor less side reactions as shown by the fact that *p*-dimethylaminobenzoic acid nitrates normally at 5–10° in mixed acid to give 3-nitro-4-dimethylaminobenzoic acid, but at 60–70° gives a complicated mixture containing *p*-nitrodimethylaniline[81]. Even at low temperatures, though, phenolic aldehydes are prone to undergo nitrodecarbonylation, piperonal, vanillin, and anisaldehyde, all giving about 30% of the nitrodecarbonylated product when nitrated at 0° [82].

Use of a sulfonyl group to block positions in order to obtain desired orientation, in electrophilic substitution, e.g., the preparation of 2,6-dinitroaniline from *p*-chlorobenzenesulfonic acid, is common textbook knowledge. This procedure, however, is not always successful because of halo- or nitrodesulfonation, particularly with phenols[83]. Along with the stabilization that an electron-withdrawing group in a phenol can give against oxidation by concentrated nitric acid, nitrodesulfonation can be useful. Thus, picric acid can be obtained from trinitration of phenol-2,4-disulfonic acid; similarly, sulfonation is used prior to the dinitration of 1-naphthol (equation 16).

OH → OH, SO_3H, SO_3H → OH, NO_2, NO_2 (16)

II. NITRATIONS UNDER NON-ACIDIC CONDITIONS

Nitrations which are predominately nucleophilic in nature are possible with suitably electron-rich aromatics. Thus, phenoxides can be nitrated with tetranitromethane[84]. Both water and pyridine are useful solvents. Azulene has been nitrated with tetranitromethane in pyridine in high yield.[85] Earlier reports purported that tetranitromethane nitrated anilines, but recent work with amino acids has shown that tetranitromethane does not nitrate tryptophan, but it is specific for tyrosine[86].

Tyrosine is quantitatively converted to 3-nitrotyrosine with tetranitromethane (equation 17). The optimum conditions are between pH 8 and 9. At higher pH, hydroxide causes breakdown of the tetranitromethane, and below pH 7 no nitration occurs*.

* The reaction of tetranitromethane with phenols has recently been studied. There is strong indication of a radical process[167].

Ferrocene is so easily oxidized in acidic medium that its nitro derivative remained elusive. The nitroferrocene, however, was

$$\text{HO}-C_6H_4-CH_2CH(\overset{\oplus}{N}H_3)COO^{\ominus} + C(NO_2)_4 \longrightarrow \text{HO}-C_6H_3(NO_2)-CH_2(\overset{\oplus}{N}H_3)COO^{\ominus} + HC(NO_2)_3 \quad (17)$$

prepared from its lithium derivative which was subsequently treated with propyl nitrate or dinitrogen tetroxide at $-70°$ [87,88] (equation 18).

$$\begin{aligned} Fe(C_5H_5)_2 + C_4H_9Li &\longrightarrow Fe(C_5H_5)(C_5H_4Li) + C_4H_{10} \\ Fe(C_5H_5)(C_5H_4Li) + PrONO_2 &\longrightarrow Fe(C_5H_5)(C_5H_4NO_2) + PrOLi \end{aligned} \quad (18)$$

It has long been held that 3-nitropyrrole could be prepared in a similar manner from the sodium derivative of pyrrole and isoamyl nitrate[89]. Morgan and Morrey[90] have shown this to be false; only 1 % of 2-nitropyrrole, not 3-nitro, was obtained from the treatment of pyrrole with sodium and isoamyl nitrate.

Low acidity is achieved with nitrate salts. Anhydrous pyridinium nitrate[91], applied in presence of excess pyridine, has been used to nitrate naphthalene (40 % yield of 1-nitronaphthalene) and anthracene (70 %, 9-nitroanthracene). Urea nitrate[92] has been used to nitrate azulene.

The use of copper nitrate in acetic anhydride has already been mentioned as a useful reagent for nitrating silanes[79] which undergo facile protodisilylation. The preparation of acyl nitrates from the acyl chloride and silver nitrate in acetonitrile has been used to nitrate the cyclopentadienyldiazonium zwitterion[51a].

The recently discovered addition compounds of picoline and lutidines with nitronium tetrafluoroborate[166] offer a new method for effecting homogeneous nitrations at room temperature and under essentially neutral conditions (equation 18a).

$$\left[\text{2,6-}(H_3C)_2C_5H_3\overset{+}{N}-NO_2 \right]^+ BF_4^- + ArH \xrightarrow[25°]{CH_3CN} \left[C_5H_5\overset{+}{N}-H \right]^+ BF_4^- + ArNO_2 \quad (18a)$$

It seems conceivable that the *N*-nitropyridinium ion may be a better reagent than alkyl nitrates for reaction with organometallics (*vide supra*).

III. OXIDATION OF AMINO AND NITROSO COMPOUNDS TO NITROARENES

Ring nitrosation of phenols and dimethylanilines is a useful process because of its marked propensity in aqueous medium to occur *para*. Subsequent oxidation with dilute nitric acid gives the nitro compound. Phenol, when nitrated in dilute nitric acid, is both directly nitrated and nitrosated, if urea has not been added to destroy nitrous acid. Since oxidation of the nitroso derivative results in the formation of more nitrous acid, the oxidative nitrosation becomes autocatalytic[37].

Nitrous acid, also when concentrated, undergoes disproportionation to nitric acid and nitric oxide. A process known as 'Zinke Nitration,' which consists of treating phenols with sodium or potassium nitrite in glacial acetic acid probably depends on the disproportionation for production of nitric acid. The phenolic ether, 2,3,6-tribromo-4-methoxyphenol, with this reagent gives the 6-nitro ether[93] (equation 19). Whether this reaction consists of initial

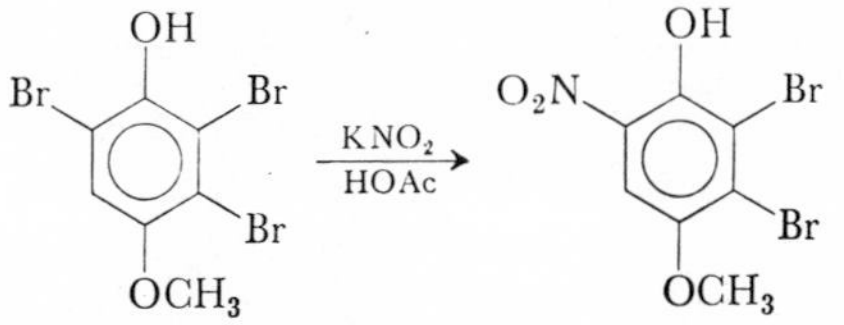

$$\text{2,3,6-Br}_3\text{-4-OCH}_3\text{-C}_6\text{H(OH)} \xrightarrow[\text{HOAc}]{\text{KNO}_2} \text{2-O}_2\text{N-3,6-Br}_2\text{-4-OCH}_3\text{-C}_6\text{H(OH)} \qquad (19)$$

nitration or oxidative nitrosation is not known although nitrosodebromination seems unlikely.

The 3-nitration of 2-methylindole in hot nitric acid seems to proceed through nitrosation to give the tautomeric oxime which is then oxidized to the 3-nitro-2-methylindole[67].

The oxidation of amines to nitroso compounds can be effected with Caro's acid or hydrogen peroxide. The preparation of 2,5-dinitrobenzoic acid, in which the nitro groups are *para* to each and with one *ortho* and the other *meta* to the carboxyl group—a difficult arrangement to achieve by direct nitration—is readily achieved by oxidizing 5-nitro-2-aminotoluene with Caro's acid first to the nitroso compound; concomitant oxidation of the nitroso and methyl groups with acid dichromate gives the nitro acid[94] (equation 20).

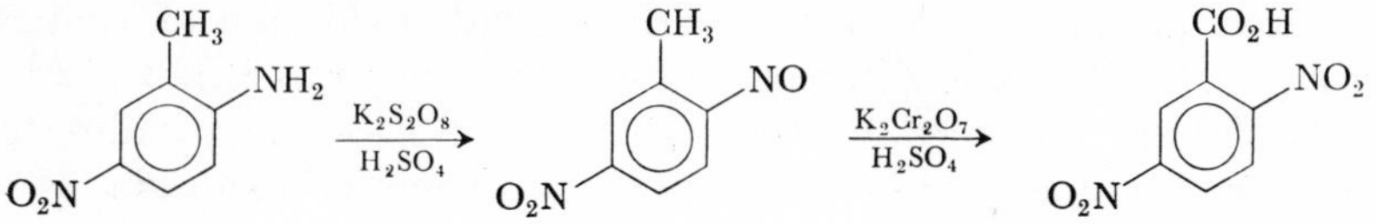

$$\text{CH}_3\text{-C}_6\text{H}_3(\text{NH}_2)(\text{NO}_2) \xrightarrow[\text{H}_2\text{SO}_4]{\text{K}_2\text{S}_2\text{O}_8} \text{CH}_3\text{-C}_6\text{H}_3(\text{NO})(\text{NO}_2) \xrightarrow[\text{H}_2\text{SO}_4]{\text{K}_2\text{Cr}_2\text{O}_7} \text{CO}_2\text{H-C}_6\text{H}_3(\text{NO}_2)(\text{NO}_2) \qquad (20)$$

Holmes and Bayer[95a] have shown that 30% hydrogen peroxide in acetic acid oxidizes amines to the nitroso compound if the reagents are simply allowed to stand at room temperature; at 70–80° in the presence of a larger excess of 30% hydrogen peroxide the nitroso group is further oxidized to the nitro compound.

Although the method of Holmes and Bayer using 30% hydrogen peroxide in acetic acid may give somewhat lower yields of oxidized

TABLE 11. Oxidation of aromatic amines with peroxy acids to nitro and nitroso compounds.

Anilines	90% H_2O_2–$(CF_3CO)_2O$ Yield $ArNO_2$, %[95c]	90% H_2O_2–Ac_2O Yield $ArNO_2$, %[95c]	30% H_2O_2–HOAc Yield, %[95a] ArNO	ArNO$_2$
Aniline	89	83	—	—
4-Cl	87	62	—	—
4-CH_3	78	72	—	—
2,6-$(Cl)_2$-4-CH_3	—	—	22.6	—
4-CN	96	—	—	—
2,6-$(Cl)_2$-4-CN	—	—	—	83
2,6-$(Br)_2$-4-CN	—	—	—	68
2,4,6-$(Cl)_3$	98	—	73.8	?
2,4,6-$(Br)_3$	100	—	80.8	?
4-EtO_2C	99	66	—	—
2,6-$(Cl)_2$-4-EtO_2C	—	—	38.7	—
4-OCH_3	0	82	—	—

compounds, the procedures of Emmons[95b,c] which employ 90% hydrogen peroxide with acetic anhydride or trifluoroacetic anhydride are less attractive. A comparison of yields for the different procedures is given in Table 11.

With the anhydrous peroxy acids, Emmons has noticed that peroxytrifluoroacetic acid is a superior reagent to peracetic for weakly basic amines such as *p*-nitroaniline, but peracetic acid is superior to the fluoro peracid for the oxidation of *p*-anisidine, which is hydroxylated by peroxytrifluoroacetic acid. 2-Naphthylamine with peroxy acids gives an intractable mixture which also results in part from hydroxylation.

Azoxy formation is common with peroxide oxidation of amines. High acidity tends to disfavor its formation, but this is not always successful as evidenced by the formation of 3,3′-azoxypyridine in treating 3-aminopyridine with 30% hydrogen peroxide in 30% fuming sulfuric acid[96].

The reagent, 30% hydrogen peroxide dissolved in 30% fuming sulfuric acid, is the usual one employed for the oxidation of aminopyridines[96]; 60–70% yields have been obtained in the oxidation of 2- and 4-aminopyridines, 2-amino-5-bromopyridine, and from all the 2-aminopicolines except the 5-methyl-2-aminopyridine which gave only a 30% yield of 2-nitro-5-methylpyridine.

IV. REPLACEMENT OF DIAZONIUM ION WITH THE NITRO GROUP

Diazotization of aromatic amines followed by treatment with sodium nitrite, generally in the presence of a copper sulfite catalyst, is a companion method to peroxide oxidation for the conversion of an amino group to a nitro group. Whereas aminonaphthalenes give complex mixtures on attempted peroxide oxidation, the Sandmeyer-type process has been extensively used to prepare the nitro derivatives of naphthalene inaccessible by direct nitration[125] (equation 21). In

$$\text{1-}NH_2\text{-4-}NO_2\text{-naphthalene} \xrightarrow[HOAc-H_2SO_4]{NaNO_2} \text{1-}N_2^{\oplus}\text{-4-}NO_2\text{-naphthalene} \xrightarrow[Cu^{2+}-Cu^{+}]{NaNO_2} \text{1,4-}(NO_2)_2\text{-naphthalene} \quad (21)$$

contrast, 2- and 4-aminopyridines can be diazotized only under special conditions[97], but oxidation of the 2- and 4-aminopyridines gives good yields of the nitropyridines[98].

The replacement of the diazonium group by nitrite ion can only be effected in neutral or basic media. To achieve neutrality or slight alkalinity various methods are used: addition of calcium carbonate[99a] or sodium bicarbonate[99b], or precipitation (and washing free of acid) of the diazonium salts as the sulfates[99c], fluoroborates[100], or colbaltinitrites[99d].

Although the use of diazonium fluoroborates is described in *Organic Syntheses*[100] for the preparation of 2- and 4-dinitrobenzenes, much better yields can be obtained by the method of Ward and coworkers[99b] by adding the solution of diazonium sulfate to a solution of excess sodium nitrite containing excess sodium bicarbonate.

Nitro formation by means of the Sandmeyer process is complicated by several factors. Some amines are difficult to diazotize; the combination of concentrated sulfuric acid plus glacial acetic acid as a medium for diazotization will overcome the solubility problem with the amine, and the combination also overcomes the slow diazotizing ability of nitrosylsulfuric acid when sulfuric acid is used alone. Where diazo-oxide formation complicates the use of a neutral solution of the diazonium salt, the precipitation of a solid sulfate or cobaltinitrite is necessary. Most diazonium salts need the presence of a cupric–cuprous catalyst to be effectively replaced by nitrite; the 2- and 4-nitrobenzenediazonium ions do not require the copper catalyst indicating that the presence of electron-withdrawing groups make the catalyst unnecessary; but the corresponding compounds in the naphthalene series give low yields without it. A mixture of cuprous oxide with copper sulfate and the greenish yellow-brown precipitate formed from equal weights of sodium sulfite and hydrated copper sulfate are effective catalysts. In all cases, the diazonium ion is added to the catalyst and sodium nitrite, both in *excess*, the copper sulfate at 400 mole % and the sodium nitrite at 2000–6000 mole %.

To isolate the diazonium sulfates as solid compounds ether is added in large amounts to the diazonium ion prepared in sulfuric–acetic acid mixture; the solid cobaltinitrites are precipitated from aqueous medium by adding a small excess of sodium cobaltinitrite to the solution of diazonium sulfate or diazonium chloride previously neutralized with calcium carbonate and filtered.

The success of the various methods for specific types of substituted amines is shown in Table 12[99].

TABLE 12. Yields (%) of nitro compounds obtained by use of neutralized solutions and isolated-solid diazonium salts[99].

	Solution		Solid	
Diazotized amine	$Ca(CO_3)$[99a]	$NaHCO_3$[99b]	$(ArN_2)_2SO_4$[99c]	$(ArN_2)_3Co(NO_2)_6$[99d]
Aniline	35	—	—	75.5
2-Nitroaniline	70	97	—	67.4
4-Nitroaniline	76	97	—	75
4-Chloroaniline	35	—	—	82.5
4-Anisidine	16	—	—	68
2-Anisidine	—	—	—	63
4-Toluidine	—	—	—	69
2-Toluidine	—	—	—	61
2-Naphthylamine	15	—	57	60
4-Nitro-1-naphthylamine	25	50	65	—
5-Nitro-2-naphthylamine	15	—	55	—
Benzidine	10	—	16	—

An unusual nitro formation has been reported during the deamination with nitrous acid of guanosine; a 5% yield of 2-nitroinosine[101] was isolated as its ammonium salt (**17**).

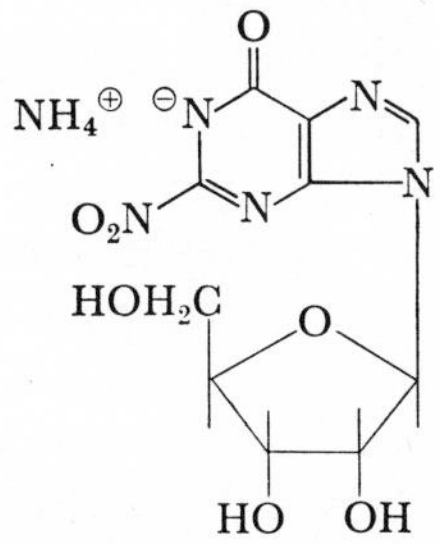

(**17**)

V. NITRATIONS WITH OXIDES OF NITROGEN

The oxides, N_2O_3, NO_2, N_2O_4, and N_2O_5, when dissolved in sulfuric acid or combined with certain Lewis acids, give rise to nitronium ion which obviously will nitrate aromatic compounds. The products are typical of electrophilic nitration in mixed acid and the only significance might be an economic one in the utilizatoin of by-product nitrogen dioxide.

In non-ionizing solvents or in the gas phase the use of the oxides of nitrogen in the higher oxidation states may offer some advantage over the normal methods of nitration.

Although dinitrogen pentoxide is not readily available, it has been suggested for nitration of easily hydrolyzed compounds such as benzoyl chloride[102] (equation 22).

$$C_6H_5COCl \xrightarrow[CCl_4]{N_2O_5} m\text{-}O_2NC_6H_4COCl \qquad (22)$$

Dinitrogen tetroxide is very accessible and nitrations with it are numerous, but often messy. Riebsomer[103] has done an extensive review of its use.

In the vapor state dinitrogen tetroxide is extensively dissociated to the radical, nitrogen dioxide. Accordingly, Titov[104] has proposed a radical mechanism similar to ionic nitration (equation 23).

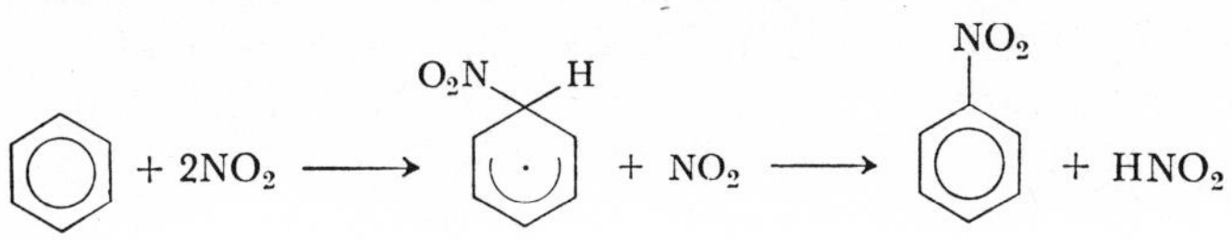

$$C_6H_6 + 2NO_2 \longrightarrow C_6H_6(NO_2)^{\cdot} + NO_2 \longrightarrow C_6H_5NO_2 + HNO_2 \qquad (23)$$

Compounds of the same or lower reactivity than benzene need high temperatures or photoactivation to obtain significant reaction. Because of the low discrimination by radicals and the oxidative properties, alkylarenes give complex mixtures with nitrogen tetroxide. The more active polycyclic aromatic hydrocarbons present a more favorable picture. Naphthalene has been quantitatively converted to 1-nitronaphthalene by heating equimolar portions of napththalene and nitrogen tetroxide at 150°.

Pyridine is said to give a 10% yield of 3-nitropyridine with nitrogen dioxide at 120° [106]. It is questionable whether pyridine undergoes electrophilic nitration since the high temperatures, 300–450°, necessary to effect only slight nitration (6–15%) with potassium nitrate in sulfuric acid[107], suggest radical nitration. Nitrobenzene has been nitrated photochemically with nitric acid[108]. The production of picric acid among the products is similar to the products obtained with irradiated mixtures of benzene and nitrogen tetroxide[109].

Liquid phase reactions present another problem. Dinitrogen tetroxide can undergo two heterolyses (equation 24). Even if the

$$\begin{aligned} N_2O_4 &\longrightarrow NO^{\oplus} + NO_3^{\ominus} \\ N_2O_4 &\longrightarrow NO_2^{\oplus} + NO_2^{\ominus} \end{aligned} \tag{24}$$

initial medium is non-ionizing, a small amount of reaction will replace hydrogen from the aromatic giving rise to nitrous acid, nitric acid, and water as possible products. The nitrations thus become, most probably, ionic. That acidity is not great, however, has recommended its use with organometallics, and some unusual isomer distributions are observed in liquid phase nitration with dinitrogen tetroxide.

The use of N_2O_4 at $-70°$ with ferrocenyllithium gives a small yield of nitroferrocene[110]. The oxynitration of benzene to picric acid with nitric acid and mercuric nitrate has been shown to involve formation of phenylmercuric nitrate which is then converted to nitrosobenzene[111]. Some diaryl mercury compounds also have been shown to give nitroso derivatives with nitrogen tetroxide.[112,113] If the nitroso compound can be produced, oxidation will readily transform it to the nitro compound.

Tropolone[114] has been nitrated with nitrogen tetroxide in petroleum ether at 10–15° in 3 hours. The proportion of the 5-nitro to 3-nitro derivative is 5:1 from the nitrogen tetroxide but only 2:1 when nitration is effected with nitric acid in acetic acid. Quinoline, which is nitrated with mixed acid in the 5 and 8 positions[65], is nitrated with nitrogen tetroxide[106] in the 7 position. This result is analogous to nitration with lithium and copper nitrate in acetic anhydride[66], which also gives 7-nitroquinoline, and suggests that nitrogen tetroxide might be used more often where acidity must be kept low. The overall effect on isomer distribution is not completely straightforward since liquid nitrogen tetroxide[105] reacts with biphenyl at room temperature to give essentially the same proportion of 2- and 4-nitrobiphenyls (35:65) as mixed acid[16e], although the less protic medium, nitric acid in acetic anhydride[16f], gives a higher amount of 2-nitrobiphenyl (68%).

VI. REARRANGEMENT OF *N*-NITROAMINES

The acid-catalyzed rearrangement of nitramines has been studied a lot but is not used synthetically. The avoidance reflects the esoteric procedures for preparing the *N*-nitroamines.

The ease of preparation and the stability of nitramines depend on the particular amine. Aromatic amines, whose rings are strongly deactivated with nitro groups, such as 2,4-dinitro-*N*-methylaniline, are *N*-nitrated with 70% nitric acid in 2 hours at room temperature[35] (equation 25). With 4-nitromethylaniline, a suspension of the amine

$$\text{2,4-}(NO_2)_2C_6H_3NHCH_3 \xrightarrow[\text{2 hr, }-25^\circ]{70\%\ HNO_3} \text{2,4-}(NO_2)_2C_6H_3N(NO_2)CH_3 \tag{25}$$

in acetic acid is *N*-nitrated with anhydrous nitric acid dissolved in acetic anhydride[36] (equation 26). A more generally useful procedure

$$\text{4-}NO_2C_6H_4NHCH_3 \xrightarrow[\text{HOAc, 1 hr, }-25^\circ]{HNO_3-Ac_2O} \text{4-}NO_2C_6H_4N(NO_2)CH_3 \tag{26}$$

employs nitrate esters in the presence of a base or metal to convert amines to the nitramine anion; the free nitramine is obtained by slightly acidifying an aqueous solution of the nitramine salt at 0° and extracting it into ether. The barium salts of nitramines are precipitated from aqueous solutions of the potassium or lithium salt by barium chloride; the barium salts show considerable stability and storage of the nitramines is best as the barium salts[115].

N-Nitroaniline has been prepared with potassium ethoxide and ethyl nitrate[116] (equation 27).

$$C_6H_5NH_2 \xrightarrow[EtOH]{K} C_6H_5NHK \xrightarrow[Et_2O]{EtONO_2} C_6H_5KNNO_2 \xrightarrow[H_2O]{HCl} C_6H_5HNNO_2 \qquad (27)$$

Butyl- and phenyllithium have been found to be useful bases for proton extraction and amyl as well as ethyl nitrate can be employed as the nitrating agent[115,117] (equation 28). For liberation of the free

$$ArNH_2 \xrightarrow{PhLi} ArNHLi \xrightarrow{AMONO_2} (ArNNO_2)^{\ominus}Li^{\oplus} \qquad (28)$$

nitramine, saturation of the salt with carbon dioxide in the presence of water has been found useful for the more unstable amines. *N*-Nitro-1-naphthylamine has been obtained this way[115] (equation 29).

$$\begin{gathered} NaphNH_2 \xrightarrow{BuLi} NaphNHLi \xrightarrow{EtONO_2} NaphN(NO_2)Li \\ NaphN(NO_2)Li \xrightarrow{sat.\ aq\ BaCl_2} (NaphNNO_2)_2Ba \xrightarrow[H_2O]{CO_2} NaphNHNO_2 \end{gathered} \qquad (29)$$

Neither the early method of Bamberger[118], alkaline oxidation of diazonium ions, nor direct *N*-nitration with nitrogen pentoxide in carbon tetrachloride are too practical. Direct *N*-nitration with nitronium tetrafluoroborate in sulfolane might be useful[33]. The nitrate ester of acetone cyanohydrin[119] has been found useful for preparing aliphatic nitramines[120] but seems to have not been used with aromatic amines.

The addition of a nitramine or its metal salt to aqueous acid will cause rearrangement to ring nitration. The *o*-nitro compound predominates but some *p*-nitro derivative is found; the amount of *para* isomer increases with decreasing acidity[116]. This is shown in Table 13.

Rearrangement products are recovered in yields greater than 95% at high acidities, but only 60% yields are obtained from 3 *M* acid. Much tar is formed at the low acidity. Concentrated sulfuric acid (98%) is too vigorous: flames are reported with the addition of *N*-nitroaniline to 18 *M* sulfuric acid[116].

Labeling with isotopic nitrogen and cross-nitration experiments have shown the rearrangement to be intramolecular. This is true for both the *ortho* and *para* products[115,121]. That the *para* derivative also is produced intramolecularly has lead to a controversy[115,116,122] as to mechanism. Migration by a π-complex all the way to the *para* position is felt unlikely and rather elaborate mechanisms have been proposed to overcome the problem of spanning the large distance to the *para* position, intramolecularly.

It would seem that a study of the rearrangement of 2-(*N*-nitroamino)pyridine would be informative since this compound is

TABLE 13. Percentage of *o*-nitroaniline from rearrangement of *N*-nitroaniline at 0° [116].

Acid concn, w/w	% *ortho*
74% H_2SO_4–H_2O	95
52% H_2SO_4–H_2O	88
25% H_2SO_4–H_2O	78
37% $HClO_4$–H_2O	72
26% $HClO_4$–H_2O	70

peculiar: the predominant product is the *para*-like, 5-nitro-2-aminopyridine[123] (equation 30).

2-(NHNO$_2$)pyridine $\xrightarrow{H^+}$ 5-(O_2N)-2-(NH_2)pyridine (30)

VII. PROBLEMS IN ORIENTATION

Some positions in aromatic nuclei are not accessible by direct nitration. The illustration of some of the paths whereby nitro compounds with unusual orientation are prepared seems necessary.

That aromatic nitro compounds are susceptible to nucleophilic substitution should not be forgotten. The nitro group deactivates the ring which holds it and causes further nitration to occur in the unsubstituted ring. However, 1,4-dinitronaphthalene can be prepared through nucleophilic amination of 1-nitronaphthalene with hydroxylamine followed[124] with diazotization and replacement of the diazonium group with sodium nitrite[125] (equation 21).

Nucleophilic displacement of nitro groups can also occur with di- and polynitro compounds permitting the formation of unusual orientations[126,127] (equations 31 and 32).

$$1,3-(O_2N)_2C_6H_4 \xrightarrow[MeOH]{KCN} 2-CN-1-CH_3O-3-NO_2-C_6H_3 \quad (31)$$

$$1,3,5-(NO_2)_3C_6H_3 \xrightarrow{CH_3ONa} 1-OCH_3-3,5-(O_2N)_2C_6H_3 \quad (32)$$

A long-involved synthesis is necessary to prepare 2,3-dinitronaphthalene[128]; the overall yield was 14%. This most inaccessible of the ten dinitronaphthalenes is obtained by starting with 6-acetyltetralin (equation 33).

$$\text{6-}COCH_3\text{-tetralin} \xrightarrow{H_2NOH} \text{6-}C(CH_3)NOH\text{-tetralin} \xrightarrow[P_2O_5]{H_3PO_4} \text{6-}NHCOCH_3\text{-tetralin} \xrightarrow[Ac_2O]{HNO_3} \text{6-NHAc-7-}NO_2\text{-tetralin} \longrightarrow$$

$$\text{6-}NH_2\text{-7-}NO_2\text{-tetralin} \xrightarrow[NaNO_2]{HNO_2} \text{6,7-}(NO_2)_2\text{-tetralin}$$

$$\text{6,7-}(NO_2)_2\text{-tetralin} \xrightarrow[100^\circ]{Br_2} \xrightarrow{170-190^\circ} \text{2,3-}(NO_2)_2\text{-naphthalene} \quad (33)$$

Nitration of anthracene occurs in the 9 and 10 positions. Mono- and polynitration of 9,10-dehydro-9,10-ethanoanthracene gives only β substitution (equation 34). Pyrolysis of the ethanoanthracenes

$$\text{9,10-dihydro-9,10-ethanoanthracene} \xrightarrow[CH_3NO_2,\ 0-3^\circ]{HNO_3-Ac_2O} \text{2-}NO_2 \xrightarrow[CH_3NO_2,\ 0-3^\circ]{HNO_3-Ac_2O}$$

$$\text{2,6-}(NO_2)_2 + \text{2,7-}(O_2N)_2 \xrightarrow[CH_3NO_2,\ 2-5^\circ]{HNO_3-H_2SO_4}$$

$$\text{2,3,6-}(NO_2)_3 \xrightarrow[CH_3NO_2,\ 23^\circ]{HNO_3-H_2SO_4} \text{2,3,6,7-}(O_2N)_4 \quad (34)$$

causes a reverse Diels–Alder thereby giving β-substituted anthracenes[129]. β-Substitution seems to be usual with benzocyclenes[130].

Thiophene directs to the 2 position when nitrated with nitric acid in acetic anhydride. About 5% of the 3 isomer is produced but isolation is not possible. The only feasible preparation of 3-nitrothiophene is from the chlorosulfonation of thiophene[131] (equation 35).

S $\xrightarrow{ClSO_3H}$ S SO_2Cl $\xrightarrow{HNO_3}$

O_2N S SO_2Cl $\xrightarrow{H_2O}$ NO_2 S

(35)

The 2,4-dinitrothiophene can be obtained from either the 2-nitro- or 3-nitrothiophene; 2,5-dinitrothiophene is produced as a by-product (15%) in the nitration of 2-nitrothiophene. Heating the mixture of 2,4- and 2,5-dinitrothiophenes with mixed acid will destroy the 2,4-dinitro compound while the 2,5 isomer is unaffected[131].

The preparation of 3,4-dinitrothiophene has been achieved by using halogens as blocking groups. Dinitration of 2,5-dibromothiophene gives the 2,5-dibromo-3,4-dinitrothiophene[132]. One of the bromine atoms is replaced by hydrogen on treatment with either sodium iodide in acetic acid or with hypophosphorous acid; the remaining halogen can only be removed with copper in refluxing butyric acid[133] (equation 36). Nitration of the monobromo derivative

O_2N NO_2 Br S Br $\xrightarrow{H_3PO_2}$ O_2N NO_2 S Br $\xrightarrow[C_3H_7CO_2H]{Cu}$ O_2N NO_2 S

(36)

with mixed acid provides 2-bromo-3,4,5-trinitrothiophene which can be reduced to the trinitrothiophene with hypophosphorous acid.

The preparation of 2,3-diaminopyridine also makes use of bromine as a blocking agent prior to nitration; the bromine is removed by catalytic hydrogenation with palladium[134] (equation 37).

N NH_2 $\xrightarrow{Br_2}$ Br N NH_2 $\xrightarrow[H_2SO_4]{HNO_3}$ Br NO_2 N NH_2 $\xrightarrow[HCl]{Fe}$

Br NH_2 N NH_2 $\xrightarrow[Pd]{H_2}$ NH_2 N NH_2

(37)

The *N*-oxides of pyridines provide a means of preparing 4-nitropyridines. The *N*-oxide is reduced with phosphorous trichloride; phosphorous tribromide also reduces the *N*-oxide but replaces the nitro group with bromine[136] (equation 38).

$$\text{3-methylpyridine} \xrightarrow[\text{HOAc}]{H_2O_2} \text{3-methylpyridine } N\text{-oxide} \xrightarrow[H_2SO_4]{KNO_3} \text{4-nitro-3-methylpyridine } N\text{-oxide} \xrightarrow{PCl_3} \text{4-nitro-3-methylpyridine} \quad (38)$$

Carboxyl groups may be used as blocking agents. Carbazole is nitrated in the 3 position predominately but only to a slight extent (4%) in the 1 position[137]. To prepare the 1-nitrocarbazole, 3,6-carbazoledicarboxylic acid is first nitrated to 1-nitro-3,6-carbazoledicarboxylic acid which is then decarboxylated[138] (equation 39).

$$\text{1-nitro-3,6-carbazoledicarboxylic acid } (HO_2C, CO_2H, NO_2, N\text{–}H) \xrightarrow[\text{quinoline}]{\text{Cu}} \text{1-nitrocarbazole} \quad (39)$$

A similar decarboxylation with copper chromite and quinoline of 4-nitropyrrole-2-carboxylic acid provides the most convenient route to 3-nitropyrrole. However, in this case, which is common with many heterocyclic compounds, the substituted pyrrole nucleus must be prepared through ring closure; the 4-nitropyrrole-2-carboxylic acid is made by the condensation of nitromalondialdehyde with glycine ethyl ester[139]. Ring closure, also is the best method to prepare 2-nitrocarbazole, a carcinogen. The synthesis employs an interesting deactivation of one ring of biphenyl with acid to nitrate the other ring[140] (equation 40).

$$\text{2-nitrobiphenyl} \xrightarrow[\text{PdC}]{H_2} \text{2-aminobiphenyl} \xrightarrow[H_2SO_4]{HNO_3} \text{2-amino-4'-nitrobiphenyl} \xrightarrow[H_2SO_4]{NaNO_2} \text{4'-nitrobiphenyl-2-diazonium } (N_2^{\oplus}) \xrightarrow{NaN_3} \text{2-azido-4'-nitrobiphenyl } (N_3) \xrightarrow{170°} \text{2-nitrocarbazole} \quad (40)$$

VIII. MISCELLANEOUS ELECTROPHILIC NITRATING REAGENTS

Numerous combinations of Lewis acids with varied carriers of an incipient nitronium ion have been reported. A comprehensive discussion of these miscellaneous nitrating reagents has been presented by Olah and Kuhn[16b]. Aside from the sulfuric acid catalyzed nitration with alkyl nitrates, these reagents give poor yields and offer no advantage over the methods of nitration already discussed. The use of nitrate esters should be investigated more, since there is indication[59] that lower temperatures are needed to effect nitration and large amounts of sulfuric acid, such as are used to compensate for the water content of nitric acid, can be avoided. Because the comprehensive survey[16b] of these various nitrating agents has been done and since they are mainly of academic interest, only a tabulation with pertinent references is given in Table 14.

TABLE 14. Miscellaneous nitrating agents.

NO_2^+ carrier	Lewis acid catalyst	Ref.
HNO_3	HF	141
HNO_3	BF_3	142
$AgNO_3$, $Ba(NO_3)_2$, $NaNO_3$, KNO_3, NH_4NO_3, $Pb(NO_3)_2$	$FeCl_3$, BF_3, $SiCl_4$, $AlCl_3$	143
$RONO_2$	H_2SO_4	59, 144–147
$EtONO_2$	$AlCl_3$, $SnCl_4$, $SbCl_5$, $FeCl_3$	143
NO_2F	—	148
NO_2F	BF_3, PF_5, AsF_5, SbF_5	149
NO_2Cl	HF, $AlCl_3$	150
	$TiCl_4$	149
N_2O_3	BF_3	151
N_2O_4	H_2SO_4	152, 153
N_2O_4	$AlCl_3$, $FeCl_3$	154–157
N_2O_4	BF_3	143, 158–161
N_2O_4	SbF_5, AsF_5, IF_5	162
N_2O_5	—	4
N_2O_5	BF_3	163

A novel nitrating agent must be mentioned, however, because its use may provide a means of decreasing the amount of *ortho* nitration as the results of the steric requirements of the reagent[164]. This novel reagent is derived from a sulfonated resin anhydrous polystyrene polysulfonic acid (Rohm and Haas amberlite IR-120).

Dehydrated (azeotropic distillation with toluene) sulfonated resin behaves like concentrated sulfuric acid toward 90% nitric acid producing nitronium ion which is ion paired with the resin (equation 41). The ion-pair salt, thus formed, is considerably larger

$$HNO_3 + 2ResSO_3H \longrightarrow NO_2^{\oplus} + H_3O^{\oplus} + 2ResSO_3^{\ominus} \quad (41)$$

than a 'free' nitronium ion. Nitration of toluene with this reagent at 65–70° gives nitrotoluenes with an *ortho–para* ratio of 0.68, much lower than the *ortho*:*para* ratio of 1.65 obtained when nitration is effected with 90% nitric acid alone.

Finally—although it may not be essentially electrophilic—mention is given to the photochemical nitration of quinoline 1-oxide in the *3 position* with nitrosyl chloride and butyl nitrite, as unexpected nitrating agents and an unusual product orientation[165].

Cognizance should be given to the fact that many of the procedures in the preparation of aromatic nitro compounds may be explosive. Safety shields ought to be standard operating procedure when working with acetyl nitrate (or acetic anhydride plus nitric acid), tetranitromethane, or hydrogen peroxide. The toxicity of the oxides of nitrogen ought to be recalled, and the knowledge that aromatic nitro compounds are very poisonous and readily absorbed through the skin should be impressed on all. Note should also be taken that some nitro heterocyclic compounds and nitro derivatives of fused ring systems may be carcinogenic.

IX. REFERENCES

1. P. C. Myhre and M. Beug, *J. Am. Chem. Soc.*, **88**, 1569 (1966).
2. C. K. Ingold, *Structure and Mechanism in Organic Chemistry*, Cornell University Press, Ithaca, N.Y., 1953.
3. R. J. Gillespie and D. J. Millen, *Quart. Rev.* (London), **2**, 277 (1948).
4. P. B. D. De La Mare and J. H. Ridd, *Aromatic Substitution, Nitration and Halogenation*, Academic Press, New York, 1959.
5. J. H. Ridd in *Studies on Chemical Structure and Reactivity* (Ed. J. H. Ridd), John Wiley and Sons, New York, 1966.
6. M. A. Paul, *J. Am. Chem. Soc.*, **80**, 5329 (1958).
7. P. B. D. De La Mare and R. Koenigsberger, *J. Chem. Soc.*, **1964**, 5327.
8. G. Wheland, *J. Am. Chem. Soc.*, **64**, 900 (1942).
9. G. A. Olah and S. J. Kuhn, *J. Am. Chem. Soc.*, **80**, 6541 (1958).
10. M. J. S. Dewar, *The Electronic Theory of Organic Chemistry*, Clarendon Press, Oxford, 1949.
11. H. C. Brown and J. D. Brady, *J. Am. Chem. Soc.*, **74**, 3570 (1952).
12. R. E. Rundle and J. H. Goring, *J. Am. Chem. Soc.*, **72**, 5337 (1950).
13. G. A. Olah, S. J. Kuhn, and S. H. Flood, *J. Am. Chem. Soc.*, **83**, 4571 (1961).
14. G. S. Hammond, *J. Am. Chem. Soc.*, **77**, 334 (1955).
15. R. O. C. Norman and R. Taylor, *Electrophilic Substitution in Benzenoid Compounds*, Elsevier Publishing Co., New York, 1965.

16. (a) Ref. 13.
(b) G. A. Olah and S. J. Kuhn in *Friedel-Crafts and Related Reactions*, Vol. III, Part 2 (Ed. G. A. Olah), Interscience Publishers, New York, 1964, p. 1461.
(c) G. A. Olah, S. J. Kuhn, S. H. Flood, and J. C. Evans, *J. Am. Chem. Soc.*, **84,** 3687 (1962).
(d) A. Fisher, J. Packer, J. Vaughan, and G. J. Wright, *J. Chem. Soc.*, **1964**, 3687.
(e) R. L. Jenkins, R. McCullough, and C. F. Booth, *Ind. Eng. Chem.*, **22**, 31 (1930).
(f) O. Semamura and Y. Mizuno, *J. Chem. Soc.*, **1958**, 3875.
(g) E. Hayashi, K. Inana, and T. Ishikawa, *J. Pharm. Soc. Japan*, **79**, 972 (1959).
(h) J. R. Knowles, R. O. C. Norman, and G. K. Radda, *J. Chem. Soc.*, **1960**, 4885.
(i) A. F. Holleman and B. R. deBruyn, *Rec. Trav. Chim.*, **19**, 188 (1900).
(j) M. A. Paul, *J. Am. Chem. Soc* , **80**, 5332 (1958).
(k) J. D. Roberts, J. K. Sanford, F. L. J. Sixma, H. Cerfontain, and R. Zagt, *J. Am. Chem. Soc.*, **76**, 4525 (1954).
(l) F. Arnall and T. Lewis, *J. Soc. Chem. Ind.*, **48**, 159T (1929).
(m) P. Kovacic and J. J. Hiller, Jr., *J. Org. Chem.*, **30**, 2871 (1965).
(n) P. H. Griffiths, W. A. Walkey, and H. B. Watson, *J. Chem. Soc.*, **1934**, 631.
(o) K. Halvarson and L. Melander, *Arkiv Kemi*, **11**, 77 (1957).
17. (a) Ref. 15, p. 66.
(b) Ref. 5, p. 141.
18. Ref. 16h.
19. E. D. Hughes, C. K. Ingold, and R. I. Reed, *J. Chem. Soc.*, **1950**, 2400.
20. L. L. Craccio and R. A. Marcus, *J. Am. Chem. Soc.*, **84**, 1838 (1962).
21. Ref. 16d.
22. A. G. Anderson, J. A. Nelson, and J. Tazuma, *J. Am. Chem. Soc.*, **75**, 4980 (1953).
23. H. C. Brown and R. A. Wirkkala, *J. Am. Chem. Soc.*, **88**, 1447 (1966).
24. (a) H. C. Brown and L. M. Stock, *J. Am. Chem. Soc.*, **79**, 1421 (1957).
(b) Ref. 23.
(c) H. Cohen, E. D. Hughes, M. H. Jones, and M. A. Peeling, *Nature*, **169**, 291 (1951).
(d) C. K. Ingold, A. Lapworth, E. Rothstein, and D. Ward, *J. Chem. Soc.*, **1931**, 1959.
25. Ref. 6.
26. W. S. Tolgyesi, *Can. J. Chem.*, **43**, 343 (1965).
27. L. M. Stock, *J. Org. Chem.*, **26**, 4120 (1961).
28. (a) Ref. 16o.
(b) R. O. C. Norman and G. K. Radda, *J. Chem. Soc.*, **1961**, 3030.
(c) Ref. 4, p. 76.
29. Ref. 16m.
30. (a) P. H. Griffiths, W. A. Walkey, and H. B. Watson, *J. Chem. Soc.*, **1961**, 3610.
(b) Ref. 29.
(c) Ref. 28a.
(d) Ref. 16l.
(e) A. F. Holleman, *Chem. Rev.*, **1**, 187 (1925).
(f) Ref. 28b.
(g) D. R. Harvey and R. O. C. Norman, *J. Chem. Soc.*, **1962**, 3822.
(h) Ref. 16e.
(i) F. Bell, J. Kenyon, and P. H. Robinson, *J. Chem. Soc.*, **1926**, 1239.
(j) C. J. Billings and R. O. C. Norman, *J. Chem. Soc.*, **1961**, 3885.
31. (a) Ref. 27.
(b) Ref. 16h.

32. N. Kornblum, R. J. Berrigan, and W. J. LeNoble, *J. Am. Chem. Soc.*, **85**, 1141 (1963).
33. G. A. Olah, J. A. Olah, and N. A. Overchuk, *J. Org. Chem.*, **30**, 3373 (1965).
34. (a) E. Bamberger and K. Landsteiner, *Chem. Ber.*, **26**, 485 (1893).
(b) E. Bamberger, *Chem. Ber.*, **27**, 359, 384 (1894); **28**, 399 (1895); **30**, 1248 (1897).
35. (a) J. Glager, E. D. Hughes, C. K. Ingold, A. T. James, G. T. Jones, and E. Roberts, *J. Chem. Soc.*, **1950**, 2657.
(b) E. D. Hughes and G. T. Jones, *J. Chem. Soc.*, **1950**, 2678.
36. A. F. Holleman, J. C. Hartogs, and T. van der Linden, *Chem. Ber.*, **44**, 704 (1911).
37. C. A. Bunton, E. D. Hughes, C. K. Ingold, D. I. H. Jacobs, M. H. Jones, G. J. Minkoff, and R. I. Reed, *J. Chem. Soc.*, **1950**, 2628.
38. K. A. Kobe and H. Levin, *Ind. Eng. Chem.*, **42**, 352, 356 (1950).
39. R. D. Kleene, *J. Am. Chem. Soc.*, **71**, 2259 (1949).
40. (a) Ref. 30e.
(b) J. P. Wibaut and R. van Strik, *Rec. Trav. Chim.*, **77**, 317 (1958).
(c) J. W. Baker and W. G. Moffitt, *J. Chem. Soc.*, **1931**, 314.
(d) K. E. Cooper and C. K. Ingold, *J. Chem. Soc.*, **1927**, 836.
41. E. R. Ward, C. D. Johnson, and L. A. Day, *J. Chem. Soc.*, **1959**, 487.
42. G. S. Hammond and M. F. Hawthorne in *Steric Effects in Organic Chemistry* (Ed. M. S. Newman), John Wiley and Sons, New York, 1956, p. 180.
43. G. S. Hammond and K. J. Douglas, *J. Am. Chem. Soc.*, **81**, 1184 (1959).
44. (a) C. W. McGary, Y. Okamoto, and H. C. Brown, *J. Am. Chem. Soc.*, **77**, 3037 (1955).
(b) Y. Yukawa and Y. Tsuno, *Bull. Chem. Soc. Japan*, **32**, 971 (1959).
(c) Ref. 16h.
45. (a) L. M. Stock, *J. Org. Chem.*, **26**, 4120 (1961).
(b) Y. Okamoto and H. C. Brown, *J. Am. Chem. Soc.*, **80**, 4976 (1958).
(c) L. M. Stock and H. C. Brown, *J. Am. Chem. Soc.*, **84**, 1242 (1962).
46. J. Hine, *Physical Organic Chemistry*, McGraw-Hill, New York, 1956, p. 72.
47. P. E. Fanta and D. S. Tarbell in *Organic Syntheses*, Coll. Vol. III (Ed. E. C. Horning), John Wiley and Sons, New York, 1960, p. 661.
48. R. C. Fuson, *Reactions of Organic Compounds*, John Wiley and Sons, New York, 1962, p. 39.
49. (a) Ref. 16h.
(b) Ref. 16j.
(c) J. D. Roberts, *J. Am. Chem. Soc.*, **76**, 4525 (1954).
(d) J. R. Knowles and R. O. C. Norman, *J. Chem. Soc.*, **1961**, 2938.
50. R. Bonnett and G. F. Stephenson, *J. Org. Chem.*, **30**, 2791 (1965).
51. (a) D. J. Cram and R. D. Partos, *J. Am. Chem. Soc.*, **85**, 1273 (1963).
(b) F. Sondheimer and A. Shani, *J. Am. Chem. Soc.*, **86**, 3168 (1964).
(c) J. W. Cook, J. D. Loudon, and D. K. V. Steel, *J. Chem. Soc.*, **1954**, 530.
(d) A. G. Anderson, R. Scotoni, E. J. Cowles, and G. Fritz, *J. Org. Chem.*, **22**, 1193 (1957).
(e) C. E. Braun, C. D. Cook, C. Merritt, Jr., and J. E. Rousseau in *Organic Syntheses*, Coll. Vol. IV (Ed. N. Rabjohn), John Wiley and Sons, New York, 1963 p. 711.
(f) J. C. Howard in *Organic Syntheses*, Coll. Vol. IV (Ed. N. Rabjohn), John Wiley and Sons, New York, 1963, p. 42.
(g) R. E. Buckles and M. P. Bellis in *Organic Syntheses*, Coll. Vol. IV (Ed. N. Rabjohn), John Wiley and Sons, New York, 1963, p. 722.
(h) Ref. 47.
(i) C. A. Fetscher in *Organic Syntheses*, Coll. Vol. IV (Ed. N. Rabjohn), John Wiley and Sons, New York, 1963, p. 735.

(j) K. A. Kobe and T. F. Doumain in *Organic Syntheses*, Coll. Vol. III (Ed. E. C. Horning), John Wiley and Sons, New York, 1960, p. 653.
(k) G. R. Robertson, *Laboratory Practice of Organic Chemistry*, Macmillan, New York, 1954, p. 252.
(l) H. M. Fitch in *Organic Syntheses*, Coll. Vol. III (Ed. E. C. Horning), John Wiley and Sons, New York, 1960, p. 658.
(m) R. N. Icke, C. E. Redemann, B. B. Wisegarver, and G. A. Alles in *Organic Syntheses*, Coll. Vol. III (Ed. E. C. Horning), John Wiley and Sons, New York, 1960, p. 644.
(n) Ref. 51k, p. 255.
(o) E. C. Taylor, Jr., and A. J. Crovetti in *Organic Syntheses*, Coll. Vol. IV (Ed. N. Rabjohn), John Wiley and Sons, New York, 1963, p. 654.
(p) R. Q. Brewster, B. Williams, and R. Phillips in *Organic Syntheses*, Vol. III (Ed. E. C. Horning), John Wiley and Sons, New York, 1960, p. 337.
(q) H. P. Schultz in *Organic Syntheses*, Coll. Vol. IV (Ed. N. Rabjohn), John Wiley and Sons, New York, 1963, p. 364.
(r) E. O. Woolfolk and M. Orchin in *Organic Syntheses*, Coll. Vol. III (Ed. E. C. Horning), John Wiley and Sons, New York, 1960, p. 837.
52. F. C. Whitmore, *Organic Chemistry*, Vol. II, Dover Publications, New York, 1961, p. 633.
53. L. G. Radcliffe and A. A. Pollitt, *J. Soc. Chem. Ind.* (London), **40**, 45T (1921).
54. G. Powell and F. R. Johnson in *Organic Syntheses*, Coll. Vol. II (Ed. A. H. Blatt) John Wiley and Sons, New York, 1943, p. 449.
55. G. Illuminati and M. P. Illuminati, *J. Am. Chem. Soc.*, **75**, 2159 (1953).
56. R. C. Miller, D. S. Noyce, and T. Vermeulen, *Ind. Eng. Chem.*, **56**, 43 (1964).
57. N. D. Chernois and J. B. Entrikin, *Semimicro Qualitative Organic Analysis*, Interscience Publishers, New York, 1961.
58. R. J. Gillespie and D. G. Norton, *J. Chem. Soc.*, **1953**, 971.
59. E. Plaźak and S. Roupuszyński, *Rocz. Chem.*, **32**, 681 (1958); *Chem. Abstr.*, **53**, 3111 (1959).
60. W. M. Lauer, G. W. Matson, and G. Stedman, *J. Am. Chem. Soc.*, **80**, 6439, (1958).
61. V. Gold and F. L. Tye, *J. Chem. Soc.*, **1952**, 2172, 2184.
62. (a) C. McLean, J. H. van der Waals, and E. L. Mackor, *Mol. Phys.*, **I**, 247 (1958).
(b) G. Dallinga and G. Ter Maten, *Rec. Trav. Chim.*, **79**, 737 (1960).
63. L. Sternbach, R. I. Fryer, O. Keller, W. Metlesics, G. Sach, and N. Steiger, *J. Med. Chem.*, **6**, 261 (1963).
64. R. I. Fryer, J. V. Early, and L. H. Sternbach, *J. Org. Chem.*, **30**, 521 (1965).
65. F. H. S. Curd, W. Graham, D. N. Richardson, and F. L. Rose, *J. Chem. Soc.*, **1947**, 1613.
66. G. Bacharach, A. H. Haut, and L. Caroline, *Rec. Trav. Chim.*, **52**, 413 (1933).
67. W. E. Noland, L. R. Smith, and K. R. Rush, *J. Org. Chem.*, **30**, 3457 (1965).
68. K. Schofield, *Quart. Rev.*, **4**, 382 (1950).
69. (a) W. E. Noland, L. R. Smith, and D. C. Johnson, *J. Org. Chem.*, **28**, 2262 (1963).
(b) W. E. Noland, K. R. Rush, and L. R. Smith, *J. Org. Chem.*, **31**, 65 (1965).
(c) W. E. Noland and K. R. Rush, *J. Org. Chem.*, **31**, 70 (1965).
70. J. H. Ridd and J. H. P. Utley, *Proc. Chem. Soc.*, **1964**, 24.
71. A. N. Nesmeyanov, T. P. Tolstaya, L. S. Isaeva, and A. V. Grib, *Dokl. Akad. Nauk SSSR*, **133**, 602 (1960).
72. K. Brown and A. R. Katritzky, *Tetrahedron Letters*, **1964**, 803.
73. D. V. Nightingale, *Chem. Rev.*, **40**, 117 (1947).

74. H. H. Hodgson and J. Nixon, *J. Chem. Soc.*, **1930**, 1085.
75. G. A. Olah and S. J. Kuhn, *J. Am. Chem. Soc.*, **86**, 1067 (1964).
76. P. C. Myhre and M. Beug, *J. Am. Chem. Soc.*, **88**, 1568 (1966).
77. T. F. Doumani and K. A. Kobe, *J. Org. Chem.*, **7**, 1 (1942).
78. F. B. Deans and C. Eaborn, *J. Chem. Soc.*, **1957**, 498.
79. R. A. Benkeser and H. Landesman, *J. Am. Chem. Soc.*, **76**, 904 (1954).
80. A. Newton, *J. Am. Chem. Soc.*, **65**, 2434 (1943).
81. (a) F. Reverdin, *Chem. Ber.*, **40**, 2442 (1907).
    (b) F. Reverdin, *Bull. Soc. Chim.* [4] **1**, 618 (1907).
82. (a) W. B. Bentley, *Am. Chem. J.*, **24**, 171 (1900).
    (b) M. P. DeLange, *Rec. Trav. Chim.*, **45**, 19 (1926).
83. C. M. Suter, *Organic Chemistry of Sulfur*, John Wiley and Sons, New York, 1945.
84. E. Schmidt and H. Fisher, *Chem. Ber.*, **53**, 1529 (1920).
85. D. H. Reid, W. H. Stafford, and W. L. Stafford, *J. Chem. Soc.*, **1958**, 1118.
86. J. F. Riordan, M. Sokolovsky, and B. L. Vallee, *J. Am. Chem. Soc.*, **88**, 4104 (1966).
87. H. Grubert and K. L. Rinehart, *Tetrahedron Letters*, **12**, 16 (1959).
88. J. F. Helling and H. Shechter, *Chem. Ind.* (*London*), 1157 (1959).
89. A. Angeli and L. Alessandri, *Atti. Acad. Real. Lincei.* [5] **20**, 311 (1910).
90. K. J. Morgan and D. P. Morrey, *Tetrahedron*, **22**, 57 (1966).
91. M. Battegay and P. Brandt, *Bull. Soc. Chim.* [4] **31**, 910 (1922).
92. W. Treibs, *Angew. Chem.*, **67**, 76 (1955).
93. (a) M. Kohn and S. Grün, *Monatsh. Chem.*, **45**, 663 (1924).
    (b) G. Schill, *Chem. Ber.*, **99**, 714 (1966).
94. W. D. Langley in *Organic Syntheses*, Coll. Vol. III (Ed. E. C. Horning), John Wiley and Sons, New York, 1964, p. 334.
95. (a) R. R. Holmes and R. P. Bayer, *J. Am. Chem. Soc.*, **82**, 3454 (1960).
    (b) W. D. Emmons, *J. Am. Chem. Soc.*, **79**, 5528 (1957).
    (c) W. D. Emmons, *J. Am. Chem. Soc.*, **76**, 3470 (1954).
96. R. H. Wiley and J. L. Hartman, *J. Am. Chem. Soc.*, **73**, 494 (1951).
97. L. C. Craig, *J. Am. Chem. Soc.*, **56**, 231 (1934).
98. A. Kirpal and W. Bohm, *Chem. Ber.*, **64**, 767 (1931); **65**, 680 (1932).
99. (a) H. H. Hodgson, F. Heyworth, and E. R. Ward, *J. Chem. Soc.*, **1948**, 1512.
    (b) E. R. Ward, C. D. Johnson, and J. G. Hawkins, *J. Chem. Soc.*, **1960**, 894.
    (c) H. H. Hodgson, A. P. Mahadevan, and E. R. Ward, *J. Chem. Soc.*, **1947**, 1392.
    (d) H. H. Hodgson and E. J. Marsden, *J. Chem. Soc.*, **1944**, 22.
100. E. B. Starkey in *Organic Syntheses*, Coll. Vol. II (Ed. A. H. Blatt), John Wiley and Sons, New York, 1943, p. 225.
101. R. Shapiro, *J. Am. Chem. Soc.*, **86**, 2948 (1964).
102. K. E. Cooper and C. K. Ingold, *J. Chem. Soc.*, **1927**, 836.
103. J. L. Riebsomer, *Chem. Rev.*, **36**, 157 (1945).
104. A. I. Titov, *Zh. Obshch. Khim.*, **18**, 190 (1948); *Chem. Abstr.*, **44**, 1044 (1950).
105. P. P. Shorygin, A. V. Topchiev, and V. A. Anan'ina, *Zh. Obshch. Khim.*, **8**, 981 (1938); *Chem. Abstr.*, **33**, 3781 (1939).
106. P. P. Shorygin and A. Topchiev, *Chem. Ber.*, **69B**, 1874 (1936).
107. H. J. den Hertog, Jr., and J. Overhoff, *Rec. Trav. Chim.*, **49**, 552 (1930).
108. C. S. Foote, P. Engel, and T. W. Del Pesco, *Tetrahedron Letters*, **31**, 2669 (1965).
109. D. Avanesov and I. Vyatskin, *Khim. Referat. Zh.*, **2**, 43 (1939); *Chem. Abstr.*, **34**, 2262 (1940).
110. Ref. 88.
111. F. H. Westheimer, E. Segel, and R. Schramm, *J. Am. Chem. Soc.*, **69**, 773 (1947).

112. E. Bamberger, *Chem. Ber.*, **30**, 506 (1897).
113. J. Kunz, *Chem. Ber.*, **31**, 1528 (1898).
114. J. W. Cook, J. D. Loudon, and D. K. V. Steel, *J. Chem. Soc.*, **1954**, 530.
115. D. V. Banthrope, J. A. Thomas, and D. L. H. Williams, *J. Chem. Soc.*, **1965**, 6135.
116. D. V. Banthrope, E. D. Hughes, and D. L. H. Williams, *J. Chem. Soc.*, **1964**, 5349.
117. W. N. White, E. F. Wolfarth, J. R. Klink, J. Kindig, C. Hathaway, and D. Lazdins, *J. Org. Chem.*, **26**, 4124 (1961).
118. E. Bamberger, *Chem. Ber.*, **26**, 472 (1893); **27**, 363 (1894).
119. J. P. Freeman and I. G. Shepard, *Org. Syn.*, **43**, 83 (1963).
120. W. D. Emmons and J. P. Freeman, *J. Am. Chem. Soc.*, **77**, 4387 (1955).
121. S. Brownstein, C. A. Bunton, and E. D. Hughes, *J. Chem. Soc.*, **1958**, 4354.
122. W. N. White, D. Lazdins, and H. S. White, *J. Am. Chem. Soc.*, **86**, 1517 (1964); W. N. White, C. Hathaway, and D. Huston, *J. Org. Chem.*, **35**, 737 (1970).
123. A. E. Bradfield and K. J. P. Orton, *J. Chem. Soc.*, **1929**, 915.
124. C. C. Price and Sing-Tuh Voong in *Organic Syntheses*, Coll. Vol. III (Ed. E. C. Horning), John Wiley and Sons, New York, 1964, p. 664.
125. H. H. Hodgson, A. P. Mahadevan, and E. R. Ward, in *Organic Syntheses*, Coll. Vol. III (Ed. E. C. Horning), John Wiley and Sons, New York, 1960, p. 341.
126. A. Russell and W. G. Tebbens in *Organic Syntheses*, Coll. Vol. III (Ed. E. C. Horning), John Wiley and Sons, New York, 1960, p. 293.
127. F. Reverdin in *Organic Syntheses*, Coll. Vol. I (Eds. H. Gilman and A. H. Blatt), John Wiley and Sons, New York, 1941, p. 219.
128. E. R. Ward and T. M. Coulson, *J. Chem. Soc.*, **1954**, 4545.
129. H. Tanida and H. Ishitobi, *Tetrahedron Letters*, **15**, 807 (1964).
130. H. Tanida and R. Muneyuki, *J. Am. Chem. Soc.*, **87**, 4794 (1965).
131. A. H. Blatt, S. Bach, and L. W. Kresch, *J. Org. Chem.*, **22**, 1693 (1957).
132. R. Mozingo, S. A. Harris, D. E. Wolf, C. E. Hoffhine, Jr., N. R. Easton, and K. Folkers, *J. Am. Chem. Soc.*, **67**, 2092 (1945).
133. A. H. Blatt, N. Gross, and E. W. Tristram, *J. Org. Chem.*, **22**, 1588 (1957).
134. F. A. Fox and T. L. Threlfall, *Org. Syn.*, **44**, 34 (1964).
135. W. Herz and L. Tsai, *J. Am. Chem. Soc.*, **76**, 4184 (1954).
136. T. B. Lee and G. A. Swan, *J. Chem. Soc.*, **1956**, 771.
137. H. Lindemann, *Chem. Ber.*, **57**, 555 (1924).
138. S. H. Tucker, R. W. G. Preston, and J. M. L. Cameron, *J. Chem. Soc.*, **1942**, 500.
139. W. J. Hale and W. Hoyt, *J. Am. Chem. Soc.*, **37**, 2551 (1915).
140. G. D. Mendenhall and P. A. S. Smith, *Org. Syn.*, **46**, 85 (1966).
141. J. H. Simons, J. M. Passino, and S. Archer, *J. Am. Chem. Soc.*, **63**, 608 (1941).
142. R. J. Thomas, W. F. Anzelotti, and C. F. Hennion, *Ind. Eng. Chem.*, **32**, 908 (1940).
143. A. V. Topchiev, *Nitration of Hydrocarbons and Other Organic Compounds*, Pergamon Press, New York, 1959.
144. H. Raudnitz, *Chem. Ber.*, **60**, 738 (1927).
145. H. R. Wright and W. J. Donalson, U.S. Pat. 2,416,974; *Chem. Abstr.*, **41**, 3485 (1947).
146. H. Colonna, *Pubbl. Ist. Chim. Univ. Bologna*, **2**, 3 (1943); *Chem. Abstr.*, **41**, 754 (1947).
147. H. Colonna and R. Andrisano, *Pubbl. Ist. Chim. Univ. Bologna*, **3**, 3 (1944); **4**, 3 (1945); *Chem. Abstr.*, **41**, 754 (1947).
148. G. Hetherington and P. L. Robinson, *J. Chem. Soc.*, **1954**, 3512.
149. S. J. Kuhn and G. A. Olah, *J. Am. Chem. Soc.*, **83**, 4564 (1961).
150. C. C. Price and C. S. Sears, *J. Am. Chem. Soc.*, **75**, 3276 (1953).
151. G. B. Bachman and T. Hokamo, *J. Am. Chem. Soc.*, **79**, 4370 (1957).

152. L. A. Penck, *J. Am. Chem. Soc.*, **49**, 2536 (1927).
153. M. Battegay, *Bull. Soc. Chim. France*, **43**, 109 (1928).
154. A. Schaarschmidt, *Chem. Ber.*, **57**, 2065 (1924).
155. A. Schaarschmidt, *Angew. Chem.*, **39**, 1457 (1926).
156. A. Schaarschmidt, H. Balzerkiewicz, and J. Gante, *Chem. Ber.*, **58**, 499 (1925).
157. A. I. Titov, *J. Gen. Chem. USSR*, **7**, 591, 667 (1937).
158. R. W. Sprague, A. B. Garrett, and H. H. Sisler, *J. Am. Chem. Soc.*, **82**, 1059 (1960).
159. J. L. Andrews and R. M. Keefer, *J. Am. Chem. Soc.*, **73**, 4169 (1951).
160. G. B. Bachman, H. Feuer, B. R. Bluestein, and C. M. Vogt, *J. Am. Chem. Soc.*, **77**, 6188 (1955).
161. G. B. Bachman and C. M. Vogt, *J. Am. Chem. Soc.*, **80**, 2987 (1958).
162. E. E. Aynsley, G. Hetherington, and P. L. Robinson, *J. Chem. Soc.*, **1954**, 1119.
163. G. B. Bachman and J. L. Dever, *J. Am. Chem. Soc.*, **80**, 5871 (1958).
164. O. L. Wright, J. Teipel, and D. Thoennes, *J. Org. Chem.*, **30**, 1301 (1965).
165. T. Kosuge, M. Yokota, and H. Sawanishi, *Chem. Pharm. Bull.* (Tokyo), **13**, 1480 (1965).
166. C. A. Kupas and R. L. Pearson, *J. Am. Chem. Soc.*, **90**, 4742 (1968).
167. T. C. Bruice, M. J. Gregory, and S. L. Wolters, *J. Am. Chem. Soc.*, **90**, 1612 (1968).

CHAPTER 2

Directing effects of the nitro group in electrophilic and radical aromatic substitutions

TADEUSZ URBAŃSKI

Warsaw Institute of Technology (Politechnika), Warszawa, Poland

I. ELECTROPHILIC SUBSTITUTION

A. Introduction

It is now accepted that nitration is due to the action of a positively charged ion (i.e., a cation) NO_2^+. Therefore a nitration by ionic substitution (or displacement) is an *electrophilic substitution*, according to the well-known nomenclature introduced by Ingold[1] ('Kationoid' substitution according to Lapworth[2] and Robinson[3]). In electrophilic substitution the two electrons which form the new covalent bond (i.e., the bond between the aromatic compound and the electrophilic reagent) are both supplied by the aromatic compound (equation 1).

$$\text{(1)} \xrightarrow[k_1]{NO_2^+} \text{(2)} \xrightarrow[k_2]{-H^+} \text{(3)} \qquad (1)$$

(1) (slow) (2) (fast) (3)

In **1** two 'free' electrons are supplied by the aromatic system (π-electron sextet). The addition of the electrophilic agent NO_2^+ results in the formation of the intermediate σ-complex **2** ('Wheland intermediate')[4]. On the basis of experiments, particularly those of Melander[5], the formation of **2** constitutes the rate-determining step. The overall reaction is of second order and follows the S_E2 mechanism, i.e., is a bimolecular electrophilic substitution.

B. Directing Effect of the Nitro Group

1. Historical review[6]

The systematic study of aromatic substitution became possible only after the correct structural relationship was established between the *ortho*, *meta*, and *para* isomers of benzene. The first attempts to formulate the orientation rules were made as early as 1875 by Hübner[7] and in the following year by Noelting[8].

They found that the substitution of benzene derivatives in the *ortho* and *para* positions occurred without simultaneous substitution in the *meta* position and that the mode of substitution depended largely on the substituent group already present in the benzene ring. Further, Noelting tried to establish a relation between the directing effect of a substituent and its chemical character. He pointed out that *meta*-directing groups such as NO_2, SO_3H, and COOH are *acidic*, whereas *basic* (e.g., NH_2) or *neutral* groups (e.g., CH_3 and/or Cl) are *ortho* and *para* directing.

The rule could not explain why the phenolic OH group is *ortho* and *para* directing, although it should be considered as acidic.

Later, Armstrong[9] drew attention to the fact that simple substituents containing double or triple bonds, e.g.,

$$-C(=O)OH, \quad -N(=O)\rightarrow O, \quad \text{or } C{\equiv}N,$$

are *meta* directing. This was supported by Vorländer[10]. The rule was accepted for some time, but eventually a number of exceptions were found and this reduced its value.

Crum Brown and Gibson[11] advanced an original concept of the substitution rule. Regarding substituent X in C_6H_5X as a derivative of HX, they stated that X will be a *meta*-directing group if HX can be oxidized to HOX in a one-step process. Thus NO_2 should be *meta* directing, for HNO_2 can be oxidized to $HONO_2$.

The most important systematic collection of experimental facts related to substitution was given by Hollemann[12,13]. He also examined substitution reactions of disubstituted benzene derivatives, C_6H_4XY, and was able to show which of the two groups had the stronger orienting effect. Further a considerable number of experiments were carried out by Hollemann using kinetic measurements. On the basis of the relative speed and the yield of substitution, Hollemann classified substituents not only according to their directing effect, but also their 'directing powers' (Table 1).

TABLE 1. 'Directing powers' of substituent directing groups.

ortho–para	$OH > NH_2 > NR_2 >$ NHAcyl $>$ Cl $>$ Br $> CH_3 >$ higher alkyls $>$ I
meta	$COOH > SO_3H > NO_2$

Although the rule is empirical, it did help to predict the nature of the product obtained on substitution of most aromatic compounds. Thus, if *m*-nitrotoluene is further nitrated, the nitro group enters the *ortho* and *para* positions relative to the methyl group. It follows from this experimental fact that the directive effect of the nitro group is less than that of the methyl group.

Hollemann also pointed out that *ortho–para*-directing groups increase the rate of aromatic (electrophilic in present day terminology) substitution, whereas *meta*-orienting groups greatly decrease it. Thus phenol, containing the *ortho–para*-directing OH group, can readily be nitrated even with dilute nitric acid, but nitrobenzene with its *meta*-directing nitro group requires a mixture of concentrated nitric and sulfuric acids. Nitrobenzene also requires vigorous conditions (e.g., high temperature) in order to be chlorinated or sulfonated. Generally speaking, the rate of the reactions of nitrobenzene is about 10^{-7} lower than that of benzene.

Additional attempts to establish general rules for the directing effect in substitution were due to Hammick and Illingworth[14], and to Mason and coworkers[15].

Several important attempts to explain the substitution rule in more modern terms were initiated in 1902 by Flürscheim[16] who introduced a concept of alternately strong and weak distribution of 'chemical affinity' around an aromatic ring. The difference of reactivity distribution due to the influence of *meta-* (NO_2) and *ortho–para-* (Cl) directing groups is shown in formulas **4** and **5**, respectively. The thick and thin lines indicated large and small

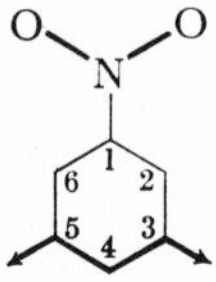

(**4**) 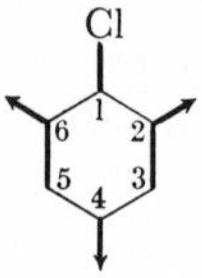 (**5**)

quantities of 'chemical affinity,' respectively, and arrows indicated 'residual' or 'free affinity.' Although the term 'affinity' is rather meaningless in the present state of chemical theories, the Flürscheim bonds of strong and weak 'affinity' may be compared with bonds of different orders between positions 1 and 2, and the arrows should indicate 'affinity forces' or 'free affinity'[6] available at reactive carbon atoms[17].

An important feature of Flürsheim theory was an introduction of the concept that alternations in degree of chemical reactivity can be transmitted from a substituent group situated in a relatively distant part of a molecule.

The alternation in chemical character of the atoms of any substituted benzene derivative was also considered later by Fry[18]. He suggested that positive and negative charges resided upon the atoms constituting the aromatic molecule and he suggested formula **6** for benzene. Thus the nitro group would induce a positive charge at

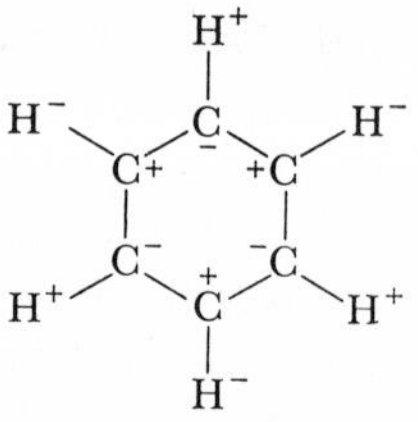

(**6**)

the hydrogen atoms in the *meta* positions (structure **7**), whereas chlorine, being essentially electronegative, would induce the positive charges at the hydrogen atoms in *ortho* and *para* positions (structure **8**).

(**7**) (**8**)

Fry also stated that positively charged hydrogen atoms are readily substituted. Further development of the idea of alternation was due to Lowry[10]. He brought the formula of Fry into accord with the Lewis–Langmuir theory of valency.

Vorländer[20] subsequently combined the theories of Flürscheim and Fry, and gave a general rule of substitution which was in agreement with experimental facts. According to Vorländer, the difference between *meta-* and *ortho–para*-directing substituents as presented in structures **9** and **10** is expressed as follows:

(**9**) (**10**)

'Bei der Bildung der Benzol-Disubstitutionsprodukte durch Halogenierung und Nitrierung von Benzol-Monosubstitutionsprodukten wird der eintretende zweite Substituent durch vorhandene positive Elemente der Seitenkette $C_6H_5E^+$ überwiegend nach der *meta*-Stellung, durch negative Elemente $C_6H_5E^-$ überwiegend nach der *para–ortho*-Stellung gelenkt.'

The importance of the polarity of the bond between the ring carbon atom and the 'key atom' of the substituent in building the directing effect was more recently pointed out by Latimer and Porter[21]. Almost simultaneously, Sutton[22] on the basis of dipole moment measurements of differently substituted benzene derivatives, introduced a concept of the *induced dipole moment* $\Delta\mu = \mu_{arom} - \mu_{aliph}$ where μ_{arom} and μ_{aliph} are dipole moments of the aromatic and *tert*-aliphatic compound, respectively. $\Delta\mu$ is negative (-0.18 to -0.88) when a *meta*-directing substituent (structure **11**), is present, and positive (0.21–0.88) when an *ortho–para*-directing substituent is present (structure **12**).

The induced moment for the nitro group is −0.88 (determined from $\mu_{arom} = -3.93$ and $\mu_{aliph} = -3.05$).

More recent data on dipole moments[23] give a value of −0.52 for the nitro group. This value was obtained from measurements of nitrobenzene and *tert*-nitrobutane in the gas phase.

Svirbely[24] further developed this idea, and stated that if the dipole moment of a monosubstituted benzene derivative were larger than 2.07 D, further substitution would occur in the *meta* position. The dipole moment of nitrobenzene is 4.08 D and hence is *meta* directing.

Subsequently this line of thought led Eyring and Ri[25] to calculate dipole moments and charge distribution from rates of nitration of substituted benzene derivatives. Titov[91] has given an original approach to the problem of the directing effect of substituents in electrophilic substitution. He divided all substituents into two classes: (1) those facilitating the oxidation of the benzene ring to the quinonoid one, and (2) those inhibiting the formation of the quinonoid system.

Electron donating groups which are ortho-para directing belong to class (1) and electron attracting, meta directing (hence the nitro group) belong to class (2).

2. Modern theories

The Fry and Vorländer concepts can be regarded as precursors of the more modern electronic theory which was developed by Lapworth[26] and Robinson[27].

Lapworth applied his earlier theories[28] of polarity and chemical changes to aromatic substitutions. He explained the reactivity at particular points in a benzene ring by the action of substituents of polar character which induce the electrical polarity. An electrical polarization can be transmitted within an aromatic ring at the moment of reaction, just as the alternate polarity was induced in conjugated systems[28].

An important feature of Lapworth's views[26] was that he stressed the importance of the presence of a *key atom* of a definite polar character in a directing group. The more pronounced the polar character of the key atom, the more the substitution is restricted to one type. Thus, the methyl group in toluene favors *ortho–para* substitution, but *meta* substitution occurs to the extent of 4%. This is due to a very weak electronegativity of the carbon atom and very weak electropositivity of hydrogen atoms in the methyl group. On the other hand, phenol is substituted exclusively in the *ortho* and *para* positions due to a very strongly negative oxygen atom in phenol.

Robinson[27] amplified and developed Lapworth's views, and Ingold[29,30], together with other British authors introduced a special terminology and symbolism.

According to this terminology an electrophilic group, like the nitro group produces a *negative inductive* effect '−I.' This is in accordance with the presentation of Victor Meyer[31] who described the nitro group as being a negative substituent.

Conversely, nucleophilic substituents are represented by the symbol '+I' (a positive inductive effect).

British workers also introduced the concept of the mesomeric effect 'M' which can be of a different sign than the inductive effect, but which is negative for the nitro group. The effect can be considered as a form of a permanent displacement of the charge. It can be related to the concept of induced dipole $\Delta\mu$ mentioned previously.

The Hammett substituent constant, σ, of the nitro group is positive in both *meta* and *para* positions, and its value is relatively high. This is typical of electron-attracting substituents.

3. *meta*-Directing effect of the nitro group

The nitro group is a substituent with a dipolar structure in which the positive end of the formal dipole is attached to the nucleus.

$$-\overset{+}{N}(=O)-O^{-} \longleftrightarrow -\overset{+}{N}(-\bar{O})=O$$

Other *meta*-directing groups have similar features, e.g.

nitrile $-\overset{+}{C}{=}\overset{-}{N}$, sulfone $R-\overset{O^-}{\underset{O^-}{S^{++}}}-$, and carbonyl $>\overset{+}{C}-O^-$

Norman and Taylor[32] pointed out recently that substituents with dipolar double or triple bonds, and among them the nitro group, are electron withdrawing through both their −I effect (which reduces the electron density at each nuclear carbon) and their −M effect which results in further electron withdrawal from *ortho* and *para* carbons. Both effects act in the same sense, giving rise to *deactivation* at all nuclear positions and particularly the *ortho* and *para* positions.

Consequently the isolated molecule with a nitro group (or any other *meta*-directing group) should be polarized as shown in structure **13**; structure **14** shows the polarization by the action of an *ortho–para*-directing group.

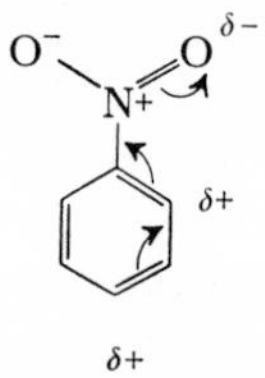

$\delta-$ $\delta+$ $\delta+$ $\delta+$ $\delta-$ $\delta-$

(**13**) 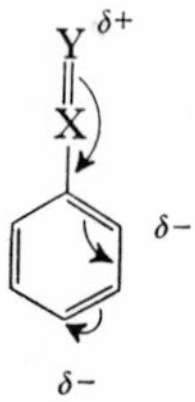 (**14**)

The mechanism of electrophilic substitution of monosubstituted benzene derivatives in which the *meta* position is least deactivated, is shown in equation 2.

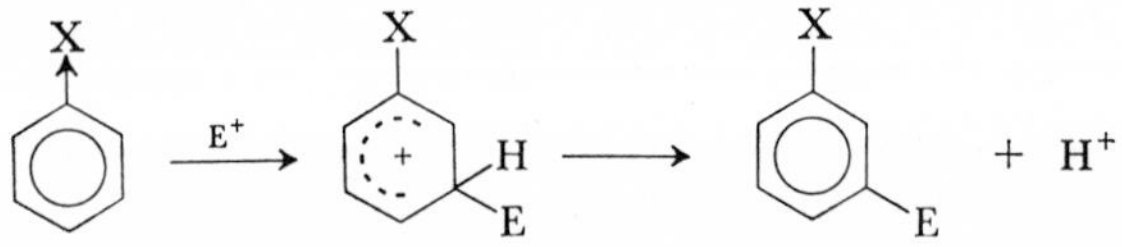

$$E^+ \longrightarrow \qquad \longrightarrow \qquad + \; H^+ \tag{2}$$

(**15**) (**16**) (**17**)

With molecular orbital calculations as a basis, 'Wheland intermediates,' such as **16** are sometimes represented by formula **18**, where the positive charge is equally shared between three carbon atoms. However, nuclear magnetic resonance measurements of the proton shifts in pentamethylcyclohexadienyl cation do not agree with an equal distribution of the positive charge[33].

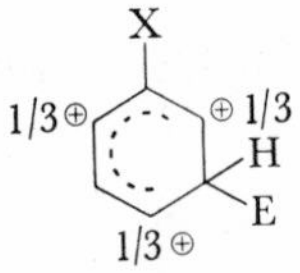

1/3⊕ ⊕1/3 1/3⊕

(**18**)

Structures **13** and **14** give a qualitative estimate of the electron density distribution around the aromatic ring.

A more modern quantitative representation of charge distribution in aromatic rings is based on the theory of molecular orbital and simplified wave mechanical calculations[34,35].

The results of π-electron density calculations[36] for nitrobenzene are presented in diagram **19** (for comparison, the π-electron distribution in aniline[35] is shown in diagram **20**).

1.94
O O
N 1.32
0.70
0.79
0.95
0.61
(**19**)

H H
N 1.92
1.02
1.03
1.00
1.02
(**20**)

Substitution of **19** in the *meta* position by positively charged species (electrophiles) is thus substantiated.

Also, the calculation of localization energy introduced by Wheland[4] can be helpful in establishing the directing effect of substituents.

4 *ortho–para*-Directing effect of the nitro group

Although the nitro group is *meta* directing, a considerable amount of *o*- and a much smaller amount of *p*-dinitrobenzenes are formed in the nitration of nitrobenzene. According to Holleman[13], the nitration of nitrobenzene yields *m*-dinitrobenzene (93.2%), *o*-dinitrobenzene (6.4%), and *p*-dinitrobenzene (0.3%). The ratio of 1/2 *ortho*:*para* is 11:0. *ortho* Substitution is even more pronounced in the chlorination of nitrobenzene (by Cl^+) leading to *m*-chloronitrobenzene (80.9%), *o*-chloronitrobenzene (17.6%), and *p*-chloronitrobenzene (1.5%). Here the ratio of 1/2 *ortho*:*para* is lower (5.9), but the absolute values are much higher than in the previous example.

There has been some controversy over the reasons why the nitro group leads to such a high 1/2 *ortho*:*para* ratio. According to De la Mare and Ridd[37] the three main views on the subject are:

(a) that the nitro group and other *meta*-directing groups specifically deactivate the *para* position[38];

(b) that *meta*-directing groups provide an additional reaction path for *ortho* substitution by the prior addition of the reagent to the group, followed by rearrangement to the *ortho* position, i.e., nitration occurs indirectly[39]; and

(c) that the lower 1/2 *ortho*:*para* ratios are a consequence of steric hindrance at the *ortho* position[40,41].

However, since correlation of orientational data for unsaturated substituents conjugated with the aromatic ring have shown that the 1/2 *ortho*:*para* and 1/2 *meta*:*para* decrease together in the same order, De la Mare and Ridd[37] suggest that electronic rather than steric factors determine the ratios of substitution.

It should be added to all these views that the calculated values of π-electron density shown in diagram **19** predict a high ratio of 1/2 *ortho*:*para* in the electrophilic substitution of nitrobenzene. Baciocchi and Illuminati[42] examined the rates of bromination and chlorination of 3-nitrodurene (**21**) into the position *para* to the nitro

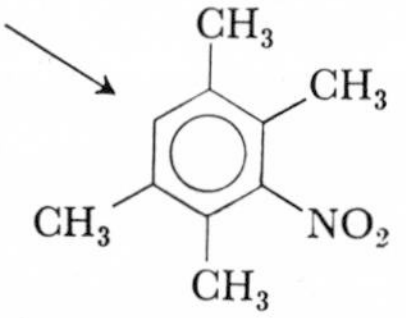

CH_3 CH_3 CH_3 NO_2 CH_3

(**21**)

group. They found a strong deactivating action of the nitro group, of the order of 10^6–10^7, and this was also the case with 2-nitromesitylene and 3-nitroisodurene. However, the reactivity of **21** with molecular halogen was found to be higher than that predicted from the electrical effects in electrophilic substitution. The authors ascribed the higher rate to a steric inhibition of resonance of the nitro group. On the other hand, with 2-nitromesitylene and 3-nitroisodurene, where *meta* substitution occurred, a slight decrease of deactivation was observed. Here the expected minor effect of steric inhibition of resonance might be overshadowed by increased hindrance to the approach of the reagent at the reaction center.

It should also be mentioned that in the homologs of benzene, the influence of the alkyl groups should not be neglected. Thus Norman and Radda[43] have pointed out that when nitrotoluenes are subjected to electrophilic substitution, the electron-donating methyl group acts contrary to the nitro group: the nitro group destabilizes intermediate **16** in the order *meta* $<$ *ortho* $<$ *para*, whereas the methyl group stabilizes it in the same order.

5. Directing effect of a nitro group placed in the side chain or ring

The *meta*-directing effect of the nitro group placed in a side chain is reduced. It decreases with an increase of the distance between the nitro group and the aromatic ring subjected to substitution. The corresponding data are collected in Table 2.

TABLE 2. Effect of a nitro group in the side chain on electrophilic substitution (nitration).

Compound	*m*-Nitro derivative, % yield	Ref
Nitrobenzene	93	44
Phenylnitromethane	67	
	48	45
Phenyl-ω-nitroethane	13	44
ω-Nitrostyrene	2	44, 46

The low yield of *meta* substitution in ω-nitrostyrene shows that the nitro group on the vinyl group, which is known to be *ortho–para* directing, has practically no directing effect in electrophilic substitutions.

The influence of the side-chain nitro group upon *meta* substitution (e.g., in nitration) can also be altered by other side-chain substituents, as indicated in structures **22–24**[46].

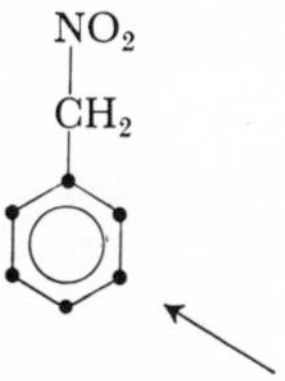

$C_6H_5CH_2NO_2$ **(22)** 67%

$C_6H_5C(CH_3)_2NO_2$ **(23)** 29%

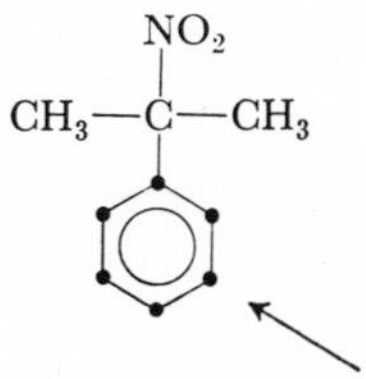

$C_6H_5CBr_2NO_2$ **(24)** 84%

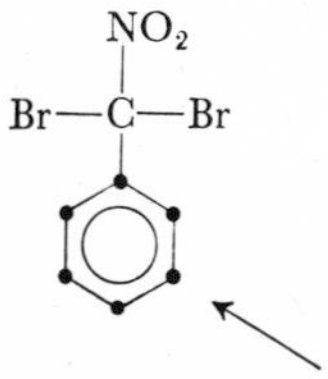

An interesting problem of the directing effect of *o*-, *m*-, and *p*-nitrophenyl groups upon the electrophilic substitution has been investigated by Mizuno and Simamura[47]. They examined the nitration of mononitrodiphenyl with nitric acid and acetic anhydride at 0° and found that the nitrophenyl substituent is deactivating but mainly *ortho–para* orienting. The partial rate factors and *ortho*:*para* ratios are indicated in diagrams **25–28**.

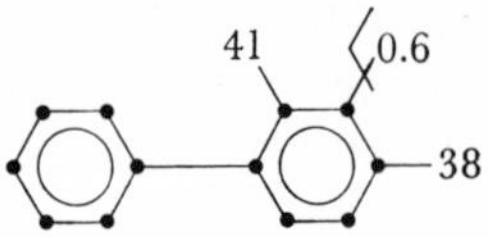

41 0.6 38

ortho:*para* = 2.2

(**25**)

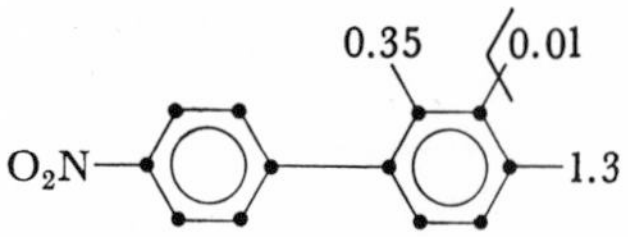

0.35 0.01 O_2N 1.3

ortho:*para* = 0.54

(**26**)

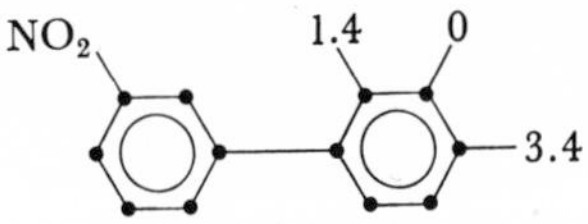

NO_2 1.4 0 3.4

ortho:*para* = 0.82

(**27**)

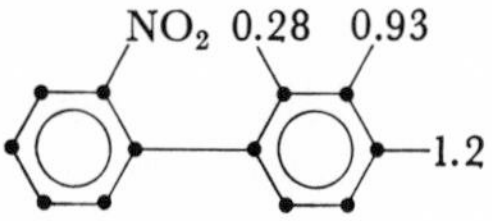

NO_2 0.28 0.93 1.2

ortho:*para* = 0.46

(**28**)

The authors concluded that the polar effect of the nitrophenyl groups is similar to that of halogen atoms.

C. Indirect Substitution

Bamberger[48] suggested that the substitution (including nitration) of aromatic compounds containing the NH_2 group is mainly an indirect process. The idea was based on experimental facts which he had observed earlier[49]: aniline nitrate is transformed into phenylnitroamine (through the loss of water), and the latter undergoes an isomerization 'Bamberger rearrangement' to yield *o*-nitroaniline along with a smaller proportion of *p*-nitroaniline.

Blanksma[50] extended Bamberger's hypothesis on the indirect nitration of aromatic amines. He suggested that a similar mechanism might exist in the instance of nitration of phenols: nitrate esters of phenols can be formed as a first step of the reaction, and then they are transformed into *C*-nitro compounds.

Further experiments on Bamberger's rearrangement were carried out by Orton and coworkers[51]. They came to the conclusion that the rearrangement is an acid-catalyzed reaction.

Holleman and coworkers[52] attempted to test Bamberger's hypothesis by comparing the orientation when aniline was nitrated (1) by a direct substitution and (2) by a rearrangement.

They found that the orientation through a rearrangement may differ considerably from the orientation through direct substitution. Thus the action of sulfuric acid upon anilinium nitrate yielded mainly *p*- and *m*-nitroaniline with a very small proportion of the *ortho* derivative, whereas the rearrangement of phenylnitroamine yielded (as in Bamberger's experiments) *o*-nitroaniline as the main product.

More recently a number of papers were published on Bamberger's rearrangement.

Thus Hughes and Jones[53] came to the following conclusions:

(1) the acid-catalyzed rearrangement of phenylnitromethane and the nitration of aniline, under comparable conditions, involve two very different types of orientations;

(2) nitric acid or any other nitrating agent are not intermediates of any importance in the Bamberger rearrangement and;

(3) the acid-catalyzed rearrangement in polar solvents is essentially intramolecular.

The rearrangement is slow, and of first order[54]. Isotopic tests[55–57] seem to confirm an intramolecular mechanism. For no ionic or radical fission was noticed which would be followed by intermolecular recombination of the counterfragments. However, more recently, splitting off of nitrite ions from nitramines was noticed by White and coworkers[58] and confirmed by Banthorpe, *et al.*[56,59]

White[58] used the isotope dilution method. He found that *N*-nitro-*N*-methylaniline-^{14}C in 0.1 *N* HCl at 40° produced: 52.1% *o*-nitro-*N*-methylaniline, 30.9% *p*-nitro-*N*-methylaniline, 9.9% *N*-methylaniline, and no *m*-nitro-*N*-methylaniline.

Banthorpe, *et al.*[56,59], found much evidence against a π-complex and a radical cage mechanism, and explained the rearrangement as shown in equation 3. They also suggested that *C*-nitrites can be

$$\underset{\textbf{(29)}}{C_6H_5NHNO_2} \underset{}{\overset{H^+}{\rightleftharpoons}} \underset{\textbf{(30)}}{C_6H_5\overset{+}{N}H_2\overset{-}{N}O_2} \longrightarrow \underset{\textbf{(31)}}{\overset{+}{N}H_2{=}C_6H_5^{+}} + NO_2^- \qquad (3)$$

formed as intermediates of the rearrangement.

Banthorpe and Thomas[60] also found that *N*-methyl-*N*-nitro-1-naphthylamine, in solvents such as toluene, rapidly rearranges to form 2- and 4-nitro isomers on heating to 100°, and also on exposure to ultraviolet irradiation at room temperature. The reaction appears to be more complex than the acid-catalyzed process, but no evidence was found for a mechanism involving homolytic or heterolytic fission. Consequently, as before, an intramolecular migration is suggested as in the acid-catalyzed rearrangement.

The Bamberger rearrangement was found to be responsible for the migration of the nitro group in *C*-nitro derivatives of aromatic amines described by Pausacker and Scroggie[61]. They found that heating of 2,3-dinitroacetanilide with sulfuric acid yielded 2,5-dinitroaniline (46 %), 3,4-dinitroaniline (23 %), and a small quantity of 2,3-dinitroaniline (5 %).

It was originally suggested that the nitration reaction may be reversible. However, more recent studies of these workers[62] showed that the mechanism of the reaction involves a reversed Bamberger rearrangement and the mechanism can be expressed as shown in equation 4.

NHR, NO_2, NO_2 (**32**)
Δ H_2SO_4
N, NO_2, R, NO_2 (**33**) $\overset{H^+}{\rightleftharpoons}$ NHR, NO_2 (**34**) + NO_2^+ ⟶ NHR, NO_2, NO_2 (**36**)
(**34**) ⟶ NHR, O_2N, NO_2 (**35**)

(4)

R = H or $COCH_3$

It was found by the same authors that heating 2,3-dinitrophenol with sulfuric acid leads to partial isomerization to 2,5-dinitrophenol. This probably could be explained in terms of the Blanksma[50] hypothesis, i.e., through the formation of an intermediate nitrate ester of phenol (**38**) (equation 5).

Other dinitro compounds, viz., those substituted in the 2,5 and 3,4 positions do not undergo such rearrangement. This fact is evidence that only that group can migrate which is subjected to steric hindrance, i.e., the nitro group in *ortho* position to the adjacent groups.

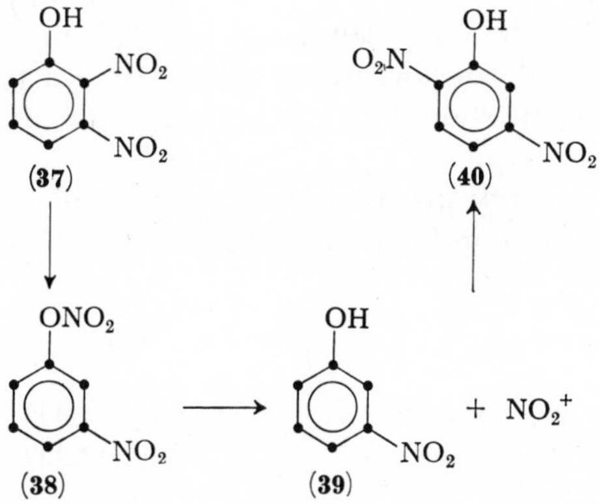

(5)

The reversibility of *C*-nitration seems to be limited to the above-mentioned cases. In general the reversibility is possible when a group such as NHR or OH is present, which can form an intermediate with the mobile nitro group.

The mobility of a *m*-nitro group in nitro derivatives of toluene was recently verified by Urbański and Ostrowski[63]. These workers kept solutions of various nitro derivatives of toluene in concentrated sulfuric acid at 90–95° for *ca.* 60 hours. *o*-Nitrotoluene (**41**), *m*-nitrotoluene (**42**), *p*-nitrotoluene (**43**), 2,4,6-trinitrotoluene (**44**), and 2,4,5-trinitrotoluene (**45**) were examined.

No change was found in the boiling points of **41** and **42**, and in the melting points of **43**–**45**. It was, however, found that solutions containing **42** or **45** eventually produced a slight blue color with diphenylamine.

However, Gore[64] reported that on heating 9-nitroanthracene in a mixture of sulfuric and trichloroacetic acids at 65–95° for 25 minutes 'the odor of nitrous fumes was noticeable.' Work-up of the reaction mixture gave nitric acid (81 %) and anthraquinone (21 %), the latter probably formed by oxidative action of nitric acid. The hydrolysis of the nitro group in 9-nitroanthracene on acid treatment is perhaps not so surprising if one considers the high reactivity of this position[36].

II FREE-RADICAL SUBSTITUTION

A. Directing Effect of the Nitro Group

When a '*meta*-directing group' such as the nitro group is present in the aromatic ring and the ring is attacked by a free radical, the homolytic substitution does not occur in the *meta* but in *ortho* and *para* positions, i.e., in a way similar to nucleophilic substitutions.

From theoretical considerations Wheland[4] has pointed out that any radical reagent should attack preferentially the *ortho* and *para* positions.

Data for the energy distribution in nitrobenzene when subjected to electrophilic and radical substitutions are shown in structures **46** and **47**.

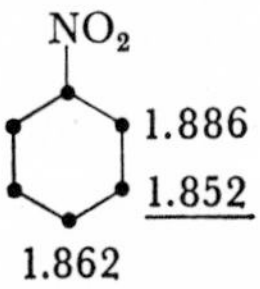

NO_2
1.886
1.852
1.862

electrophilic
substitution
(**46**)

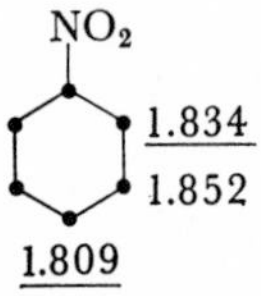

NO_2
1.834
1.852
1.809

radical
substitution
(**47**)

The relative reaction rates for both, electrophilic and radical substitutions as calculated from atom localization energies at 18° and 80°, respectively[4], are given in structures **48** and **49**.

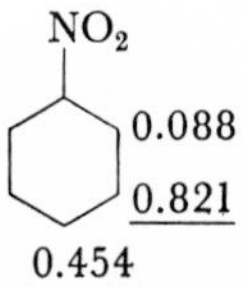

NO_2
0.088
0.821
0.454

electrophilic
substitution
(**48**)

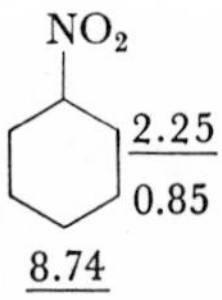

NO_2
2.25
0.85
8.74

radical
substitution
(**49**)

The differences between the directing effects in the two types of substitution can be clearly seen (at least qualitatively) although the experimental values may be somewhat different.

Wheland[65] also drew attention to the possible formation of quinonoid-type intermediates **50–52**. This view was later emphasized By Weiss and coworkers[66]. They concluded from hydroxylation experiments of nitrobenzene under free-radical conditions (*vide infra*) that the influence of the nitro group is due (1) to the greater availability of an unpaired electron at the *ortho* and *para* positions and (2) to the higher stability of quinoid structures **50** and **51** than of **52**.

(**50**) (**51**) (**52**)

Although experimental observations are not very numerous, these conclusions seem to be valid.

Thus Fieser and coworkers[67] have shown that aromatic nitro compounds can be methylated when heated in acetic acid to 90–95° with lead tetraacetate (equations 6 and 7).

$$\text{1,3,5-}(NO_2)_3C_6H_3 \xrightarrow{(CH_3COO)_4Pb} \text{2,4,6-}(NO_2)_3C_6H_2CH_3 + \text{2,4,6-}(NO_2)_3C_6H(CH_3)_2 \quad (6)$$

yield ca. 10% yield ca. 4%

$$\text{2,4,6-}(NO_2)_3C_6H_2CH_3 \xrightarrow{(CH_3COO)_4Pb} \text{2,4,6-}(NO_2)_3C_6H(CH_3)_2 \quad (7)$$

yield 32%

Thermal decomposition of lead tetraacetate probably liberates the acetyloxy radical (**53**), which furnishes the methyl radical (**54**) upon decarboxylation (equation 8).

$$(CH_3COO)_4Pb \longrightarrow \underset{(\mathbf{53})}{CH_3COO\cdot} \longrightarrow \underset{(\mathbf{54})}{\cdot CH_3} + CO_2 \quad (8)$$

Interaction of **53** with *sym*-trinitrobenzene leads to radical **55** which then reacts with the methyl radical (**54**) to give 2,4,6-trinitrotoluene (equation 9).

$$1,3,5\text{-}(O_2N)_3C_6H_3 + \mathbf{53} \longrightarrow \dot{C}_6H_2(NO_2)_3\ (\mathbf{55}) + CH_3COOH \quad (9)$$

$$\mathbf{55} + \mathbf{54} \longrightarrow 2,4,6\text{-}(O_2N)_3C_6H_2CH_3$$

Similarly, phenylation has been accomplished with lead tetrabenzoate[68].

Kharash, *et al.*[69], confirmed the formation of free radicals from lead tetraacetate, but Mosher and Kehr[70] regarded the methylating action of this reagent as the result of an ionic reaction leading to the formation of carbonium ions.

Waters and coworkers[71] studied the decomposition of *tert*-butyl peroxide in various aromatic solvents. When nitrobenzene was used at 143°, nitrotoluenes resulted. The proportion of *ortho*, *meta*, and *para*-isomers was 65.5, 6, and 28.5 %, respectively.

Weiss and coworkers[66] investigated the reaction of nitrobenzene with hydroxyl radicals (**56**) produced by the hydrogen peroxide–ferrous salt reaction (equation 10), and obtained *o*-nitrophenol (25–30 %), *m*-nitrophenol (20–25 %), and *p*-nitrophenol (50–55 %).

$$Fe^{2+} + H_2O_2 \longrightarrow Fe^{3+} + HO^- + \underset{(\mathbf{56})}{HO\cdot} \quad (10)$$

The formation of these compounds can be explained via the free-radical intermediates **50, 51,** and **52** (R = OH).

Klapproth and Westheimer[72] studied the reaction of mercuration of nitrobenzene (equation 11).

$$R\begin{matrix}NO_2\\H\end{matrix} \xrightarrow[\text{or } Hg(OOCCH_3)_2]{HgClO_4} R\begin{matrix}NO_2\\HgX\end{matrix} \quad (11)$$

$$X = ClO_4^- \text{ or } CH_3COO^-$$

The reaction had been known for some time[73], but the results were rather inconsistant. Jackson and Frant[74] studied the so-called 'classical' mercuration which consists of heating aromatic compounds with mercuric acetate in a non-polar medium. 'Classical' mercuration of nitrobenzene at 150° gave 53, 32, and 15% of *ortho*, *meta*, and *para* isomers, respectively, suggesting that a homolytic substitution may be involved.

Klapproth and Westheimer have shown that many of the apparent anomalies in the orientation in aromatic mercuration can be better understood if attention is paid to mercuration either with ionized mercuric salts in strong acid solution or with largely undissociated mercuric acetate in non-polar solvents.

With mercuric perchlorate in aqueous perchloric acid and particularly at low temperature, the orientation is typical of electrophilic substitution (*meta* in the case of nitrobenzene). In the reaction with mercuric acetate, considered to be partly radical substitution, the orientation is largely *ortho–para* (see Table 3).

TABLE 3. Orientation effects in the mercuration of nitrobenzene[72].

Exptl conditions	Isomer, %	
	ortho + *para*	*meta*
$Hg(ClO_4)_2$ in 60% $HClO_4$ at 25°	11	89
$Hg(ClO_4)_2$ in 40% $HClO_4$ at 95°	37	63
$Hg(OOCCH_3)_2$ in nitrobenzene at 95°	52	48
$Hg(OOCCH_3)_2$ in nitrobenzene at 150°	57	43

Similar results were obtained by Ogata and Tsuchida[75]. Thus, the proportions of *o*-, *p*-, and *m*-mercurated nitrobenzene were up to 37.5, 6.5, and 57.0%, respectively, when mercuration was carried out by mercuric nitrate in nitric acid at 90°.

The large proportion of *meta*-substitution product is probably due to the lack of a sharp demarcation line between the electrophilic and the radical substitution in the discussed reactions.

A partial radical substitution may be responsible for the nitration of nitrobenzene to dinitrobenzenes by nitric acid in the presence of mercuric oxide, as reported by Ogata and Tsuchida[76]. They found as much as 26% *o*- and only 24% *m*-dinitrobenzenes. These authors did not report the formation of the *para* isomer.

Hey and coworkers[77–82] studied the arylation of nitrobenzene through the action of various sources of the aryl radicals p-$RC_6H_4\cdot$, generated from such sources as diazotates, p-$RC_6H_4N_2ONa$, nitrosoacylarylamines, p-$RC_6H_4N(NO)COCH_3$, and acyl peroxides, $(p$-$RC_6H_4CO_2)_2$.

The average substitution in the *meta* position for R = Br and CH_3 was only 12.1 and 8.6%, respectively, and was essentially independent of the source of the aryl radical.

The phenylation of nitrobenzene gave the figures collected in Table 4.

TABLE 4. Substitution of nitrobenzene with phenyl radical.

Source of phenyl radical	Nitrophenyls, %		
	ortho	*meta*	*para*
Sodium benzenediazoate	54 ± 4	9 ± 2	37 ± 4.4
Benzoyl peroxide	59.5 ± 4	8.5 ± 2	32 ± 4

The high proportion of *ortho* and *para* substituents could be explained in terms of quinonoid structures **50** and **51**. It should, however, be born in mind that an aromatic radical of the type $XC_6H_4\cdot$ may acquire a polar character owing to the electron-attracting or electron-repelling properties of the substituent X.

TABLE 5. Arylation of benzotrichloride (80°)[84].

Radical	Isomer, %		
	ortho	*meta*	*para*
$C_6H_5\cdot$	12	49	39
p-$NO_2C_6H_4\cdot$	0	73	27

Thus the radical *p*-$NO_2C_6H_4\cdot$, if considered as somewhat electrophilic in character, might be expected to react most readily at nucleophilic sites. On the other hand, the *p*-tolyl radical should be considered as somewhat nucleophilic in character.

The electrophilic character of the *p*-nitrophenyl radical was demonstrated by Hey, *et al.*[78], Dannley and Sternfeld[83], and more recently by Saunders[84] (Table 5).

Chang, Hey, and Williams[80] studied the *p*-nitrophenylation of chloro- and bromobenzenes, and compared the results with the phenylation of chloro-, bromo-, and nitrobenzenes (Table 6). The products were analyzed by infrared spectroscopy and isotope dilution methods.

The electrophilic character of the *p*-halogenophenyl radicals is demonstrated by the increased proportion of the substitution at the *meta* position, in agreement with the directing influence of the nitro group in electrophilic substitution.

TABLE 6. *p*-Halogenophenylation of nitrobenzene (80°)[80].

Compound	Radical	Isomer, %		
		ortho	*meta*	*para*
$C_6H_5NO_2$	$C_6H_5\cdot$	62.5	9.8	27.7
$C_6H_5NO_2$	*p*-$ClC_6H_4\cdot$	59.0	13.8	27.2
$C_6H_5NO_2$	*p*-$BrC_6H_4\cdot$	57.7	13.2	29.1

B. Activating Effect of the Nitro Group

Hey and Grieve[85] found already in 1934 that the nitro group activates the aromatic ring toward homolytic substitution. For instance the competitive phenylation of toluene and nitrobenzene by phenyl radicals showed that the yield of nitrodiphenyls was about four times greater than the yield of methyldiphenyls. This is the result of the general character of free-radical reactions which are free of the powerful electrostatic forces which dominate heterolytic reactions.

Hey and his coworkers[77,79,81] gave a quantitative analysis of the rate of homolytic attack on nitrobenzene in terms of partial rate factors[86]. The rates of the substitution by the phenyl radical of nitrobenzene as compared with chlorobenzene, and the partial rate factors are given in Tables 7 and 8, respectively.

TABLE 7. Rate ratios and isomer distributions for phenylation of nitrobenzenes and chlorobenzenes.

Compound	X	Rate ratio PhX/PhH[81]	Isomer, %[77]		
			ortho	*meta*	*para*
Nitrobenzene	NO_2	2.94	58	10	32
Chlorobenzene	Cl	1.06	62	24	14

TABLE 8. Partial rate factors for phenylation of nitrobenzenes and chlorobenzenes[81].

Partial rate factors	F_o	F_m	F_p
$PhNO_2/PhH$	5.5	0.86	4.9
PhCl/PhH	1.6	1.0	1.2

Hey[77] also calculated the partial rate factors for homolytic substitution in nitrobenzene from data given by Wheland[4]. He obtained for $PhNO_2/PhH$: $F_o = 2.25$, $F_m = 0.85$, and $F_p = 8.7$.

There are, however, exceptions to this rule. Thus Hey and coworkers[82] found that the *p*-chlorophenyl radical attacks nitrobenzene less rapidly (rate ratio 0.53) than benzene (rate ratio 2.94) (Table 7). Moreover, *o*- and *p*-nitrophenyl radicals are less reactive toward nitrobenzene than toward benzene. Also with these negatively substituted radicals the proportion of *meta* substitution in nitrobenzene is increased, because of the electrophilic character of the substituent. In other words, the radicals with a nitro group (and possibly also with chlorine) acquire an electrophilic character.

According to Dannley and Gippin[88] the thermal decomposition of benzoyl peroxide in α-nitronaphthaline (and also in α-chloro- and α-bromonaphthalines) leads to monosubstitution by a benzoyloxy group in the 2, 4, and 5 positions (equation 12). In addition,

NO₂ — $\xrightarrow{(C_6H_5CO)_2O_2}$ — NO₂, $OCOC_6H_5$ (3%) + NO₂, $OCOC_6H_5$ (18%) + NO₂, $OCOC_6H_5$ (10%) (12)

benzoic acid, carbon dioxide, benzene, and esters are also formed.

The substituents have the following relative activating influence toward attack of the benzoyloxy radical:

NO_2 >	Br >	Cl >	H
19	2.0	1.2	1.0

On the other hand, the attack of triphenylmethyl radical on aromatic substrates in the presence of benzoyl peroxide indicates the following order of reactivity according to Benkeser and Schroeder[89]: $C_6H_5OCH_3 > C_6H_5Cl > C_6H_5H > C_6H_5COOCH_3 > C_6H_5CF_3 > C_6H_5NO_2$. Nitrobenzene failed to react in this system, and the experimental results indicate the electrophilic nature of the triphenylmethyl radical.

An excellent review of homolytic aromatic substitution has been given by Williams[90]. It includes a description of some unpublished work.

III. REFERENCES

1. G. K. Ingold, *Chem. Rev.*, **15**, 265 (1934).
2. A. Lapworth, *Nature*, **115**, 625 (1925).
3. R. Robinson, *Solway Reports*, **1931**, 434.
4. G. W. Wheland, *J. Am. Chem. Soc.*, **64**, 900 (1942).
5. L. Melander, *Nature*, **163**, 599 (1949); *Acta Chem. Scand.*, **3**, 95 (1949).
6. A more detailed historical review is presented by: F. Henrich in *Theorien der Organischen Chemie*, Vieweg and Sohn, Braunschweig, 1921; W. A. Waters in *Physical Aspects of Organic Chemistry*, 5th ed., Routledge and Kegan Paul, London, 1953; and P. H. Hermans in *Introduction to Theoretical Organic Chemistry* (Ed. R. E. Reeves), Elsevier Publishing Co., Amsterdam, 1954.
7. H. Hübner, *Chem. Ber.*, **8**, 873 (1875).
8. E. Noelting, *Chem. Ber.*, **9**, 1797 (1876).
9. H. E. Armstrong, *J. Chem. Soc.*, **51**, 258 (1887).
10. D. Vorländer, *Ann. Chem.*, **320**, 122 (1902).
11. A. Crum Brown and J. Gibson, *J. Chem. Soc.*, **61**, 367 (1892).
12. A. F. Hollemann, *Die direkte Einführung von Substituenten in den Benzolkern*, Veit and Co., Leipzig, 1910.
13. A. F. Hollemann, *Chem. Rev.*, **1**, 187 (1925).
14. D. L. Hammick and W. S. Illingworth, *J. Chem. Soc.*, **1930**, 2358.
15. S. F. Mason, E. Race, and F. E. Pounder, *J. Chem. Soc.*, **1935**, 1673.
16. B. Flürscheim, *J. Prakt. Chem.*, **66**, 321 (1902); **71**, 497 (1905); *Chem. Ber.*, **39**, 2015 (1906).
17. C. A. Coulson and G. S. Rushbrooke, *Proc. Cambridge Phil. Soc.*, **36**, 193 (1940).
18. H. S. Fry, *Z. Phys. Chem.*, **76**, 385, 396, 591 (1911); *J. Am. Chem. Soc.*, **34**, 664 (1912); **36**, 248, 1035 (1914).
19. T. M. Lowry, *J. Chem. Soc.*, **123**, 826 (1923).
20. D. Vorländer, *Chem. Ber.*, **52**, 263 (1919); **58**, 1893 (1925).
21. W. M. Latimer and C. W. Porter, *J. Am. Chem. Soc.*, **52**, 206 (1930).
22. L. E. Sutton, *Proc. Roy. Soc.*, **A133**, 668 (1931).
23. 'Landolt-Börnstein Zahlenwerte und Funktionen aus Physik, Chemie, Astronomie, Geophysik und Technik,' 6 Aufl., Band I, Teil 3, Springer Verlag, Berlin, 1951, pp. 460–463; O. A. Osipov, V. I. Minkin, and Yu. B. Kletnik, *Spravochnik po Dipolnym Momentam* (reference book on dipole moments), University Press, Rostov, 1961.
24. W. J. Svirbely and L. C. Warner, *J. Am. Chem. Soc.*, **57**, 655 (1935); W. J. Svirbely, *J. Am. Chem. Soc.*, **61**, 2555 (1939).
25. H. Eyring and T. Ri, *J. Chem. Phys.*, **8**, 433 (1940).
26. A. Lapworth, *J. Chem. Soc.*, **121**, 416 (1922).
27. W. O. Kermack and R. Robinson, *J. Chem. Soc.*, **121**, 427 (1922); J. Allan, A. E. Oxford, and R. Robinson, *J. Chem. Soc.*, **1926**, 401; R. Robinson, *Two Lectures on an Outline of an Electrochemical (Electronic) Theory of the Course of Organic Reactions*, Institute of Chemistry of Great Britain and Ireland, London, 1932; R. Robinson, *J. Soc. of Dyers and Colourists*, **50**, 65 (1934).
28. A. Lapworth, *J. Chem. Soc.*, **73**, 445 (1898).
29. C. K. Ingold and E. H. Ingold, *J. Chem. Soc.*, **1926**, 1310.
30. C. K. Ingold, *Structure and Mechanism in Organic Chemistry*, Cornell University Press, Ithaca, N.Y., 1953.
31. V. Meyer, *Chem. Ber.*, **20**, 534, 2994 (1887); **21**, 1295, 1306, 1331, 1334 (1888).

32. R. O. C. Norman and R. Taylor, *Electrophilic Substitution in Benzenoid Compounds*, Elsevier Publishing Co., Amsterdam, 1965, p. 71.
33. L. P. Colpa, C. MacLean, and E. L. Mackor, *Tetrahedron*, **19**, Suppl. 2, 65 (1963).
34. J. D. Roberts, *Notes on Molecular Orbital Calculations*, W. A. Benjamin, New York, 1961.
35. A. Streitwieser, Jr., *Molecular Orbital Theory for Organic Chemistry*, John Wiley and Sons, New York, 1961.
36. K. Higasi, H. Baba, and A. Rembaum, *Quantum Organic Chemistry*, Interscience Publishers, New York, 1965.
37. P. B. D. de la Mare, and J. H. Ridd, *Aromatic Substitution*, Butterworths Publication Ltd., London, 1959, p. 82.
38. C. K. Ingold, *Ann. Rept. Progr. Chem.*, **23**, 140 (1926).
39. A. Lapworth and R. Robinson, *Memoirs Proc. Manchester Lit. Phil. Soc.*, **72**, 43 (1928).
40. J. D. Roberts and A. Streitwieser, Jr., *J. Am. Chem. Soc.*, **74**, 4723 (1952).
41. R. D. Brown, *J. Am. Chem. Soc.*, **75**, 4077 (1953).
42. E. Baciocchi and G. Illuminati, *J. Am. Chem. Soc.*, **86**, 2677 (1964).
43. R. O. C. Norman and G. K. Radda, *J. Chem. Soc.*, 3610 (1961).
44. J. W. Baker, *J. Chem. Soc.*, **1929**, 2225; J. W. Baker and I. S. Wilson, *J. Chem. Soc.*, **1927**, 872.
45. T. Urbański, *Compt. Rend.*, **206**, 122 (1938); T. Urbański and Gedroyć, *Rocz. Chem.*, **18**, 125 (1938).
46. J. W. Baker and C. K. Ingold, *J. Chem. Soc.*, **1926**, 2462; J. W. Baker, *J. Chem. Soc.*, **1929**, 2257.
47. Y. Mizuno and O. Simamura, *J. Chem. Soc.*, **1958**, 3875.
48. E. Bamberger, *Chem. Ber.*, **27**, 584 (1894); **28**, 399 (1895); **30**, 1248 (1897); E. Bamberger and E. Hoff, *Ann. Chem.*, **311**, 91 (1900).
49. E. Bamberger, *Chem. Ber.*, **26**, 471, 485 (1893); **27**, 359 (1894).
50. J. J. Blanksma, *Rec. Trav. Chim.*, **21**, 281 (1902); **23**, 202 (1904).
51. K. J. P. Orton, *J. Chem. Soc.*, **81**, 490, 806 (1902); K. J. P. Orton and A. E. Smith, *J. Chem. Soc.*, **87**, 389 (1905); **91**, 146 (1907); K. J. P. Orton and C. Pearson, *J. Chem. Soc.*, **93**, 725 (1908); A. E. Bradfield and K. J. P. Orton, *J. Chem. Soc.*, **1929**, 915.
52. A. F. Hollemann, J. C. Hartogs, and T. van der Linden, *Chem. Ber.*, **44**, 704 (1911).
53. E. D. Hughes and G. T. Jones, *J. Chem. Soc.*, **1950**, 2678.
54. E. D. Hughes, C. K. Ingold, and R. B. Pearson, *J. Chem. Soc.*, **1958**, 435.
55. S. Brownstein, C. A. Burton, and E. D. Hughes, *J. Chem. Soc.*, **1958**, 4354.
56. D. V. Banthorpe, E. D. Hughes, and D. L. H. Williams, *J. Chem. Soc.*, **1964**, 5349.
57. D. V. Banthorpe, J. A. Thomas, and D. L. H. Williams, *J. Chem. Soc.*, **1965**, 6135.
58. W. N. White, J. R. Klink, D. Lazdins, C. Hathaway, J. T. Golden, and H. S. White, *J. Am. Chem. Soc.*, **83**, 2024 (1961); **86**, 1517 (1964).
59. D. V. Banthorpe and J. A. Thomas, *J. Chem. Soc.*, **1965**, 7149.
60. D. V. Banthorpe and J. A. Thomas, *J. Chem. Soc.*, **1965**, 7158.
61. K. H. Pausacker and J. G. Scroggie, *Chem. Ind.* (London), **1954**, 1290.
62. K. H. Pausacker and J. G. Scroggie, *J. Chem. Soc.*, **1955**, 1897.
63. T. Urbański and T. Ostrowski, according to T. Urbański, *Chemistry and Technology of Explosives*, Vol. I, Pergamon Press, Oxford–PWN, Warszawa, 1964, p. 41.
64. P. H. Gore, *J. Chem. Soc.*, **1957**, 1437.
65. G. W. Wheland, *The Theory of Resonance*, John Wiley and Sons, New York, 1945.
66. H. Loebl, G. Stein, and J. Weiss, *J. Chem. Soc.*, **1949**, 2074.
67. L. F. Fieser, R. C. Klapp, and W. H. Daudt, *J. Am. Chem. Soc.*, **64**, 2052 (1942).
68. D. H. Hey, C. J. N. Stirling, and G. H. Williams, *J. Chem. Soc.*, **1954**, 2747.

69. M. S. Kharash, H. N. Friedlander, and W. H. Urry, *J. Org. Chem.*, **16,** 553 (1951).
70. W. A. Mosher and C. L. Kehr, *J. Am. Chem. Soc.*, **75,** 3172 (1953).
71. W. A. Waters, B. R. Cowley, and R. O. C. Norman, according to G. H. Williams, Ref. 90.
72. W. J. Klapproth and F. H. Westheimer, *J. Am. Chem. Soc.*, **72,** 4461 (1950).
73. J. Jürgens, *Rec. Trav. Chim.*, **45,** 61 (1926); S. Coffey, *J. Chem. Soc.*, **1926,** 3215.
74. G. R. Jackson and M. S. Frant, *J. Am. Chem. Soc.*, **77,** 5625 (1955).
75. Y. Ogata and M. Tsuchida, *J. Org. Chem.*, **20,** 1637 (1955).
76. Y. Ogata and M. Tsuchida, *J. Org. Chem.*, **21,** 1065 (1956).
77. D. H. Hey, *J. Chem. Soc.*, **1952,** 1974.
78. D. H. Hey, A. Nechvatal, and T. S. Robinson, *J. Chem. Soc.*, **1951,** 2892.
79. D. R. Augood, D. H. Hey, and G. H. Williams, *J. Chem. Soc.*, **1952,** 2094.
80. Chang Shih, D. H. Hey, and G. H. Williams, *J. Chem. Soc.*, **1958,** 1885.
81. D. H. Hey, S. Orman, and G. H. Williams, *J. Chem. Soc.*, **1961,** 565.
82. J. I. G. Cadogan, D. H. Hey, and G. H. Williams, *J. Chem. Soc.*, **1955,** 1425.
83. R. L. Dannley and M. Sternfeld, *J. Am. Chem. Soc.*, **76,** 4543 (1954).
84. F. C. Saunders, Ph.D. Thesis, London, 1958, according to G. H. Williams, Ref. 90.
85. W. S. M. Grieve and D. H. Hey, *J. Chem. Soc.*, **1934,** 1797.
86. C. K. Ingold, A. Lapworth, E. Rothstein, and D. Ward, *J. Chem. Soc.*, **1931,** 1959; C. K. Ingold and M. S. Smith, *J. Chem. Soc.*, **1938,** 905; M. L. Bird and C. K. Ingold, *J. Chem. Soc.*, **1938,** 918.
87. D. H. Hey, H. N. Moulden, and G. H. Williams, *J. Chem. Soc.*, **1960,** 3769.
88. R. L. Dannley and M. Gippin, *J. Am. Chem. Soc.*, **74,** 332 (1952).
89. R. A. Benkeser and W. Schroeder, *J. Am. Chem. Soc.*, **80,** 3314 (1958).
90. G. W. Williams, *Homolytic Aromatic Substitution*, International Monographs Vol. 4, Pergamon Press, Oxford, 1960.
91. A. I. Titov, Voprosy Reaktsionnoi Sposobnosti i Orientatsii v Teorii Nitrovania Aromaticheskikh Soedinenii po Ionno Kompleksnomu Tipu (Problems of reactivity and orientation in theory of nitration of aromatic compounds according to an ionic complex type) in Organicheskie Poluprodukty i Krasiteli (Organic intermediates and dyes), Vol. 2, ed. A. I. Korolev, p. 46, Goskhimizdat, Moskva, 1961.

CHAPTER **3**

Activating and directing effects of the nitro group in aliphatic systems

HANS H. BAER and LJERKA URBAS

Department of Chemistry,
University of Ottawa,
Ottawa, Ontario, Canada

I. THE HENRY ADDITION AND RELATED REACTIONS

A. General Features and Reaction Conditions

Shortly before the turn of the century, L. Henry discovered the aldol-type, alkali-catalyzed addition of nitromethane[1] and homologous primary nitroalkanes[2] to aliphatic aldehydes. The reaction, which was later extended to include secondary nitroalkanes and many substituted nitroalkane derivatives on the one hand, and ketones as well as aromatic aldehydes on the other, has since become one of the main avenues to a variety of more complicated nitro compounds. Generally proceeding with facility and affording good yields, the reaction has made accessible a large number of new nitro alcohols and nitro glycols, which, in turn, have served as starting points for the synthesis of nitro olefins, amino alcohols, oximes, hydroxycarbonyl compounds, and many other products.

Although the Henry addition represents but another example of the more general Knoevenagel reaction, i.e., the addition of reactive methylene to carbonyl compounds, its scope is sufficiently wide to warrant a special treatment within that broader classification. Early work has been reviewed by Hass and Riley[3], Levy and Rose[4], and Shvekhgeimer, Piatakov, and Novikov[5]; a useful monograph containing extensive tables has been presented by Perekalin[5a]; accounts of more recent applications, particularly in the field of sugars and cyclitols, have been given by Sowden[6], Lichtenthaler[7], and Baer[8]; and implications with special reference to the synthesis of aliphatic polynitro compounds have been outlined in a review by Noble, Jr., Borgardt, and Reed[9].

In its most general form, the Henry reaction is illustrated by equation 1. It requires the presence of a hydrogen atom on the

$$RCHO + RCH_2NO_2 \longrightarrow RCHOHCH(R)NO_2 \qquad (1)$$

carbon that carries the nitro group, and hence it occurs with primary and secondary nitroalkanes, while tertiary ones do not react. The first step undoubtedly is the dissociation of the α hydrogen which is promoted by the activating effect of the nitro group. Consequently, polar solvents such as water or alcohols are generally used, and the function of the catalyst is to make the nitronate ion available for nucleophilic attack at the carbonyl carbon of the aldehyde (or ketone) reaction partner. A wide variety of basic catalysts can be used. These include alkali metal hydroxides, carbonates, bicarbonates, and alkoxides as perhaps the most commonly employed bases[10]. Calcium hydroxide[11,12], aluminum ethoxide[13], and magnesium aluminum ethoxide[13] have been recommended as efficient catalysts. The conversion of the latter into insoluble salts upon completion of the reaction may facilitate their removal from the reaction mixture[3], although this point would appear to be of diminished importance since cation-exchange resins have been introduced into general practice for such purposes. Catalysis of the Henry reaction by anion-exchange resins appears to be very promising[14,15]. A great many organic bases have proved to be suitable condensing agents. Thus, triethylamine[16,17] is a useful catalyst, while primary and secondary aliphatic amines, though effective, may promote side reactions because of their reactivity toward the aldehyde components. On the other hand, this reactivity can be taken advantage of when the goal of the nitroalkane–aldehyde reaction is not the nitro alcohol itself but products arising from its dehydration and further transformation, e.g., in Mannich reactions (see section II). Ammonium acetate has also been recommended[18,19]. Several instances have been reported where the choice of the catalyst influenced the nature of the products obtained; this will be dealt with in subsequent sections.

Generally, the reaction proceeds with sufficient speed at or slightly above or below room temperature. Secondary nitroalkanes usually react somewhat more sluggishly than primary ones, and may require catalysts of stronger basicity. Because of the tendency inherent in aldehydes to undergo intermolecular aldol additions or Cannizzaro reactions in alkaline medium, it is essential that the reaction conditions be carefully controlled, especially with regard to the alkalinity of the medium. This precaution is also necessary because the nitroalkanes employed may enter into side reactions. Thus, nitromethane is known to give, in the presence of alkali, methazonic and nitroacetic acids whereas higher homologs may form trialkylisoxazoles.

Primary nitroalkanes may react with more than one molecule of aldehyde, as illustrated in equation 2. For this reason, the proportion of the reagents must be chosen so as to direct the course of the reaction, as much as possible, toward the formation of the desired product. When the nitro alcohol produced contains a primary or

$$RCH_2NO_2 + RCHO \longrightarrow R\overset{\displaystyle OH}{\overset{|}{C}}H{-}CH(R)NO_2 \xrightarrow{RCHO} R\overset{\displaystyle OH}{\overset{|}{C}}H{-}\overset{\displaystyle NO_2}{\overset{|}{C}}(R){-}\overset{\displaystyle OH}{\overset{|}{C}}HR \quad (2)$$

secondary nitro group, and when a molar quantity of basic condensing agent was employed, the product will be present as a nitronate salt (equation 3). The nitro alcohol must then be liberated by

$$RCHO + RCH_2NO_2 + NaOH \longrightarrow R\overset{\displaystyle OH}{\overset{|}{C}}H{-}C(R){=}NO_2Na + H_2O \quad (3)$$

acidification, which is to be done carefully, preferably with dilute, weak acids in the cold, because of the possibility of the Nef reaction with concomitant loss of nitrous oxide occurring at that stage. Another reaction that may take place at this point is the dehydration to a nitro olefin, and this occurs with special ease in the case of aromatic nitro alcohols. In fact, the spontaneous formation of nitrostyrenes from β-nitro-α-hydroxyphenylalkanes is difficult, though not impossible, to avoid. If the sodium nitronate of a secondary nitroparaffin is condensed with an aldehyde so as to produce an alcohol containing a tertiary nitro group, the alkalinity of the medium builds up during the reaction because the product, an alkoxide, is the salt of a much weaker acid than the nitronic acid which is consumed (equation 4).

$$R_2C{=}NO_2Na + R'CHO \longrightarrow R_2C(NO_2)CH(R')ONa \quad (4)$$

This tends to retard the reaction, but yields can be improved[20,21] by the addition of sodium bisulfite, sodium bisulfate, carbon dioxide, or acetic acid which will reduce the alkalinity and suppress reversal of the reaction.

B. *Individual Reactions*

1. Nitromethane and formaldehyde

One molecule of nitromethane can react with up to three molecules of formaldehyde[22]. When equimolar amounts of the reactants are used, a mixture of 2-nitroethanol (9%), 2-nitro-1,3-propanediol (13%), and 2-hydroxymethyl-2-nitro-1,3-propanediol (78%) is produced[23] (equation 5).

$$CH_3NO_2 + HCHO \longrightarrow CH_2OHCH_2NO_2 \xrightarrow{HCHO} (CH_2OH)_2CHNO_2 \xrightarrow{HCHO} (CH_2OH)_3CNO_2 \quad (5)$$

When a 5 *M* excess of nitromethane is employed, the yield of 2-nitroethanol is increased to about 42%[23]. Its isolation by distillation from the bis- and trismethylol derivatives is attended with some explosion hazard, and for preparative purposes alternative ways[24,25] may be preferable. The preparations by this method of 2-nitro-1,3-propanediol[26–28] and of its sodium nitronate[29] have been described, and improvements on the preparation of 2-hydroxymethyl-2-nitro-1,3-propanediol have been reported[30]. The trishydroxymethyl compound can be demethylolated by treatment with sodium alkoxide to give the diol[31].

2. Nitromethane and homologous aliphatic aldehydes

The tendency of nitromethane to add more than one molecule of aldehyde decreases as the chain length of the latter increases. The main products are monohydric nitro alcohols of the type $RCHOHCH_2NO_2$. Thus, for instance, 1-nitro-2-propanol[2], 1-nitro-2-butanol[32], 1-nitro-2-pentanol[32], 3-methyl-1-nitro-2-butanol[13,33,34], 4-methyl-1-nitro-2-pentanol[35], 1-nitro-2-octanol[35–37], and some higher homologs[38] have been obtained from acetaldehyde, propionaldehyde, butyraldehyde, isobutyraldehyde, isovaleraldehyde, *n*-heptaldehyde, and higher aldehydes, respectively. Cyclohexanecarboxaldehyde furnished 1-cyclohexyl-2-nitroethanol[39]. It is possible, however, to produce diols of the type $RCHOHCHNO_2CHOHR$ by proper choice of the reaction conditions[40–43]. Understandably, a large excess of aldehyde will favor diol formation[40]. The pH of the reaction medium also influences the product composition. Thus, Eckstein and Urbański[42] obtained 1-nitro-2-propanol in a 75–80% yield when they allowed equimolar amounts of nitromethane and acetaldehyde to interact at pH 7.5–8.0, but they produced 3-nitro-2,4-pentanediol in yields of 80–95% by employing the reactants in a molar ratio of 1:2.25 at pH 6.5–7.0.

Three molecules of aldehydes higher than formaldehyde do not seem to add to nitromethane, but in diols of the type just described the remaining α hydrogen can be replaced by a hydroxymethyl group by the addition of formaldehyde (equation 6)[10].

$$(RCHOH)_2CHNO_2 + HCHO \longrightarrow (RCHOH)_2(CH_2OH)CNO_2 \qquad (6)$$

Similarly, the interaction of nitromethane with two molecules of formaldehyde and one molecule of a homologous aldehyde may afford nitro triols. In this way, 2-hydroxymethyl-2-nitro-1,3-pentanediol, -hexanediol, -nonanediol, and -5-methylhexanediol have been prepared by the use of propionaldehyde, butyraldehyde, *n*-heptaldehyde, and isovaleraldehyde, respectively[44]. It has, however, been stated that in such reactions formaldehyde can cause a displacement, from the nitro alcohol, of a molecule of the higher aldehyde[45,46].

3. Homologous nitroalkanes and aliphatic aldehydes

Primary aliphatic nitroalkanes such as nitroethane and its homologs readily add one molecule of an aliphatic aldehyde to give nitro alcohols of the type $R{-}CHOH{-}CH(NO_2)R$. Following initial studies by Henry and his coworkers, the first systematic investigation, leading to the preparation of 13 representatives of this type was made by Vanderbilt and Hass[10]. They included in their study nitroethane, 1-nitropropane, 1-nitrobutane, and 2-methyl-1-nitropropane; and formaldehyde, acetaldehyde, and butyraldehyde. Extending the reaction to secondary nitroalkanes, they found that 2-nitropropane and 2-nitrobutane similarly added the same aldehydes furnishing nitro alcohols of the type $R{-}CHOH{-}C(NO_2)R_2$, although catalysts of greater basicity than in the case of primary nitroalkanes were required for reasonable reaction rates. This they ascribed to the lower acidity of the secondary nitro compounds. Other workers[34,38,47] later carried out many reactions along similar lines.

Primary nitroalkanes above nitromethane, but not secondary ones, can also interact with two molecules of aliphatic aldehydes. Such twofold additions have appeared for some time to be restricted to either two molecules of formaldehyde or one molecule of a given aldehyde and at least one molecule of formaldehyde[10,48–50]. Thus, Vanderbilt and Hass[10] obtained 2-alkyl-2-nitro-1,3-propanediols by double addition of formaldehyde to nitroethane, 1-nitropropane,

1-nitrobutane, and 1-nitro-2-methylpropane, but they were unable to effect double addition to these nitroalkanes of either acetaldehyde or butyraldehyde under a variety of reaction conditions. (Recently, however, nitroethane and some homologs have been found capable of undergoing twofold additions not involving formaldehyde, namely, in the cyclizations of dialdehydes discussed in sections I.B. 4 and I.B. 5.) Fieser and Gates[16] prepared from phenylnitromethane and its 3-nitro and 3,5-dinitro derivatives the corresponding 2-aryl-2-nitro-1,3-propanediols. A large number of 2-alkyl- and 2-aryl-substituted 2-nitro-1,3-propanediols were synthesized by Urbański, Eckstein, and their coworkers[17,51–54].

On the formation of γ-dinitroparaffins in amine-catalyzed condensations of nitroalkanes and formaldehyde, see section II.C.

4. Nitroalkanes and simple aliphatic dialdehydes

The nature of the products formed in the interaction of nitroalkanes and aliphatic dialdehydes much depends on the relative proportions of the reactants. When a molar ratio of at least 2:1 is employed, i.e., when one molecule of nitroalkane for each aldehyde group is available, straight-chain dinitrodiols are formed. Thus Plaut[55] has allowed nitromethane, nitroethane, and 1-nitropropane to react with glyoxal and has obtained 1,4-dinitro-2,3-butanediol, 2,5-dinitro-3,4-hexanediol, and 3,6-dinitro-4,5-octanediol, respectively. Similarly, phenylnitromethane and glyoxal gave 1,4-dinitro-1,4-diphenyl-2,3-butanediol. A large excess of nitroalkane expectedly favors this reaction, as has been shown by Novikov and coworkers[56] who were able to prepare 1,4-dinitro-2,3-butanediol in 80% yield and, moreover, to separate the product into two stereoisomers that arose in nearly equal amounts. Carroll[57] obtained isopropylidene ketals from the two stereoisomers and used these for the assignment of configuration by NMR spectroscopy. In addition he reduced the dinitrodiols to the corresponding diaminodiols whose Schiff bases with salicyladehyde were compared with those of the 1,4-diamino-2,3-butanediols independently synthesized from *meso*- and DL-tartaric acids. It was found that the higher melting dinitrodiol (m.p. 135–135.5°) had the *meso*, and the lower melting one (m.p. 101–102°) had the DL configurations.

When, on the other hand, equimolar amounts of nitromethane and glyoxal are allowed to react in aqueous solution in the presence of sodium carbonate, a stereoisomeric mixture of 1,4-dinitro-2,3,5,6-cyclohexanetetrols (**3**) is formed (equation 6a). One of the stereoisomers has been isolated and assigned the *neo*-1,4 configuration **3a**[58].

O_2NCH_3 + OCH—CHO / OCH—CHO + CH_3NO_2 ⟶ (**1**) OCH——CHOH, O_2NCH_2, CH_2NO_2, CHOH—CHO ⟶ (**2**) CHOH—CHOH, O_2NCH, CH_2NO_2, CHOH—CHO

↓ (from O_2NCH_3 + OCH—CHO / OCH—CHO)

OCH—CHOH, CH_2NO_2 + OHC—CHO ⟶ (**4**) OCH——CHOH, $CHNO_2$, OCH——CHOH

(**4**) —CH_3NO_2⟶ (**2**)

(**2**) ↓ (**3**) CHOH—CHOH, O_2NCH, $CHNO_2$, CHOH—CHOH

(**3a**) HO, OH, NO_2, O_2N, HO, OH

(6a)

It is believed that the first step in this cyclization is the formation of 3-nitrolactaldehyde (**1**), which then undergoes intermolecular double addition head to tail[7]. In fact, 3-nitrolactaldehyde, which was synthesized in an independent way[59], has been reported to give readily upon treatment with alkali, a mixture of **3** from which isomer **3a** has been isolated[7]. Alternatively, the pathway could involve 2,4-dihydroxy-3-nitroglutaric dialdehyde (**4**)[7] as the intermediate. In either event, the cyclization would pass through intermediate **2**, a mixture of stereoisomeric 3,6-dideoxy-3,6-dinitrohexoses, and it has been well established that 6-deoxy-6-nitrohexoses easily undergo internal Henry addition to form nitroinositols (see section I.B. 5a).

From succinic dialdehyde and nitroethane a straight-chain diol, 2,7-dinitro-3,6-octanediol, has been prepared[55], while no cyclization reactions appear to have been reported.

Glutaric dialdehyde and nitromethane react to give a six-membered ring[60–62]. Here, too, a mixture of stereoisomers is obtained, from which one of the three theoretically possible isomers (*trans–trans*, *cis–cis*, and DL-*cis–trans*) was isolated in pure form in 51% yield and was proved by NMR studies to be *trans,trans*-2-nitrocyclohexane-1,3-diol (**5**)[61,62] (equation 6b).

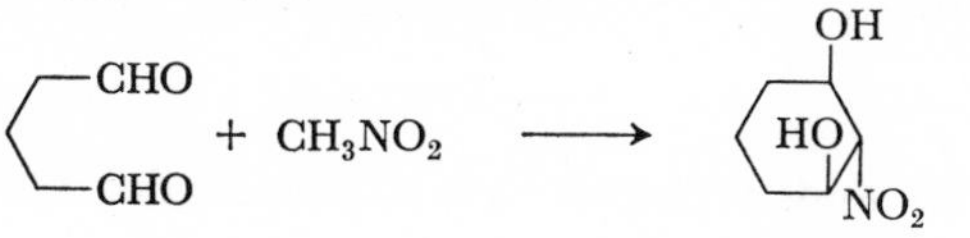

CHO, CHO + CH_3NO_2 ⟶ OH, HO, NO_2 (6b)

(**5**)

It is interesting to note that isomer **5**, with *all*-equatorial disposition of the substituents, was formed in a proportion higher than would be expected if all three isomers had an equal chance of formation. As will be pointed out in section I.B. 5b, the preponderance of certain stereoisomers due to conformational factors is frequently encountered in similar cyclizations. However, substitution of homologous nitroalkanes for nitromethane in cyclizations with glutaric dialdehyde led to individual stereoisomers in lesser yields. Thus, nitroethane gave 27% of the *trans–trans* compound **6**, and 1-nitropropane and phenylnitromethane gave 34 and 20%, respectively, of the *cis–trans* compounds **7** and **8**[63].

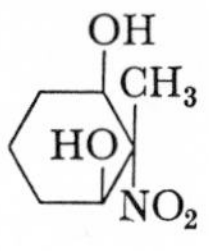

OH, CH_3, HO, NO_2

(**6**)

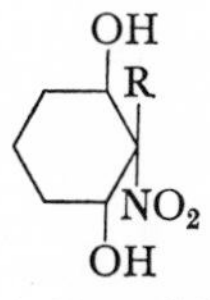

OH, R, NO_2, OH

(**7**), R = C_2H_5
(**8**), R = C_6H_5

An analogous nitromethane cyclization of adipic dialdehyde, which should have led to a seven-membered ring **9**, could not be achieved[64]. By the use of equimolar amounts of reactants under a variety of conditions, complex mixtures of products were produced among which there was always present what appeared to be cyclopenten-1-aldehyde (**10**). Evidently, **10** arose by a competing intramolecular aldol condensation of the adipic aldehyde, and thus far all attempts to prevent this occurrence have failed. When a 10 *M*

$$\text{(10) cyclopentenyl–CHO} \longleftarrow \text{OHC–}CH_2CH_2CH_2CH_2\text{–CHO} \not\longrightarrow \text{(9) cyclo-}[CH_2CH_2CH_2CH_2\text{–CHOH–CHNO}_2\text{–CHOH}]$$

$$\downarrow$$

$$O_2NCH_2CHOH(CH_2)_4CHOHCH_2NO_2 \quad \textbf{(11)} \qquad (6c)$$

excess of nitromethane was employed, a 40% yield of the straight-chain bis adduct, 1,8-dinitrooctane-2,7-diol (**11**), was obtained (equation 6c).

5. Nitroalkanes and aliphatic hydroxyaldehydes including sugars and "sugar dialdehydes"

a. Aliphatic hydroxyaldehydes and sugars. Early attempts[65] to condense nitromethane with glycolaldehyde, glyceraldehyde, and aldose sugars were recorded in 1921 but later were criticized[6] as having been inconclusive. In the years following World War II, the Henry reaction was reintroduced into carbohydrate chemistry, this time with great success, by H. O. L. Fischer and his associates, especially J. C. Sowden. It has since developed in this field to an exceedingly fruitful method, rivaling and in some respects indeed surpassing in versatility Emil Fischer's classical cyanohydrin synthesis.

Preliminary studies were intended to find out whether the alkaline degradation of aldonic acid nitriles, when performed in the presence of nitromethane, would give rise to 1-deoxy-1-nitroalditols according to equation 7, which amounts to a Henry reaction of an aldehyde generated *in situ*.

$$\begin{array}{c}CN\\|\\CHOAc\\|\end{array} \xrightarrow[CH_3OH]{NaOCH_3} \left[\begin{array}{c}CN\\|\\CHOH\\|\end{array} \xrightarrow{-HCN} \begin{array}{c}CHO\\|\end{array}\right] \xrightarrow{CH_3NO_2} \begin{array}{c}CH_2NO_2\\|\\CHOH\\|\end{array} \qquad (7)$$

The reaction was accomplished[66] on the example of 4,6-*O*-benzylidene-2,3,5-tri-*O*-acetyl-D-glucononitrile (**12**) which readily afforded 4,6-*O*-benzylidene-1-deoxy-1-nitro-D-mannitol (**13**) and, upon acid hydrolysis, the free nitro alcohol (equation 7a).

$$\begin{array}{l} CN \\ | \\ HCOAc \\ | \\ AcOCH \\ | \\ HC{-}O \\ | \qquad \searrow CHC_6H_5 \\ HCOAc \quad \nearrow \\ | \\ CH_2{-}O \\ (\mathbf{12}) \end{array} \xrightarrow[CH_3ONa]{CH_3NO_2} \begin{array}{l} CH_2NO_2 \\ | \\ HOCH \\ | \\ HOCH \\ | \\ HC{-}O \\ | \qquad \searrow CHC_6H_5 \\ HCOH \quad \nearrow \\ | \\ CH_2{-}O \\ (\mathbf{13}) \end{array} \qquad (7a)$$

It was soon realized that benzylidene-substituted aldoses with a free reducing group[67–71] as well as unsubstituted aldoses[69,71–76] will add nitromethane; the use of nitriles therefore has attained no practical significance.

The 1-nitroalditols are readily converted into the corresponding aldoses by treatment of their sodium salts with strong sulfuric acid (Nef reaction), so that the successive steps, viz., nitromethane addition to a given aldose and Nef reaction, constitute a lengthening of the sugar chain. Numerous aldoses less conveniently available otherwise have thus been synthesized[67,72,77–80]. The method also has proved satisfactory for the preparation[81–83] of 1-^{14}C-labeled sugars by the use of nitromethane-^{14}C. Furthermore, the 1-nitroalditols can be transformed[71,73,78,79] by acetylation followed by a Schmidt–Rutz reaction (section V.B. 2) into *O*-acetylated polyhydroxy-1-nitro-1-alkenes which are useful compounds for a variety of synthetic purposes. Thus, these nitro olefins may be selectively hydrogenated with a palladium catalyst to furnish 1,2-dideoxy-1-nitroalditols which, in turn, give 2-deoxyaldoses when subjected to the Nef reaction[70,71,84,85]. The nitro olefins may serve also as intermediates for the synthesis of 2-*O*-alkyl aldoses and 2-amino-2-deoxyaldoses (see section VI). The addition to aldoses of 2-nitroethanol[86] (or of nitromethane followed by formaldehyde[47]) and subsequent Nef reaction leads to ketoses possessing two more carbon atoms. These relationships are presented in Scheme 1.

The addition of nitromethane and nitroethanol to aldoses theoretically allows the formation of two epimeric 1-deoxy-1-nitroalditols and four epimeric 2-deoxy-2-nitroalditols, respectively. Epimers do in fact occur in most cases, but it is difficult at present

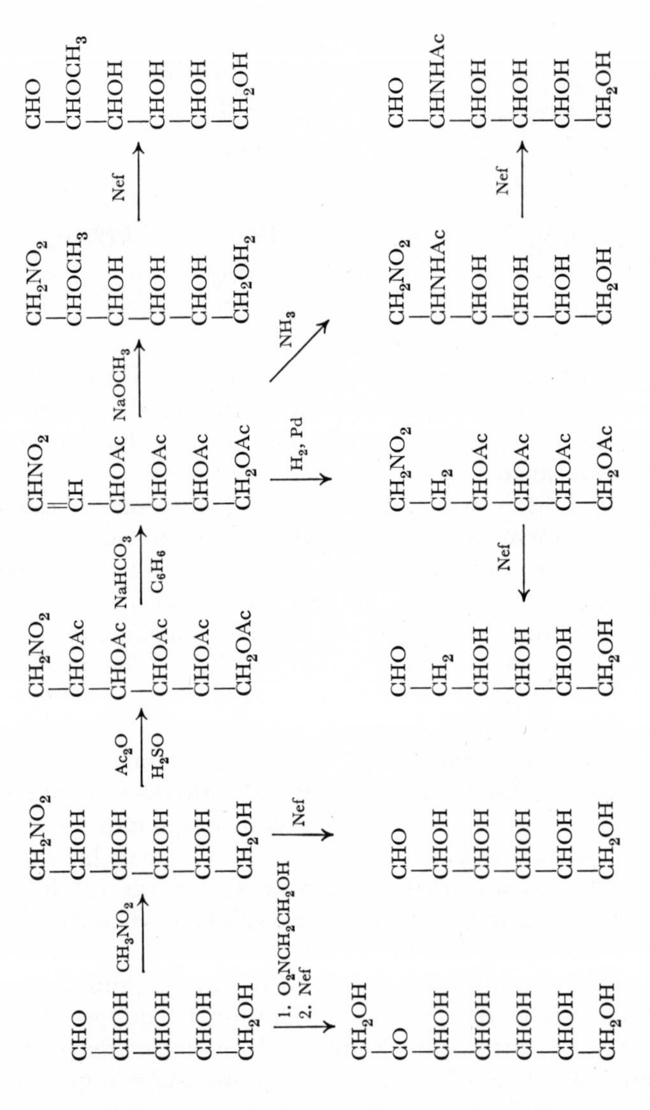
CH3NO2
Ac2O
H2SO
NaHCO3
C6H6
NaOCH3
Nef
H2, Pd
NH3
1. O2NCH2CH2OH
2. Nef

Scheme 1

to predict their ratios. From the preparative results reported so far it appears that, in general, there is no great tendency for one epimer to predominate, which is in line with expectations for these acyclic compounds. Whatever preferences in the formation of individual epimers have been observed, do not necessarily reflect thermodynamic stabilities, since factors such as fortuitous crystallization of products often have played a role, and the composition of the reaction mother liquors seldom has been examined. Although no systematic studies with a view to establishing general rules for the dependency of epimer distribution upon the reaction conditions and configuration of the starting sugars have yet been recorded, an explanation correlating at least some of the stereochemical observations has been proposed[8].

It is also worth mentioning that 2-hydroxytetrahydropyran, which may react as tautomeric 5-hydroxypentanal, undergoes sodium hydroxide catalyzed nitromethane addition. The expected adduct, 1-nitro-2,6-hexanediol, was not isolated, but on steam distillation, the acidified reaction mixture yielded its dehydration product, 2-nitromethyltetrahydropyran[87] (equation 8). Such cyclodehydrations to tetrahydropyran derivatives also occur with 1-deoxy-1-nitroalditols when they are boiled in aqueous solution, neutral or acidic,

O OH ⇌ OH CHO $\xrightarrow{CH_3NO_2}$

[OH CH—CH_2NO_2 | OH] $\xrightarrow{-H_2O}$ O —CH_2NO_2 (8)

or when heated above their melting points, or even on prolonged standing in syrupy condition. In alkaline medium, too, the formation of anhydrodeoxynitroalditols (aldosylnitromethanes) has been observed, and it is considered that they arise via intermediate α-nitroalkenes[88–90] (equations 9a and b). To a minor extent, ring closure occurred between C_2 and C_5 so as to form furanoid anhydrides[89].

An interesting application of the nitromethane reaction has enabled Grosheintz and Fischer to synthesize 6-deoxy-6-nitroaldoses[91] and deoxynitroinositols[92]. 1,2-*O*-Isopropylidene-α-D-*xylo*-pentodialdo-1,4-furanose (**14**), obtained by lead tetraacetate cleavage of 1,2-*O*-isopropylidene-α-D-glucofuranose, gave with nitromethane a mixture of the blocked 6-deoxy-6-nitro sugars **15** and **16**. Separation

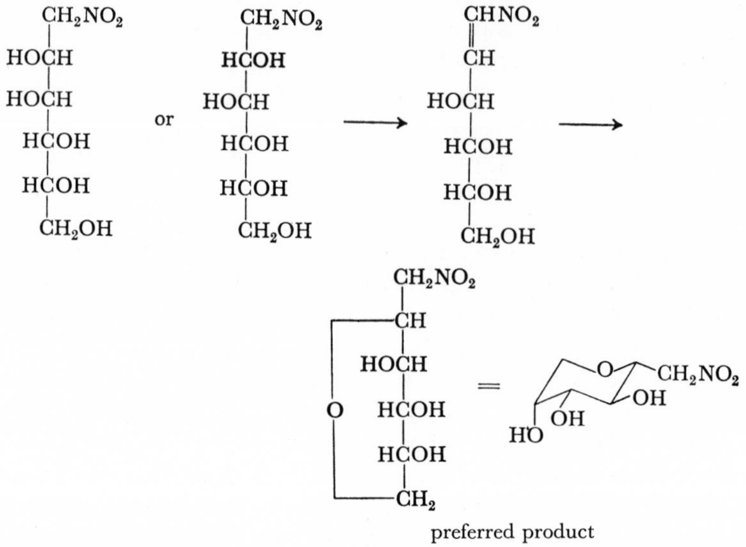

CH_2NO_2
HOCH
HOCH
HCOH
HCOH
CH_2OH

or

CH_2NO_2
HCOH
HOCH
HCOH
HCOH
CH_2OH

→

$CHNO_2$
CH
HOCH
HCOH
HCOH
CH_2OH

→

CH_2NO_2
CH
HOCH
O HCOH
HCOH
CH_2

= CH_2NO_2, OH, OH, HO, O (9a)

preferred product

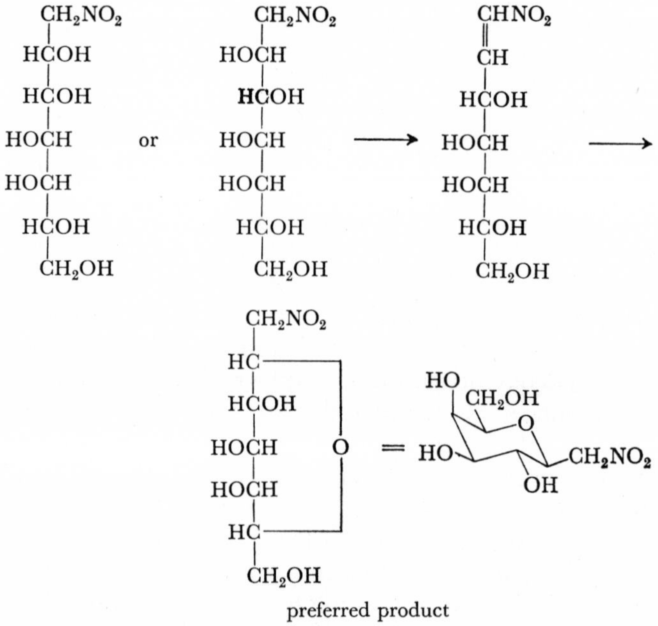

CH_2NO_2
HCOH
HCOH
HOCH
HOCH
HCOH
CH_2OH

or

CH_2NO_2
HOCH
HCOH
HOCH
HOCH
HCOH
CH_2OH

→

$CHNO_2$
CH
HCOH
HOCH
HOCH
HCOH
CH_2OH

→

CH_2NO_2
HC
HCOH
HOCH O
HOCH
HC
CH_2OH

= HO, CH_2OH, O, HO, CH_2NO_2, OH (9b)

preferred product

of **15** and **16** was effected by preferential 3,5-acetonation of **16** (see section V.A. 1). Hydrolysis of the acetone derivative then furnished 6-deoxy-6-nitro-D-glucose (**17**) and -L-idose (**18**), respectively. With aqueous barium hydroxide, either of these sugars incurred internal Henry addition to give the same mixture of optically inactive, stereoisomeric deoxynitroinositols **19**, **20**, and **21** (equation 10). The configurations of the products were later established by other investigators[93–95]. The sequence of reactions may be regarded as a stepwise nitromethane cyclization of *xylo*-trihydroxyglutaric dialdehyde (**22**), of which **14** represents a partially blocked derivative. A direct cyclization of **22** with nitromethane has also been realized[96,97], as has an analogous reaction employing nitroethane and leading to 1-deoxy-1-*C*-methyl-1-nitro-*scyllo*-inositol[63].

(**14**) $\xrightarrow{CH_3NO_2,\ OH^-}$ (**15**) + (**16**)

(**14**) $\xrightarrow{H^+}$ (**22**); (**15**), (**16**) $\xrightarrow{H^+}$ (**17**), (**18**)

(**22**) $\xrightarrow{CH_3NO_2,\ Ba(OH)_2}$; (**17**) $\xrightarrow{Ba(OH)_2}$; (**18**) $\xrightarrow{Ba(OH)_2}$ (**19**) ⇌ (**20**) ⇌ (**21**) (10)

(**19**) *scyllo*; (**20**) *myo*-1 (as racemate); (**21**) *muco*-3

The three deoxynitroinositols **19–21** are interconvertible through epimerization of their alkali salts[92]. At equilibrium, the *scyllo* (**19a**) and *myo*-1 (**20a**) stereoisomers exist in comparable amounts while the *muco*-3 stereoisomer is strongly disfavored, presumably because of diaxial substituent interaction in either chair form (**21a,b**)[94,97,98]. Replacement of the secondary nitro group in these compounds by a carbonyl function by means of the Nef reaction has not been successful[99].

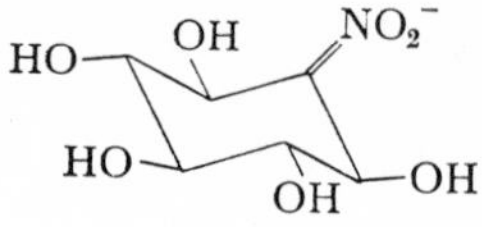

(**19a**)

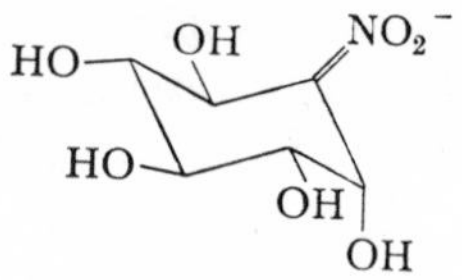

(**20a**)

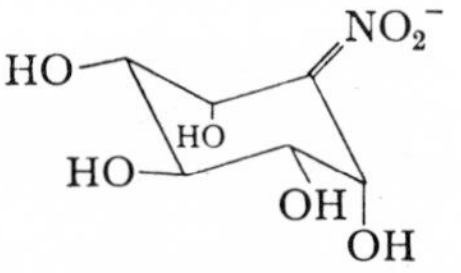

(**21a**)

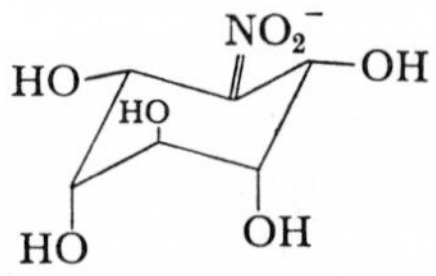

(**21b**)

Although the two nitro hexoses **17** and **18** are rapidly cyclized by the action of base to the same mixture of inositol derivatives, the L-*ido* derivative **18** reacts more readily and does so, if slowly, even in neutral to slightly acidic media (e.g., at pH 5–6)[100]. Presumably the marked degree of conformational instability in the idopyranose ring **18a**, which contrasts with the stable glucopyranose **17a**, is responsible for this difference.

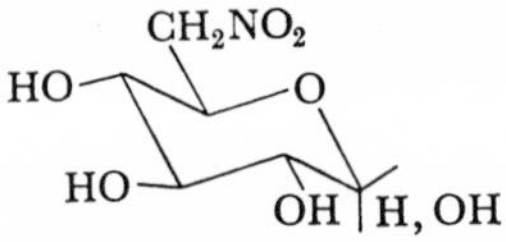

(**17a**), pyranose form

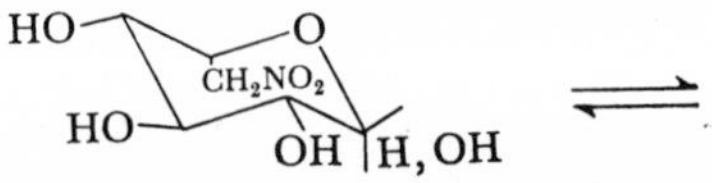

(**18a**), pyranose form

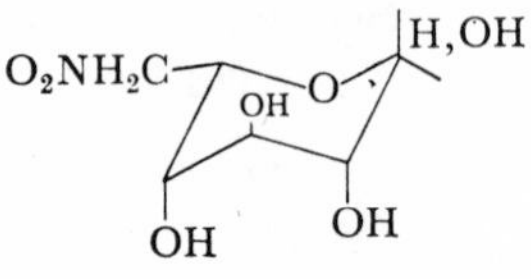

In analogy to the cyclization of *xylo*-trihydroxyglutaric dialdehyde (**22**), a derivative of *meso*-tartaric dialdehyde, namely *cis*-3,4-cyclohexylidenedioxy-2,5-dihydroxytetrahydrofuran (**23**), has recently been shown to form with nitromethane a mixture of diastereomeric 5-nitro-2,3-O-cyclohexylidenecyclopentane-1,2,3,4-tetrols (**24**),[101] (equation 11).

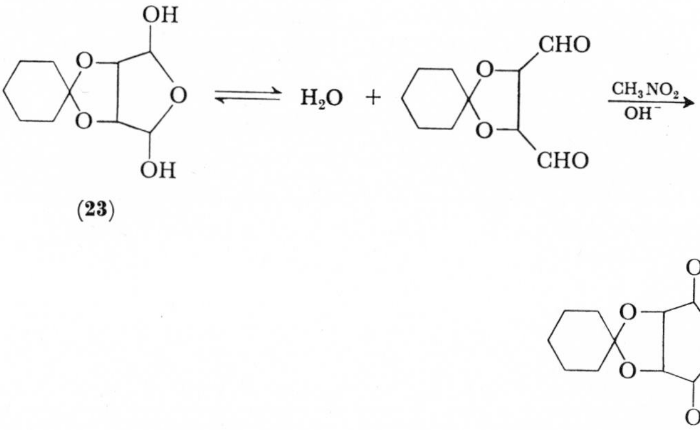

OH O O O OH ⇌ H_2O + O O CHO CHO $\xrightarrow[OH^-]{CH_3NO_2}$

(**23**)

OH O O NO_2 OH (11)

(**24**)

By applying the principle of intramolecular Henry addition to a derivative of D-glucosamine, Wolfrom and associates[102] were able to synthesize streptamine, a fragment of the antibiotic streptomycine. The pathway is represented in Scheme 2.

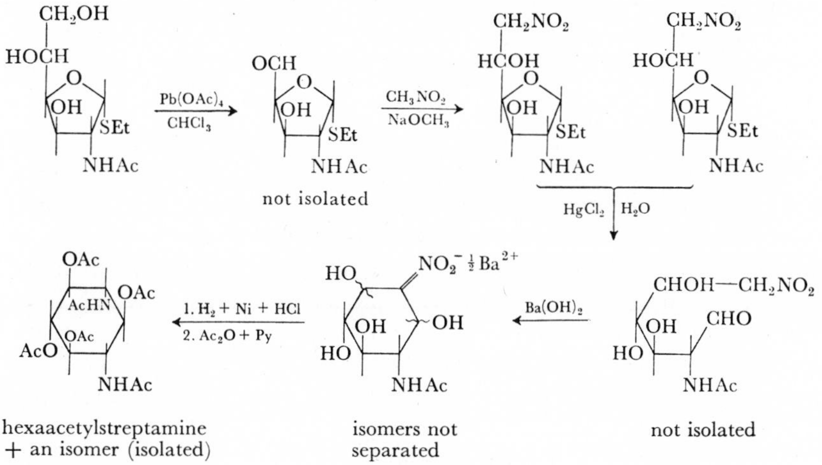

CH_2OH HOCH O OH SEt NHAc $\xrightarrow[CHCl_3]{Pb(OAc)_4}$ OCH O OH SEt NHAc (not isolated) $\xrightarrow[NaOCH_3]{CH_3NO_2}$ CH_2NO_2 HCOH O OH SEt NHAc + CH_2NO_2 HOCH O OH SEt NHAc

$\downarrow$ $HgCl_2$ | H_2O

$CHOH—CH_2NO_2$ OH CHO HO NHAc (not isolated) $\xrightarrow{Ba(OH)_2}$ HO NO_2^- $\frac{1}{2}Ba^{2+}$ OH OH HO NHAc (isomers not separated) $\xrightarrow[2.\ Ac_2O + Py]{1.\ H_2 + Ni + HCl}$ OAc OAc AcHN OAc AcO NHAc

hexaacetylstreptamine
+ an isomer (isolated)

isomers not
separated

not isolated

SCHEME 2

Novikov and coworkers performed Henry additions of nitromethane to 3-hydroxy-2,2-dimethylpropanal and -butanal[102a], and of nitromethane, nitroethane, and 2-nitropropane to acetaldol[102b]. Although not all of the nitrodiols produced were well characterized, they could be subjected to dehydration leading to various unsaturated nitro compounds.

b. 'Sugar dialdehydes.' In 1958, Baer and Fischer[103] extended the nitromethane–aldehyde addition to the sugar 'dialdehydes' which are readily available from methyl glycosides by periodate or lead tetraacetate fission. Just as aldoses, under the conditions of the Henry reaction, behave like free aldehydes although they exist predominantly in cyclic hemiacetal forms, the sugar 'dialdehydes' react as true dialdehydes although in solution they, too, assume cyclic hemiacetal or hemialdal structures[104] (equations 12a and b).

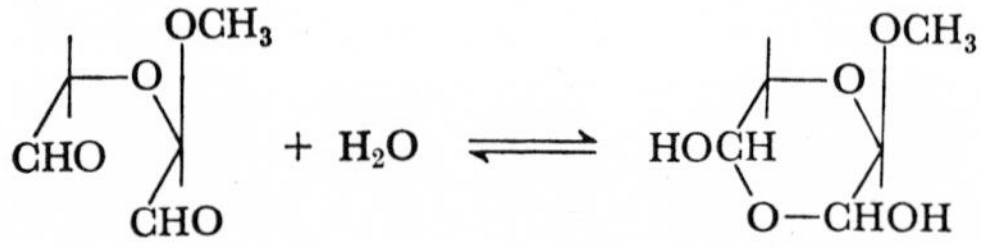

OCH$_3$ / O / CHO / CHO + H_2O ⇌ HOCH / OCH$_3$ / O / O—CHOH (12a)

(**25**)

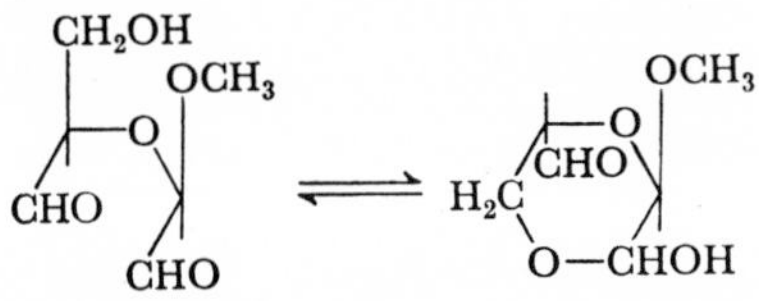

CH$_2$OH / OCH$_3$ / O / CHO / CHO ⇌ OCH$_3$ / O / CHO / H$_2$C / O—CHOH (12b)

(**26**)

Cyclization of these dialdehydes with nitromethane provides a general and facile method of introducing nitrogen into the position 3 of aldoses. The synthesis commands particular interest because of the discovery in recent years of numerous antibiotics that contain, as building units, various rare 3-amino sugars, some of which have not previously been accessible by simple synthetic approaches[105]. L′-Methoxydiglycolic aldehyde (**25**), which can be obtained from methyl β-D-xylopyranoside or from any of the methyl β-D- or α-L-pentopyranosides, was cyclized giving a mixture of stereoisomeric methyl 3-deoxy-3-nitropentopyranoside sodium salts[106,107]. One of the nitronates formed, the β-D-*erythro* salt **27**, crystallized in 43%

yield, while relatively small amounts of the α-L-*threo* and β-D-*threo* salts **28** and **29** were shown to exist in the mother liquor beside additional **27**. No evidence for the formation of the fourth possible isomer, the α-L-*erythro* salt **30**, was detected. The nitronates had originally been assumed, without proof, to exist in the C-1 chair conformation as depicted[106]. The stability of **27** and instability of **30**, with **28** and **29** occupying intermediate positions, thus appeared plausible in the light of the conformational tenets prevailing at the time. However, recent NMR data revealed that these nitro glycosides adopt, in part, inverted chair conformations[107a]. A satisfactory explanation of the stereochemical results must await further study, and due consideration must probably be given to unfavorable steric interaction[107b,c] between equatorial hydroxyl groups and adjacent nitronate groupings.

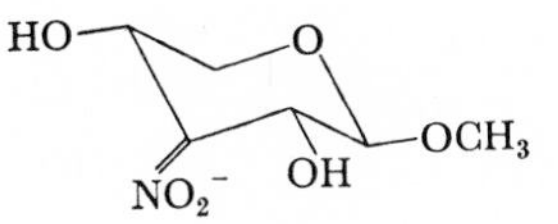

HO, O, NO_2^-, OH, OCH_3 **(27)**

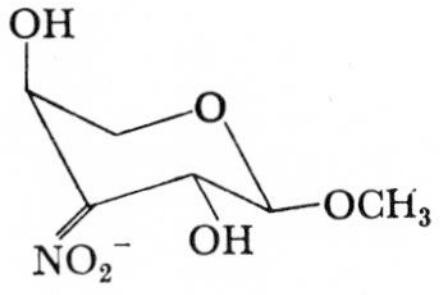

OH, O, NO_2^-, OH, OCH_3 **(28)**

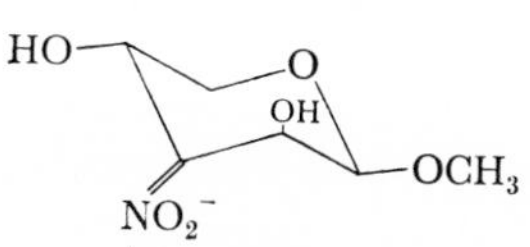

HO, O, OH, NO_2^-, OCH_3 **(29)**

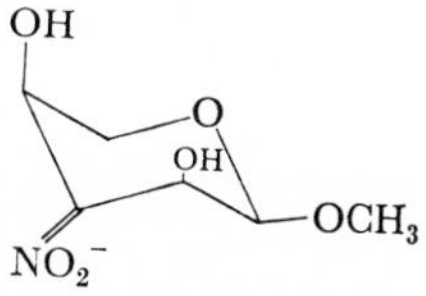

OH, O, OH, NO_2^-, OCH_3 **(30)**

The enantiomer of **25**, D′-methoxydiglycolic aldehyde, gave rise to the expected enantiomers of **27**, **28**, and **29**[106,107].

In the hexose series the dialdehyde L′-methoxy-D-hydroxymethyl-diglycolic aldehyde (**26**) was cyclized to form chiefly the nitronates **31** and **32**, and **33** as a minor component. Again, no evidence was obtained for the presence of the fourth possible product (**34**)[109–111].

Each of the nitronates **31**, **32**, and **33** in aqueous solution at room temperature epimerized within 5 hours to an equilibrium mixture containing all three of them, with **31** and **32** predominating over **33**. When the mixture was allowed to stand in solution for an extended period of time, or was heated briefly, dehydration occurred giving

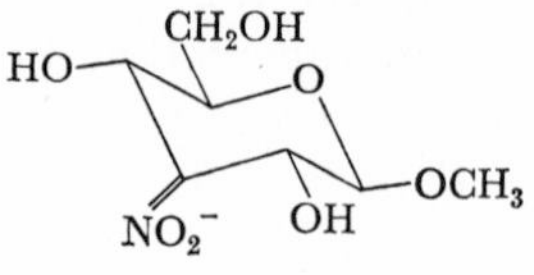

(**31**) 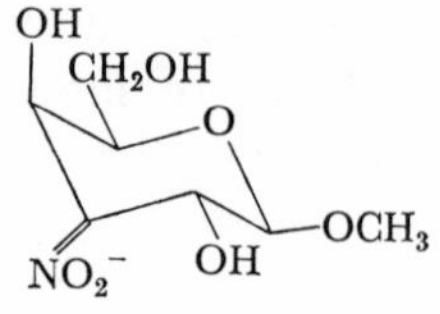 (**32**)

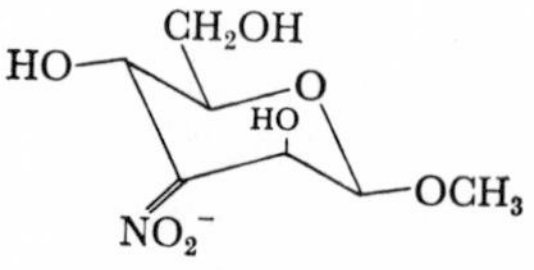

(**33**) 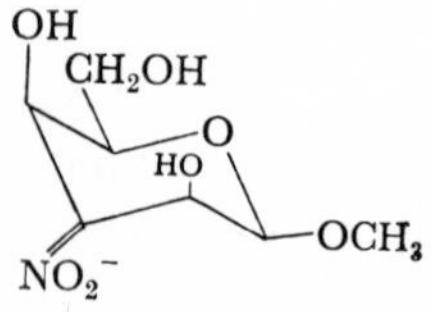 (**34**) not formed

an olefinic nitronate (equation 13). Similar reactions take place with other 3-deoxy-3-nitroglycosides in the presence of alkali[108].

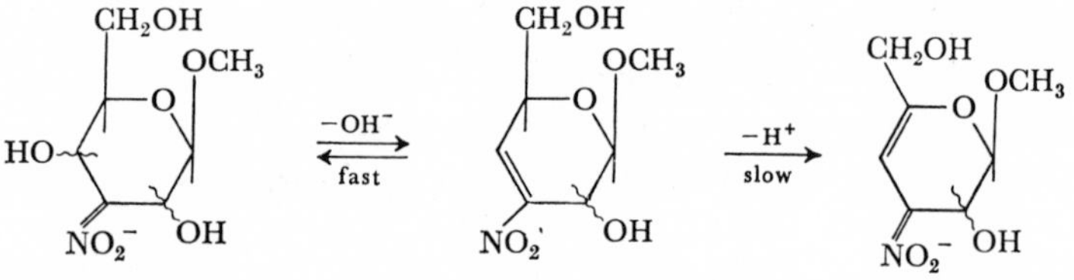

$$-\mathrm{OH}^- \text{ (fast)} \qquad -\mathrm{H}^+ \text{ (slow)} \qquad (13)$$

The diastereoisomer of dialdehyde **26**, D′-methoxy-D-hydroxymethyldiglycolic aldehyde (**35**), furnished a mixture of nitronates in which **36** and **37** strongly predominated when the reaction was kinetically controlled (equation 14). When this mixture was allowed to stand in aqueous solution it epimerized, and at equilibrium the nitronates **38** ($>40\%$) and **39** ($\sim 30\%$) were the chief components, with only 10–12 % of **36** and even less of **37** remaining[112,113] (equation 15).

It is noteworthy that in this α-hexopyranoside series configuration **38** is the most stable one, whereas its counterparts **30** and **34** in the pentopyranoside and β-hexopyranoside series, respectively, seem to be entirely unstable.

Other sugar dialdehyde–nitromethane cyclizations that have led to novel, nitrogenous carbohydrate derivatives were performed with the dialdehydes **40**[114,115] (and its enantiomorph)[116], **41**[117], **42**[118,119], **43**[120], and **44**[121] (equations 16–19). Application of the synthesis to the disaccharidic dialdehyde **45**, which is derived from sucrose,

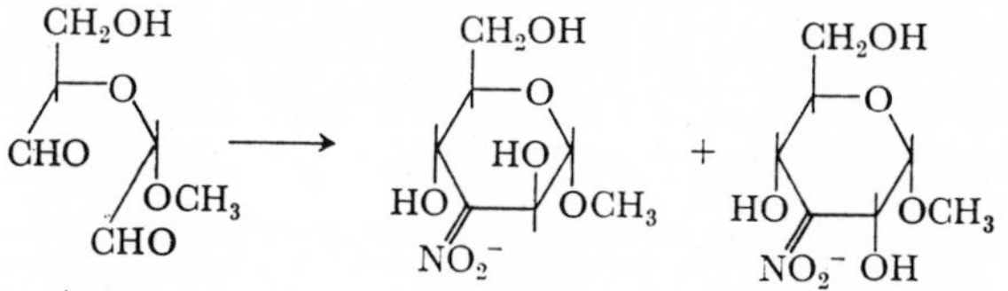

(35) → (36) + (37) (14)

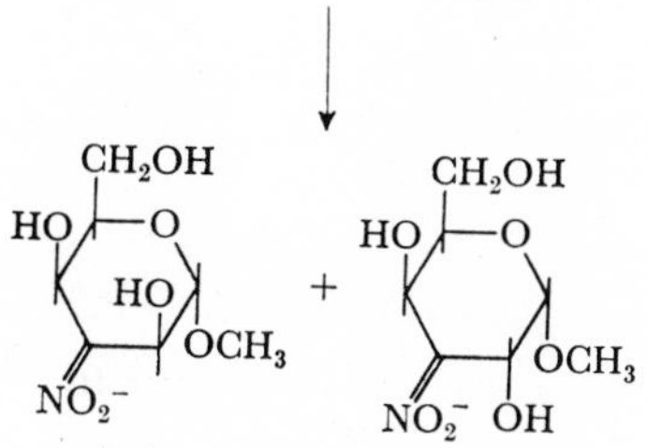

(38) + (39) (15)

furnished the first disaccharide containing a nitrogenous seven-carbon sugar[122] (equation 19a). From nucleoside dialdehydes **46** were obtained nucleosides that possess 3-nitro- (and 3-amino-) 3-deoxy-hexose moieties[123–127] (equation 20). It should be noted that in those cases where the dialdehyde is made by a glycol cleavage that does not involve the removal of a carbon atom (as formic acid), the glycoside resulting from nitromethane cyclization has one more carbon atom than the starting glycoside, and the reaction sequence then constitutes a method of lengthening the sugar chain 'from within'[106,109]. Examples include the use of a methyl pentofuranoside

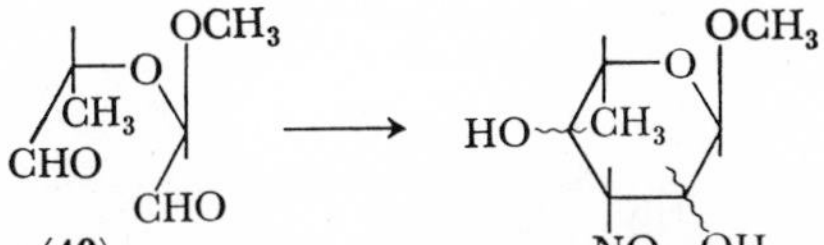

(40) → methyl 3,6-dideoxy-3-nitro-α-L-hexopyranosides (16)

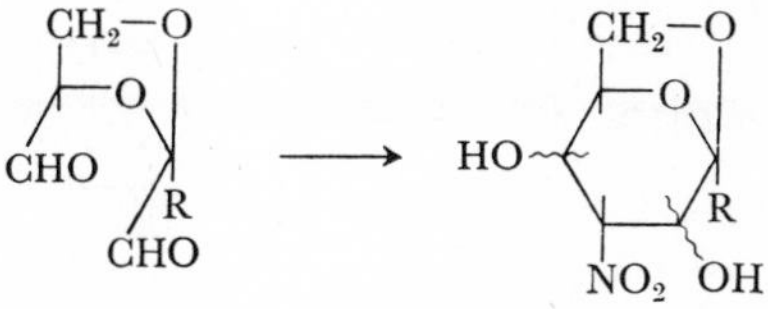

(17)

(**41**), R = H → 1,6-anhydro-3-deoxy-3-nitro-β-D-hexopyranoses

(**42**), R = CH_2OH → 2,7-anhydro-4-deoxy-4-nitro-β-D-heptulopyranoses

(18)

(**43**) → methyl 3-deoxy-3-nitro-α-D-heptoseptanosides

(19)

(**44**) → benzyl 4-deoxy-4-nitro-hexulopyranosides

(19a)

(**45**) → α-D-glucopyranosyl 4-deoxy-4-nitro-β-D-heptulopyranosides

(20)

(**46**) (also with other pyrimidine and purine moieties) → 1-(3′-deoxy-3′-nitro-β-D-hexopyranosyl)pyrimidines and -purines instead of a methyl hexopyranoside for generating the dialdehyde **47**[128] (equation 21), and also the cyclizations of **43**, **45**, and **46**.

$\xrightarrow{NaIO_4}$ $\xrightarrow{CH_3NO_2}$ (21)

methyl β-L-arabinofuranoside

(**47**)
D′-methoxy-L-hydroxymethyl-diglycolic aldehyde

methyl 3-deoxy-3-nitro-β-L-hexopyranosides

Similarly, the action of nitromethane and alkali upon periodate-oxidized cellulose has been reported[129] to result in a nitro polysaccharide which contains, in part, seven-carbon sugar units. The structure of the polymeric material has not been conclusively proved, however.

In all the sugar dialdehyde–nitromethane cyclizations, mixtures of stereoisomers were formed, but a marked selectivity in favor of one or two of the possible isomers was invariably observed, and preparative separations were accomplished either on the sodium nitronate stage or upon deionization on the nitro stage or, failing that, after hydrogenation to the corresponding amines.

An empirical observation regarding the stereochemical course of the cyclization can be stated. Under conditions of kinetic control the reaction tends to yield preferentially the nitronates in which the hydroxyl group that is formed at C_2 is arranged *trans* to the neighboring glycosidic group. This can be explained on the basis of an attack of the carbonyl group from the least hindered side (Cram's rule). Thus, when the dialdehyde **25** is assumed to react in its most probable conformation **25a** and the first step is thought to be an addition of methanenitronate ion to the carbonyl group A, the ion would approach A from the left-hand side and therefore would generate at C_2 the configuration depicted (Scheme 3). Ring closure between C_3 and the carbonyl group B then leads chiefly to **27** (and to some **28**). Alternatively, if the first reaction step consists of nitromethane addition to the carbonyl group B, the adduct would cyclize by approach of the nitronate grouping to A in the direction shown, and the resulting configuration at C_2 would be the same as before. In similar fashion the formation of products in many of the cyclizations quoted can be rationalized. Difficulties of interpretation

SCHEME 3

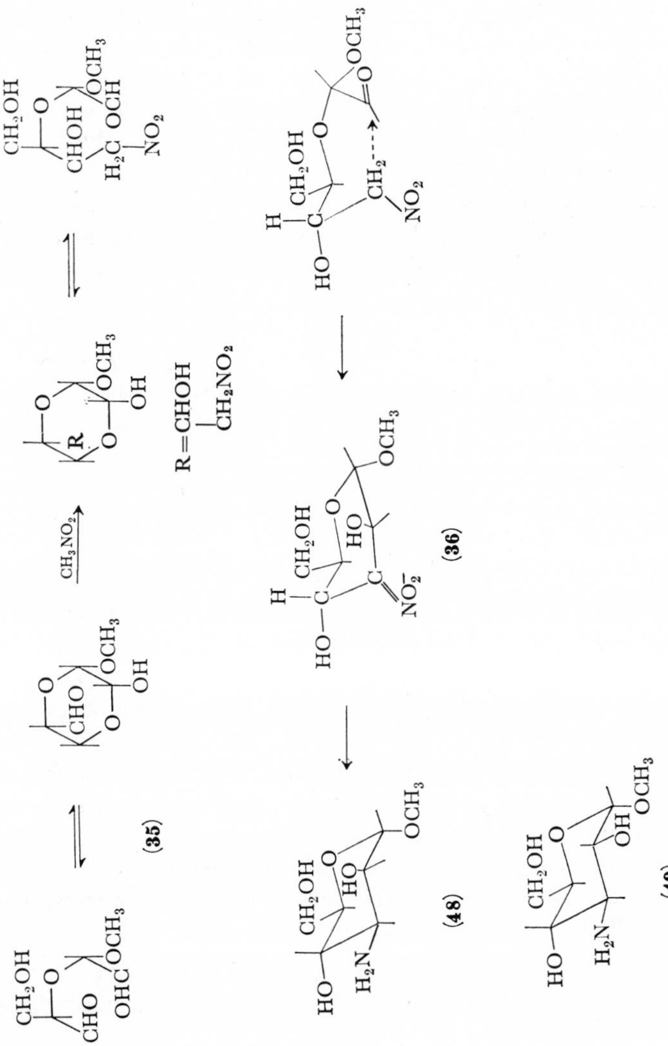

CH_2OH O CHO OHC OCH_3 ⇌ CHO O O OCH_3 OH $\xrightarrow{CH_3NO_2}$ R O O OCH_3 OH ⇌ CH_2OH O CHOH OCH_3 H_2C OCH NO_2

(35)

$R = CHOH$
CH_2NO_2

H HO—C CH_2OH O OCH_3 CH_2 O NO_2 ⟶ H HO—C CH_2OH O HO C OCH_3 NO_2^- ⟶ HO CH_2OH O HO H_2N OCH_3

(36)

(48)

HO CH_2OH O H_2N OH OCH_3

(49)

SCHEME 4

arise on occasion, however. Thus, one would predict D′-methoxy-D-hydroxymethyldiglycolic aldehyde (**35**) to yield preferentially the product **36** as depicted in Scheme 4. This product is in fact formed to a large extent as evidenced by the isolation, upon acidification and catalytic hydrogenation, of 32–36% of methyl 3-amino-3-deoxy-α-D-mannopyranoside (**48**). However, the corresponding glucopyranoside **49** is produced in preponderance (an estimated 60%), which indicates that caution must be exercised in making generalizations on this basis*. Application of the '1,2-*trans*' rule to the dialdehyde **40** would lead to the prediction that methyl 3,6-dideoxy-3-nitro-α-L-mannopyranoside (**50**) will be the favored product, but actually the 1,2-*cis* compound, methyl 3,6-dideoxy-3-nitro-α-L-glucopyranoside (**51**) predominated more than twofold over **50** (equation 22)[115].

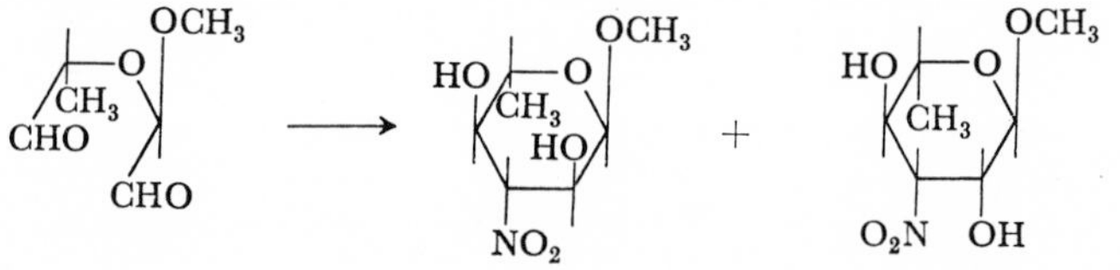

(**40**) → (**51**) + (**50**) (22)

Another stereochemical rule that holds widely though not without exception has emerged. This concerns the acidification (or deionization) of the nitronate cyclization products to give the nitro glycosides. In the majority of cases the nitro group is found to be placed in equatorial position. Thus, the salts **31**, **32**, and **33** give methyl 3-deoxy-3-nitro-β-D-glucopyranoside (**52**), -galactopyranoside (**53**), and -mannopyranoside (**54**), respectively, and not the corresponding C_3 epimers (equations 23–25).

In the α-D-hexopyranoside series the *manno*, *gluco*, and *talo* configurations (**55**, **56**, and **57**) arise from the corresponding nitronates **36**, **37**, and **38** (equations 26–28).

The nitronates derived from β-1,6-anhydrohexoses and from β-2,7-anhydroheptuloses give rise to products with *allo*, *altro*, *gulo*,

* The preponderance of **49** could be due to a subsequent, thermodynamically controlled epimerization on the nitronate stage but, as has been stated above, in the equilibrium the α-D-*gluco* salt is even less stable than the α-D-*manno* salt. There is no evidence for the occurrence of C_2 epimerization during the acidification of the nitronate, or during the hydrogenation of the nitro compound to the amine, which is done in the presence of an equimolar amount of acid.

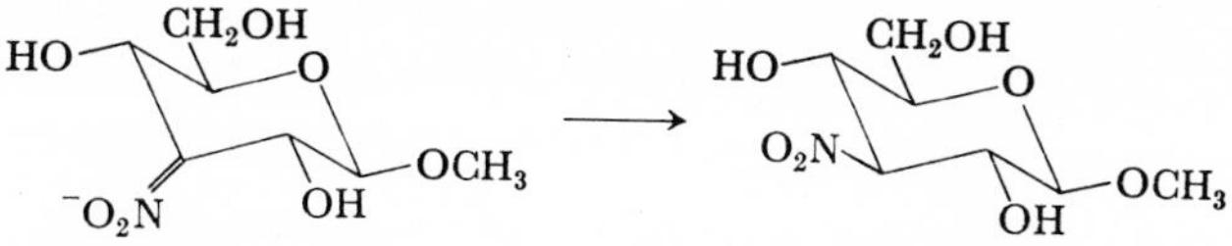

(31) (52) (23)

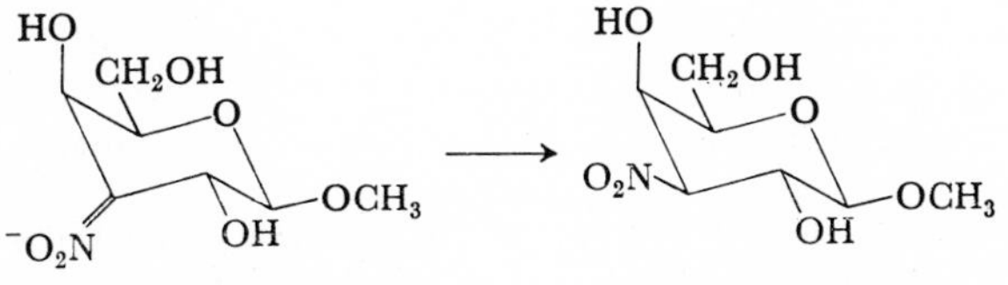

(32) (53) (24)

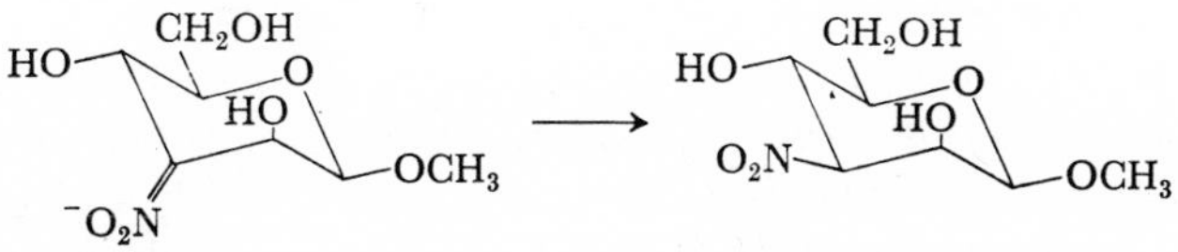

(33) (54) (25)

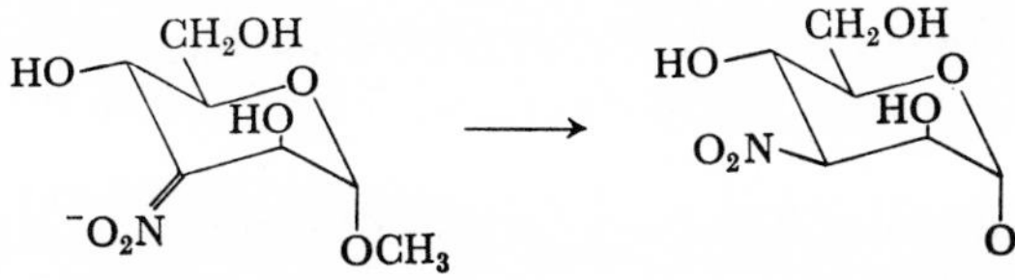

(36) (55) (26)

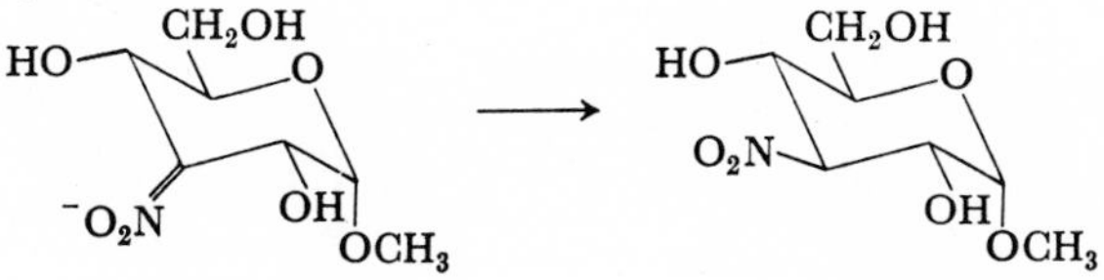

(37) (56) (27)

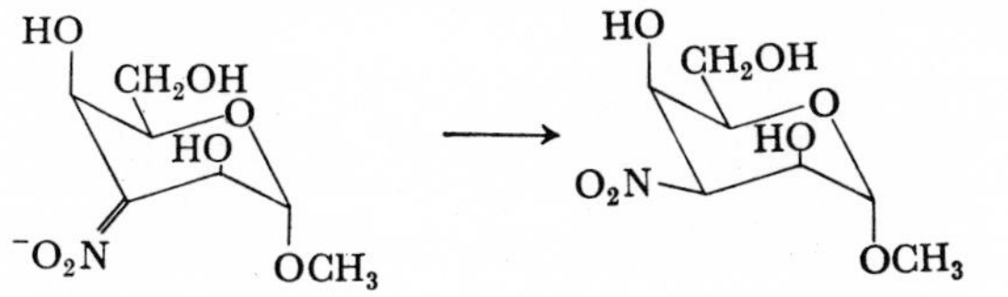

(38) (57) (28)

and *ido* configurations which likewise result from equatorial placement of the nitro group (e.g., **58** → **59**) (equation 29). Similarly,

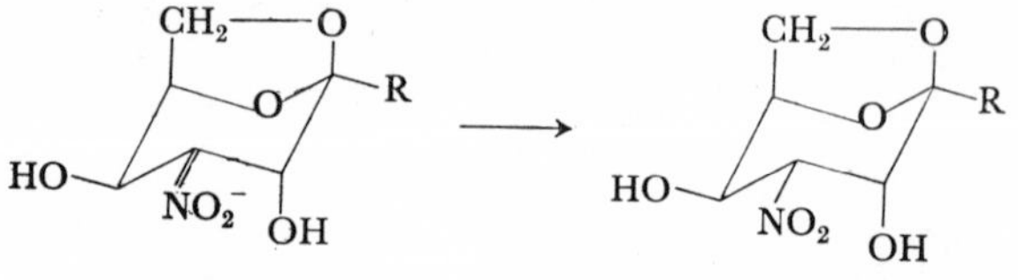

(29)

(**58**)
R = H or CH_2OH

(**59**)
β-D-*gulo* configuration

the deoxynitroinositols that are produced by cyclization of *xylo*-trihydroxyglutaric dialdehyde or of 6-deoxy-6-nitro-D-glucose (or -L-idose, see section I.B. 5a) possess equatorial nitro groups[94].

A notable exception to this fairly general rule seemed to occur in the deionization of the pentoside nitronate **27** which gave two products, the β-D-riboside **60** and the β-D-xyloside **61** (equation 30). The exact ratio is not known, but upon hydrogenation of the mixture the amine from **60** was isolated in over 50% yield, and the amine from **61** in only about 10% yield. On the other hand, from the nitronate **28** was obtained only the α-L-arabinoside **62**, and its 3-epimer, if present, could not be detected (equation 31)[107]. The nitronate **29** also furnished but one product, the β-D-arabinopyranoside **63** (equation 32). The β-D-riboside **60** had originally been thought to exist in the C-1 conformation (with NO_2 axially, and all the other substituents equatorially, oriented), and its preferential formation from **27** was therefore difficult to reconcile with the general tendency of the nitro group, in these glycosides, to adopt an equatorial orientation. However, NMR spectra have indicated the conformations shown for **60**–**63**, which all do possess an equatorial nitro group[107a].

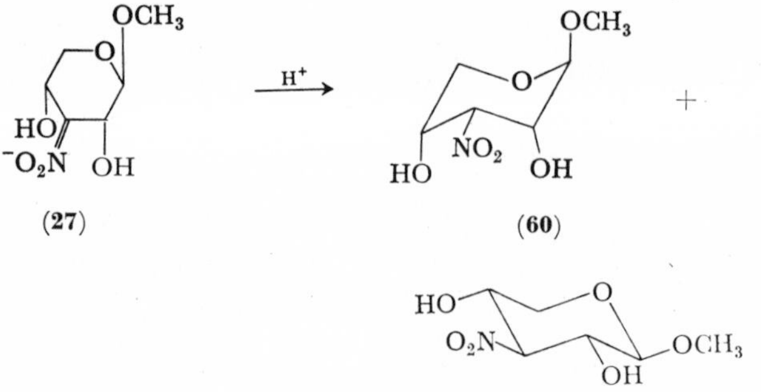

(**27**) $\xrightarrow{H^+}$ (**60**) + (**61**) (30)

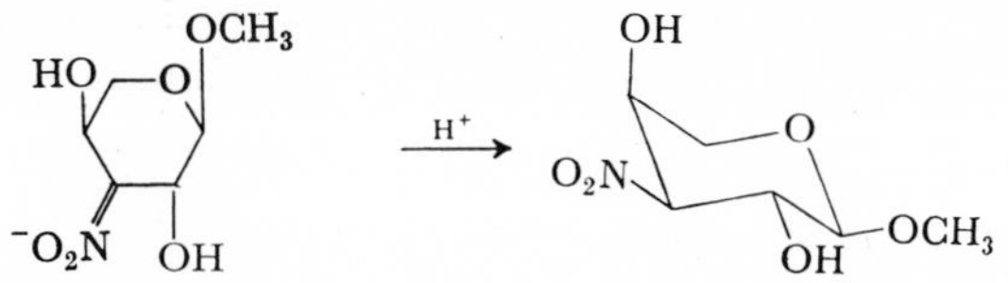

(31)

(28) (62)

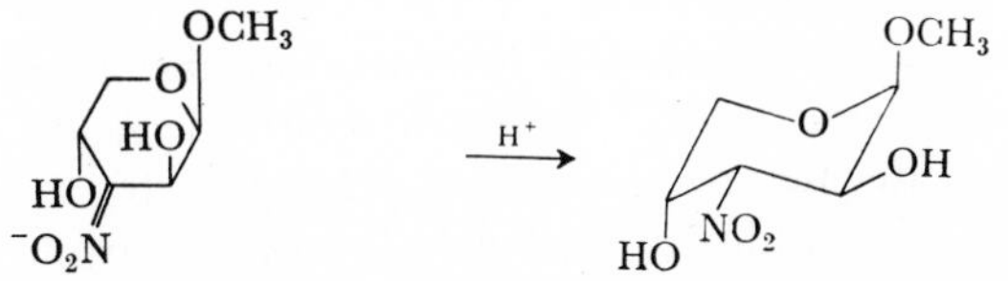

(32)

(29) (63)

Substitution of nitroethane for nitromethane has led to 3-deoxy-3-*C*-methyl-3-nitro sugar derivatives. The dialdehyde **40** gave a mixture of stereoisomers of which one component, **64**, could be isolated as a diacetate in 12.5% yield[130] (equation 33).

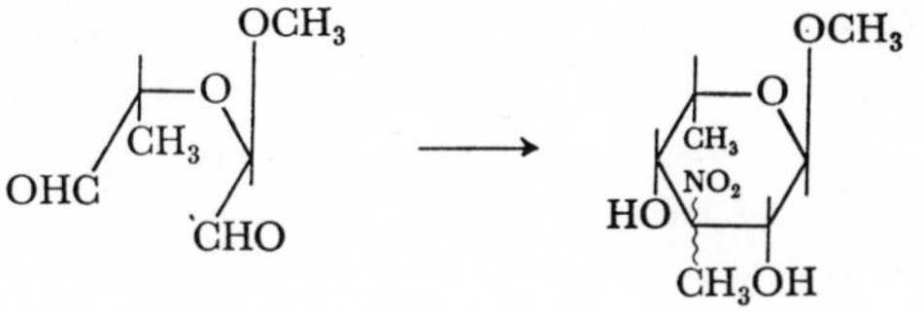

(33)

(40) (64)

An interesting observation was made in nitroethane cyclizations of the dialdehyde **26** and its diastereoisomer **35**[131]. Either dialdehyde

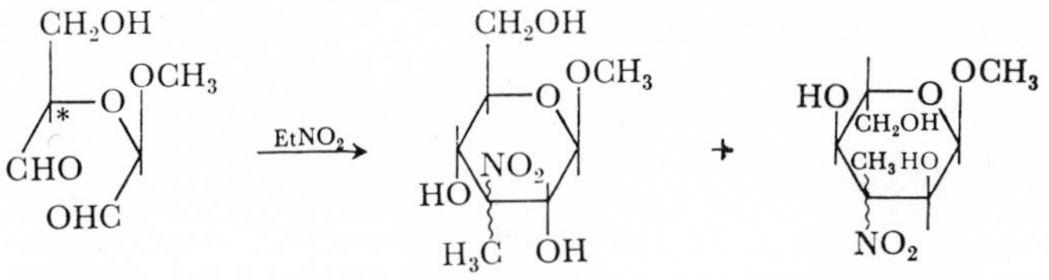

(34a)

(26) β-D α-L

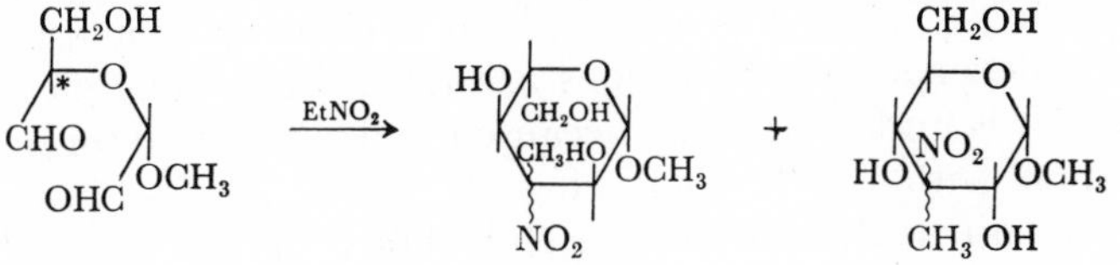

(34b)

(35) β-L α-D

furnished in 20% yield a crystalline product that was shown to be a molecular complex of α- and β-glycosides, the glycosides obtained from **26** being the enantiomorphs of those from **35**. Clearly the reaction must involve in its course a partial epimerization at the carbon atom marked with an asterisk (equations 34a and b).

6. Nitroalkanes and aromatic aldehydes

The first reactions between nitroalkanes and benzaldehyde that seem to have been carried out were reported by Priebs[132] in 1883, more than a decade before Henry published his first papers concerning aliphatic aldehydes. Priebs allowed various aldehydes to react with nitromethane and nitroethane, using zinc chloride as a catalyst and a reaction temperature of 160°. Under these conditions, condensation rather than addition occurred, and the products were β-nitrostyrene derivatives. Later on, Thiele[133,134] and Holleman[135] used basic catalysts or sodium methanenitronate and noticed the formation of the sodium salts of the nitro alcohols, just as in the cases of the aliphatic aldehydes that had meanwhile been studied. However, acidification led directly to the dehydration products (Scheme 5). Numerous β-nitrostyrene derivatives have since been prepared in this fashion or, alternatively, by amine-catalyzed condensations[136,137] of the reactants, which also lead to nitro olefins (see Hass and Riley[3], Schales and Graefe[138] and Eckstein and coworkers[19,139]).

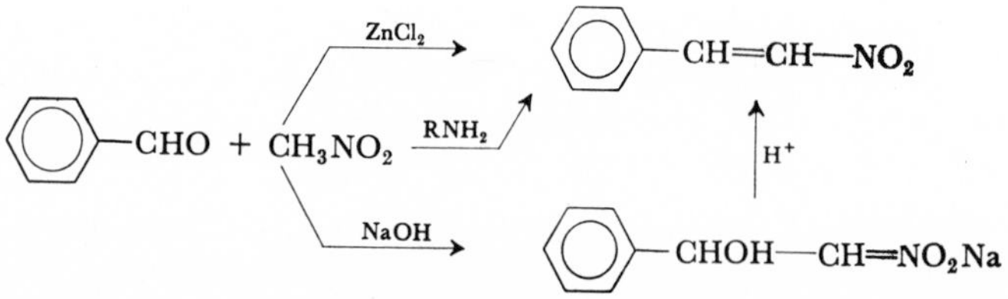

$$C_6H_5\text{—}CHO + CH_3NO_2 \xrightarrow{ZnCl_2 \text{ or } RNH_2} C_6H_5\text{—}CH{=}CH\text{—}NO_2$$

$$C_6H_5\text{—}CHO + CH_3NO_2 \xrightarrow{NaOH} C_6H_5\text{—}CHOH\text{—}CH{=}NO_2Na \xrightarrow{H^+} C_6H_5\text{—}CH{=}CH\text{—}NO_2$$

SCHEME 5

The tendency of α-phenyl-substituted β-nitro alcohols to dehydrate spontaneously is quite general, and the alcohol cannot usually be isolated unless special precautions are taken in the acidification of its salt. This is understandable because the double bond formed is in conjugation with the benzene ring. It is nevertheless possible to obtain the nitro alcohols, by careful choice of the work-up procedures[140]. If the nitro alcohol produced possesses two asymmetric centers, as for instance 2-nitro-1-phenyl-1-propanol from benzaldehyde and nitroethane, diastereomers are formed[141,142]. A fruitful application was found in the synthesis of chloramphenicol, for which benzaldehyde was allowed to react with nitroethanol to give 2-nitro-1-phenyl-1,3-propanediol, which was then used for further synthetic steps eventually furnishing the antibiotic[143].

The method of Kamlet[20], in which the bisulfite addition compound of an aldehyde is allowed to react with a sodium alkanenitronate, has been reported[142] to be particularly desirable for the synthesis of aryl-substituted nitro alcohols. In a recent example, the *o*-, *m*-, and *p*-fluorobenzaldehyde–bisulfite compounds and sodium ethanenitronate gave the corresponding 1-(fluorophenyl)-2-nitro-1-propanols (equation 35), which were then used in the synthesis of potentially insecticidal, fluorinated 1,1-diphenyl-2-nitropropanes[144].

$$FC_6H_4-CH(OH)(SO_3Na) + NaO_2N{=}CHCH_3 \longrightarrow FC_6H_4-CH(OH)CH(NO_2)CH_3 + Na_2SO_3 \quad (35)$$

Phenylnitromethane and benzaldehyde condense to give nitrostilbene, of which the *cis* and *trans* forms have been isolated[145].

However, other reactions may interfere with the generation of a nitrovinyl group. Thus, while *m*- and *p*-carboxybenzaldehydes condense in normal fashion with nitromethane to give *m*- and *p*-carboxy-β-nitrostyrenes[146], respectively, *o*-carboxy-β-nitrostyrene cannot be made in the same way from *o*-carboxybenzaldehyde. The reaction leads, instead, to 3-phthalidylnitromethane (**65**)[147,148] (equation 36) (see also section V.B.2., equation 177).

$$o\text{-}C_6H_4(CHO)(CO_2H) + CH_3NO_2 \xrightarrow{NaOH} o\text{-}C_6H_4(CHOH{-}CH{=}NO_2Na)(CO_2Na) \xrightarrow{H^+}$$

$$\left[o\text{-}C_6H_4(CHOH{-}CH_2NO_2)(CO_2H)\right] \xrightarrow{-H_2O} C_6H_4\langle CH(CH_2NO_2){-}O{-}CO\rangle \ \textbf{(65)} \quad (36)$$

When nitromethane is reacted with *o*-phthalaldehyde in alcoholic solution in the presence of sodium or potassium hydroxide, the alkali salt (**66**) of 2-nitro-1,3-dihydroxyindane (**67**) is formed[149,150]. Acidification with mineral acid leads to what had originally been thought[149,151] to be 2-nitroindanone (**68a**) or tautomeric 1-hydroxy-2-nitroindene (**68b**) but later proved to be 3-hydroxy-2-nitroindene (**69**)[150,152]. Careful deionization of the nitronate **66** afforded the free diol **67**[150]. When the reaction between *o*-phthalaldehyde and nitromethane was carried out without solvent in the presence of dry sodium carbonate, the product was the hemiacetal **71** of *o*-(1-hydroxy-2-nitroethyl)benzaldehyde (**70**). The hemiacetal, upon treatment with alcoholic alkali, was immediately converted into **66**, whereas upon dichromate oxidation, it yielded 3-phthalidylnitromethane (**65**)[150] (Scheme 6).

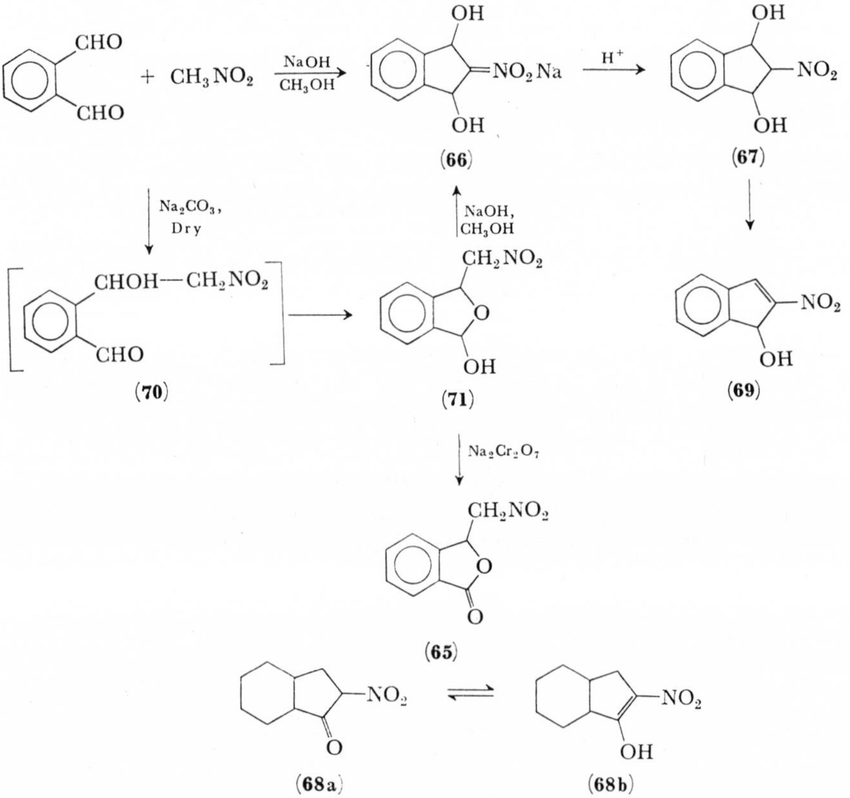

CHO, CHO + CH_3NO_2 → (NaOH, CH_3OH) (**66**) OH, OH, $=NO_2Na$ → (H^+) (**67**) OH, OH, NO_2

Na_2CO_3, Dry → (**70**) CHOH—CH_2NO_2, CHO → (**71**) CH_2NO_2, O, OH

(**71**) → (NaOH, CH_3OH) (**66**)

(**67**) → (**69**) NO_2, OH

(**71**) → ($Na_2Cr_2O_7$) (**65**) CH_2NO_2, O, O

(**68a**) NO_2, O ⇌ (**68b**) NO_2, OH

SCHEME 6

Naphthalene-2,3-dicarboxaldehyde cyclizes with nitromethane to 2-nitrobenzindene-3-ol (**72**) (equation 37), whereas homophthalic dialdehyde furnishes, through facile aromatization of intermediary 2-nitrotetralin-1,3-diol (**73**), 2-nitronaphthaline in 25% yield[152] (equation 38).

$$\text{naphthalene-2,3-(CHO)}_2 \xrightarrow[OH^-]{CH_3NO_2} \text{(72)} \quad (37)$$

(72)

$$\text{homophthalic dialdehyde (CHO, CHO)} \xrightarrow[OH^-]{CH_3NO_2} \text{(73) (OH, OH, NO}_2\text{)} \xrightarrow{-2H_2O} \text{2-nitronaphthalene (NO}_2\text{)} \quad (38)$$

(73)

Cyclizations of *o*-phthalaldehyde and naphthalene-2,3-dicarboxaldehyde with nitroethane gave 1,3-dihydroxy-2-methyl-2-nitroindan and -benzindan, respectively, whereas naphthaline-1,8-dicarboxaldehyde failed to undergo cyclization with either nitromethane or nitroethane[63].

7. Halogenated nitroalkanes and aldehydes

Chloro- and bromonitromethanes have already been recognized by Henry[153] and his school[154,155] to undergo the addition reaction with aldehyde in the usual manner, giving α-halonitro alcohols. Maas[155], and later Wilkendorf and coworkers[156,157], and Urbański and coworkers[43,158] prepared several 2-halo-2-nitro-1,3-alkanediols by further reaction of α-halonitro alcohols with aliphatic aldehydes.

α-Halonitro alcohols that have a hydrogen atom at the α-carbon may be dehydrated to α-halonitro olefins. Thus, when the addition reaction between halonitroalkane and formaldehyde is conducted in the gas phase over a dehydrating catalyst such as alumina impregnated with phosphoric or sulfuric acid, the end products are α-halonitro olefins[159].

Recently, the addition of formaldehyde to trichloronitroalkanes **74** has been investigated[160]. Nitro alcohols **75** were obtained when R = alkyl and R′ = alkyl or alkoxyl. The same was true for R = H; that is, no bismethylol derivatives were formed. However in these instances, easy (R′ = OEt) or even spontaneous (R′ = Et or *n*-Bu) dehydration to **76** occurred. When **74** (R = H, R′ = OEt) was treated with formaldehyde and alkali in the presence of ethanol or butanol, the dialkoxy derivatives **77** arose, no doubt by way of alcohol addition to intermediate **76** (equation 39).

$$\underset{(\mathbf{74})}{Cl_3CCHR'CHRNO_2} \longrightarrow \underset{(\mathbf{75})}{Cl_3CCHR'C(R)(NO_2)CH_2OH} \xrightarrow[(R\,=\,H)]{}$$

$$\underset{(\mathbf{76})}{Cl_3CCHR'C(NO_2){=}CH_2} \xrightarrow[(R'\,=\,OEt)]{R''OH} \underset{(\mathbf{77})}{Cl_3CCHR'CH(NO_2)CH_2OR''} \quad (39)$$

$$R'' = Et \text{ or } n\text{-Bu}$$

The reaction of formaldehyde with 1,1,1-trifluoro-2-nitroethane and 1,1,1,2-tetrafluoro-2-nitroethane led to bis and monomethylolation, respectively[161].

8. Nitroalkanes and halogenated aldehydes

Chloral and bromal combine easily with nitromethane in the presence of salts of weak acids, yielding 3-nitro-1,1,1-trihalo-2-propanols[162,163]. Chloral similarly reacts with nitroethane[13,164] and nitropropane[13], and also with phenylnitromethane though not as readily[164]. With ethyl nitroacetate it gives ethyl 4,4,4-trichloro-3-hydroxy-2-nitrobutyrate[165]. Fluoral and nitromethane afforded 3-nitro-1,1,1-trifluoro-2-propanol, and to heptafluorobutanal hydrate has been added nitromethane, nitroethane, and 1-nitropropane to give the corresponding fluorine-containing nitro alcohols[166]. 1,1-Dichloroacetaldehyde and nitroethane, and chloroacetaldehyde and nitromethane furnished, respectively, 1,1-dichloro-3-nitro-2-butanol and 1-chloro-3-nitro-2-propanol[167].

9. Miscellaneous nitro compounds and aldehydes

α,β-Unsaturated aldehydes may react with nitroalkanes by Henry addition at the carbonyl group or by Michael addition at the C=C double bond. Examples for the former type of addition include the reactions of nitromethane, nitroethane, 1- and 2-nitropropanes, and 1- and 2-nitrobutanes with crotonaldehyde to give unsaturated nitro alcohols[13,168,169] (equation 40).

$$MeCH{=}CHCHO + NO_2CH(R)(R') \longrightarrow MeCH{=}CHCH(OH)CR(NO_2)(R') \quad (40)$$

$$R = H;\ R' = H, Me, Et \text{ or } Pr$$

$$R = Me;\ R' = Me \text{ or } Et$$

Just as intermediate β-nitro-α-phenylethanol easily dehydrates to ω-nitrostyrene when benzaldehyde adds to nitromethane, the vinyl homolog, cinnamic aldehyde, gives rise to 1-phenyl-4-nitro-1,3-butadiene[170]. Furfural, in analogy to benzaldehyde, condenses with nitromethane to give α-(2-furyl)-β-nitroethylene[171,172], the highest yields being obtained with anhydrous methanol as the solvent[173]. Addition, without subsequent dehydration, has been achieved by Kanao who prepared a number of furyl-substituted nitro alcohols[174] (equation 41). Thiophene aldehyde[175] and 1-methyl-

$$\text{(2-furyl)CHO} + O_2NCH_2R \longrightarrow \text{(2-furyl)}CH(OH)CH(NO_2)R \quad (41)$$

2-formylbenzimidazole[176] gave nitro alcohols and, upon dehydration, nitrovinyl derivatives with nitromethane, nitroethane, or phenylnitromethane (equation 42).

$$\text{(1-methylbenzimidazol-2-yl)}CHO \xrightarrow{RCH_2NO_2} \text{(1-methylbenzimidazol-2-yl)}CHOH{-}CH(R)NO_2 \longrightarrow \text{(1-methylbenzimidazol-2-yl)}CH{=}C(R)NO_2 \quad (42)$$

$R = H, Me, \text{ or } Ph$

Pyridine-3-carboxaldehyde condensed with nitromethane, by catalysis with methylamine, to give 1-nitro-2-(3-pyridyl)ethylene, whereas phenylnitromethane under the same conditions formed 1,3-dinitro-1,3-diphenyl-2-(3-pyridyl)propane[177] (equation 43).

$$\text{(3-pyridyl)}CH{=}CHNO_2 \xleftarrow{CH_3NO_2} \text{(3-pyridyl)}CHO \xrightarrow{PhCH_2NO_2} \text{(3-pyridyl)}CH[CH(NO_2)Ph]_2 \quad (43)$$

Ethoxyacetaldehyde and nitromethane produced 1-ethoxy-3-nitro-2-propanol. Attempts to methylolate this product with formaldehyde led, by way of reversal of the addition, to the regeneration of ethoxyacetaldehyde, and the nitromethane liberated was converted into 2-nitro-1,3-propanediol and 2-hydroxymethyl-2-nitro-1,3-propanediol[45] (equation 44).

$$\begin{array}{l} EtOCH_2CHO + CH_3NO_2 \longrightarrow \\ \quad EtOCH_2CHOHCH_2NO_2 \xrightarrow[HCHO]{/\!/} EtOCH_2CHOHCH(NO_2)CH_2OH \\ \quad \downarrow HCHO \\ \quad EtOCH_2CHO + CH_3NO_2 \\ \quad\quad \downarrow \\ \quad\quad (CH_2OH)_2CHNO_2 + (CH_2OH)_3CNO_2 \end{array} \tag{44}$$

Similarly, when β-nitrolactic acid was treated with formaldehyde, the latter displaced glyoxylic acid[45] (equation 45).

$$O_2NCH_2{-}CHOH{-}CO_2H + 3CH_2O \longrightarrow CHOCO_2H + (CH_2OH)_3CNO_2 \tag{45}$$

Nitro olefins whose double bond does not involve the carbon atom that bears the nitro group have been reported to be capable of aldehyde addition. Thus, 1-nitromethylcyclohexene and acetaldehyde furnished 1-nitro-1-(cyclohexen-1-yl)-2-propanol[178], and 3-nitropropene added two molecules of formaldehyde giving 2-hydroxymethyl-2-nitro-3-buten-1-ol[179] (equation 46).

$$CH_2{=}CHCH_2NO_2 + 2CH_2O \longrightarrow CH_2{=}CHC(NO_2)(CH_2OH)_2 \tag{46}$$

Aldehyde addition reactions of ethyl nitroacetate have been extensively studied. Using aliphatic aldehydes with piperidine as a catalyst, Weisblat and Lyttle[180] obtained α-nitro-β-hydroxy esters which they converted into α-amino acids by dehydration, reduction, and hydrolysis. Dornow and Frese[181], who employed stoichiometric amounts of diethylamine as condensing agent and ligroin as the solvent, isolated α-nitro-β-hydroxy esters as crystalline diethylammonium salts **78** but noted that these in general are unstable, and are spontaneously transformed, in solution as well as in the dry state, into monodiethylammonium salts of α,α'-dinitroglutaric esters (**79**) (equation 47). This must without doubt be ascribed to

$$\begin{array}{l} RCHO + O_2NCH_2CO_2Et + NHEt_2 \rightleftharpoons RCHOHC(=NO_2^-)CO_2Et\ \overset{+}{N}H_2Et_2 \ (\mathbf{78}) \\ \quad \downarrow -H_2O,\ -NHEt_2 \\ RCH{=}C(NO_2)CO_2Et\ (\mathbf{80}) \xrightarrow[Et_2NH]{O_2NCH_2CO_2Et} RCH[CH(NO_2)CO_2Et][C(=NO_2^-)CO_2Et]\ \overset{+}{N}H_2Et_2\ (\mathbf{79}) \end{array} \tag{47}$$

R = alkyl

the reversibility of the addition on the one hand, and to a facile dehydration of **78** to the unsaturated ester **80** on the other hand. A Michael addition of regenerated nitroacetate to intermediate **80** then produces **79.** The stability of compounds **78** was found to depend on the structure of the residue R. Thus, a trichloromethyl group or aryl groups (especially those with electronegative substituents) exerted a stabilizing effect in comparison to alkyl groups[181]. However, aromatic aldehydes in an alcoholic reaction medium at low temperature gave dinitroglutaric esters (**79**, R = Ar); on heating in alcohol, the latter were converted into derivatives **81** of isoxazoline oxide[165] (equation 48).

$$\begin{array}{c} ArCHCH(NO_2)CO_2Et \\ | \\ EtO_2CC{=}NO_2^- \ Et_2{}^+NH_2 \end{array} \xrightarrow[-EtOH]{-HNO_2} \text{Ar, } CO_2Et, \ Et_2NCO \text{ isoxazoline ring with } N \rightarrow O \qquad (48)$$

(**79**) (**81**)*

The use of an equimolar amount of ethylamine instead of diethylamine in the nitroacetate–aldehyde addition led to the formation of α-nitro-β-ethylamino esters (**82**) in reactions presumed to proceed via Schiff bases (equation 49). Excess amine was found to convert

$$RCHO + H_2NEt \longrightarrow RCH{=}NEt \xrightarrow{O_2NCH_2CO_2Et} RCH(HNEt)CH(NO_2)CO_2Et \qquad (49)$$

(**82**)

these nitroamino esters, like the hydroxy esters **78**, into dinitroglutaric esters and isoxazoline oxides[181,182].

Under different reaction conditions, formaldehyde[183] and other aliphatic aldehydes[183a] combined with ethyl nitroacetate to yield ethyl β-hydroxy-α-nitroalkanoates. Glutaraldehyde was cyclized[183b] to 2-ethoxycarbonyl-2-nitro-1,3-cyclohexanediol.

Of considerable interest is also the behavior of nitroacetone[184]. Whereas aromatic aldehydes in the presence of alkali condense with the methyl group of the ketone to give unsaturated nitro ketones of type **83**, Schiff bases in the presence of acetic anhydride react with the nitromethyl group to form compounds of type **84** (equation 50).

With aromatic *o*-amino aldehydes or ketones in an acid medium, nitroacetone readily condenses to form 3-nitroquinaldines in a Friedländer-type reaction (equation 51).

* It was not established which of the two ester groupings became an amide.

$$\underset{(\mathbf{83})}{ArCH{=}CHCOCH_2NO_2} \xleftarrow[OH^-]{ArCHO} CH_3COCH_2NO_2 \xrightarrow{ArCH{=}NR} CH_3COCH(NO_2)CH(NHR)Ar \xrightarrow[(-RNH_2)]{Ac_2O} \underset{(\mathbf{84})}{CH_3COC(NO_2){=}CHAr} \quad (50)$$

$$\text{2-}(RCO)C_6H_4NH_2 + H_2CNO_2{-}C({=}O){-}CH_3 \longrightarrow \text{4-R-3-nitro-2-methylquinoline} \quad (51)$$

10. Nitroalkanes and ketones

The nature of the products formed in the interaction of nitroalkanes and ketones depends to a large extent upon the catalyst employed. With alkali hydroxides or alkoxides, quaternary ammonium hydroxides, and primary or tertiary aliphatic amines it is possible to conduct addition reactions that stop at the nitro alcohol stage (equation 52).

$$RCOR + NO_2CH_2R \longrightarrow (R)_2COHCH(NO_2)R \quad (52)$$

Nitro alcohols of this type dehydrate easily, and the resulting nitro olefins are liable to react further, either with excess of nitroalkane giving dinitroalkanes, or with excess ketone giving nitro ketones (see section III). When the ketone–nitroalkane reaction is catalyzed by secondary amines such as diethylamine, piperidine, piperazine, or morpholine, little or no nitro alcohol is isolated and nitro olefins, dinitroalkanes, or nitro ketones are formed in proportions influenced by the ratio of reactants. The first study of the nitroalkane–ketone interaction was reported by Fraser and Kon[185] who obtained 1-nitromethylcyclohexanol from cyclohexanone and nitromethane in the presence of sodium ethoxide, and analogous products with nitroethane and 1-nitropropane, while they observed formation of dinitroalkanes from acetone and some of its homologs with nitromethane in the presence of piperidine. For the preparation of 1-nitromethylcyclohexanol various modifications and improvements were reported by later workers[178,186–188]. Hass and coworkers[189,190] extended considerably the general scope of the reaction, especially with view to the preparation of nitro olefins, dinitroalkanes, and nitro ketones. Lambert and Lowe[191] were able to isolate 2-methyl-1-nitro-2-propanol, in a 62% yield in the sodium methoxide catalyzed addition of nitromethane to acetone. The latter authors also established that 2-methyl-1-nitro-2-propanol when allowed to stand at room temperature for several days in the presence of diethylamine, partly suffers cleavage into acetone and nitromethane; nitromethane so liberated adds to the double bond of 2-methyl-1-nitro-propene which is generated from part of the starting material by loss of water, the final product being 2,2-dimethyl-1,3-dinitropropane (equation 53).

$$2(CH_3)_2C(OH)CH_2NO_2 \xrightarrow{Et_2NH} \begin{cases} (CH_3)_2CO + CH_3NO_2 \\ H_2O + (CH_3)_2C{=}CHNO_2 \end{cases} \longrightarrow (CH_3)_2C(CH_2NO_2)_2 \qquad (53)$$

The occurrence of this reaction affords a general method of preparing α,γ-dinitroalkanes, in which 1 mole of ketone, 2 moles of nitromethane, and 1 mole of diethylamine are mixed together and reaction intermediates are not isolated[192].

The reversal of a Henry addition can also be promoted by the presence of excess formaldehyde which binds the nitromethane that is liberated from a nitro alcohol. Thus, Urbański and coworkers[158] have shown that cyclohexanone is formed when 1-nitromethyl- or 1-halonitromethylcyclohexanol is treated with formaldehyde and alkali (equation 54).

$$\text{1-HO-1-}(CHXNO_2)\text{cyclohexane} \xrightarrow[NaOH]{2(\text{or }3)HCHO} \text{cyclohexanone} + X'C(NO_2)(CH_2OH)_2 \qquad (54)$$

$X = H, Cl, \text{ or } Br; \quad X' = CH_2OH, Cl, \text{ or } Br$

Acetaldehyde also caused the liberation of cyclohexanone from the chloro- and bromonitromethylcyclohexanols; from the former, 3-chloro-3-nitro-2,4-pentanediol was produced and from the latter, 1-bromo-1-nitro-2-propanol (equation 55).

Nightingale and associates studied the nitroalkane additions of alkyl-substituted cyclohexanones, cyclohexenones, and cyclohexanediones. They obtained[193] alkyl-substituted 1-nitromethylcyclohexanols from nitromethane and the 3- and 4-methylcyclohexanones by catalysis with sodium ethoxide, whereas 2-methylcyclohexanone did

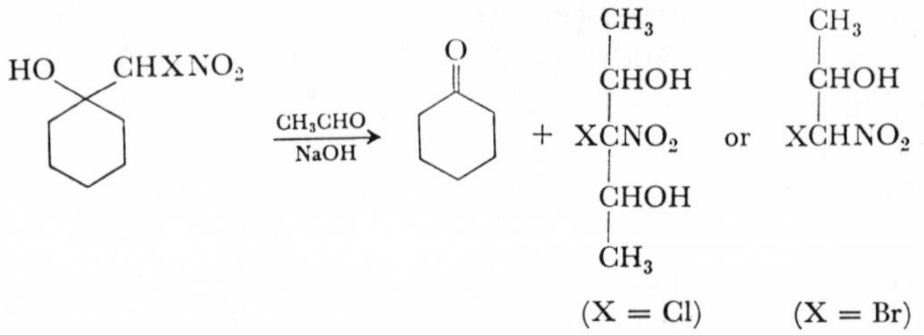

$$\text{1-(CHXNO}_2\text{)-1-hydroxycyclohexane} \xrightarrow[\text{NaOH}]{\text{CH}_3\text{CHO}} \text{cyclohexanone} + \underset{(\text{X = Cl})}{\text{CH}_3\text{CHOH–CX(NO}_2\text{)–CHOHCH}_3} \text{ or } \underset{(\text{X = Br})}{\text{CH}_3\text{CHOH–CHXNO}_2} \quad (55)$$

not react, probably because of steric hindrance. Higher nitroalkanes gave low yields of nitro alcohols from the 3- and 4-methylcyclohexanones when piperidine was used as the catalyst. Later, the authors[194] isolated the corresponding alkyl-substituted 1-nitromethyl-1-cyclohexenes from the piperidine-catalyzed interaction of nitromethane with the alkylcyclohexanones mentioned. They also found that 3-methyl-2-cyclohexen-1-one (and some related unsaturated ketones) add nitromethane across the carbon–carbon double bond rather than the carbonyl group, giving 3-methyl-3-nitromethylcyclohexanone (and related saturated ketones). Interestingly, 1,2- as well as 1,4-cyclohexanedione reacted, in the presence of potassium carbonate, with only one molecule of nitromethane even when a large excess of the latter was employed. The products were the 2- (and 4-) hydroxy-2- (and 4-) nitromethylcyclohexanones. Similarly, one carbonyl group only of acenaphthenequinone and of phenanthrenequinone underwent nitroalkane addition; no identifiable products were obtained from anthraquinone or the naphthoquinones[194].

Homologous nitromethylcycloalkenes containing seven- and eight-membered rings were synthesized from cycloheptanone and cyclooctanone by Eckstein and coworkers[195–197]. When they tried to condense cycloheptanone or cyclooctanone with nitroethane or 1-nitropropane in the presence of piperidine, they isolated the enamines *N*-(1-cycloheptenyl)piperidine and *N*-(1-cyclooctenyl) piperidine[197].

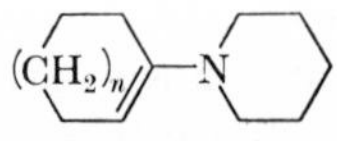

$(CH_2)_n$ ring–N(piperidine)

$n = 2$ or 3

N-(1-Cyclohexenyl)piperidine ($n = 1$), which was prepared by condensing cyclohexanone with piperidine, was converted into 1-cyclohexenylnitromethane by boiling with nitromethane in dioxane solution[197].

In the reaction between nitromethane and cyclohexanone when catalyzed by piperidine or other secondary aliphatic amines an interesting by-product of the composition $C_{14}H_{20}N_2O_3$ is formed[191,197,198]. Its structure was established independently by Noland and Sundberg[199] and by House and Magin[200] to be 14-hydroxy-14-azadispiro[5.1.5.2]pentadec-9-ene-7,15-dione-7-oxime (**85**). Although the mechanism of its formation has not been securely elucidated, plausible pathways have been suggested.

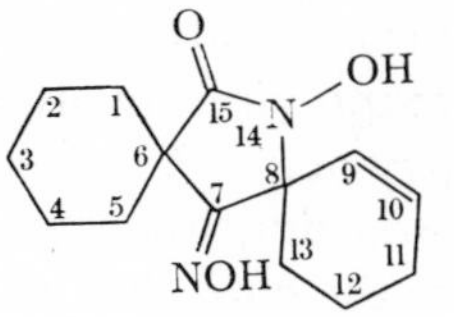

(**85**)

Nightingale and coworkers demonstrated that similar products may also arise from nitromethane and other cyclic ketones[201].

Two examples of cyclizations of diketones with nitromethane, analogous to those of dialdehydes, have been reported. Bicyclo-[1.3.3]nonane-3,7-dione and its 9-dichloromethyl-9-methyl derivative were cyclized in the presence of sodium methoxide to the corresponding 2-nitro-1,3-dihydroxyadamantanes[202,203] (equation 56).

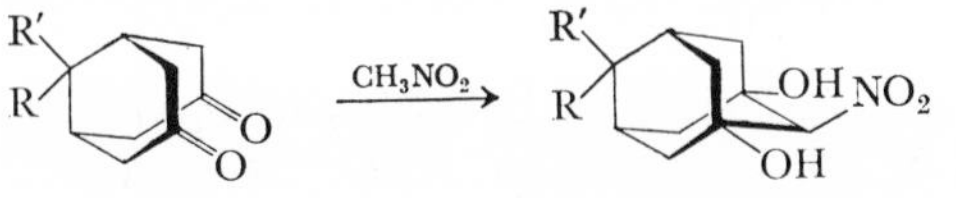

$$\xrightarrow{CH_3NO_2} \tag{56}$$

$R = R' = H$
$R = CH_3;\ R' = CHCl_2$

II. Polynitroalkanes and aldehydes

Geminal dinitroalkanes that possess a hydrogen atom on the carbon bearing the nitro groups are capable of undergoing the Henry addition. Thus, the dimethylol derivative of dinitromethane, 2,2-dinitropropanediol, was obtained by Feuer and associates[204] by acidifying an aqueous mixture of potassium dinitromethane and formaldehyde. Similarly, 1,1,3,3-tetranitropropane gave a bis-methylol derivative, 2,2,4,4-tetranitro-1,5-pentanediol[205], whereas bis(potassium 2,2-dinitroethyl)amine or its *N*-substituted derivatives did not add to formaldehyde[206]. Action of base upon methylol derivatives of dinitroalkanes results in a reversion of the Henry addition (demethylolation). Thus 2,2-dinitropropanediol upon treatment at room temperature with one mole of base affords the salt of 2,2-dinitroethanol, and heating of the latter with excess base slowly leads to the salt of dinitromethane[204]. On the other hand, the potassium salt of 2,2-dinitroethanol will react with formaldehyde to furnish 2,2-dinitropropanediol[207], while *gem*-dinitro alcohols of the type $RC(NO_2)_2CH_2OH$ can be obtained also by monomethylolation of *gem*-dinitroalkanes[208,209].

Dinitromethane and 1,1-dinitroethane added glyoxylic acid to furnish the expected α-hydroxy acids, while with glyoxal the reaction took an unexpected course in that it involved a partial oxidation and yielded the same α-hydroxy acids[210].

Dinitroacetonitrile [211] and dinitroacetamide[212] have been treated with formaldehyde to give the corresponding β-hydroxy-α,α-dinitropropionic acid derivatives.

From 1,4-dinitrobutane there have been obtained the monomethylol, bismethylol, and bisdimethylol derivatives **86**, **87**, and **88** by addition of one, two, and four molecules of formaldehyde, respectively[12,213,214], but there was no evidence for the formation of the unsymmetrical diol or the triol (equation 57).

$$O_2NCH_2CH_2CH_2CH_2NO_2 \longrightarrow \begin{cases} NO_2CH_2CH_2CH_2CH(NO_2)CH_2OH & (\mathbf{86}) \\ HOCH_2CH(NO_2)CH_2CH_2CH(NO_2)CH_2OH & (\mathbf{87}) \\ (HOCH_2)_2C(NO_2)CH_2CH_2C(NO_2)(CH_2OH)_2 & (\mathbf{88}) \end{cases} \qquad (57)$$

Similar additions were carried out with 1,5-dinitropentane[12,213,215], 1,6-dinitrohexane[12], and 1,4-dinitrocyclohexane[216], and some of the symmetrical bismethylol derivatives were used[217] in oxidative nitrations leading ultimately to α,α,ω,ω-tetranitroalkanes. 1,1,3,3-Tetranitropropane gave 2,2,4,4-tetranitro-1,5-pentanediol[205], and 1,1,3,3-tetranitrobutane gave 2,2,4,4-tetranitropentanol[218].

By condensing aromatic aldehydes with 1,4-dinitrobutane, Perekalin and coworkers[214,219] obtained 1,6-diaryl-2,5-dinitro-1,5-hexadienes, and from 2,3-dimethyl- (and diaryl-) 1,4-dinitro-2-butenes they synthesized a number of 1,6-diaryl-3,4-dimethyl- (or diaryl-)2,5-dinitro-1,3,5-hexatrienes. Similar work was reported by Rembarz and Schwill[220].

Dinitromethane has been reported to react with glyoxal and with succinic dialdehyde to give, respectively, 1,1,4,4-tetranitro-2,3-butanediol and 1,1,6,6-tetranitro-2,5-hexanediol[55]. The product from glyoxal, stated to be a brown liquid, would seem to have been of dubious purity, and it has already been mentioned above that other workers[210] observed a different course of this reaction.

When the potassium salt of dinitromethane was treated with chloroacetaldehyde, the intermediate 3-chloro-2-hydroxy-1,1-dinitropropane potassium cyclized to 4-hydroxy-3-nitroisoxazoline *N*-oxide[221] (equation 58).

$$ClCH_2CHO + HC(NO_2)NO_2K \longrightarrow$$

$$ClCH_2CHOH—C(NO_2)NO_2K \longrightarrow \text{[4-hydroxy-3-nitroisoxazoline } N\text{-oxide: ring with HO, } NO_2\text{, N}\rightarrow\text{O]} \quad (58)$$

II. MANNICH REACTIONS

A. Reactions with Secondary Amines

In the Mannich reaction, formaldehyde reacts with a secondary amine and a compound possessing a reactive α-hydrogen to give dialkylaminomethyl derivatives (equation 59). A comprehensive discussion of the reaction mechanism and conditions has been given by Hellmann and Opitz[222].

$$R_2NH + CH_2O + H—CR_2—CR_2Y \longrightarrow R_2N—CH_2—CR_2—CR_2Y \quad (59)$$
$$Y = \text{activating group}$$

Henry was the first to show that similar reactions will occur with nitroalkanes. He established that *N*-hydroxymethylpiperidine condenses with nitromethane and nitroethane to yield, respectively, 2-nitro-1,3-di-*N*-piperidinopropane and 2-nitro-1,3-di-*N*-piperidino-2-methylpropane[223,224] (equation 60).

$$2\ C_5H_{10}NCH_2OH + CH_2RNO_2 \longrightarrow O_2NCR(CH_2NC_5H_{10})_2 \quad (60)$$
$$R = H \text{ or } CH_3$$

Later investigators[225–228] found that higher primary nitroalkanes chiefly condense with only one molecule of N-hydroxymethylated secondary amines, and that from nitromethane and nitroethane, mono-condensation products can be obtained at lower temperatures (equation 61).

$$C_5H_{10}NCH_2OH + CH_2RNO_2 \longrightarrow C_5H_{10}NCH_2CHRNO_2 \qquad (61)$$
$$R = H \text{ or alkyl}$$

Dornow and coworkers obtained Mannich bases by the interaction of formaldehyde and secondary amines with ketones such as ω-nitroacetophenone[229], nitroacetone[184], and benzylidene- and furfurylidenenitroacetones[230] (equation 62).

$$RCOCH_2NO_2 + CH_2O + NHR'R'' \longrightarrow RCOCH(NO_2)CH_2NR'R'' \qquad (62)$$
$$R = Ph, CH_3, PhCH{=}CH, C_4H_3OCH{=}CH; \quad R', R'' = \text{alkyl}$$

On the other hand, ethyl nitroacetate did not give a stable Mannich base with methylenebisdiethylamine; the reaction proceeded to the diethylammonium salt of ethyl α,α'-dinitroglutarate[231] (equation 63).

$$2\,CH_2NO_2CO_2Et + Et_2NCH_2NEt \longrightarrow \left[EtO_2C{-}\underset{\underset{NO_2}{\|}}{C}{-}CH_2{-}\underset{\underset{NO_2}{\|}}{C}{-}CO_2Et\right]^{2-} 2EtNH_2^+ \qquad (63)$$

The secondary nitroparaffins, 2-nitropropane and 2-nitrobutane, were condensed with formaldehyde and a variety of secondary aliphatic amines that included dimethyl-, diethyl-, dibutyl-, and bis(2-ethylhexyl)amines, piperidine, morpholine, 2,5-dimethylpyrrolidine, and diethanolamine[232] (equation 64).

$$R_2NH + CH_2O + (CH_3)_2CHNO_2 \longrightarrow R_2N{-}CH_2{-}C(CH_3)_2NO_2 + H_2O \qquad (64)$$

Analogous Mannich bases were also obtained from 2-nitropropane and *N*-phenylpiperazine as well as from diallylamine and dimethallylamine, whereas unsubstituted piperazine and 3,5-dimethylpiperazine gave the corresponding bis condensation products[233] (equation 65).

$$HN\langle\rangle NH + 2\,CHO + 2\,(CH_3)_2CHNO_2 \longrightarrow H_3C\underset{NO_2}{\overset{CH_3}{C}}CH_2N\langle\rangle NCH_2\underset{NO_2}{\overset{CH_3}{C}}CH_3 \qquad (65)$$

vic-Nitro alcohols, diols, and triols interact with secondary amines to give nitroamines[227,232–234]. Thus 2-nitroethanol, the methylol derivative of nitromethane, and piperidine afforded the same diamine that was obtained with nitromethane and hydroxymethylpiperidine[227] (equation 66).

$$2\ O_2NCH_2CH_2OH + 2\ C_5H_{11}N \longrightarrow (C_5H_{11}NCH_2)_2CHNO_2 + CH_3NO_2 + 2H_2O \quad (66)$$

2-Nitro-2-methyl-1-propanol, the methylol derivative of 2-nitropropane, was shown to react with piperazine and some related secondary amines to give the same Mannich products that were obtained from these amines with 2-nitropropane and formaldehyde. Likewise, interaction of the nitro alcohol with the methylenediamines that form readily upon mixing the secondary amines with formaldehyde afforded the same Mannich bases[233].

Similarly, the bismethylol derivative of nitroethane, 2-methyl-2-nitro-1,3-propanediol, yielded 2-methyl-2-nitro-1,3-dipiperidinopropane[227] (equation 67) and 2-methyl-2-nitro-1,3-bis(dimethylamino)propane[232]; and tris(hydroxymethyl)nitromethane gave, with diethylamine, 3,3′-bis(diethylamino)-2-nitroisobutyl alcohol[234] (equation 68).

$$CH_3C(NO_2)(CH_2OH)_2 + 2\ C_5H_{11}N \longrightarrow CH_3C(NO_2)(CH_2NC_5H_{10})_2 + 2H_2O \quad (67)$$

$$O_2NC(CH_2OH)_3 + 2\ NH(C_2H_5)_2 \longrightarrow [(C_2H_5)_2NCH_2]_2C(NO_2)CH_2OH + 2H_2O \quad (68)$$

Whereas it was claimed[234] that 1-nitro-2-octanol condensed with formaldehyde and diethylamine to form 1-diethylamino-2-nitro-3-nonanol (equation 69), an analogous reaction between 1-nitro-2-propanol, piperidine, and formaldehyde resulted[227] in the liberation of acetaldehyde and the production of 2-nitro-1,3-dipiperidinopropane (equation 70).

$$CH_3(CH_2)_5CHOHCH_2NO_2 + CH_2O + NH(C_2H_5)_2 \longrightarrow CH_3(CH_2)_5CHOHCHNO_2CH_2N(C_2H_5)_2 + H_2O \quad (69)$$

$$CH_3CHOHCH_2NO_2 + 2\ CH_2O + 2\ C_5H_{11}N \longrightarrow CH_3CHO + O_2NCH(CH_2NC_5H_{10})_2 + 2H_2O \quad (70)$$

B. *Reactions with Primary Amines*

Although in a previous study[225] hydroxymethylmonoalkylamines had failed to undergo Mannich reactions with nitroalkanes,

Senkus[235,236] has successfully carried out such reactions using methylamine, isopropylamine, 1-butylamine, 2-butylamine, benzylamine, 1-phenylethylamine, 2-amino-1-butanol, and 2-amino-2-methyl-1-propanol as monoalkylamine components, and nitroethane, 1- and 2-nitropropane, and 2-nitrobutane as nitro components. Again, the products could be obtained either by allowing the amine to react with formaldehyde and thereafter adding the nitroalkane, or by first generating the methylol derivative of the nitroalkane, which was then treated with the amine. The latter reaction is the slower, and it is believed that the first step in it is a demethylolation of the nitro alcohol, after which this process becomes identical with the first one (equation 71).

$$\mathrm{RNHCH_2OH + R_2CHNO_2 \longrightarrow RNHCH_2CR_2NO_2 + H_2O} \qquad (71a)$$

$$\mathrm{2\,RNHCH_2OH + RCH_2NO_2 \longrightarrow (RNHCH_2)_2CRNO_2 + 2H_2O} \qquad (71b)$$

alternatively

$$\mathrm{RNH_2 + HOCH_2CR_2NO_2 \longrightarrow RNHCH_2CR_2NO_2 + H_2O} \qquad (71a')$$

$$\mathrm{2\,RNH_2 + (HOCH_2)_2CRNO_2 \longrightarrow (RNHCH_2)_2CRNO_2 + 2H_2O} \qquad (71b')$$

With aromatic amines the reaction is more difficult, and a basic catalyst such as a quaternary ammonium hydroxide is required for its success. Nevertheless, good to excellent yields have been achieved with 2-nitropropane, formaldehyde, and a variety of primary arylamines and aromatic diamines. *N*-Methylaniline reacted also, but *N*,*N*-diphenylamine did not[237].

With excess formaldehyde, the nitrodiamines obtained in equation 71 (b or b′), are cyclized to hexahydropyrimidine derivatives[238–240] (equation 72).

$$\mathrm{RNHCH_2C(NO_2)(R)CH_2NHR + CH_2O \longrightarrow} \text{ 1,3-di-R-5-R-5-nitrohexahydropyrimidine (ring: } \mathrm{CH_2}\text{, RN, NR, } \mathrm{H_2C}\text{, } \mathrm{CH_2}\text{, C(R)(NO}_2\text{))} \qquad (72)$$

On the other hand, reaction between a primary amine, a primary nitroalkane, and formaldehyde in a molar ratio 1:1:3 leads to 3,5-dialkyl-5-nitrotetrahydro-1,3-oxazines[241], which is understandable since the medium contains nitro alcohol and hydroxymethylalkylamine, as intermediates, which combine according to equation 73. The bismethylol derivative of the nitroalkane together with 1 mole each of formaldehyde and amine may be used instead.

$$RNHCH_2OH + RCH(NO_2)CH_2OH \longrightarrow (R)(O_2N)C(CH_2OH)(CH_2NHR) \xrightarrow[-H_2O]{CH_2O} (R)(O_2N)C\langle CH_2{-}O{-}CH_2{-}NR{-}CH_2 \rangle \quad (73)$$

Numerous tetrahydrooxazine derivatives of this type have been prepared in this way.[51,53,241–252] It has also been shown that, for the preparation of *N*-(arylmethyl)tetrahydrooxazines, hexahydro-*sym*-triazines may act as a source of amine and part of the formaldehyde[253,254] (equation 74). It is of interest that 3-nitro-2,4-pentanediol

$$3\ (R')(O_2N)C(CH_2OH)_2 + \text{(RN)}_3(CH_2)_3 \xrightarrow{-3H_2O} 3\ R'(O_2N)C\langle CH_2{-}O{-}CH_2{-}NR{-}CH_2 \rangle \quad (74)$$

R = CH_2Ar; R′ = alkyl, or some other group

did not give heterocyclic products on treatment with primary amines. Instead, acetaldehyde was split off and resinification took place[42].

The synthesis and reactions of the oxazine derivatives have been reviewed by Eckstein and Urbański[255].

C. C-Alkylations with Mannich Bases

Mannich bases derived from primary nitroalkanes have been found capable of alkylating reactive methylene or methine compounds. Actually, an elimination of amine from the Mannich base is thought to occur, so that the latter behaves as a potential nitro olefin which will add to the reactive methylene compound in a Michael-type reaction. No external catalyst is required[256] (equation 75).

$$CH_3CH_2CH(NO_2)CH_2NR_2 \xrightarrow{-HNR_2} [CH_3CH_2C(NO_2){=}CH_2] \xrightarrow{R'R''CHNO_2} CH_3CH_2CH(NO_2)CH_2C(NO_2)R'R'' \quad (75)$$

$R = CH_3$ or C_2H_5; $R' = C_2H_5$, $R'' = H$ or $R' = R'' = CH_3$

For the same reason, an amine exchange occurred when the Mannich base was heated with excess piperidine.

N-(2-Nitroisobutyl)dimethylamine, a Mannich base derived from a secondary nitroparaffin, is structurally incapable of amine elimination and therefore does not undergo these reactions[256].

The same mechanism has been invoked for the formation of γ-dinitroparaffins from primary nitroalkanes and formaldehyde when catalysis by secondary amines is used. It was found that secondary amines are considerably better catalysts than triethylamine or sodium carbonate, and this was attributed to a more facile formation of the nitro olefin intermediate from a Mannich base precursor than from a nitroalkane–methylol derivative[257] (equation 76).

$$R'CH_2NO_2 + CH_2O \begin{cases} \xrightarrow{R_2NH} R'CH(NO_2)CH_2NR_2 \xrightarrow[\text{fast}]{-R_2NH} \\ \xrightarrow[\text{or } Na_2CO_3]{R_3N} R'CH(NO_2)CH_2OH \xrightarrow[\text{slow}]{-H_2O} \end{cases} R'CH(NO_2){=}CH_2 \quad (76)$$

Prior to these investigations, 2-nitroalkenes had actually been prepared, in yields of 50–75%, by the pyrolysis at reduced pressure of 1-diethylamino-2-nitroalkane hydrochlorides which had been obtained from 1-nitroalkanes and hydroxymethyldiethylamine[228] (equation 77).

$$RCH(NO_2)CH_2NEt_2{\cdot}HCl \xrightarrow[50\text{–}100\text{ mm}]{100\text{–}175°} RC(NO_2){=}CH_2 + Et_2NH_2Cl \quad (77)$$

Various Mannich bases derived from non-nitro compounds have been used to *C*-alkylate nitromethane and ethyl nitroacetate. Thus, Reichert and Posemann[258] allowed ω-dimethylaminopropiophenone (**89**) to react with nitromethane in the presence of sodium methoxide and obtained the γ-nitro ketones **91**, **92**, and **93**, presumably via the intermediate olefin **90** (equation 78). Dornow and coworkers[231]

$$\underset{(\mathbf{89})}{PhCOCH_2CH_2N(CH_3)_2} \xrightarrow{-HN(CH_3)_2} \underset{(\mathbf{90})}{PhCOCH{=}CH_2} \xrightarrow{CH_3NO_2} \underset{(\mathbf{91})}{PhCOCH_2CH_2CH_2NO_2}$$

$$\mathbf{91} + \mathbf{90} \longrightarrow \underset{(\mathbf{92})}{(PhCOCH_2CH_2)_2CHNO_2} \quad (78)$$

$$\mathbf{92} + \mathbf{90} \longrightarrow \underset{(\mathbf{93})}{(PhCOCH_2CH_2)_3CNO_2}$$

prepared, for instance, ethyl 4-benzoyl-2-nitrobutyrate by the interaction of ethyl nitroacetate with **89** (equation 79), and similarly, ethyl 3-(2-ketocyclohexyl)-2-nitropropionate by the reaction with 2- diethylaminomethylcyclohexanone (equation 80). Later, they also treated **89** with ethyl nitromalonate and obtained a condensation product that was partially hydrolyzed and decarboxylated to furnish the same 4-benzoyl-2-nitrobutyrate[259]. The use of nitroacetanilide gave the corresponding anilide[260].

$$\mathbf{89} + CH_2(NO_2)CO_2Et \longrightarrow PhCOCH_2CH_2CH(NO_2)CO_2Et \qquad (79)$$

$$\text{2-}(CH_2N(Et)_2)\text{cyclohexanone} + CH_2(NO_2)CO_2Et \longrightarrow \text{2-}(CH_2CH(NO_2)CO_2Et)\text{cyclohexanone} \qquad (80)$$

The Mannich base **94** from benzylideneacetone, formaldehyde, and diethylamine gave, with ethyl nitroacetate, the α,α′-dinitro diester **95**[231] (equation 81).

$$\underset{(\mathbf{94})}{PhCH{=}CHCOCH_2CH_2N(Et)_2} + CH_2(NO_2)CO_2Et_2 \xrightarrow{-HNEt_2}$$

$$PhCH{=}CHCOCH_2CH_2CH(NO_2)CO_2Et \xrightarrow{O_2NCH_2CO_2Et}$$

$$\underset{(\mathbf{95})}{EtO_2CCH(NO_2)CH(Ph)CH_2COCH_2CH_2CH(NO_2)CO_2Et} \qquad (81)$$

Gramine (**96**), which can be regarded as a Mannich base derived from indole, underwent reaction with ethyl nitroacetate to form ethyl 3-(3-indolyl)-2-nitropropionate (**97**) (equation 82), the reduction of which furnished D,L-tryptophan[261]. Later, ethyl nitromalonate was used with advantage to synthesize **97**[262].

$$\underset{(\mathbf{96})}{\text{3-}(CH_2NEt_2)\text{indole}} \xrightarrow{-HNEt_2} \left[\text{3-methyleneindolenine } (=CH_2)\right] \xrightarrow{O_2NCH_2CO_2Et} \underset{(\mathbf{97})}{\text{3-}(CH_2CH(NO_2)CO_2Et)\text{indole}} \qquad (82)$$

C-Alkylation of ethyl nitroacetate failed with a tertiary Mannich base that is structurally incapable of amine elimination, namely diethylaminoethylantipyrine (**98**); only the salt of the base with the nitro ester was produced. The reaction succeeded, however, when the methiodide of **98** was employed[231,260], since quaternization weakens the bond between the nitrogen atom and the heterocycle which can dissociate as a resonance-stabilized cation[222] (equation 83).

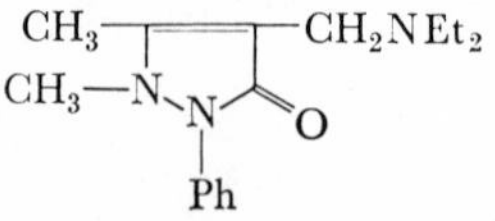

(**98**)

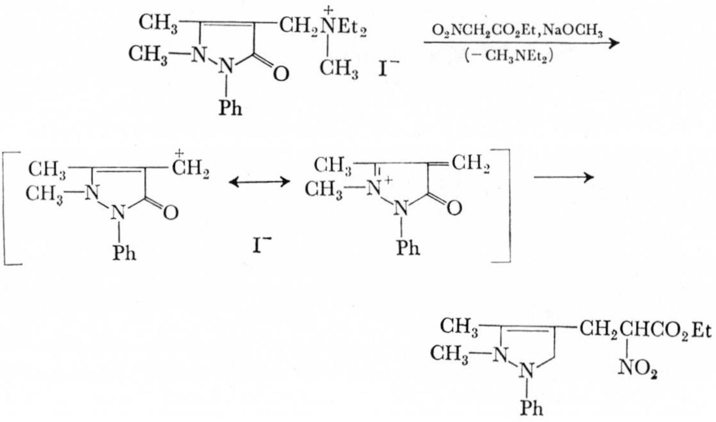

$$\xrightarrow[(-CH_3NEt_2)]{O_2NCH_2CO_2Et,\ NaOCH_3}$$

(83)

D. Reactions with Ammonia

Heterocyclic products may also arise in reactions between primary nitroalkanes, formaldehyde, and ammonia. Hirst and coworkers[263] investigated the case of 1-nitropropane and obtained, along with resinous material, condensation products that were formulated as tetrahydrooxazine derivatives **99** and **100**, and oxaazacyclooctane **101** (equations 84–86).

$$EtCH_2NO_2 + 3\ CH_2O + NH_3 \longrightarrow [HOCH_2C(Et)(NO_2)CH_2NHCH_2OH] \longrightarrow \quad (84)$$

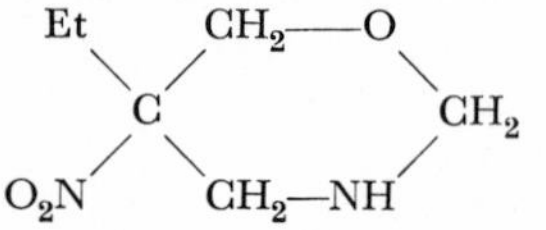

(**99**)

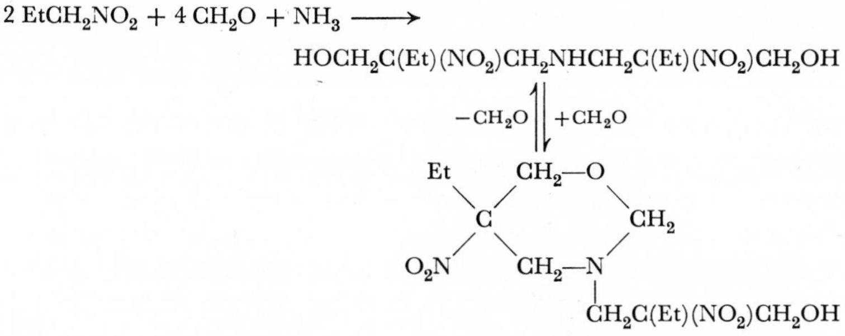

$$2\,EtCH_2NO_2 + 4\,CH_2O + NH_3 \longrightarrow HOCH_2C(Et)(NO_2)CH_2NHCH_2C(Et)(NO_2)CH_2OH \underset{+CH_2O}{\overset{-CH_2O}{\rightleftharpoons}} \textbf{(100)} \quad (85)$$

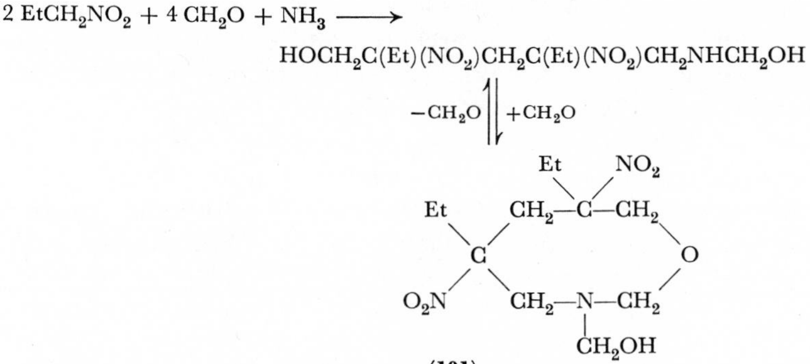

Et CH₂—O CH₂ C O₂N CH₂—N CH₂C(Et)(NO₂)CH₂OH

(100)

$$2\,EtCH_2NO_2 + 4\,CH_2O + NH_3 \longrightarrow HOCH_2C(Et)(NO_2)CH_2C(Et)(NO_2)CH_2NHCH_2OH \underset{+CH_2O}{\overset{-CH_2O}{\rightleftharpoons}} \textbf{(101)} \quad (86)$$

Et NO₂ Et CH₂—C—CH₂ C O O₂N CH₂—N—CH₂ CH₂OH

(101)

In later studies[264,265] of the same system, however, 3,7,10-trinitro-3,7,10-triethyl-1,5-diazabicyclo[3.3.3]undecane (**102**) and a little 5-nitro-5-ethylhexahydropyrimidine (**103**) were isolated. Acid hydrolysis of **102** caused cleavage to formaldehyde, nitropropane, and 3,7-dinitro-3,7-diethyl-1,5-diazacyclooctane (**104**).

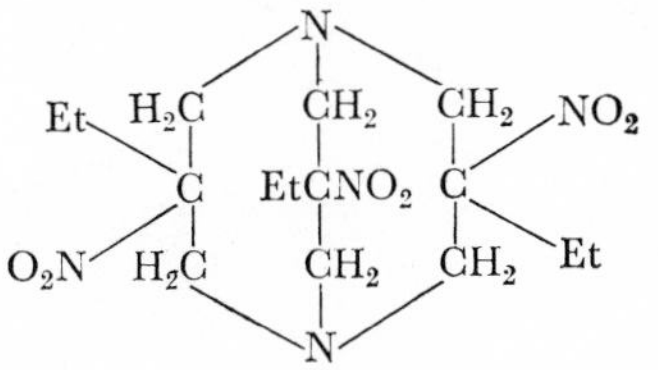

(102) 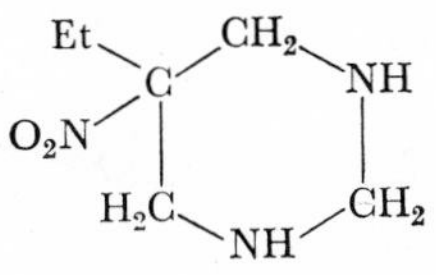 **(103)**

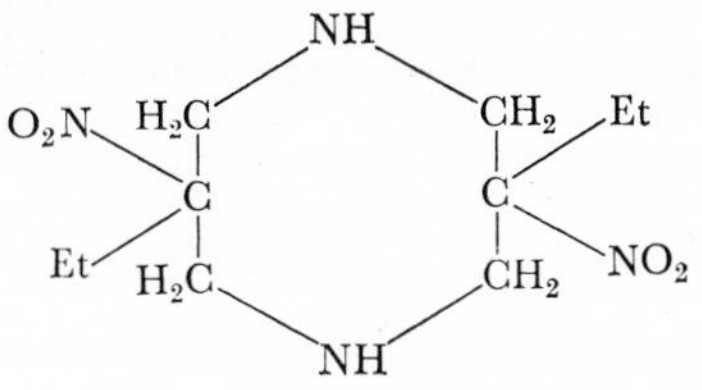

(104)

Urbański and coworkers also studied analogous reactions with 1-nitrobutane[266], 2-methyl-1-nitropropane[267], 1-nitropentane and 1-nitrohexane[51], and arylnitromethanes[250]. Thus, for instance, 5-nitro-5-propyltetrahydro-1,3-oxazine (**105**, R = H) was formed along with 2-nitro-2-hydroxymethylpentylamine (**106**) when 1-nitrobutane, formaldehyde, and ammonia were allowed to react in molar proportions 1:3:1. If an excess of ammonia was employed the product was 5-nitro-5-propylhexahydropyrimidine (**107**) which could be cleaved by aqueous ethanol to 1,3-diamino-2-nitro-2-propylpropane (**108**). The products **105** (R = H), **106**, and **107** were also obtained, in better yields, when the bismethylol derivative **109** of 1-nitrobutane was first prepared and then treated with the appropriate amounts of formaldehyde and ammonia. With ammonia alone the diol **109** furnished 3,7-dinitro-3,7-dipropyl-1,5-diazacyclooctane (**110**) (Scheme 7). The amino alcohol **106**, on warming with formaldehyde, was readily converted into **105** (R = H), whereas other aliphatic aldehydes gave 2-substituted oxazine derivatives (R = alkyl) in poor yields (5–7%). On the other hand, aromatic aldehydes gave **105** (R = aryl) in yields of 60–90%.

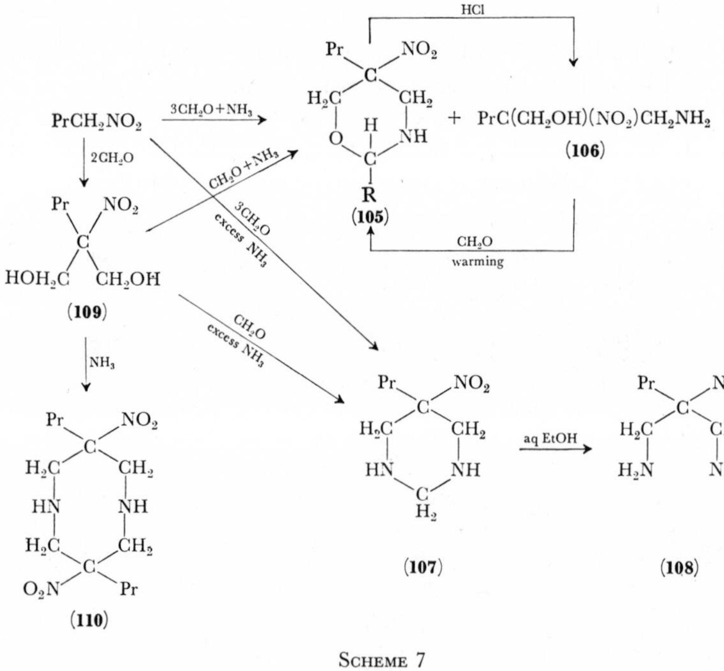

$PrCH_2NO_2$ →($3CH_2O + NH_3$)→ **105** + $PrC(CH_2OH)(NO_2)CH_2NH_2$ (**106**)

106 →(HCl)← **105**; **106** →(CH_2O, warming)→ **105**

$PrCH_2NO_2$ →($2CH_2O$)→ $PrC(NO_2)(CH_2OH)_2$ (**109**)

109 →($CH_2O + NH_3$)→ **105**

$PrCH_2NO_2$ →($3CH_2O$, excess NH_3)→ **107**

109 →(CH_2O, excess NH_3)→ **107**

107 →(aq EtOH)→ **108**

109 →(NH_3)→ **110**

SCHEME 7

Hydrochloric acid hydrolysis reconverted the oxazine derivatives into **106**[268]. When *N*-alkyl-substituted amino alcohols of this type, which were prepared analogously[246,247], were treated with phosgene in the presence of pyridine, cyclization occurred to give 3,5-dialkyl-5-nitrotetrahydro-1,3-oxazine-2-ones[269].

The condensation of nitroethane with formaldehyde and ammonia differed from those of the higher nitroalkanes in that a bisoxazine derivative **111** was formed[240].

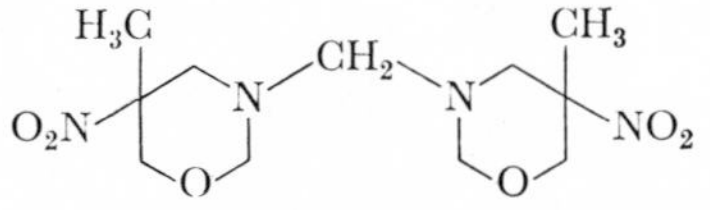

H_3C CH_3 CH_2 N N O_2N NO_2 O O

(**111**)

No 5-nitrotetrahydrooxazine derivative could be obtained from nitromethane, ammonia, and formaldehyde[239].

Application of the reaction to the system 1-nitropropane–formaldehyde–ethylenediamine gave 1-(2-nitro-1-butyl)-6-ethyl-6-nitro-1,4-diazacycloheptane (**112**) and 3,7-diethyl-3,7-dinitro-1,5-diazabicyclo[3.3.2]decane (**113**)[270].

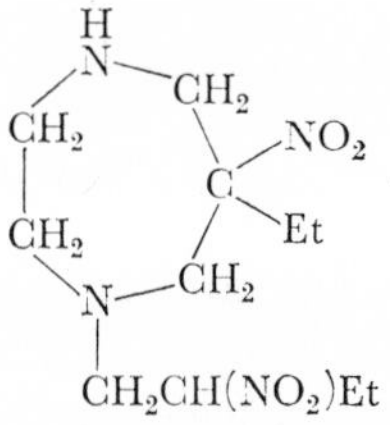

H N CH_2 CH_2 NO_2 C CH_2 Et N CH_2 $CH_2CH(NO_2)Et$

(**112**)

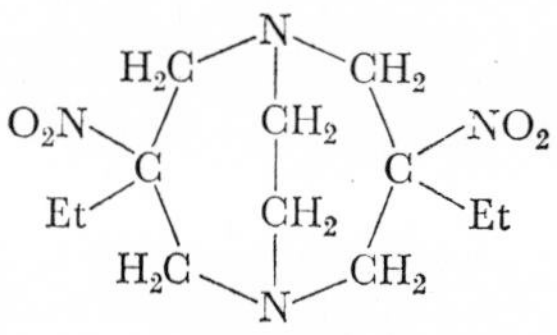

N H_2C CH_2 O_2N CH_2 NO_2 C C Et CH_2 Et H_2C CH_2 N

(**113**)

E. Stereochemistry

The stereochemistry of some of the heterocycles mentioned above has also been examined by the Polish workers[271]. Based on considerations of molecular models and dipole measurements, they[272] assigned to the diazacyclooctane **104**, which occurs in a *cis* and a *trans* form, the conformations **104a** and **104b**. The diazabicycloundecane **102** appears to exist in the asymmetrical conformation **102a**.

5-Nitrohexahydropyrimidines were shown[273] to exist in the chair conformation, with the nitro group axial when there is an alkyl substituent at C_5 (**114a**), or equatorial when C_5 carries hydrogen (**114b**).

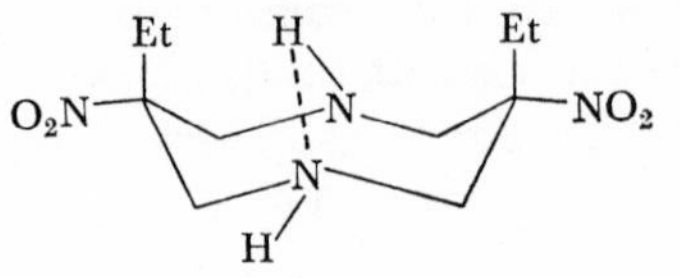

cis
(**104a**)

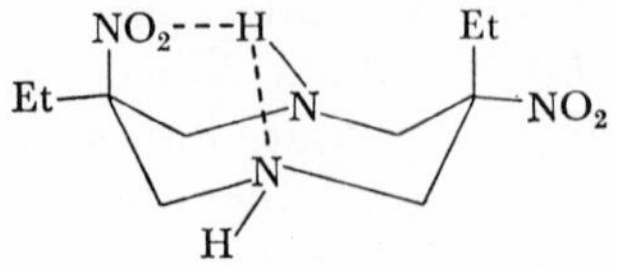

trans
(**104b**)

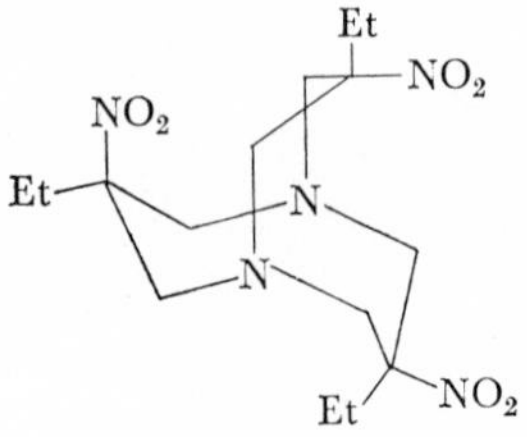

(**102a**)

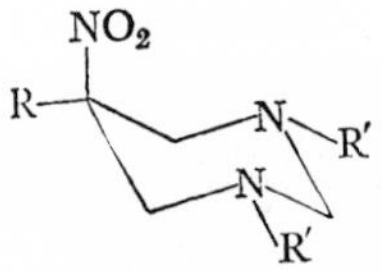

(**114a**) 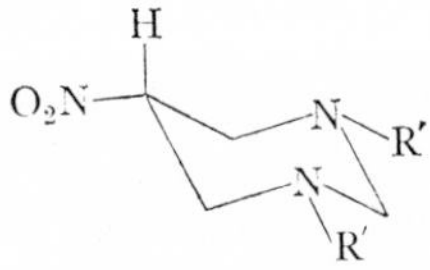 (**114b**)
R′ = CH_2Ph or cylohexyl; R = Me, Et, or *n*-Pr

Similarly, dipole moment and nuclear magnetic resonance measurements have revealed the conformations of several 5-nitro-tetrahydro-1,3-oxazines[248,274,275]. If C_5 carries an alkyl substituent, the latter is equatorial and the nitro group is axial. The disposition of an alkyl substituent at the ring nitrogen varies; it is equatorial for cyclohexyl and *t*-butyl (**115a**), but axial for methyl, ethyl, *n*-butyl, and benzyl (**115b**). Replacement by hydrogen of the alkyl group at C_5 results in conformational instability with rapid inter-conversion of chair forms.

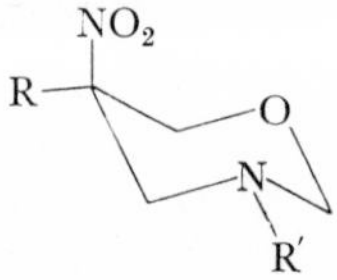

(**115a**)
R′ = cyclohexyl or *t*-Bu

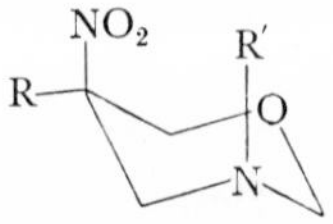

(**115b**)
R′ = Me, Et, *n*-Bu, or CH_2Ph

By quaternization of the tetrahydrooxazines with alkyl halides, pairs of diastereoisomers may be obtained[276] (equations 87 and 88).

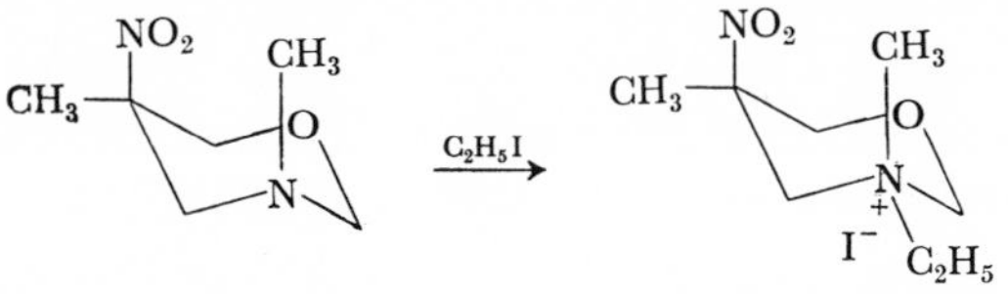

$$\text{(5-methyl-5-nitro-3-methyl-tetrahydro-1,3-oxazine)} \xrightarrow{C_2H_5I} \text{N-ethyl quaternary salt, } I^- \quad (87)$$

$$\text{(5-methyl-5-nitro-3-ethyl-tetrahydro-1,3-oxazine)} \xrightarrow{CH_3I} \text{N-methyl quaternary salt, } I^- \quad (88)$$

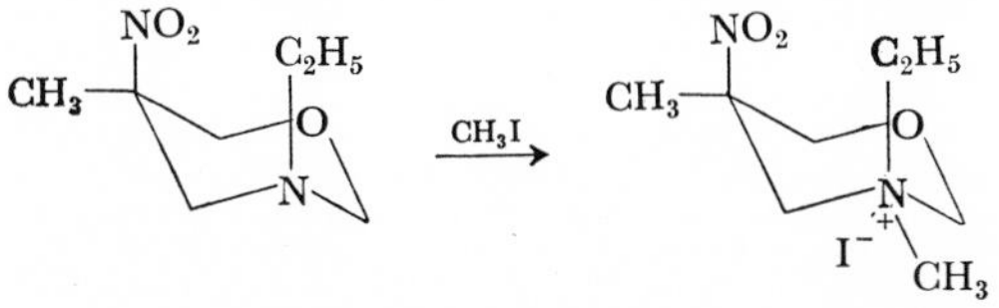

F. Reactions Involving Polynitro Compounds

The Mannich reaction with *gem*-dinitro compounds was first studied by Feuer, Bachman, and May[277], who found that 2,2-dinitro-1,3-propanediol or sodium 2,2-dinitroethanol undergo condensation with glycine at pH 4 to yield 3,3,5,5-tetranitropiperidinoacetic acid (**116**) (equation 89). Ethyl glycine hydrochloride gave the corresponding ethyl ester **117**. The conditions of the reaction,

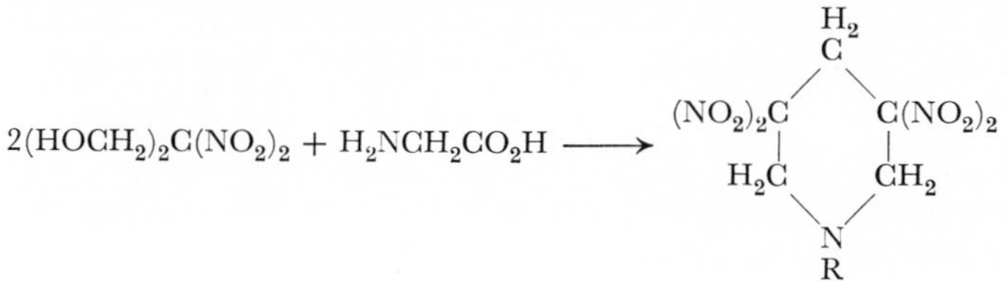

$$2(HOCH_2)_2C(NO_2)_2 + H_2NCH_2CO_2H \longrightarrow \text{3,3,5,5-tetranitropiperidine, } N\text{-}R \quad (89)$$

(**116**), R = CH_2CO_2H
(**117**), R = CH_2CO_2Et
(**118**), R = H, CH_2CF_3, $CH_2Si(CH_3)_3$, or $CH_2CH_2N(NO_2)CH_3$

particularly its dependence on the pH and the ratio of reactants were studied in detail, and a possible mechanism was proposed that could explain a stepwise formation of the cyclic product. Interestingly, condensation of ethanolamine with either *gem*-dinitro compound did not lead to cyclization but gave *N*-(2,2-dinitroethyl)ethanolamine. Related tetranitropiperidines **118** were prepared by Hamel[278] from 2,2,4,4-tetranitro-1,5-pentanediol and various amines; in similar fashion, 2,4-dinitraza-1,5-pentanediol afforded a hexahydrotriazine **119** (equation 90). Many further examples of the use of the Mannich reaction for the preparation of polynitroaliphatic amines and nitroamines have been reported since Feuer's work in 1954[277], notably by Frankel and Klager[206,279–281], Novikov[282],

$$HOCH_2N(NO_2)CH_2N(NO_2)CH_2OH + H_2NR \longrightarrow \text{(119)} \quad (90)$$

(119): ring $O_2N{-}N\,(CH_2)\,N{-}NO_2$, H_2C, CH_2, $N{-}R$

$R = CH_2CH_2C(NO_2)_2CH_3$

and Parker[211,212], and the reactants used and products obtained have been reviewed by Noble, Borgardt, and Reed[9].

More recently, Ungnade and Kissinger[283] investigated the condensation of 2,2-dinitropropanol (**120**) with a series of aliphatic and aromatic amines (equation 91). The resulting (2,2-dinitropropyl)-amines **121** could be isolated as such in some cases, or as crystalline

$$\underset{(\mathbf{120})}{CH_3C(NO_2)_2CH_2OH} \longrightarrow \underset{(\mathbf{121})}{CH_3C(NO_2)_2CH_2NRR'} \quad (91)$$

$R = H, R' = \text{alkyl or aryl}; R = R' = \text{alkyl}$

nitrates or nitroamines (if R = H) in others. It was noted that electron-withdrawing substituents tend to inactivate the amines. Thus, amides, urethans, and ureas did not react with **120**. Similarly, while *m*- and *p*-nitroanilines reacted slowly with **120**, 2,4-dinitroaniline and picramide did not react. An abnormal reaction was observed with aminoguanidine bicarbonate, the product being the aminoguanidine salt of 1,1-dinitroethane. Another abnormal reaction occurred with 1,3-diamino-2-propanol, inasmuch as the expected product, 2,2,10,10-tetranitro-4,8-diaza-6-undecanol (**122**), disproportionated in the aqueous reaction medium as shown in equation 92. Formaldehyde and 1,1-dinitroethane were then

$$\underset{(\mathbf{122})}{[CH_3C(NO_2)_2CH\,NHCH_2]_2CHOH} + H_2O \longrightarrow$$
$$\mathbf{120} + \underset{(\mathbf{123})}{H_2NCH_2CH(OH)CH_2NHCH_2C(NO_2)_2CH_3} \quad (92)$$

provided by **120**, and the formaldehyde condensed with remaining **122** to a hexahydropyrimidine derivative while the 1,1-dinitroethane formed a salt with the disproportionation product **123**.

III. MICHAEL ADDITIONS

A. General Features and Reaction Conditions

The Michael reaction, which has been exhaustively reviewed in 1959 by Bergmann, Ginsburg, and Pappo[284], is the base-catalyzed addition of an addend or donor (A) containing a reactive α-hydrogen atom to an activated carbon–carbon double bond in an acceptor (B) (equation 93). X and Y may represent a wide variety of activating substituents. As far as the chemistry of aliphatic nitro compounds

$$\underset{\text{A}}{X{-}\overset{R'}{\underset{R''}{C}}H} + \underset{\text{B}}{\overset{R}{\underset{R}{C}}{=}\overset{R}{C}{-}Y} \longrightarrow X{-}\overset{R}{\underset{R}{C}}{-}\overset{R}{\underset{R}{C}}{-}\overset{R}{\underset{H}{C}}{-}Y \qquad (93)$$

is concerned, primary and secondary nitroalkanes can serve as donors as can many of their derivatives such as nitroalkyl ethers, nitroalkyl sulfides, nitroalkyl sulfones, nitro esters, arylnitroalkanes, and dinitro compounds. On the other hand, α-nitro olefins, in which the nitro group and olefinic double bond form a conjugated system, can act as acceptors. For a qualitative comparison of the efficacy of various activating groups in the Michael reaction one may presume that the acidity of the donor and the polarity of the carbon–carbon double bond in the acceptor are the important factors to be considered. Since the acidity of the hydrogen in A decreases in the sequence $X = NO_2 > SO_3R > CN > CO_2R > CHO > COR$[285], one may expect nitroalkanes to be excellent donors. Conversely, since the electromeric effects of the activating groups which produce polarity in B diminish in the sequence $Y = CHO > COR > CN > CO_2R > NO_2$[284], α-nitroalkenes should be relatively poor acceptors. Nevertheless, a large number of Michael additions using nitroethylenic acceptors have been reported in the literature and listed in tabular form up to the year 1955[284]. These examples, to be sure, include all the many cases where both reactants were activated by nitro groups, so that lack of reactivity in the acceptor may have been compensated by the high reactivity of the donor. Furthermore, acceptors have frequently been employed whose double bonds were activated not only by a nitro group but, in addition, by substituents such as CO_2R, CHO, or C_6H_5. Examples are alkyl α-nitroacrylates, -crotonates, and -cinnamates, hydroxymethylenenitroacetaldehyde (the tautomeric form of nitromalonaldehyde), and nitrostyrenes.

The catalysts employed in Michael reactions that involve nitro compounds are: most frequently, sodium ethoxide or occasionally, other alkoxides; less frequently, metallic sodium and aqueous or alcoholic sodium hydroxide; often, diethylamine; rarely, other amines. The use of basic ion-exchange resins has also been investigated[286]. In some special instances no catalyst was required. Solvents used most often are alcohols, but it must be remembered that nitro olefins easily add alkoxide ion, and when such competition between the catalyst and the donor anion for the acceptor molecule exists, a non-hydroxylic solvent such as ether, benzene, or dioxane is preferred. The examples of Michael additions described herein are mostly taken from the more recent literature; for older work the reader is referred to the comprehensive review by Bergmann, Ginsburg, and Pappo[284], which covers the field up to 1955, and to the monograph by Perekalin[5a] which provides detailed tables incorporating also literature from several subsequent years.

B. Nitroalkane Donors and Non-Nitro Acceptors

The addition of nitroalkanes to α,β-unsaturated ketones, a reaction first investigated by Kohler and his school[287] and summarized by Hass[3], has been employed for the preparation of γ-nitro ketones which can serve as starting points for a route leading into the cyclopropane series. Several adducts of nitroalkanes and methyl vinyl ketone were obtained also by von Schickh[46].

Mesityl oxide in the presence of catalytic amounts of diethylamine combines with nitromethane to 4,4-dimethyl-5-nitro-2-pentanone[288], and with nitroethane to 4,4-dimethyl-5-nitro-2-hexanone[289].

Benzylideneacetone was shown to react with nitroethane giving 5-nitro-4-phenyl-2-hexanone, and with 1,1-dinitroethane giving 5,5-dinitro-4-phenyl-2-hexanone[290]. Analogous reactions with nitromethane and the nitropropanes had previously been carried out[288].

The interactions between phenyl vinyl ketone and nitromethane, nitroethane, dinitromethane, and 1,1-dinitroethane were studied by Novikov and Korsakova[291] who obtained, depending upon the reaction conditions, mono adducts or bis adducts, the latter from the three first-mentioned nitroalkanes (equation 94).

$$\text{PhCOCH{=}CH}_2 \xrightarrow{\text{CH}_2(\text{NO}_2)_2} \text{PhCOCH}_2\text{CH}_2\text{CH}(\text{NO}_2)_2 \xrightarrow{\text{PhCOCH{=}CH}_2} (\text{PhCOCH}_2\text{CH}_2)_2\text{C}(\text{NO}_2)_2 \qquad (94)$$

Many Michael additions of nitroalkanes to chalcones are described in the earlier literature; recently, an extension of these reactions to include thiophene analogs of chalcones has been reported[292].

2-Methyl-1-nitropropene (**124**) and 2-methyl-3-nitropropene (**125**) which are known to isomerize into each other by alkaline catalysis (see section V.B. 2b), have been shown to do so also by heating in toluene at 110°. When either olefin (or a mixture of both) was treated under these conditions with acrolein, a mixture of the adduct **126** (from **125**) and an isomerization product **127** was produced[293] (equation 95).

$$\underset{(\mathbf{124})}{CH_3C(CH_3){=}CHNO_2} \rightleftharpoons \underset{(\mathbf{125})}{CH_2{=}C(CH_3)CH_2NO_2} \xrightarrow{CH_2=CHCHO}$$

$$\underset{(\mathbf{126})}{CH_2{=}C(CH_3)CH(NO_2)CH_2CH_2CHO} + \underset{(\mathbf{127})}{CH_3C(CH_3){=}C(NO_2)CH_2CH_2CHO} \quad (95)$$

Nitroalkanes also add α,β-unsaturated esters and nitriles, whereby depending on the conditions mono, bis, and (with nitromethane) tris adducts may be obtained. von Schickh[46] has reported numerous reactions of this kind, in which alkyl acrylates, crotonates, and methacrylates, and acrylonitrile were combined with a variety of nitroalkanes in the presence of alkali. A more recent example was described by Vita and Bucher[294].

Basic ion-exchange resins may be used as catalysts, as has been shown, for instance, in reactions between 2-nitropropane and methyl methacrylate[286,295], diethyl ethylidenemalonate[295], and benzyl acrylate[295]. Excess acrylonitrile[296] reacted with nitromethane and a strongly basic resin to give mainly tris(β-cyanoethyl)nitromethane, and analogously with nitroethane to yield chiefly α,α-bis(β-cyanoethyl)nitroethane (γ-methyl-γ-nitropimelodinitrile).

It had been found by Kloetzel[297] that the addition of 2-nitropropane to esters of fumaric or maleic acid in the presence of diethylamine leads to dialkyl 3-methyl-3-nitrobutane-1,2-dicarboxylate (**128**) and, through loss of nitrous acid, to dialkyl isopropylidenesuccinate (**129**), with the amount of amine present governing the ratio of products (equation 96). The Michael adducts **130** and **131** formed with nitromethane and nitroethane were not isolated but lost nitrous acid to give itaconic esters (**132**) and ethylidenesuccinic esters (**133**), respectively (equation 97). von Schickh[46], however, reported the successful addition of nitroethane, 2-nitropropane, 2-nitrobutane, and nitrocyclohexane to diethyl maleate and fumarate giving products of type **128** and **131**, and more recently **130** and **131** (R = Et) were prepared in 25 and 45% yields by Polish workers[298] who used potassium fluoride as a catalyst in the addition reactions.

Since the early 1950's, extensions of the Michael addition to include aliphatic *gem*-dinitro compounds have been studied, and it

$$\begin{array}{l}CHCO_2R\\ \|\\ CHCO_2R\end{array} + O_2NCH(CH_3)_2 \xrightarrow{Et_2NH} \begin{array}{c}CH_2CO_2R\\ |\\ CHCO_2R\\ |\\ O_2NC(CH_3)_2\\ (\mathbf{128})\end{array} \xrightarrow{-HNO_2} \begin{array}{c}CH_2CO_2R\\ |\\ CCO_2R\\ \|\\ C(CH_3)_2\\ (\mathbf{129})\end{array} \quad (96)$$

$$\begin{array}{l}CH_2CO_2R\\ |\\ CCO_2R\\ \|\\ CHR'\end{array} \xleftarrow{Et_2NH} \begin{array}{l}CHCO_2R\\ \|\\ CHCO_2R\end{array} + R'CH_2NO_2 \xrightarrow{KF} \begin{array}{l}CH_2CO_2R\\ |\\ CHCO_2R\\ |\\ O_2NCHR'\end{array} \quad (97)$$

$$\begin{array}{lll}(\mathbf{132}),\ R' = H & R = CH_3 \text{ or } C_2H_5 & (\mathbf{130}),\ R' = H\\ (\mathbf{133}),\ R' = CH_3 & & (\mathbf{131}),\ R' = CH_3\end{array}$$

was found that dinitromethane, 1,1-dinitroethane, and 1,1-dinitropropane add to α,β-unsaturated esters, aldehydes, ketones, and sulfones under the catalytic influence of bases[299–301,290,291]. Potassium 1,1-dinitroethanol was shown[301] to react with methyl acrylate giving the expected adduct. When the pH of the medium is kept between 5 and 6, this adduct may be isolated, but in the presence of excess base it suffers demethylolation to the salt of methyl 4,4-dinitrobutyrate. The latter, in turn, is able to undergo further Michael additions.

Recently, Solomonovici and Blumberg[302] obtained the normal Michael adducts from 1,1-dinitropropane and 1,1-dinitrobutane with some 2,2-dinitroalkyl α,β-unsaturated esters (equation 98).

$$CH_3CH_2CH(NO_2)_2 + CH_2{=}CHCO_2CH_2C(NO_2)_2CH_2CH_3 \longrightarrow CH_3CH_2C(NO_2)_2CH_2CH_2CO_2CH_2C(NO_2)_2CH_2CH_3 \quad (98)$$

But when the mononitroalkane, 1-nitropropane, was refluxed with 2,2-dinitrobutyl acrylate in ethanol in the presence of piperidine, the normal adduct was not formed. Due to transesterification, some 2,2-dinitro-1-butanol was liberated which underwent demethylolation to 1,1-dinitropropane, and the latter in turn added to excess dinitrobutyl ester to give 2,2-dinitrobutyl 4,4-dinitrohexanoate (equation 99).

$$CH_2{=}CHCO_2CH_2C(NO_2)_2CH_2CH_3 \xrightarrow{EtOH} CH_2{=}CHCO_2Et + HOCH_2C(NO_2)_2CH_2CH_3$$

$$HOCH_2C(NO_2)_2CH_2CH_3 \longrightarrow CH_2O + CH(NO_2)_2CH_2CH_3 \quad (99)$$

$$CH_2{=}CHCO_2CH_2C(NO_2)_2CH_2CH_3 + CH(NO_2)_2CH_2CH_3 \longrightarrow CH_3CH_2C(NO_2)_2CH_2CH_2CO_2CH_2C(NO_2)_2CH_2CH_3$$

Michael additions with α,ω-dinitroalkanes have also been achieved, notably by Feuer and coworkers[303–305]. When 1,4-dinitrobutane was treated with 2 moles of methyl vinyl ketone, the symmetrical bis adduct was obtained. With 4 moles of the ketone, however, the expected tetrakis adduct could not be isolated since it underwent internal aldol addition to give the cyclic derivative **134** (equation 100). No such cyclizations occurred with methyl

$$O_2NCH_2CH_2CH_2CH_2NO_2 + 2\ CH_3COCH{=}CH_2 \longrightarrow CH_3COCH_2CH_2CH(NO_2)CH_2CH_2CH(NO_2)CH_2CHCOCH_3$$

$$\xrightarrow{4\ CH_3COCH=CH_2} \left[(CH_3COCH_2CH_2)_2C(NO_2)CH_2{-}CH_2C(NO_2)(CH_2CH_2COCH_3)_2\right] \longrightarrow \mathbf{134} \qquad (100)$$

$$\text{(134): two cyclohexane rings, each bearing } NO_2,\ COCH_3,\ CH_3,\ OH\text{, linked by } {-}CH_2{-}CH_2{-}$$

(**134**)

acrylate or acrylonitrile as acceptors, and the tetrakis adducts could be isolated[303]. A bis adduct was obtained from 1,5-dinitropentane and methyl vinyl ketone[304].

Bis adducts have likewise been found to arise from $\alpha,\alpha,\omega,\omega$-tetranitroalkanes (or their bismethylol derivatives) and suitable acceptors[305,305a]. This has been verified with 1,1,4,4-tetranitrobutane, 1,1,5,5-tetranitropentane, and 1,1,6,6-tetranitrohexane as donors, and with such acceptors as methyl vinyl ketone, acrolein, methyl acrylate, acrylonitrile, and methyl vinyl sulfone[305]. On the other hand, the potassium salt of 1,1,3,3-tetranitropropane added only one molecule of methyl acrylate[205] even under vigorous conditions; steric hindrance in the mono adduct probably prevents its further reaction[278]. However, Novikov and coworkers[305a] reported that 1,1,3,3-tetranitropropane gives mono as well as bis adducts with phenyl vinyl ketone and with 5-methyl-1,4-hexadien-3-one. An unexpected course took the reaction between 2 moles of methyl acrylate and 1 mole of the dipotassium salt of bis(2,2-dinitroethyl)-amine (**135**). Instead of the expected bis(4-carbomethoxy-2,2-dinitrobutyl)amine (**136**), there was formed in high yield dimethyl

4,4-dinitropimelate (**137**)[206,278], (equation 101). This reaction is

$$\underset{(\mathbf{135})}{\mathrm{HN(CH_2\overset{\overset{\displaystyle NO_2}{|}}{C}{=}NO_2K)_2}} + 2\mathrm{CH_2{=}CHCO_2CH_3} \not\longrightarrow \underset{(\mathbf{136})}{\mathrm{HN(CH_2\overset{\overset{\displaystyle NO_2}{|}}{\underset{\underset{\displaystyle NO_2}{|}}{C}}CH_2CH_2CO_2CH_3)_2}} \qquad (101)$$

$$\longrightarrow \underset{(\mathbf{137})}{\mathrm{(NO_2)_2C(CH_2CH_2CO_2CH_3)_2}}$$

explained by assuming **135** to undergo, first, a reverse Mannich reaction to give the 2-hydroxy-1-nitro-1-ethanenitronate anion (**138**), which then is added to methyl acrylate forming methyl 5-hydroxy-4,4-dinitrovalerate (**139**) (equation 102). The hydroxy

$$\mathbf{135} \longrightarrow \mathrm{NH_3} + 2\ \underset{(\mathbf{138})}{\mathrm{HOCH_2\overset{\overset{\displaystyle NO_2}{|}}{C}{=}NO_2^-}} + 2\mathrm{K^+} \xrightarrow{\mathrm{CH_2{=}CHCO_2CH_3}}$$

$$\underset{(\mathbf{139})}{\mathrm{\underset{\underset{\displaystyle HO}{|}}{H_2C}{-}\overset{\overset{\displaystyle NO_2}{|}}{\underset{\underset{\displaystyle NO_2}{|}}{C}}{-}CH_2CH_2{-}\overset{\overset{\displaystyle O}{\|}}{C}{-}OCH_3}} \qquad (102)$$

ester **139** then would suffer demethylolation, and subsequent addition of the resulting anion **140** to methyl acrylate would afford **137** (equation 103).

$$\mathbf{139} \xrightarrow{\mathrm{OH^-}} \mathrm{CH_2O} + \underset{(\mathbf{140})}{\mathrm{\bar{O}_2N{=}\overset{\overset{\displaystyle NO_2}{|}}{C}{-}CH_2CH_2{-}\overset{\overset{\displaystyle O}{\|}}{C}{-}OCH_3}} \xrightarrow{\mathrm{CH_2{=}CHCO_2CH_3}} \mathbf{137} \qquad (103)$$

Michael additions of dinitroacetonitrile and dinitroacetamide to various unsaturated compounds have been studied by Parker and coworkers[211,212], and numerous further examples involving polynitro compounds are cited in a comprehensive review by Noble, Borgardt, and Reed[9].

C. Non-Nitro Donors and Nitroalkene Acceptors

A very large number of reactions between nitroalkenes and active methylene compounds such as acetylacetone, alkyl acetoacetates, dialkyl malonates, alkyl cyanoacetates, benzyl cyanide, and others have been reported in the literature. Among the nitroalkenes used as acceptors are not only simple ones such as the nitropropenes, nitrobutenes, and many of their homologs including 1-nitrocyclohexene, but also a great variety of substitution products such as nitrostyrenes, unsaturated nitro esters, and halogenated derivatives[284]. It is noteworthy that even acetone has been claimed to add to 2-methyl-1-nitropropene[189,306] (equation 104). This reaction is

$$CH_3COCH_3 + (CH_3)_2C{=}CHNO_2 \longrightarrow CH_3COCH_2C(CH_3)_2CH_2NO_2 \quad (104)$$

said to occur when acetone is treated with nitromethane; the primary product, 2-methyl-1-nitro-2-propanol, which arises by Henry addition, loses a molecule of water giving the nitro olefin to which excess acetone then may add in competition to excess nitromethane.

Extensive use of the Michael reaction has been made in recent years by Perekalin and coworkers[307,308]. They demonstrated that β-nitrostyrene reacts with ethyl malonate, ethyl acetoacetate, acetylacetone and benzoylacetone, yielding adducts as shown for example in equation 105, ($R = H$).

$$p\text{-}R{-}C_6H_4CH{=}CHNO_2 + H_2C(CO_2Et)_2 \longrightarrow p\text{-}R{-}C_6H_4\underset{\displaystyle CH(CO_2Et)_2}{\underset{|}{C}H}CH_2NO_2 \quad (105)$$

Similar results were obtained[309] with a number of *para*-substituted β-nitrostyrenes, which, with ethyl acetoacetate and triethylamine as catalysts, gave ethyl α-acetyl-β-aryl-γ-nitrobutyrates ($R = NO_2$, CH_3, OCH_3, $N(CH_2)_2$). Acetylacetone and α-nitro-β-(p-tolyl)-ethylene afforded, similarly, 3-acetyl-1-nitro-2-(p-tolyl)-4-pentanone. Furthermore p-bis(β-nitrovinyl)benzene has been shown to give bis adducts with dimethyl and diethyl malonate, ethyl acetoacetate, ethyl cyanoacetate, and 1-phenyl-3-methyl-5-pyrazolone[310] (equation 106).

$$O_2NCH{=}CH{-}C_6H_4{-}CH{=}CHNO_2 + 2\ \text{(1-phenyl-3-methyl-5-pyrazolone)} \longrightarrow$$

$$\text{(pyrazolonyl)}{-}\underset{\displaystyle CH_2NO_2}{CH}{-}C_6H_4{-}\underset{\displaystyle O_2NCH_2}{CH}{-}\text{(pyrazolonyl)} \quad (106$$

1,3-Indandione and 2-aryl-1,3-indandiones gave adducts with β-nitrostyrenes[311–313] (equation 107).

$$\text{1,3-indandione} + 2\,\text{CHNO}_2{=}\text{CHAr} \longrightarrow \text{2,2-bis}(\text{CH(Ar)–CH}_2\text{NO}_2)\text{-1,3-indandione}$$

$$\text{2-Ar-1,3-indandione} + \text{CHNO}_2{=}\text{CHAr} \longrightarrow \text{2-Ar-2-(CH(Ar)–CH}_2\text{NO}_2\text{)-1,3-indandione} \quad (107)$$

2,5-Dinitro-1,6-diphenyl-1,5-hexadiene (**141**) added 2 moles of dimethyl malonate to give the adduct **142**, which upon catalytic hydrogenation over Raney nickel produced the bislactam **143**[214] (equation 108).

$$\underset{(\mathbf{141})}{\text{PhCH}{=}\text{C(NO}_2)\text{CH}_2\text{CH}_2\text{C(NO}_2){=}\text{CHPh}} \xrightarrow{2\text{CH}_2(\text{CO}_2\text{Me})_2}$$

$$\underset{(\mathbf{142})}{\text{PhCH[CH(CO}_2\text{Me)}_2]\text{–CH(NO}_2)\text{CH}_2\text{CH}_2\text{CH(NO}_2)\text{–CH[CH(CO}_2\text{Me)}_2]\text{Ph}} \xrightarrow{\text{H}_2}$$

$$\underset{(\mathbf{143})}{\text{Ph–CH–CH–CH}_2\text{CH}_2\text{–CH–CH–Ph (each CH(Ph)–CH(CO}_2\text{Me)–CO–NH–CH lactam ring)}} \quad (108)$$

Interestingly, when the conjugated triene 3,4-dimethyl-2,5-dinitro-1,6-diphenyl-1,3,5-hexatriene (**144**) was treated in the same manner with dimethyl malonate, it behaved as though it were non-conjugated and added two molecules of the methylene component (equation 109).

$$\underset{(\mathbf{144})}{\text{PhCH}{=}\text{C(NO}_2)\text{–C(CH}_3){=}\text{C(CH}_3)\text{–C(NO}_2){=}\text{CHPh}} \longrightarrow \text{PhCH[CH(CO}_2\text{Me)}_2]\text{–CH(NO}_2)\text{–C(CH}_3){=}\text{C(CH}_3)\text{–CH(NO}_2)\text{–CH[CH(CO}_2\text{Me)}_2]\text{Ph} \quad (109)$$

Raman spectra showed that conjugation is indeed absent in **144**, which can be seen from molecular models to be the result of steric interaction of the substituents, disturbing the coplanarity of the system.

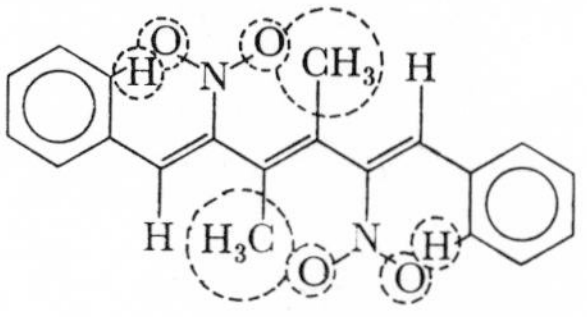

(**144**)

Perekalin and coworkers also synthesized 1,4-dinitro-1,3-butadiene (**145**)[314,315], 2,3-dimethyl-1,4-dinitrobutadiene (**146**)[315,316], 1,4-dinitro-2,3-diphenyl-1,3-butadiene (**147**) (*cis–cis* and *trans–trans* isomers)[315], and 1,4-dinitro-1,4-diphenyl-1,3-butadiene (**148**) (*cis–cis* and *trans–trans* isomers)[315]. Investigation of the structure of these compounds by means of Raman spectroscopy[317] showed that the conjugated system of **145** includes the double bonds and nitro groups, and that the phenyl groups in **148** make a considerable contribution to conjugation. On the other hand, introduction of substituents in the 2,3 position (derivatives **146** and **147**) sterically disturbs the coplanarity and conjugation of the system. These steric factors as well as the effects of the substituents on the electron distribution in the diene system influence the course of Michael additions[318]. The unsubstituted dinitrobutadiene (**145**), although very sensitive toward base and prone to polymerization, was treated with the sodium salt of dimedone (which is neutral in character) and gave a 1,2 adduct that subsequently underwent an allylic shift of the double bond (equation 110). The same behavior was observed with the 1,4-diphenyl derivatives **148**, with malononitrile, dimethyl malonate, and dimethyl methylmalonate being used in the latter instance. Under analogous conditions, the 2,3-dimethyl derivative **146** did not add methylene compounds but underwent allylic rearrangement (equation 111). It is believed that the rearranged diene is less sterically strained than the *trans*-fixed **146**, which may provide the driving force for this reaction. In contrast to **148**, the 2,3-phenyl-substituted dienes (**147**) reacted with malononitrile in the 1,4 position giving an intermediate which by 1,4 elimination of nitrous acid was converted into **149**. The reaction was much faster with *cis–cis*-**147** than with *trans–trans*-**147** (equation 112).

A novel synthesis of the furan ring, based on the Michael reaction,

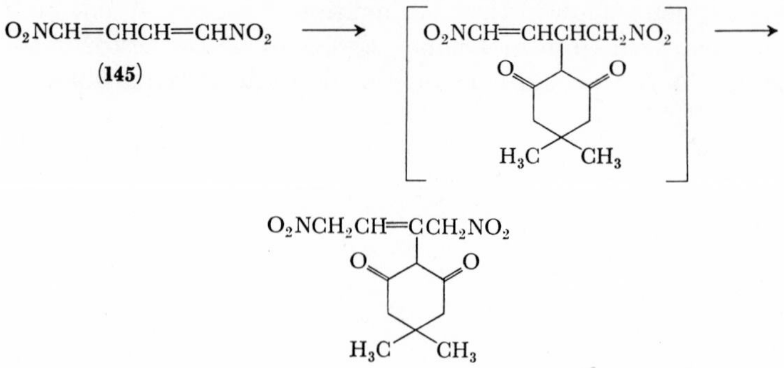

$$O_2NCH{=}CHCH{=}CHNO_2 \ (\mathbf{145}) \longrightarrow \left[O_2NCH{=}CHCHCH_2NO_2 \text{ (with 5,5-dimethyl-1,3-cyclohexanedione-2-yl; } H_3C, CH_3) \right] \longrightarrow \quad (110)$$

$$O_2NCH_2CH{=}CCH_2NO_2 \text{ (with 5,5-dimethyl-1,3-cyclohexanedione-2-yl; } H_3C, CH_3)$$

Ph, NO_2, O_2N, Ph

cis–cis

O_2N, Ph, Ph, NO_2

trans–trans

(**148**)

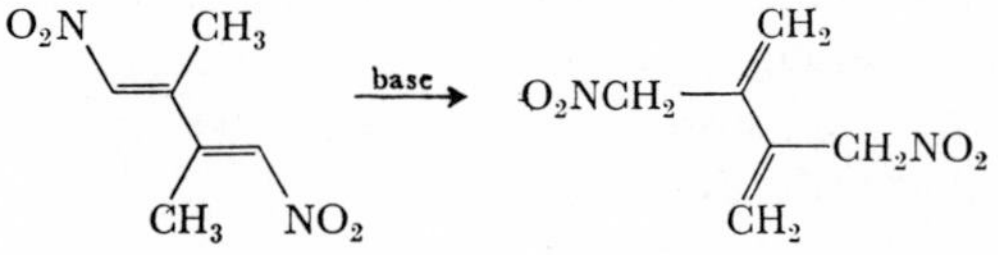

O_2N, CH_3, CH_3, NO_2 (**146**) $\xrightarrow{\text{base}}$ O_2NCH_2—C(=CH_2)—C(=CH_2)—CH_2NO_2 (111)

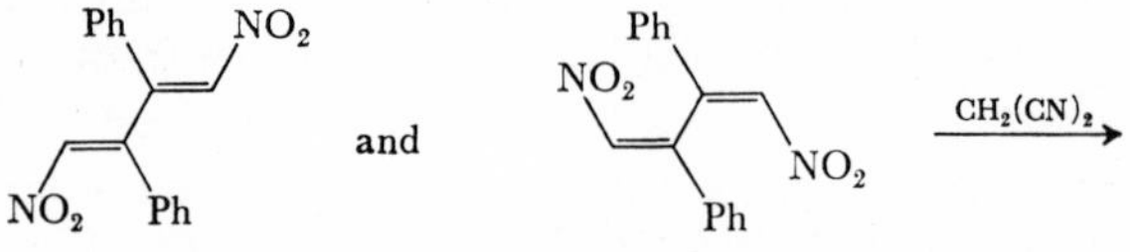

Ph, NO_2, NO_2, Ph

trans–trans

and

Ph, NO_2, NO_2, Ph

cis–cis

(**147**)

$\xrightarrow{CH_2(CN)_2}$

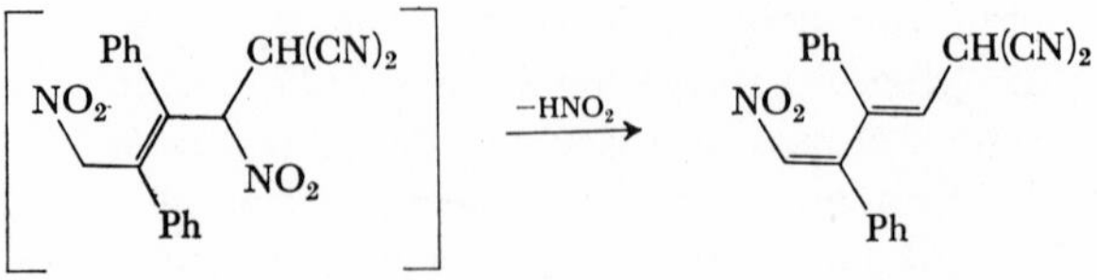

[Ph, $CH(CN)_2$, NO_2, NO_2, Ph] $\xrightarrow{-HNO_2}$ Ph, $CH(CN)_2$, NO_2, Ph (112)

(**149**)

was devised by Boberg and Schultze[319]. Addition of sodium ethyl acetoacetate to β-methyl- and β-ethyl-β-nitrostyrene followed by Nef reaction of the adducts and cyclization of the resulting diketones afforded tetrasubstituted furans (equation 113). The reaction proved

$$\mathrm{PhCH{=}C(R)NO_2} + {}^{-}\mathrm{CH(CO_2Et)C(=O)CH_3} \longrightarrow \mathrm{PhCH[C(R){=}NO_2^-]{-}CH(CO_2Et)C(=O)CH_3} \xrightarrow[\text{(Nef)}]{H^+}$$

$$\mathrm{PhCH[C(=O)R]{-}CH(CO_2Et)C(=O)CH_3} \xrightarrow[(-H_2O)]{H^+} \text{furan: Ph–C}{=}\text{C(R)–O–C(CH}_3){=}\text{C–CO}_2\text{Et} \quad (113)$$

$$R = CH_3 \text{ or } C_2H_5$$

to be of general applicability. With other β-keto esters as donors and with cyclic nitro olefins (1-nitrocyclohexene, 1-nitrocycloheptene, and 1-nitrocyclooctene) as acceptors, furan derivatives **150** were obtained[320].

$$\text{(CH}_2)_n\text{-fused furan with } CO_2Et \text{ and } R$$

(**150**)

$R = CH_3$, C_2H_5, or C_5H_{11}

$n = 4, 5, \text{ or } 6$

When adducts from 1-nitro-1-alkenes and dimethyl malonate sodium were reduced, carbomethoxypyrrolidinones **151** arose which could be converted into γ-amino acids either by acid hydrolysis, or by alkaline hydrolysis, followed by decarboxylation and repeated hydrolysis[321] (equation 114).

$$\mathrm{O_2NCH{=}CHR} + \mathrm{CH_2(CO_2CH_3)_2} \longrightarrow \mathrm{O_2NCH_2CH(R)CH(CO_2CH_3)_2} \xrightarrow{H_2} \text{(151)}$$

$$\text{(151): R–CH–CH–CO}_2\text{CH}_3\text{, H}_2\text{C, C=O, NH ring} \xrightarrow[\text{2. heat}]{\text{1. KOH}} \text{R–CH–CH}_2\text{, H}_2\text{C, C=O, NH ring} \quad (114)$$

$$\text{pyrrolidinone} \xrightarrow{KOH} \mathrm{H_2NCH_2CH(R)CH_2CO_2H} \xleftarrow{HCl} \text{(151)}$$

Interesting results were obtained by Perekalin and coworkers who examined reactions of β-bromo-β-nitrostyrene with several compounds containing reactive methylene groups[322,323]. With dimethyl malonate in the presence of sodium methoxide at 0–5°, the normal adduct **152**, ($R = R' = OCH_3$) was produced. Heating **152** in benzene solution with triethylamine led to dehydrobromination. On evidence adduced from Raman spectra the product was first thought to be the furan derivative **154** which was presumed to arise via the enol **153.** When other β-dicarbonyl compounds (ethyl acetoacetate, acetylacetone, benzoylacetone, dimedone) were heated with β-bromo-β-nitrostyrene in the presence of triethylamine, dehydrobrominated products believed to be dihydrofurans **154** were formed immediately (equation 115). Later on, Perekalin and his

$$PhCH{=}C(NO_2)Br + H_2C(COR)(COR') \longrightarrow \underset{(152)}{Ph{-}CH(CH(NO_2)Br){-}CH(COR)(COR')} \rightleftharpoons$$

$$\underset{(153)}{Ph{-}CH(CH(NO_2)Br){-}C(COR){=}C(OH){-}R'} \xrightarrow{-HBr} \underset{(154)}{Ph{-}CH{-}C(COR){=}C(R'){-}O{-}CH(NO_2)} \quad (115)$$

For dimethyl malonate: $R = R' = OCH_3$
ethyl acetoacetate: $R = OEt$, $R' = CH_3$
acetylacetone: $R = R' = CH_3$
benzoylacetone: $R = CH_3$, $R' = Ph$
dimedone: $R, R' = CH_2C(CH_3)_2CH_2$

associates[324] included in their studies β-bromo-β-nitro-*p*-nitrostyrene and some additional cyclic donors. The product **155** that was formed with dimedone in boiling benzene in the presence of triethylamine was found to lose, on prolonged heating, a molecule of nitrous acid, and structure **156** was proposed for the end product (equation 116). From β-bromo-β-nitro-*p*-nitrostyrene and the sodium derivatives of,

$$\underset{(155)}{p\text{-}NO_2{-}Ph,\ O_2N\text{-dihydrofuran fused cyclohexanone},\ CH_3,\ CH_3} \longrightarrow \underset{(156)}{p\text{-}NO_2{-}Ph\text{-furan fused cyclohexanone},\ CH_3,\ CH_3} \quad (116)$$

respectively, 2-carbethoxycyclopentanone, 1,3-indanone, and 3-methyl-1-phenyl-5-pyrazolone, the bromine-containing adducts of type **152** were obtained when the reactions were carried out at 0–5°. Analogous reactions were performed[325,326] using 1-bromo-1-nitropropene. Adducts of type **152** (with CH_3 instead of C_6H_5) could be isolated in reactions, performed at −70°, with the sodium derivatives of diethyl malonate and ethyl acetoacetate in absolute ethanol. On the other hand, refluxing of the dicarbonyl compounds and the bromonitroalkene in 95% ethanol in the presence of potassium acetate resulted in dehydrobromination and cyclization.

In a subsequent communication[327], however, it was reported that the dehydrobromination products encountered in the previous studies were not dihydrofuran derivatives as originally believed but were actually cyclopropane derivatives **157**. This was concluded from the fact that **157** was obtained not only from **158** but also by an alternative synthesis which started from methyl α-bromo-α-carbomethoxy-β-alkyl- (or aryl-)γ-nitrobutyrate (**159**) (equation 117). Hence, revision of the earlier structural formulations was necessary. This

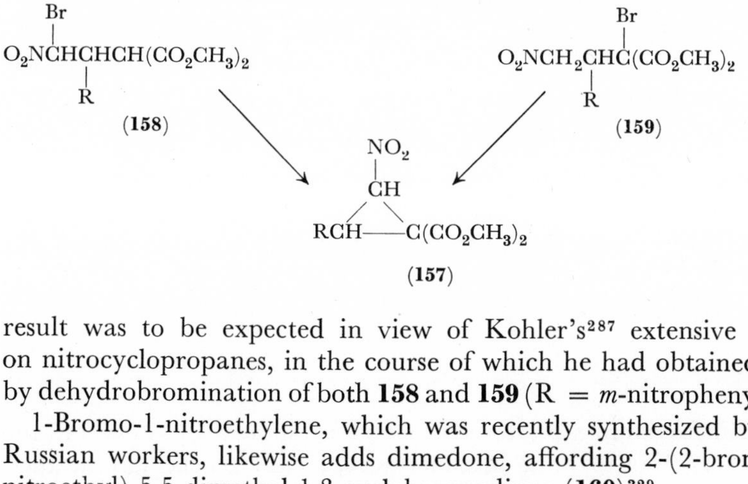

$$O_2NCH(Br)CH(R)CH(CO_2CH_3)_2 \; (\mathbf{158}) \longrightarrow \mathbf{157} \longleftarrow O_2NCH_2CH(R)C(Br)(CO_2CH_3)_2 \; (\mathbf{159}) \qquad (117)$$

157: cyclopropane ring $\overset{NO_2}{CH}$ / RCH—$C(CO_2CH_3)_2$

(**158**) (**159**) (**157**)

result was to be expected in view of Kohler's[287] extensive work on nitrocyclopropanes, in the course of which he had obtained **157** by dehydrobromination of both **158** and **159** (R = *m*-nitrophenyl)[328].

1-Bromo-1-nitroethylene, which was recently synthesized by the Russian workers, likewise adds dimedone, affording 2-(2-bromo-2-nitroethyl)-5,5-dimethyl-1,3-cyclohexanedione (**160**)[329].

As another example, involving a heterocyclic donor, may be

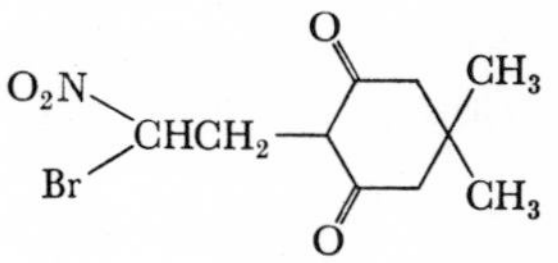

$$(O_2N)(Br)CHCH_2\text{-[5,5-dimethyl-1,3-cyclohexanedione-2-yl]}$$

(**160**)

mentioned the syntheses of some nitroalkylbarbituric acids[330] (equation 118).

$$\mathrm{OC}\begin{matrix}\mathrm{NH{-}CO}\\ \\ \mathrm{NH{-}CO}\end{matrix}\mathrm{CH_2} \xrightarrow{\mathrm{RCH{=}CHNO_2}} \mathrm{OC}\begin{matrix}\mathrm{NH{-}CO}\\ \\ \mathrm{NH{-}CO}\end{matrix}\mathrm{CHCH(R)CH_2NO_2} \qquad (118)$$

R = Me, Et, or *n*-Pr

The chemistry of 1-amino-2-nitroalkenes[331] provides some interesting applications of the Michael reaction. Hurd and Sherwood[332] allowed ethyl ethoxymethylenemalonate (**161**) to react with nitromethane in the presence of certain secondary amines (piperidine or morpholine—the reaction failed with others), and they obtained dialkylaminonitroalkenes **164**. Compound **161** may be regarded as a vinylogous carbonate that gives a vinylogous urethan **162**, to which addition of nitromethane then occurs. The adduct **163** subsequently loses a molecule of ethyl malonate by way of a reverse Michael reaction, yielding the product **164** (equation 119).

$$\underset{(\mathbf{161})}{\mathrm{EtOCH{=}C(CO_2Et)_2}} \xrightarrow{\mathrm{C_5H_{11}N}} \underset{(\mathbf{162})}{\mathrm{C_5H_{10}NCH{=}C(CO_2Et)_2}} \xrightarrow{\mathrm{CH_3NO_2}}$$

$$\underset{(\mathbf{163})}{\mathrm{C_5H_{10}N\underset{|}{C}HCH(CO_2Et)_2} \atop \mathrm{CH_2NO_2}} \longrightarrow \underset{(\mathbf{164})}{\mathrm{C_5H_{10}NCH{=}CHNO_2}} + \mathrm{CH_2(CO_2Et)_2} \qquad (119)$$

Similar reactions between various ethoxymethylene β-dicarbonyl compounds and nitroalkanes in the presence of bases were carried out by Dornow and Lüpfert[333], but these authors reported the formation of nitroalkylidene derivatives **165** (equation 120).

$$\mathrm{C_2H_5OCH{=}C(COCH_3)_2} + \mathrm{RCH_2NO_2} \xrightarrow[-\mathrm{C_2H_5OH}]{} \underset{(\mathbf{165})}{\mathrm{R\underset{|}{C}HCH{=}C(COCH_3)_2} \atop \mathrm{NO_2}} \qquad (120)$$

1-Dimethylamino-2-nitroethylene was shown by Severin and coworkers[334,335] to be a Michael acceptor. It reacts, for instance, with ketones, whereby the dimethylamino group is eliminated and unsaturated nitro ketones arise (equation 121). The product obtained

$$\mathrm{RCOCH_3} + \mathrm{(CH_3)_2NCH{=}CHNO_2} \xrightarrow{\mathrm{EtOK}}$$

$$\left[\mathrm{(RCOCH_2\underset{|}{C}HCH{=}NO_2K)} \atop \mathrm{N(CH_3)_2}\right] \xrightarrow{-\mathrm{HN(CH_3)_2}}$$

$$\mathrm{RCOCH{=}CHCH{=}NO_2K} \xrightarrow{\mathrm{H^+}} \mathrm{RCOCH{=}CHCH_2NO_2} \qquad (121)$$

R = CH_3 or Ar

with benzylideneacetone (R = $C_6H_5CH{=}CH$) gave, upon borohydride reduction, 1-nitro-6-phenyl-1,3,5-hexatriene.

Benzyl cyanide, malononitrile, ethyl cyanoacetate, and alkyl malonates were found to react analogously with 1-dimethylamino-2-nitroethylene.

D. Nitroalkane Donors and Nitroalkene Acceptors

The addition of nitroalkanes to nitroalkenes is one of the most convenient routes to aliphatic dinitro and polynitro compounds. Following early work by some other authors,[145,306,336–338] intensive studies concerning this reaction were carried out by Lambert and coworkers[191,339,340] and by Bahner and Kite[341,342]. The syntheses of 1-methoxy-2,4-dinitro-2-methylpentane (**166**)[339] and of 3,5-dinitroheptane (**167**)[341] may serve as illustrations (equations 122 and 123).

$$CH_3C(NO_2){=}CH_2 + CH_3CH(NO_2)CH_2OCH_3 \xrightarrow{NaOEt} CH_3CH(NO_2)CH_2C(CH_3)(NO_2)CH_2OCH_3 \quad (122)$$

(**166**)

$$CH_3CH_2CH{=}NO_2K + CH_2{=}C(NO_2)CH_2CH_3 \longrightarrow CH_3CH_2CH(NO_2)CH_2CH(NO_2)CH_2CH_3 \quad (123)$$

(**167**)

Addition may occur between nitroalkanes and nitro olefins that are generated *in situ* by elimination of water from nitro alcohols. Thus, one of the products in the reaction between acetone and nitromethane is 2,2-dimethyl-1,3-dinitropropane[306] (equation 124).

$$(CH_3)_2CO + CH_3NO_2 \longrightarrow (CH_3)_2C(OH)CH_2NO_2 \longrightarrow$$

$$[(CH_3)_2C{=}CHNO_2] \xrightarrow{CH_3NO_2} (CH_3)_2C(CH_2NO_2)_2 \quad (124)$$

In subsequent years a large number of reactions along similar lines have been described,[165,166,177,181,182,256,343–345] and the reactants used and products obtained have been listed in tabular form[284]. More recently, several new 1,3-dinitroalkanes were prepared by the addition of nitroalkanes to 1-nitro-1-alkenes[346]. It may suffice here to cite one example from the extensive work of Dornow and coworkers[165,177,181,182,343]. Ethyl α-nitrocinnamate was prepared by heating a Schiff base of benzaldehyde, e.g., benzylidene-*n*-butylamine, with ethyl nitroacetate in the presence of acetic anhydride (equation 125).

$$\mathrm{PhCH{=}NBu\text{-}}n + \mathrm{CH_2NO_2CO_2Et} \xrightarrow{\mathrm{Ac_2O}} \left[\begin{array}{c}\mathrm{H} \\ \mathrm{PhCCH(NO_2)CO_2Et} \\ | \\ \mathrm{HNBu\text{-}}n\end{array}\right] \longrightarrow$$

$$\mathrm{PhCH{=}C(NO_2)CO_2Et} + (n\text{-}\mathrm{BuNH_2} \longrightarrow n\text{-}\mathrm{BuNHCOCH_3}) \quad (125)$$

The unsaturated nitro ester was then treated with additional ethyl nitroacetate, in the presence of diethylamine, to give diethyl α,α′-dinitro-β-phenylglutarate[343]. The same ester arose in a single operation when benzylidene–aniline and ethyl nitroacetate reacted in the absence of acetic anhydride but in the presence of diethylamine; obviously, unsaturated nitro ester was formed as an intermediate and then rapidly added a second molecule of nitroacetate[181].

Examples of the addition of 1,1-dinitroethane to nitro olefins have been reported by Shechter and Zeldin[300], Klager[344], and Novikov and coworkers[347] who used as acceptor molecules, respectively: 2-nitropropene; methyl 2-nitro-2-pentenoate; and nitroethylene, 1-nitropropene, 1-nitro-1-butene, and 1-nitro-1-pentene (equations 126 and 127).

$$\mathrm{CH_2{=}C(NO_2)CH_3} + \mathrm{CH_3CH(NO_2)_2} \longrightarrow \mathrm{CH_3C(NO_2)_2CH_2CH(NO_2)CH_3} \quad (126)$$

$$\mathrm{RCH{=}CHNO_2} + \mathrm{CH_3CH(NO_2)_2} \longrightarrow \mathrm{CH_3C(NO_2)_2CH(R)CH_2NO_2} \quad (127)$$

$$\mathrm{R} = \mathrm{H, CH_3, C_2H_5, or\ } n\text{-}\mathrm{C_3H_7}$$

1,1-Dinitropropane and 1,1-dinitrobutane were added to 1-nitropropene and homologous nitroalkenes[347a].

It is considered that 1,1-dinitroethylene (**169**) plays a role as an intermediate in certain Michael-type reactions. Thus, when an aqueous solution of 2,2-dinitroethanol potassium salt (**168**) was gradually acidified to pH 4 and then made alkaline again, the dipotassium salt of 1,1,3,3-tetranitropropane (**171**) was produced in a 56% yield[205]. Apparently part of the dinitroethanol suffered dehydration in this process, and the dinitroethylene formed added a molecule of surviving dinitroethanol, to give the alcohol **170** which subsequently was demethylolated to the final product **171** (equation 128).

$$\underset{(\mathbf{168})}{HOCH_2C(NO_2)NO_2K} \xrightarrow{H^+} HOCH_2CH(NO_2)_2 \xrightarrow{-H_2O} \underset{(\mathbf{169})}{[CH_2{=}C(NO_2)_2]}$$

$$\mathbf{169} + HOCH_2C(NO_2)NO_2K \longrightarrow$$

$$\underset{(\mathbf{170})}{HOCH_2C(NO_2)_2CH_2\underset{\displaystyle NO_2}{\underset{|}{C}}{=}NO_2K} \xrightarrow[-CH_2O,\,-H_2O]{+KOH} \underset{(\mathbf{171})}{KO_2N{=}\overset{\displaystyle NO_2}{\overset{|}{C}}CH_2\overset{\displaystyle NO_2}{\overset{|}{C}}{=}NO_2K} \quad (128)$$

A number of similar reactions have previously been postulated to involve **169** which has, however, not been isolated[218,348,349] (see also sections IV and VI.A). More recently, Feuer and Miller[350] made an important and detailed study on the Michael addition using as acceptors certain nitro olefins that are generated *in situ* from 2-nitroalkyl acetates. It was found that 2-nitroalkyl acetates react under mild conditions with 2 moles of the sodium salt of the donor nitroparaffin to afford the olefin which then undergoes the Michael addition. For hydroxylic solvents (methanol or *t*-butyl alcohol) the following reaction path was suggested (equations 129–132).

$$R'R{-}C{=}NO_2Na + R''OH \rightleftharpoons R'R{-}CHNO_2 + NaOR'' \quad (129)$$

$$R'''{-}\overset{\displaystyle NO_2}{\overset{|}{C}}HCH_2OAc + NaOR'' \rightleftharpoons R'''{-}\overset{\displaystyle NO_2}{\overset{|}{C}}{=}CH_2 + NaOAc + R''OH \quad (130)$$

$$R'RC{=}NO_2Na + R'''{-}\overset{\displaystyle NO_2}{\overset{|}{C}}{=}CH_2 \rightleftharpoons R'R{-}\overset{\displaystyle NO_2}{\overset{|}{C}}HCH_2{-}\overset{\displaystyle NO_2Na}{\overset{\|}{C}}{-}R''' \quad (131)$$

$$R'R{-}\overset{\displaystyle NO_2}{\overset{|}{C}}H{-}CH_2{-}\overset{\displaystyle NO_2Na}{\overset{\|}{C}}{-}R''' + R''OH \rightleftharpoons R'R{-}\overset{\displaystyle NO_2}{\overset{|}{C}}H{-}CH_2{-}\overset{\displaystyle NO_2}{\overset{|}{C}}H{-}R''' + NaOR'' \quad (132)$$

Under these conditions the addition product was obtained as its sodium salt. Alternatively, equivalent amounts of the salt of the donor, 2-nitroalkyl acetate and sodium acetate could be employed. The reaction path then is essentially the same, with acetate elimination induced by alkoxide ion that arises from alcoholysis of sodium acetate. This procedure affords the Michael addition product as the free nitro compound. The reaction proceeds in similar fashion in nonhydroxylic solvents (tetrahydrofuran), in which case the salt of the donor reacts directly with the 2-nitroalkyl acetate to give the olefin and sodium acetate (equation 133).

Nitro olefin precursors used in this work were 2-nitrobutyl acetate, 3-nitro-2-butyl acetate, and 1,6-diacetoxy-2,5-dinitrohexane, and

$$R'R{-}C{=}NO_2Na + R''{-}\underset{}{\overset{NO_2}{\overset{|}{C}}}HCH_2OAc \rightleftharpoons R''{-}\overset{NO_2}{\overset{|}{C}}{=}CH_2 + R'R{-}CHNO_2 + NaOAc \quad (133)$$

donors included 1- and 2-nitropropanes, 1,1-dinitroethane, 1,1-dinitropropane, 2-nitro-1,3-propanediol, and ethylenedinitramine.

Similar work was carried out by Klager[351] who obtained 2,2,4,6,6-pentanitroheptane by the reaction of 2 moles of 1,1-dinitroethane with either 3-acetoxy-2-nitro-1-propene or its precursor, 1,3-diacetoxy-2-nitropropane (equation 134). Corresponding polynitro

$$AcOCH_2{-}\overset{NO_2}{\overset{|}{C}}H{-}CH_2OAc \xrightarrow{-HOA} CH_2{=}\overset{NO_2}{\overset{|}{C}}{-}CH_2OAc \xrightarrow{CH_3CH(NO_2)_2}$$

$$CH_3\underset{NO_2}{\underset{|}{\overset{NO_2}{\overset{|}{C}}}}CH_2\overset{NO_2}{\overset{|}{C}}HCH_2OAc \xrightarrow{-HOAc} CH_3\underset{NO_2}{\underset{|}{\overset{NO_2}{\overset{|}{C}}}}CH_2\overset{NO_2}{\overset{|}{C}}{=}CH_2 \xrightarrow{CH_3CH(NO_2)_2}$$

$$CH_3\underset{NO_2}{\underset{|}{\overset{NO_2}{\overset{|}{C}}}}CH_2\overset{NO_2}{\overset{|}{C}}HCH_2\underset{NO_2}{\underset{|}{\overset{NO_2}{\overset{|}{C}}}}CH_3 \quad (134)$$

compounds were also obtained when 1,1-dinitropropane, 1,1-dinitrobutane, and methyl 4,4-dinitrobutyrate were used as donor components.

The addition of *gem*-dinitroalkanes to β-nitrostyrene and a number of its nuclear substitution products was studied by Solomonovici and Blumberg[352], who found that normal adducts were formed without the need of an external catalyst.

IV. DIELS–ALDER REACTIONS

During the first two decades following the discovery, in 1928, of the Diels–Alder reaction, relatively few examples of the use of nitroethylene and other conjugated nitroalkenes as dienophiles have been reported. It was shown by Alder and coworkers[353] that nitroethylene, 1-nitropropene, and 1-nitropentene form adducts with cyclopentadiene and that 1-nitropentene gives adducts with butadiene and 2,3-dimethylbutadiene. Several adducts of various aliphatic and cycloaliphatic dienes[354–357] with β-nitrostyrene were prepared, and whereas the latter dienophile quantitatively added to

1,3-diphenylisobenzofuran, it did not react with furan, 2-methylfuran, or 2,5-dimethylfuran[354,355]. (For a review of these older investigations see H. L. Holmes[358].)

Alder's original work was later greatly extended by other investigators, frequently with the purpose of synthesizing alicyclic amines by reduction of the nitro adducts. It has been established that the reaction between cyclopentadiene and nitroalkenes is a general one, giving substituted bicyclo[2.2.1]-2-heptenes that carry a secondary or tertiary nitro group in the 5 position, depending on the nitroalkene employed[359–365c]. For example, 2-nitro-1-butene, 2-nitro-2-butene, and 1-nitro-1-heptene gave, respectively, the 5-ethyl[362], 5,6-dimethyl[364], and 6-pentyl[365] derivatives of 5-nitrobicyclo[2.2.1]-2-heptene **172, 173**, and **174** (equation 135). Hexachlorocyclopentadiene gave adducts with nitroethylene and various nitroalkyl acrylates[365d].

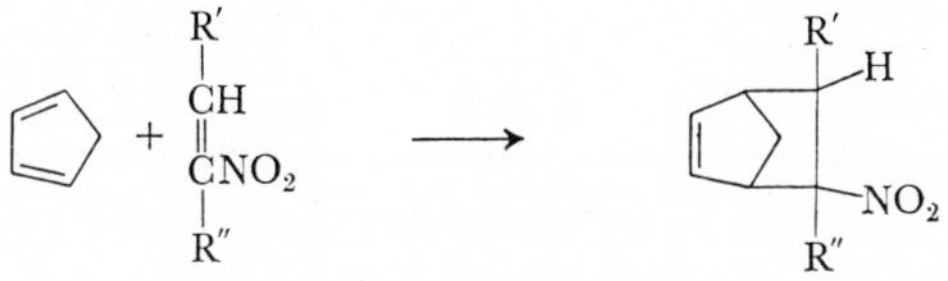

$$\text{cyclopentadiene} + R'CH{=}C(NO_2)R'' \longrightarrow \text{5-}R''\text{-6-}R'\text{-5-nitrobicyclo[2.2.1]-2-heptene} \qquad (135)$$

(**172**), $R' = H$, $R'' = C_2H_5$
(**173**), $R' = R'' = CH_3$
(**174**), $R' = n\text{-}C_5H_{11}$, $R'' = H$

In reactions with cyclopentadiene, 1,4-dinitro-1,3-butadiene behaved as a dienophile affording[318] a mono adduct **174a** and a bis adduct **174b**.

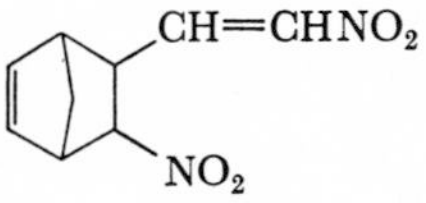

$CH{=}CHNO_2$, NO_2

(**174a**)

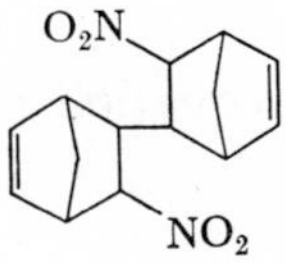

O_2N, NO_2

(**174b**)

In similar fashion, 5-nitrobicyclo[2.2.2]-2-octene was synthesized from 1,3-cyclohexadiene and nitroethylene[366].

A systematic study concerning the addition of eight homologous 1-nitroalkenes to 2,3-dimethyl-butadiene was performed by Drake and Ross[367]; the expected 4-alkyl-1,2-dimethyl-5-nitrocyclohexenes were obtained in excellent yields (equation 136). The same

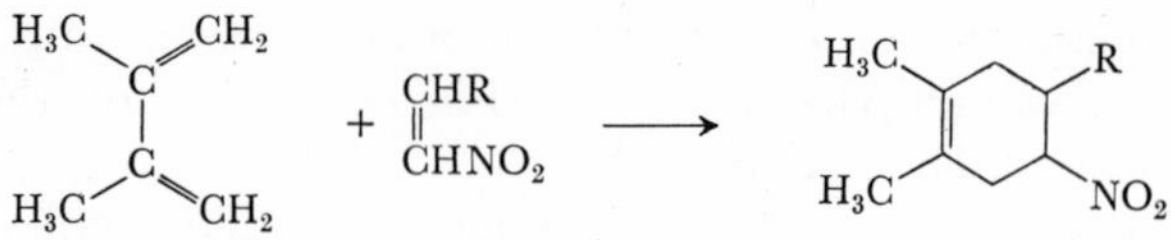

$$(136)$$

authors[368] also allowed 2-methoxy-1,3-butadiene to react with 4-ethoxy-1-nitrobutene; the resulting enol ether was readily hydrolyzed with acid to 3-(2-ethoxyethyl)-4-nitrocyclohexanone (equation 137).

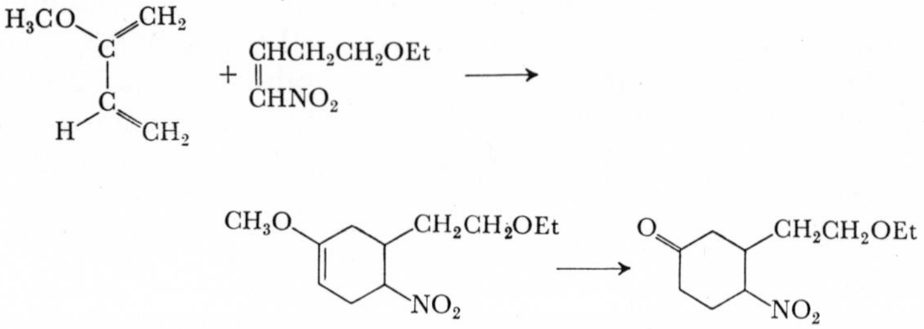

$$(137)$$

Using 1,3-butadiene[369] and *trans*-1,3-pentadiene (piperylene)[370], Novikov and colleagues synthesized several similar nitrocyclohexene derivatives. Besides nitroethylene and 1-nitropropene, they also employed as dienophiles 3,3,3-trichloro-1-nitropropene, methyl 3-nitroacrylate, and, interestingly, 3-nitropropene. The latter compound, although not an α-nitroalkene, nevertheless gave with piperylene an adduct, to which the structure of 3-methyl-4(or 5)-nitromethylcyclohexene was assigned. As far as the structural orientation in the cyclizations with piperylene is concerned, the Russian authors found[371] that nitroethylene exclusively yields the *ortho* adduct **175** (R = H), whereas substituted nitroethylenes gave mixtures of isomers **175** and **176**, in which those of type **175** predominated (equation 138).

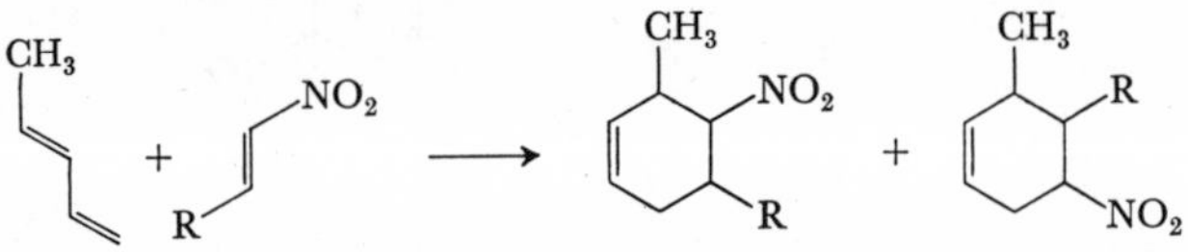

$$(138)$$

(**175**) (**176**)

R = H, CH_3, Ph, CO_2CH_3, or CCl_3 (R $\neq$ H)

The use of furans as dienophiles has also attracted further attention. Thus, it was found by Etienne and collaborators[372] that 1,3-diphenylisobenzofuran adds nitroethylene and 1-nitropropene to yield 1,4-diphenyl-1,4-epoxy-2-nitrotetralins, which can be converted into nitronaphthalenes by the action of alcoholic hydrochloric acid (equation 139).

Ph, O, Ph + CHR=CHNO$_2$ ⟶ Ph, O, R, NO$_2$, Ph $\xrightarrow{H^+}$ Ph, R, NO$_2$, Ph

R = H or CH_3 (139)

More recently, Russian workers[373] reported that the relative lack of reactivity of simple furans[354,355] is also borne out in their failure to add nitroethylene, although some of the more reactive dienophiles (3-nitroacrylic acid, 3-nitroacrylonitrile, and 1,1,1-trichloro-3-nitropropane) did give adducts with furan and various methylfurans.

With regard to the stereochemical course of these reactions, Alder and coworkers have determined that in the addition of nitroethylene to cyclopentadiene, the activating group of the dienophile is placed in *endo* position in the adduct, in accord with what is generally observed in such cases. It should be noted, however, that this does not seem to be the exclusive way of addition. Thus Roberts, Lee, and Saunders[363], who employed milder reaction conditions, assumed to have obtained a mixture of the *endo* and *exo* isomers of 5-nitronorbornene, with the former strongly preponderating. Fraser[373a] confirmed the *endo* configuration of the product by NMR spectroscopy, but apparently he did not encounter the *exo* isomer. The latter was subsequently shown to arise easily by alkaline equilibration[373b]. In a detailed investigation, Noland and coworkers[365a] examined the addition of 1-nitropropene, 1-nitrobutene, and various 2-aryl-1-nitroethylenes to cyclopentadiene, and they found *endo*/*exo* product ratios of 9:1 (one case), 6:1 (three cases), 4:1 and 3:1 (one case each). Hence, the Alder rule of *endo* addition is certainly valid if it refers to the perponderant product. van Tamelen and Thiede[360] have based on it a deduction of the configuration of 5-*endo*-nitro-6-*exo*-methylbicyclo[2.2.1]-2-heptene (**177**) previously prepared by Alder[353] from cyclopentadiene and 1-nitropropene (equation 140). The methyl group in **177** could be demonstrated to be *exo*, i.e., *trans* to the nitro group; since in diene syntheses *trans* adducts are expected to arise from *trans* dienophiles, it was concluded that the 1-nitropropene used existed in that configuration.

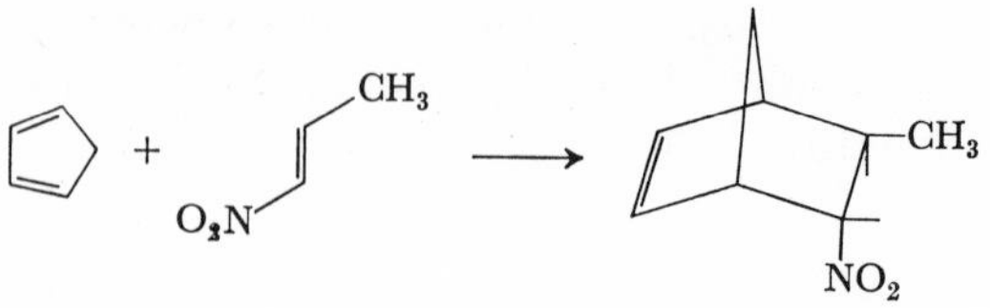

CH_3 O_2N → CH_3 NO_2 (140)

(**177**)

Similarly, Arnold and Richardson[374] argued that ω-nitrostyrene, which with butadiene afforded an adduct that was reduced to *trans*-2-phenylcyclohexylamine, possesses the *trans* configuration (see also Zimmerman and Nevins[375]) (equation 141).

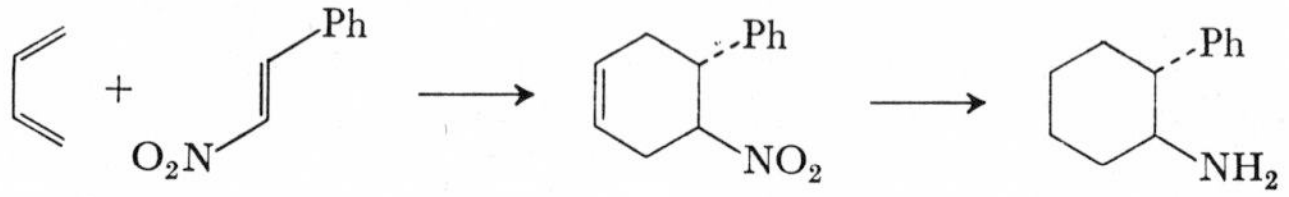

Ph O_2N → Ph NO_2 → Ph NH_2 (141)

Reactions between nitro olefin dienophiles and anthracene have been simultaneously reported by Klager[344] and Noland and coworkers[376]. Thus, nitroethylene gave a 71% yield of 9,10-dihydro-9,10-(11-nitroethano)anthracene (**178**)[344]. Fair to good yields were obtained[376] of methyl- or phenyl-substituted products **179**, **180**, and **181** when the appropriate 1- or 2-monosubstituted nitro olefins were employed (equation 142). On the other hand, 1,2- and 2,2-disubstituted nitro olefins gave little or no adduct. It may be mentioned in this connection that Hudak and Meinwald[377] have found two monosubstituted nitroethylenes, namely 5-cyano-3,3-dimethyl-1-nitro-1-pentene and 6-methoxy-3,3-dimethyl-1-nitro-1-hexene, to be inert as dienophiles. This lack of reactivity appeared to be due to steric hindrance caused by the geminal methyl groups next to the double bond.

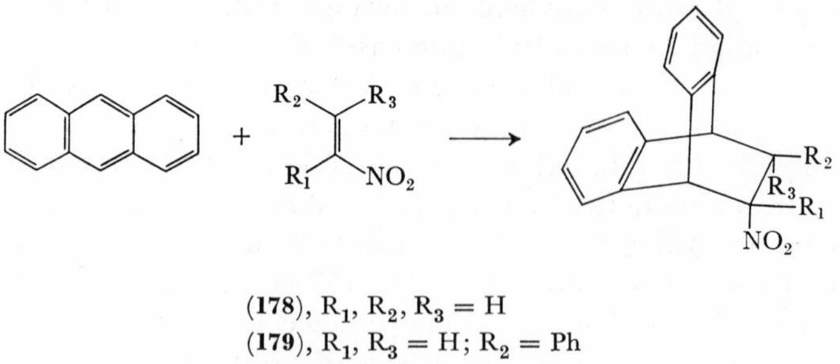

R_2 R_3 R_1 NO_2 → R_2 R_3 R_1 NO_2 (142)

(**178**), $R_1, R_2, R_3 = H$
(**179**), $R_1, R_3 = H; R_2 = Ph$
(**180**), $R_1, R_3 = H; R_2 = CH_3$
(**181**), $R_1 = CH_3; R_2, R_3 = H$

Klager has shown, moreover, that the nitroethylene adduct **178** can serve as a starting point for the preparation of α-substituted nitroethylenes[344]. Methylolation of **178** followed by acetylation furnished the acetoxymethyl derivative **182**, which by thermal decomposition regenerated anthracene and produced 3-acetoxy-2-nitropropene. Michael reaction of the sodium nitronate of **178** with methyl acrylate or acrylonitrile led to **183**, which upon pyrolysis afforded methyl 4-nitro-4-pentenoate and 4-nitro-4-pentenonitrile, respectively (equation 143).

178 $\xrightarrow{NaOCH_3}$ NO_2Na $\xrightarrow{1.\ CH_2O,\ 2.\ Ac_2O}$ O_2N CH_2OAc (**182**) $\xrightarrow{\Delta}$ anthracene + $CH_2{=}C(NO_2)CH_2OAc$

$\downarrow CH_2{=}CHR$

O_2N CH_2CH_2R (**183**) $\xrightarrow{\Delta}$ anthracene + $CH_2{=}C(NO_2)CH_2CH_2R$

$R = CO_2CH_3$ or CN

(143)

A limitation of this procedure is that the nitro olefin liberated may not withstand the drastic conditions of the thermal treatment. When the preparation of 1,1-dinitroethylene was attempted by this technique, complete degradation of the aliphatic part of the molecule took place[378]. Unfortunately, attempts to circumvent the pyrolysis by applying the principle of Diels and Thiele, which consists of a displacement of the *endo*-ethylene bridge by reaction with excess maleic anhydride, proved unsuccessful[378].

Although 1,1-dinitroethylene has not yet been isolated, it is known to play a role as an intermediate in a number of reactions (see section III.D). When a mixture of cyclopentadiene and 2,2-dinitroethanol in chlorobenzene was heated at 100–110°, 5,5-dinitrobicyclo[2.2.1]-2-heptene (**184**) was formed; it obviously arose by dehydration of the alcohol and Diels–Alder reaction of intermediate 1,1-dinitroethylene[207] (equation 144).

$$\text{cyclopentadiene} + \underset{\text{CH(NO}_2)_2}{\overset{\text{CH}_2\text{OH}}{|}} \longrightarrow \text{5,5-dinitronorbornene } \textbf{(184)} \qquad (144)$$

An analogous case was reported by Babievskii and associates[183] who synthesized ethyl β-hydroxy-α-nitropropionate, which was found to dehydrate easily. Although the dehydrated product, ethyl α-nitroacrylate, could not be isolated, its existence was verified by heating the hydroxy ester with cyclopentadiene, butadiene, or anthracene, whereby the adducts arose that were expected from reaction of the olefinic ester.

The formation of nitro olefins from 2-nitroalkyl acetates and their utilization *in situ* in diene additions has been studied in detail by Feuer, Miller, and Lawyer[379]. These workers have effected Diels–Alder reactions using as dienophile precursors 2-nitrobutyl acetate, 3-nitro-2-butyl acetate, 2-nitropentyl acetate, 1,6-diacetoxy-2,5-dinitrohexane, and 2-acetoxy-2-perfluoropropyl-1-nitroethane. All of these acetates reacted well with cyclopentadiene in the presence of sodium acetate, the solvent being ethanol, *t*-butyl alcohol, or benzene. Reaction of 2-nitrobutyl acetate with anthracene was brought about in refluxing xylene. The function of the added sodium acetate is to generate the nitro olefin from the nitroacetate in the fashion previously discussed (see section III.D).

V. SOME REACTIONS OF NITRO ALCOHOLS AND THEIR DERIVATIVES

A. Nitro Acetals and Ketals

1. Formation

β,β′-Dihydroxynitroalkanes, when treated with aldehydes or ketones in the presence of an acid catalyst, form cyclic acetals or ketals, which are substituted 5-nitro-1,3-dioxanes (equation 145).

$$\text{RCHOH–C(R)(NO}_2\text{)–CHOHR} + \text{O=CR}_2 \longrightarrow \text{O}_2\text{N(R)C}\begin{matrix}\diagup\text{CH(R)–O}\diagdown \\ \diagdown\text{CH(R)–O}\diagup\end{matrix}\text{CR}_2 \qquad (145)$$

Catalysts such as hydrochloric acid, sulfuric acid, and *p*-toluenesulfonic acid have been employed to condense such nitro alcohols as trishydroxymethylnitromethane, 2-nitro-1,3-propanediol and several of its 2-alkyl and 2-aryl derivatives, and 3-nitro-2,4-pentanediol with a variety of aliphatic aldehydes and ketones, benzaldehyde and other aromatic aldehydes, and alkyl levulinate.[42,239,250,380–386] 2-Substituted 5-bromo-5-nitro-1,3-dioxanes were obtained in analogous fashion[387].

Ketalization of 2-nitro-1,3-propanediol and its 2-substituted derivatives (including 2,2-dinitro-1,3-propanediol and trishydroxymethylnitromethane) proceeds especially well in the presence of boron trifluoride etherate[254,388]. By this method, 2,5-bis(hydroxymethyl)-2,5-dinitro-1,6-hexanediol gave the expected bis ketal with acetone[389] (equation 146).

$$HOCH_2C(CH_2OH)(NO_2)CH_2CH_2C(CH_2OH)(NO_2)CH_2OH \xrightarrow[BF_3]{2(CH_3)_2CO} (H_3C)_2C\begin{matrix}O-CH_2\\O-CH_2\end{matrix}C(NO_2)CH_2CH_2C(NO_2)\begin{matrix}CH_2-O\\CH_2-O\end{matrix}C(CH_3)_2 \tag{146}$$

It has been recognized that geometrical isomerism should occur in 2-substituted 5-nitro-1,3-dioxanes, and stereoisomeric forms have in fact been isolated in a few instances[381,385]. Conformational studies based on dipole moment values have been conducted on a number of 5-nitro-1,3-dioxanes, and it was suggested that the nitro group tends to be equatorial when C_5 carries hydrogen, and axial when C_5 carries an alkyl group[271].

Unsubstituted 2-nitro-1,3-propanediol appears to form a 1,3-dioxane with acetone less readily than do its 2-substituted derivatives, since the reaction failed to be promoted by anhydrous cupric sulfate or sulfuric acid[239] although it succeeded by boron trifluoride catalysis[388]. The original explanation[239] that intramolecular hydrogen bonding (**185a**) might interfere with the ketalization seems to be untenable in the light of the facile cyclization of substituted 2-nitro-1,3-propanediols with aldehydes and ketones.[243,381,386,387,390] It has therefore been suggested[391] that intermolecular (**185b**) rather than intramolecular hydrogen bonding might be responsible for the phenomenon.

Triol acetals and ketals of type **186** have been reported[239] to resist

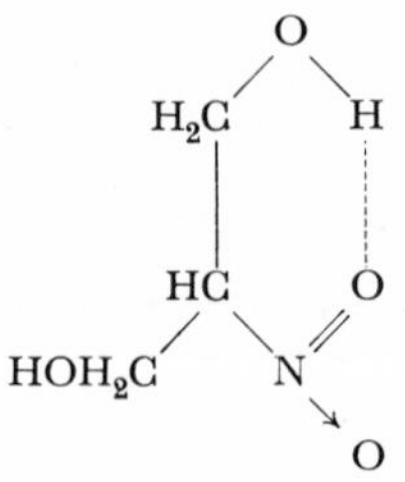

(**185a**) 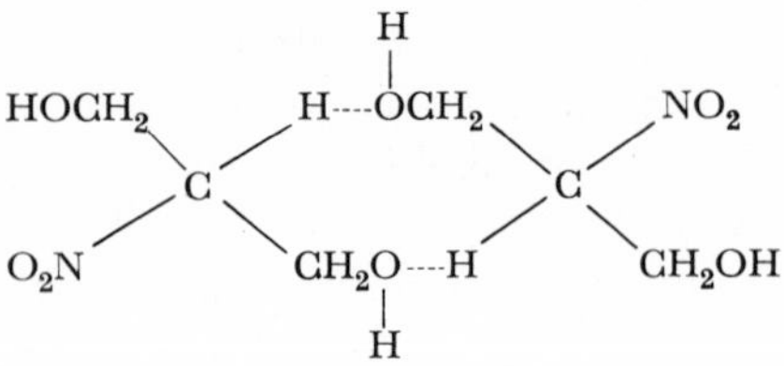 (**185b**)

methylation and oxidation and to be stable toward alkali. This, however, is difficult to reconcile with the findings of Eckstein[387], according to which 5-hydroxymethyl-5-nitro-2-phenyl-1,3-dioxane (**186**, R = C_6H_5) can be reversibly interconverted, in alkaline medium, into 5-nitro-2-phenyl-1,3-dioxane (**187**, R = C_6H_5), as was indicated by a positive response of **186** to the pseudonitrole reaction and by the actual preparation of **187** from **186** (equation 147). Moreover, when **186** was alkylated with *p*-nitrobenzyl chloride

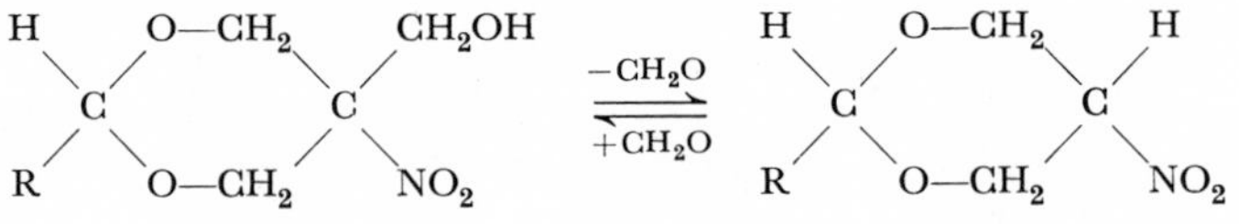

$$\mathbf{186} \underset{+CH_2O}{\overset{-CH_2O}{\rightleftharpoons}} \mathbf{187} \qquad (147)$$

and potassium hydroxide, replacement of the hydroxymethyl by the *p*-nitrobenzyl group took place and the same 5-substituted derivative was formed that also arose under similar conditions from **187**[243]. Furthermore, identical 5-arylazo derivatives were obtained from **186** and **187** by reaction with aryldiazonium salts at pH 7.5–8.5[390]. The isopropylidene ketal corresponding to **186** also underwent replacement, in alkaline medium, of the 5-hydroxymethyl group by arylazo groups[254,392].

Another way in which **187** and related dioxanes with a secondary nitro group could be obtained is the replacement of bromine by hydrogen in 2-substituted 5-bromo-5-nitro-1,3-dioxanes through the action of sodium ethyl malonate, alcoholic potassium hydroxide, benzylamine in dioxane[387], or sodium ethyl alkylmalonates[393].

The *meso* and *racemic* epimers of 1,4-dinitro-2,3-butanediol gave cyclic ketals with acetone, which differed in their NMR spectra and could thus be used to elucidate the configurations of the parent diols[57] (equation 148, see also I.B. 4).

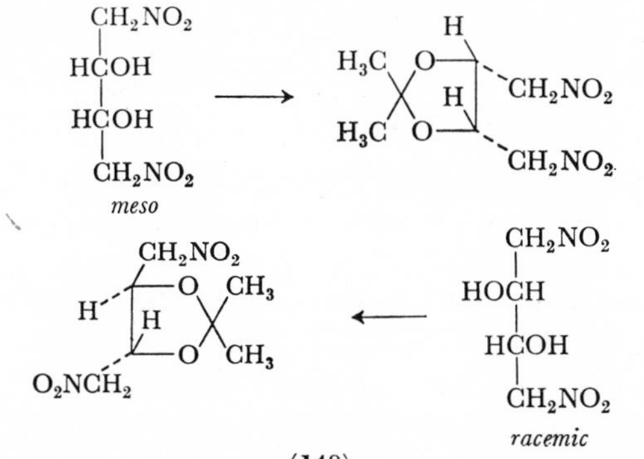

(148)

(148)

Both 6-deoxy-1,2-*O*-isopropylidene-6-nitro-α-D-glucofuranose (**188**) and -β-L-idofuranose (**189**) when treated with acetone and sulfuric acid give 1,2:3,5-di-*O*-isopropylidene derivatives (**190** and **191**) (equations 149 and 150); since **189** affords a much better yield than **188**, this acetonation can be used to achieve a partial separation of the *gluco* and *ido* isomers[91]. It has been pointed out[394] that, if the six-

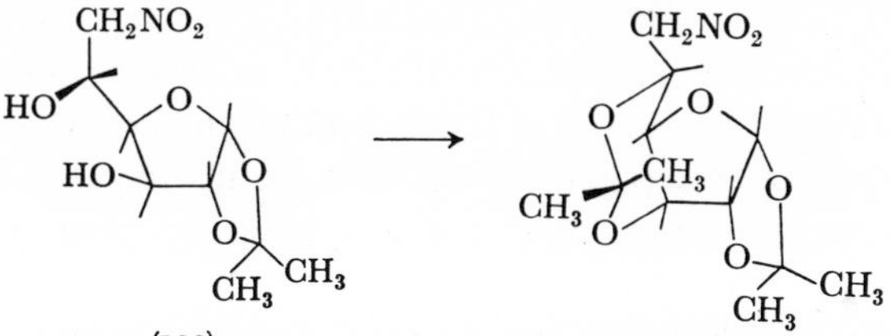

(149)

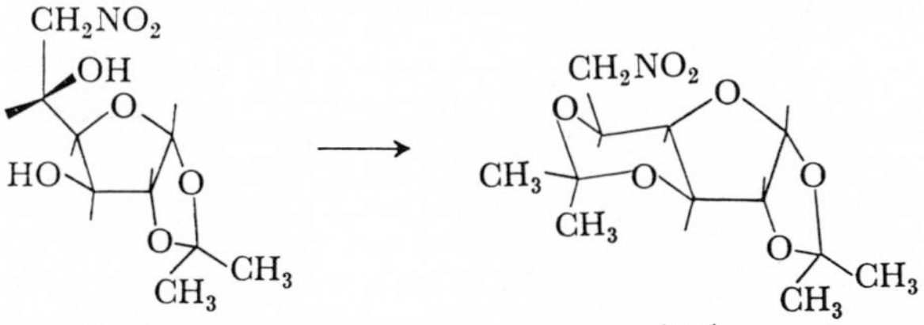

(150)

membered 3,5-*O*-isopropylidene rings exist in chair conformations as drawn for **190** and **191**, there will be a diaxial substituent interaction in **190** but not in **191** (the same is true for inverted chairs), which may be the reason for a greater stability and hence more facile formation of the latter. This view appears to be supported by the fact that **191** is more resistant than **190** against alkali (see below), but no investigations concerning the actual, preferred conformations of the compounds have yet been carried out. One would also have to consider the conformations of **188** and **189** prior to and during the reaction with acetone. Molecular models show that if the nitro groups are engaged in hydrogen bonding with the hydroxyl at C_3 then the hydroxyl at C_5 would be favorably disposed in **189**, but unfavorably so in **188**, for cyclization with acetone.

Of the three deoxynitroinositols **192**, **193**, and **194**, only **194** forms a diisopropylidene derivative **195**, whereas the others do not react with acetone under ordinary conditions[92,94]. Since *trans* vicinal hydroxyls in pyranoside and inositol systems generally are

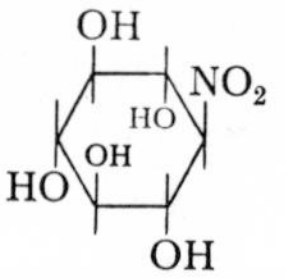

scyllo (**192**) *myo*-1 (**193**) *muco*-3 (**194**) (**195**)

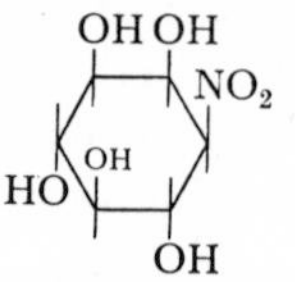

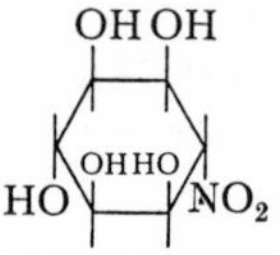

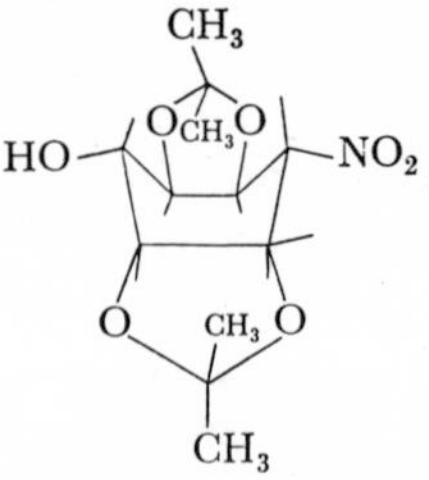

reluctant to form cyclic ketals with acetone, the behavior of **192** is understandable. On the other hand, **193** might be expected to yield at least a monoketal; possibly the nitro group in *cis* position to the *cis*-glycol grouping prevents it from reacting. For **195**, a boat conformation as depicted would seem reasonable and is supported[94] by NMR spectroscopic evidence.

1,4-Dideoxy-1,4-dinitro-*neo*-inositol (**196**) has been found to form a monoisopropylidene derivative only **197**, despite the presence of two pairs of *cis*-hydroxyls. For a diisopropylidene derivative to be possible, the boat conformation **198** with diaxial attachment of the

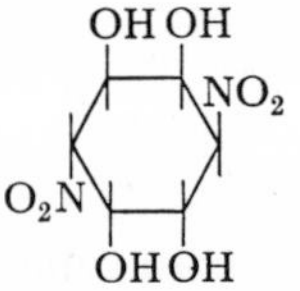

(**196**) 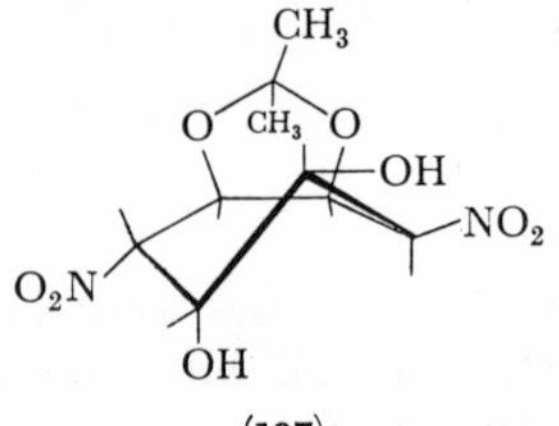 (**197**) 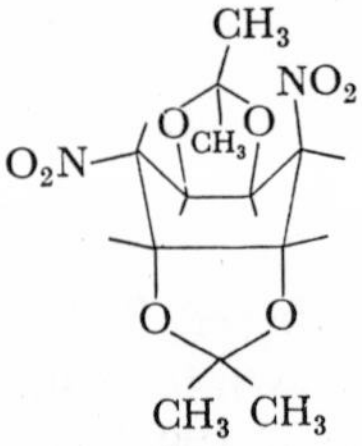 (**198**)

second ketal ring and flagstaff position of one nitro group would have to arise, which is considered energetically unfavorable[58].

Several methyl 3-deoxy-3-nitrohexopyranosides, e.g., the β-D-*gluco* (**199**) and β-D-*galacto* (**200**) compounds, have been found to give well-crystallized cyclic benzylidene acetals (**201** and **202**) that have proved valuable as intermediates in the synthesis of rare amino sugars[395–397] (equations 151 and 152). Similarly, amorphous methyl 4-deoxy-4-nitro-α-D-*gluco*-heptulopyranoside (**203**) could be obtained in crystalline condition by purification via its 1,3:5,7-di-*O*-benzylidene derivative **204**[122] (equation 153).

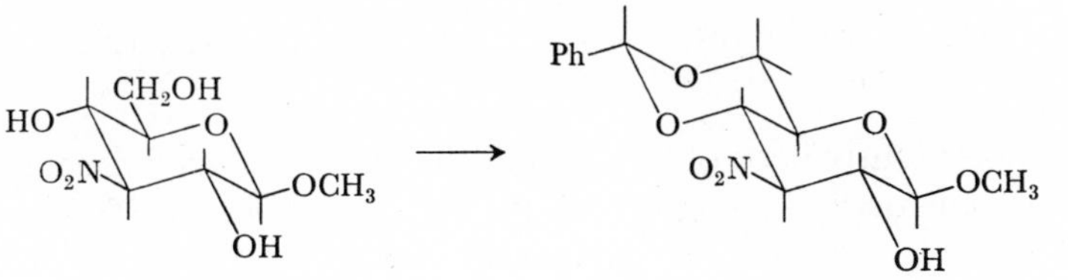

(**199**) → (**201**) (151)

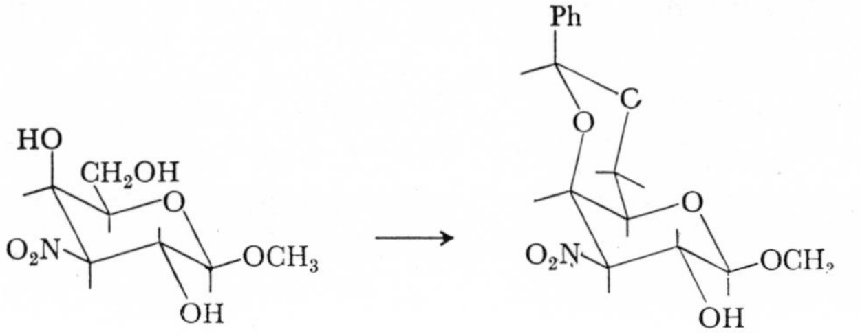

(**200**) → (**202**) (152)

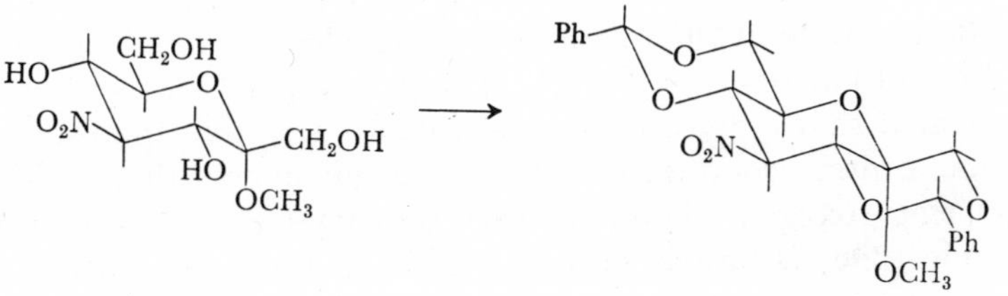

(**203**) → (**204**) (153)

2. Cleavage

Just like ordinary acetals and ketals, the nitro derivatives are easily hydrolyzed by acids under mild conditions. For instance, compounds of type **201** are debenzylidenated by short heating in 70% acetic acid or by refluxing in aqueous methanol in the presence of a cation-exchange resin. More unusual is the lability toward alkali of acetals and ketals of certain β-nitro alcohols, particularly nitro sugars. Whereas in the studies by Urbański, Eckstein, and coworkers on 5-nitro-1,3-dioxane derivatives, cited above, no alkaline cleavage of acetalic bonds seems to have been observed, Feuer and Markofsky[398] reported that bis(2-nitrobutoxy)methane is cleaved by aqueous sodium hydroxide at 38° to 2-nitro-1-butanol and formaldehyde (equation 154).

$$(CH_3CH_2CHNO_2CH_2O)_2CH_2 \xrightarrow[NaOH]{H_2O} 2\ CH_3CH_2CHNO_2CH_2OH + CH_2O \quad (154)$$

It had previously been observed that the base-catalyzed deacetylation of 2-nitroethyl β-D-glucopyranoside tetraacetate (**205**) was attended by fission of the glycosidic linkage[399] (equation 155). The nitrogenous fragment was not identified but in the light of newer experiences can be presumed to have been nitroethylene or, more likely, a product arising from it by further action of base.

CH₂OAc, OOCH₂CH₂NO₂, OAc, AcO, OAc (**205**) ⟶ CH₂OH, O, OH, HO, OH, H,OH + $(CH_2{=}CHNO_2)$ (155)

Obviously the glycosidic bond, which ordinarily is alkali stable, was rendered labile in **205** by the activating effect of the nitro group in the β position of the aglycon. Baer and Rank[98] have recently demonstrated that a glycosidic bond becomes sensitive against base also when a nitro group is located in the sugar moiety in β position to the ring oxygen. Thus, methyl 6-deoxy-6-nitro-α,β-D-glucopyranoside (**206**) is largely cleaved within 20 min when heated at 98° in the presence of 1.1 equivalents of 0.01 *N* sodium hydroxide solution. The same cleavage took place at room temperature, too, but then required several days. The reaction product was a stereoisomeric mixture of deoxynitroinositols **208** that originated from internal Henry condensation of intermediate nitrohexose **207** (equation 156).

A similar lability toward alkali had previously been observed for the 3,5-*O*-isopropylidene groups in the diisopropylidene compounds

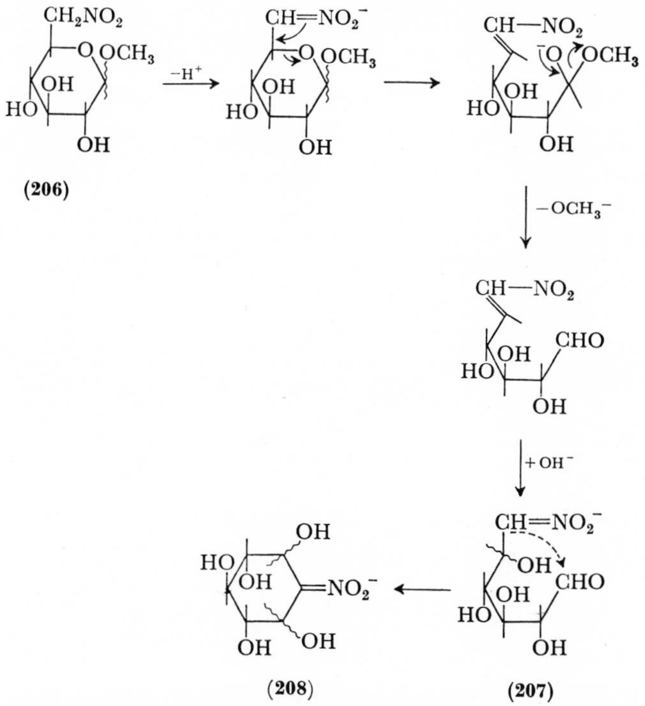

(206) (208) (207) (156)

190 and **191**. Compound **190** loses one molecule of acetone within 30 min when it is treated with 0.1 *N* sodium hydroxide in methanol at room temperature, and a mixture of the 5-*O*-methyl ethers **209** is produced[400] (equation 157). Compound **191** is somewhat more stable but is cleaved at 50° within 3 hours[401].

The 4,6-*O*-benzylidene group in the nitroglycoside **201** is split off rapidly under the influence of warm 0.1 *N* sodium hydroxide[108]. Liberation of benzaldehyde was also noticed when the 2-*O*-ethyl ether of **201** was allowed to stand for 3 hours at room temperature in ethanol solution, in the presence of a catalytic amount of sodium ethoxide[402]; however, cleavage was slight in this case and in similar ones, and it was possible to utilize the blocking function of the benzylidene group in **201** for certain synthetic purposes (see sections VI.A and C). Treatment of **201** and related compounds with acetic anhydride in pyridine, or with sodium acetate in alcohol did not cause debenzylidenation. In any event, the present evidence shows that in nitro sugars such acetal or ketal groupings which are structurally capable of β elimination, can serve as blocking groups in basic media only under careful control of conditions if at all.

$$\mathbf{190} \xrightarrow[(NaOH)]{-H^+} \text{[cyclic isopropylidene intermediate: HC–O, HC–O (}C(CH_3)_2\text{), O–CH, HC–O, HC–O, CH=NO}_2^-\text{]} \longrightarrow \text{[}H_3C, H_3C\text{ >C(O–CH)(O}^-\text{), HC–O, HC=CHNO}_2\text{]}$$

$$\xrightarrow[+CH_3OH]{-CH_3COCH_3} \text{[HC–O, HC–O (}C(CH_3)_2\text{), HOCH, HC–O, CHOCH}_3\text{, CH=NO}_2^-\text{]} \quad \textbf{(209)} \qquad (157)$$

There exist few accounts in the literature on acetals derived from *gem*-dinitro alcohols. Novikov and Shvekhgeimer[403], who studied the reaction between various nitro alcohols and vinyl ethers, obtained mixed acetals from 2,2-dinitropropanol (equation 158).

$$CH_2{=}CHOR + HOCH_2C(NO_2)_2CH_3 \longrightarrow CH_3CH(OR)(OCH_2C(NO_2)_2CH_3) \qquad (158)$$

R = Et, *i*-Pr, or Pr

Formaldehyde acetals have been prepared by Ungnade and Kissinger[404] from 2,2-dinitropropanol and 2-chloro-2,2-dinitroethanol; the cyclic acetal from 2,2-dinitro-1,3-propanediol and cyclohexane carboxaldehyde (2-cyclohexyl-5,5-dinitro-1,3-dioxane) has been described by Eckstein and coworkers[405].

Another type of cyclic ketals is represented by the ethylene glycol condensation products of α-nitroketones that were prepared by Hurd and Nilson[406] (equation 159). These compounds were found to be resistant toward hydrolysis with dilute sulfuric acid. Concentrated hydrochloric acid did cause cleavage but produced hydroximyl chlorides from those derivatives that possessed a primary nitro group (equation 160).

$$O_2NCHRCOCH_2R' + HOCH_2CH_2OH \xrightarrow[C_6H_6]{p\text{-TsOH}} \begin{matrix} CH_2\text{—}O \\ | \\ CH_2\text{—}O \end{matrix} \!\!>\!\! C \!\!<\!\! \begin{matrix} CH_2R' \\ \\ CHRNO_2 \end{matrix} \qquad (159)$$

$$R = H \text{ or } CH_3;\ R' = H \text{ or } CH_3$$

$$\begin{matrix} CH_2\text{—}O \\ | \\ CH_2\text{—}O \end{matrix} \!\!>\!\! C \!\!<\!\! \begin{matrix} R \\ \\ CH_2NO_2 \end{matrix} \xrightarrow{\text{concd HCl}} RCOC(Cl){=}NOH \qquad (160)$$

The formation of such haloisonitroso ketones from nitro ketones by the action of hydrogen halides (or of acylating agents in the presence of aluminum halide) has also been observed by Dornow and coworkers[230,407].

B. Esters of Nitro Alcohols and their Conversion into Nitro Olefins

Nitro alcohols may be esterified by acylating agents, usually under standard conditions. They may be combined with inorganic as well as organic acids. Certain esters can be obtained by the addition of acids across the double bond of α-nitro olefins, and conversely, α-nitro olefins are readily accessible through the elimination of acid from esters of β-nitro alcohols.

I. Esters of inorganic acids

Nitroalkyl nitrates have been obtained by treatment of nitro alcohols with nitric acid[408–411], or with dinitrogen tetroxide in the presence of oxygen[412]. They are also formed, along with nitrites and nitro compounds, in the addition of dinitrogen tetroxide to alkenes[413–417], and they play a role in the nitration of alkenes with nitric acid[418] or acetyl nitrate[419–421]. Nitroalkyl nitrites apparently are solvolyzed by water or methanol more readily than are 2-nitroalkyl nitrates. The latter may suffer elimination of nitric acid by the action of alkoxides, whereby 2-nitroalkyl ethers are produced[422]. There have also been described several nitrate esters of polynitro alcohols[281,423,424].

Acyclic[424,425] and cyclic[424,426] nitroalkyl sulfites have been prepared by the action of thionyl chloride upon nitro alcohols and nitro glycols, respectively, although displacement of hydroxyl by chlorine may also occur and may in fact be the main reaction[427,428] (equations 161–163).

$$2\ ClCH(NO_2)CH_2OH \xrightarrow{SOCl_2} [ClCH(NO_2)CH_2O]_2SO \quad (161)$$

$$O_2NCCH_3(CH_2OH)_2 \xrightarrow{SOCl_2} \underset{H_3C}{\overset{O_2N}{>}}C\langle\begin{matrix}CH_2-O\\CH_2-O\end{matrix}\rangle SO + O_2NCCH_3(CH_2Cl)_2 \quad (162)$$

$$(HOCH_2)_3CNO_2 \xrightarrow{SOCl_2}$$

$$(ClCH_2)_3CNO_2 + \underset{ClCH_2}{\overset{O_2N}{>}}C\langle\begin{matrix}CH_2-O\\CH_2-O\end{matrix}\rangle SO + \underset{HOCH_2}{\overset{O_2N}{>}}C\langle\begin{matrix}CH_2-O\\CH_2-O\end{matrix}\rangle SO \quad (163)$$

The cyclic esters can be designated as 5-substituted 5-nitro-1, 3-dioxathiane 2-oxides. A number of 5-arylazo derivatives were prepared recently by cyclization of 2-arylazo-2-nitro-1,3-propanediol with thionyl chloride[429], and were found to possess fungicidal properties[430].

Other heterocyclic systems, namely 5-alkyl-5-nitro-2-phenyl-2-bora-1,3-dioxanes and 2,2-dimethyl-5-alkyl-5-nitro-2-sila-1,3-dioxanes were obtained by cyclizations of 2-alkyl-2-nitro-1,3-propanediols using phenylboronic acid and diacetoxydimethylsilane, respectively, and conformational studies based on dipole moment measurements were undertaken[271,386]. An ester of silicic acid, di-*t*-butyl (2-ethyl-2-nitrotrimethylene)orthosilicate, was mentioned in a patent[431]. 2,2-Dinitropropanol was allowed to react with boron trichloride to give tris(2,2-dinitropropyl) borate, and 2,2-dinitro-1,3-propandiol gave, with dichlorodimethylsilane, a mixture believed to contain the cyclic siloxane[424].

Nitroalkyl hydrogen sulfates arose by the action of chlorosulfonic acid upon some simple nitro alcohols in dioxane, and they were characterized as their crystalline *S*-benzylthiuronium salts[432].

Nitroalkyl phosphates of the general formula $(O_2NCR_2CH_2O)_3PO$ were obtained by the interaction of nitro alcohols with phosphorus oxychloride, or with phosphorus pentachloride followed by hydrolysis[433]. The phosphate of 2,2-dinitropropanol was made by esterification with polyphosphoric acid[424,434].

2. Esters of organic acids

a. Esterification. For the preparation of esters of organic acids and nitro alcohols, standard procedures of esterification can in general be used. Esters may be made by the use of acyl halides, acid anhydrides, or acids. Catalysts such as sulfuric acid, *p*-toluene-sulfonic acid, polyphosphoric acid, boron trifluoride, aluminum chloride, and sodium acetate have been employed. Reactions may be carried out without solvent or with a solvent such as pyridine. In direct esterifications with acids, the water formed may be removed by azeotropic distillation. Finally, acid-catalyzed trans-esterifications between methyl esters and nitro alcohols have been performed.

Acetates of some simple nitro alcohols have been obtained as early as 1897[435], and since then an extensive literature on esters of various kinds has developed. It suffices here to cite only a few references[10,436–438] to older investigations, and to refer to a review by Shvekhgeimer, Piatakov, and Novikov[5] in which much of the work up to the mid-1950's has been covered. Some more recent work of special interest will be mentioned in the following paragraphs.

Considerable progress has been made in regard to the esterification of polynitro alcohols. In many instances the use of acyl halides with or without an inert solvent,[56,204,205,210,218,439] or of acid anhydride in the presence of catalytic amounts of sulfuric acid[213,440], was successful. The direct esterification, catalyzed by sulfuric acid, of dinitro alcohols with organic acids has also been achieved, for example in the preparation of *gem*-dinitroalkyl acrylates and methacrylates[281]. Difficulties were however encountered in other cases. These were overcome by the introduction of aluminum chloride, trifluoroacetic anhydride, and polyphosphoric acid as esterification catalysts[211,378,434]. A table listing a large number of esters derived from polynitro alcohols was compiled in 1964[9].

Although, in common acylations of alcohols, pyridine is one of the most extensively used agents to promote the reaction, it had found but few applications[441,442] with nitro alcohols until Kissinger and coworkers[404,424] systematically examined its suitability and worked out conditions under which good results are obtained, especially with *gem*-dinitro alcohols. The authors[443] also made a detailed investigation concerning the pyridine–nitro alcohol complexes that play a role in these acylations.

Polyhydroxynitroalkanes such as 1-deoxy-1-nitroalditols[6], deoxynitroinositols[92,99], 1,3-dihydroxy-2-nitrocyclohexane[61,62], and 1,4-dinitro-2,3,5,6-tetrahydroxycyclohexane[58] have been acetylated with acetic anhydride in the presence of sulfuric acid. With nitro sugar glycosides this procedure has usually been avoided for fear of acetolysis and (or) anomerization at the glycosidic center, and similar reservations apply to the acetylation of partially blocked nitro sugar derivatives that contain acetal or ketal groupings. The use of acetic anhydride and pyridine has been successful in some cases while in others it entailed complications. Thus, methyl 4,6-*O*-benzylidene-3-deoxy-3-nitro-β-D-glucopyranoside (**201**) and its *manno* isomer **210** afforded their 2-*O*-acetyl derivatives **211** and **212** by acetylation in the cold, whereas the *galacto* isomer **202** did not yield **213** but incurred decomposition[395,396]. Attempted acetylations of the corresponding nonbenzylidenated glycosides **199**, **214**, and **200** also failed, because of decomposition, when undiluted mixtures of acetic anhydride and pyridine were used even in the cold, although the triacetate **215** of **199** could be obtained, in less than 40% yield, when an inert diluent (tetrahydrofuran) was employed[444]. The instability of these nitro sugars is due to a tendency to form, under certain conditions, unsaturated reactive products. Recrystallization of the triacetate **215**, for example, from ethanol in the presence of pyridine gave, in part, a crystalline yellow compound (λ_{max} 345 mμ)

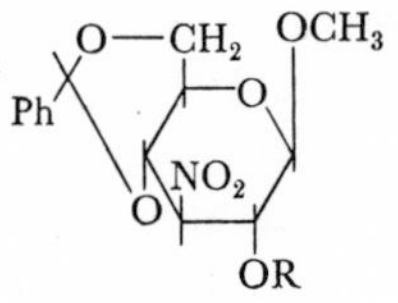

(**201**), R = H
(**211**), R = Ac

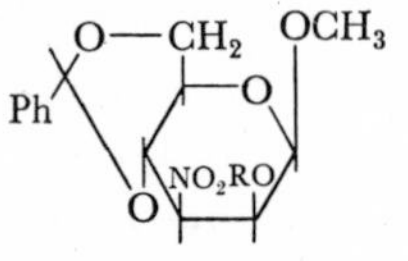

(**210**), R = H
(**212**), R = Ac

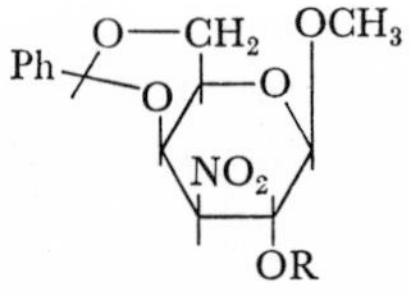

(**202**), R = H
(**213**), R = Ac

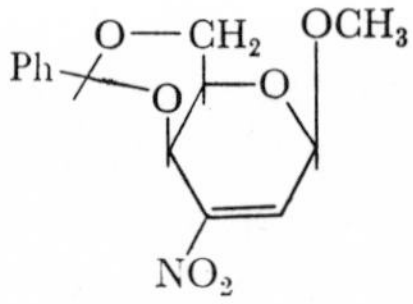

(**216**)

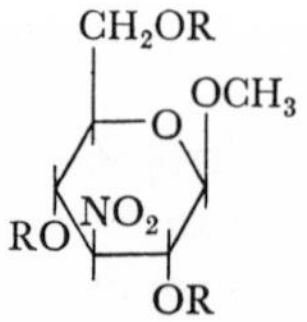

(**199**), R = H
(**215**), R = Ac

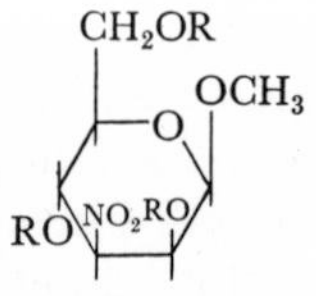

(**214**), R = H
(**222**), R = Ac

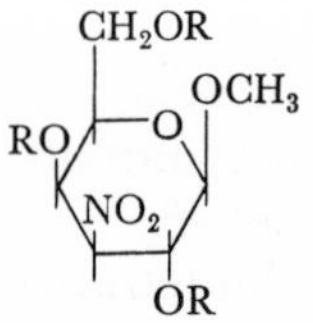

(**200**), R = H
(**223**), R = Ac

that was optically inactive and appeared to be no longer a carbohydrate[444]. This behavior is reminiscent of the conversion of acetylated nitroinositols into nitrophenols by the action of pyridine (see section V.B 2b), and it was found[446] that **215** readily suffers base-catalyzed elimination of two molecules of acetic acid to give a 2,4-diene which dimerizes immediately by way of a Diels-Alder reaction. The dimer, which has been isolated, is transformed under certain conditions into derivatives of 7-nitroisochromene.

Interesting results occurred when the three stereoisomeric benzylidene glycosides **201**, **210**, and **202** were treated with hot acetic anhydride and anhydrous sodium acetate[396]. The glucoside **201** yielded its acetate **211** as expected, whereas the mannoside **210** unexpectedly gave the same acetate **211**, and not **212**; and the galactoside **202** underwent dehydration forming the 2,3-unsaturated nitro olefin **216**. Furthermore[397], methyl 4,6-*O*-benzylidene-3-deoxy-3-nitro-α-D-glucopyranoside (**217**) (the α anomer of **201**) could be smoothly acetylated with acetic anhydride and pyridine in tetrahydrofuran to give the 2-*O*-acetyl derivative **218**. On the other hand, the α-D-*talo* isomer **219** failed to afford an acetate **220** under these conditions, and with hot acetic anhydride and sodium acetate it was dehydrated to the nitro olefin **221**. The nitro group in all of these glycosides possesses the same steric disposition, namely, it is

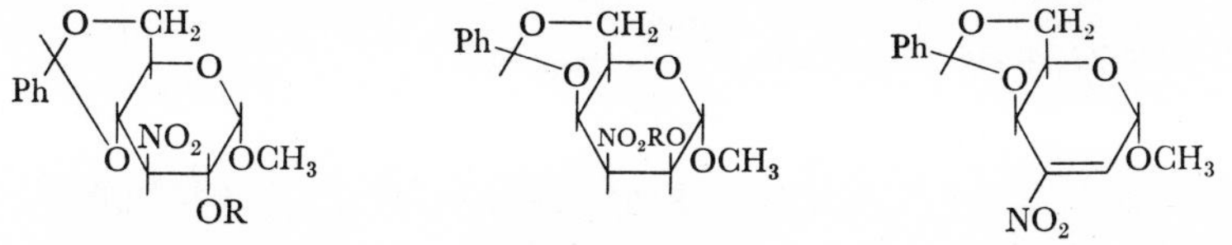

(**217**), R = H
(**218**), R = Ac

(**219**), R = H
(**220**), R = Ac

(**221**)

equatorially oriented. Obviously, the facility of acetylation and its course, that may or may not involve epimerization or give rise to olefins and other transformation products, is governed by the configurations at the carbon atoms adjacent to the nitro group.

Smooth acetylations of several nitro sugars that were difficult to acetylate otherwise were accomplished using cold acetic anhydride in the presence of boron trifluoride[444]. In this way, **215**, **222**, **223**, and **220** were made from **199**, **214**, **200**, and **219**, respectively. No displacement of the nitro by an acetoxy group was observed in these instances, although such a displacement has been reported[447] to occur when this acetylating agent is allowed to act upon nitromethane and other nitroalkanes.

By methylolation of 1,4-dinitrobutane under certain conditions, Feuer, Nielsen, and Colwell[213] obtained 2,5-dinitro-1,6-hexanediol as a separable mixture of epimers, m.p. 163–165° and 112–113°. Treatment of the low-melting epimer with acetic anhydride and sulfuric acid catalyst for 1 hour at 100° caused partial epimerization giving some of the diacetate of the high-melting epimer. The high-melting diol has been acetylated[214] with acetyl chloride.

Some special reactions pertinent to the chemistry of nitro esters deserve mention. Feuer and Gardner[448] investigated the interaction of nitro alcohols and ketene divinyl acetal. 2-Nitroethanol, 2-nitro-1-butanol, and 3-nitro-2-butanol reacted readily to form the corresponding divinyl nitroalkyl orthoacetates. Subsequent acid hydrolysis gave nitroalkyl acetates and acetaldehyde (equation 164).

$$CH_2{=}C(OCH{=}CH_2)_2 + HOCH_2CH_2NO_2 \xrightarrow{H^+} CH_3{-}\underset{\displaystyle OCH_2CH_2NO_2}{\underset{|}{C}}(OCH{=}CH_2)_2 \xrightarrow[H^+]{H_2O}$$

$$CH_3CO_2CH_2CH_2NO_2 + 2CH_3CHO \qquad (164)$$

Novikov and coworkers[449] obtained 2-nitroethyl acetate in 37% yield and 1,3-diacetoxy-2,2-dinitropropane in 19% yield, by the reaction of the corresponding alcohols with ethoxyacetylene in ether containing hydrogen chloride (equation 165).

$$HC{\equiv}COC_2H_5 + HOCH_2CH_2NO_2 \xrightarrow{HCl} H_2C{=}C\begin{matrix} \diagup OC_2H_5 \\ \diagdown OCH_2CH_2NO_2 \end{matrix} \xrightarrow{H_2O}$$

$$CH_3CO_2CH_2CH_2NO_2 + C_2H_5OH \qquad (165)$$

b. Deacylation. Whereas aromatic nitro alcohols of the type of 1-phenyl-2-nitroethanol are dehydrated to nitro olefins of the β-nitrostyrene type very easily—often spontaneously or by mild acid treatment (see section I.B. 6), aliphatic β-nitro alcohols usually require more drastic conditions for their dehydration. On the other hand, the esters of aliphatic nitro alcohols can readily be converted into α-nitroalkenes by β elimination of the acyloxy group, provided, of course, that the nitro group is not a tertiary one (equation 166).

This dehydroacylation is of considerable interest not only for the preparation of α-nitroalkenes as such, but also in view of the role that they play as intermediates in various reactions.

$$R{-}\underset{\displaystyle NO_2}{\underset{|}{\overset{\displaystyle H}{\overset{|}{C}}}}{-}\overset{\displaystyle OCOR}{\overset{|}{C}}H{-}R \xrightarrow{-HOCOR} R{-}\underset{\displaystyle NO_2}{\underset{|}{C}}{=}CH{-}R \qquad (166)$$

The preparative conversion of nitro esters into nitro olefins, often referred to as the Schmidt–Rutz reaction, can be accomplished by boiling the ester (usually the acetate) in a dry, inert solvent over an inorganic alkaline catalyst. Ether or benzene and sodium or potassium bicarbonate or carbonate are most commonly employed. The yields generally are good to excellent although losses may occur due to polymerization of the product, particularly in the case of low molecular weight olefins[450–453]. A quantitative conversion, which was performed in chloroform with potassium bicarbonate, has been reported[56], for instance, of 2,3-diacetoxy-1,4-dinitrobutane to give 1,4-dinitro-1,3-butadiene whereas treatment of 1,6-diacetoxy-2,5-dinitrohexane with potassium carbonate in benzene furnished 2,5-dinitro-1,5-hexadiene in 37 % yield[214]. The reaction has also been used to prepare chlorine-[345,436] and fluorine-containing[166] nitro olefins, but an attempted acetate elimination from 2-bromo-2-nitroethyl acetate did not succeed[329]*. This latter fact is interesting in view of the extreme ease with which 2-bromo-2,2-dinitroethyl acetate reacts with a variety of bases (see below). Another variant consists of heating the nitro ester with anhydrous sodium acetate and removing the eliminated acetic acid by vacuum distillation.[41,191,422,451,454] Pyrolytic deacylation of 1-nitro-2-benzoyloxypropane was reported to yield 1-nitropropene[455], and soon afterward was described a general, continuous process for the production of nitro olefins by vapor phase decomposition of nitro alcohol acetates over catalysts such as alkaline earth salts, silica gel, aluminum sulfate, aluminum phosphate, or zinc chloride[456,457]. Heating of 2,4-diacetoxy-3-nitropentane with sodium acetate caused elimination of two molecules of acetic acid to give 3-nitro-1,3-pentadiene[41]. Treatment of 3,3-dimethyl-1-nitro-2,4-butanediol with ketene resulted in partial acetylation and partial dehydration, the product being 4-acetoxy-3,3-dimethyl-1-nitro-1-butene[102a]. Dehydration also occurred when the Henry addition products from acetaldol and nitroalkanes (see section I.B.5a) were heated with phthalic anhydride, and various nitro dienes were obtained[102b].

Extensive use of the Schmidt–Rutz reaction has been made in the synthesis of carbohydrate nitro olefins, for which usually the benzene–sodium bicarbonate technique was employed. Numerous 1-deoxy-1-nitroalditol acetates were converted into the corresponding polyacetoxy-1-nitro-1-alkenes **224** (equation 166a) whose significance as intermediates for various syntheses is outlined in

* The desired 1-bromo-1-nitroethylene was however obtained by dehydration with phosphorus pentoxide of 2-bromo-2-nitroethanol.

section VI. 1-Deoxy-1-nitroalditols are obtained from aldoses and nitromethane, as pairs of 2-epimers (section I.B. 5a). Since either epimer will give the same nitro olefin, separation prior to the Schmidt–Rutz reaction is not necessary.[6,70,71,78,79,84,85,458] Derivatives of glucose **225** containing a terminal nitro olefin grouping have been prepared[98,459] similarly. No evidence for the existence of geometrical isomers in these nitro olefins has been uncovered yet*, although *cis–trans* isomerism has been demonstrated to occur in 1-nitropropene[460].

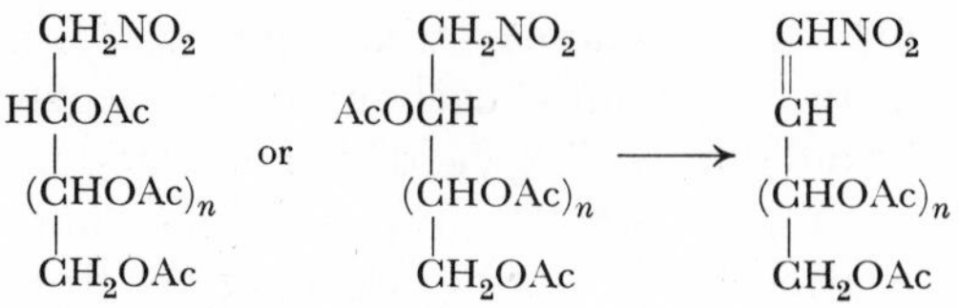

$$\begin{array}{c} CH_2NO_2 \\ | \\ HCOAc \\ | \\ (CHOAc)_n \\ | \\ CH_2OAc \end{array} \quad \text{or} \quad \begin{array}{c} CH_2NO_2 \\ | \\ AcOCH \\ | \\ (CHOAc)_n \\ | \\ CH_2OAc \end{array} \longrightarrow \begin{array}{c} CHNO_2 \\ \| \\ CH \\ | \\ (CHOAc)_n \\ | \\ CH_2OAc \\ \mathbf{(224)} \end{array} \qquad (166a)$$

CHNO₂ = CH– attached to a furanose ring bearing OAc, with O–R–O acetal

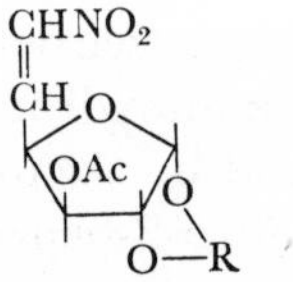

(225)

R = isopropylidene or cyclohexylidene

The ease of acetate elimination can be expected to depend upon structural and steric factors. Thus, while in the carbohydrate derivatives containing a primary nitro group the olefin formation in refluxing benzene usually is complete within a few hours, the conversion of methyl 2-*O*-acetyl-4,6-*O*-benzylidene-3-deoxy-3-nitro-*β*-D-glucopyranoside (**211**) into methyl 4,6-*O*-benzylidene-2,3-dideoxy-3-nitro-*β*-D-*erythro*-hex-2-enopyranoside (**226**) required 1 to 2 days[395]. Interestingly, no significant difference in rate was observed in the reaction of the corresponding *β*-D-*manno* derivative **212**, (equation 167)[396]. This result is not unexpected if one assumes the rate-determining step to be the abstraction of a hydrogen ion from the carbon bearing the nitro group (C_3); this carbon has the same configuration, with axial hydrogen, in **211** and **212**. On the other hand, the α-anomer **218** of **211** was converted into **227** more

* The first case of *cis–trans* isomerism in a carbohydrate nitro olefin has recently been reported by G. B. Howarth, D. G. Lance, W. A. Szarek, and J. K. N. Jones, *Can. J. Chem.*, **47**, 81 (1969).

sluggishly, which may be explained by steric hindrance, due to the axial methoxyl group, in the approach of the catalyst to the axial hydrogen that is to be abstracted from C_3[397] (equation 168).

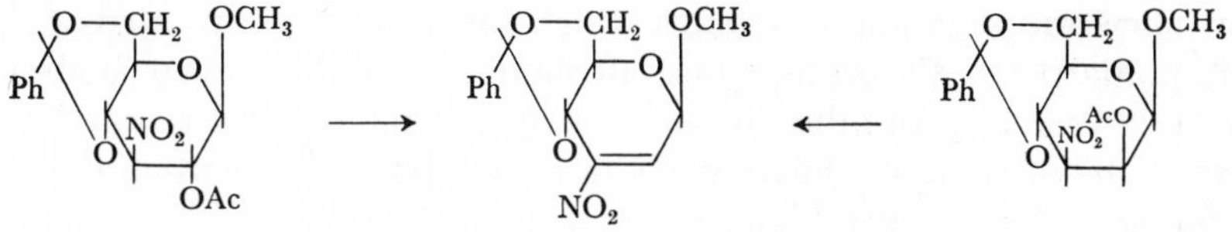

(211) → (226) ← (212) (167)

(218) → (227) (168)

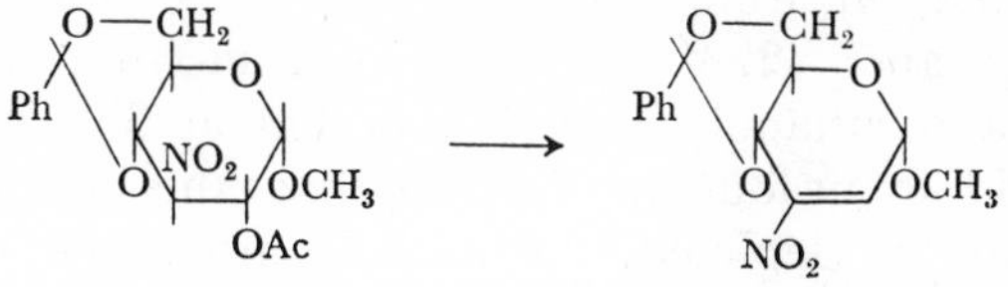

Pyridine is not normally considered to be a sufficiently strong base to bring about acetate elimination from the more stable nitro esters. There are cases, however, where the stability of the reaction product provides the necessary driving force. Thus, acetylated deoxynitroinositols[92] and 1,4-dideoxy-1,4-dinitroinositol[58] are quickly aromatized by pyridine, at or slightly above room temperature, to give diacetyl-5-nitroresorcinol and 2,5-dinitrophenyl acetate, respectively (equations 169 and 170).

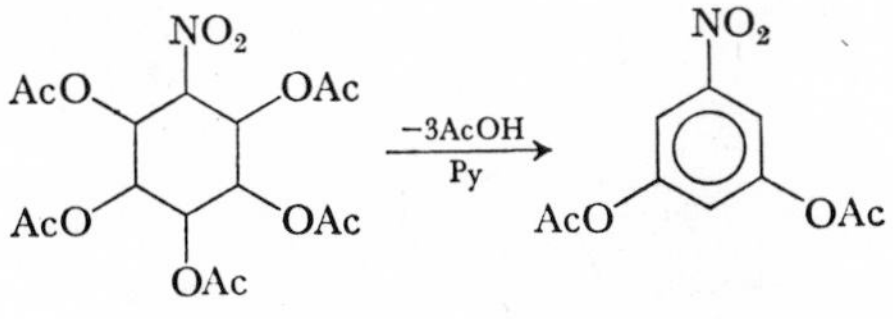

$\xrightarrow[\text{Py}]{-3\text{AcOH}}$ (169)

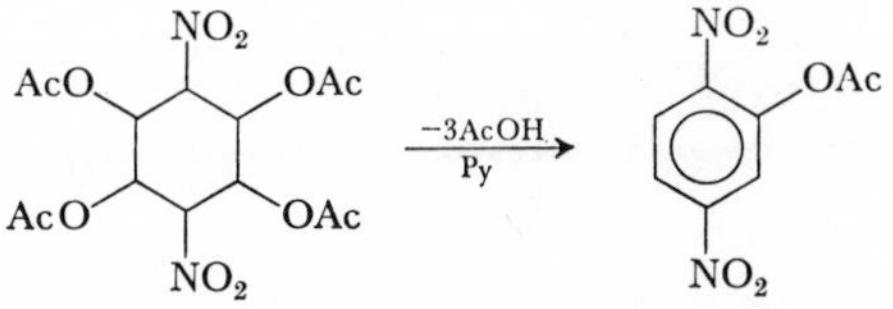

$\xrightarrow[\text{Py}]{-3\text{AcOH}}$ (170)

The difficulties encountered in attempted acetylations in pyridine of certain nitro glycosides (see V.B. 2a) were probably due to related phenomena.

In hydroxylic solvents, too, β-acyloxynitroalkanes exhibit a great susceptibility to cleavage by basic agents. It has been recognized that a β elimination of the acid moiety occurs, rather than an acyl–oxygen fission as in solvolyses of carboxylic esters of ordinary primary and secondary alcohols, when nitro esters are treated with ammonia[178,461–463], or aqueous alcoholic alkalis, or even boiling water[10,437,441]. A number of α-nitro olefins have been isolated in excellent yields upon heating esters in a mixture of methanol and aqueous 0.5 *N* sodium bicarbonate solution[34].

In a study concerning the formation and reactivity of alicyclic β-nitroacetates, Bordwell and Garbisch[464] compared the rates of acetate elimination from the 1-acetoxy-*trans*- and -*cis*-2-nitro-1-phenylcyclohexanes (**228** and **229**) in a mixture of piperidine, chloroform, and ethanol. They found that at 35°, the *trans*-ester lost acetic acid about four times as rapidly as the *cis*-ester. Assuming the conformations depicted, the authors attributed this to steric hindrance in the approach of the catalyst in the case of the *cis*-ester, in which an axial hydrogen must be abstracted. The product obtained from both esters was 6-nitro-1-phenylcyclohexene (**231**), which arose by isomerization of intermediate 2-nitro-1-phenylcyclohexene (**230**) (equation 171). The lesser stability of **230** was attributed to the steric requirement of the substituents, by which

OAc, Ph, H, NO_2 (**228**) OAc, Ph, NO_2, H (**229**)

slow | $-H^+$ → OAc, Ph, NO_2^- ← $-H^+$ | slower (171)

Fast | $-OAc^-$

Ph, NO_2 (**230**) ⟶ Ph, NO_2 (**231**)

both the nitro and the phenyl group might be prevented from effectively conjugating with the double bond. It should be pointed out, however, that base-catalyzed migrations of the double bond from α,β to β,γ position have been found to occur, to varying degrees, in several nitroalkene systems where the prerequisities of such explanation did not exist.[179,185,454,460,465–467]

It is noteworthy that neither in the piperidine-induced eliminations nor in reactions with alcoholic alkali (which gave similar results) an addition of base to intermediate **230** was observed, even though the conversion **230** → **231** was incomplete at least in the absence of excess alkali. A nucleophile, in adding to the double bond in this case, would have to enter in an axial direction, and this appears to be quite unfavored as can be seen from the stereoselectivity of nucleophilic additions to cyclic nitro olefin sugars (see section VI). It may possibly have been for a similar reason that Eckstein and associates[466,467] who obtained cyclohexylidenenitromethane (**233**) by acetate elimination from 1-nitromethylcyclohexyl acetate (**232**), effected isomerization of **233** to 1-cyclohexenylnitromethane (**235**) by treatment with aqueous diethylamine, but did not observe any nucleophilic addition. The yield of **235** was only 35%, however, and whether an adduct **234** was in fact formed as a by-product, or as a reaction intermediate as suggested by the authors, is difficult to assess. The formation of **235** could well be initiated by abstraction of an allylic proton in **233**, and protonation of the mesomeric anion **236** would then lead to **235**, alone or together with **233** (equation 172) in the manner proposed by Shechter and Shepherd[454] for the system 2-methyl-1-nitropropene–2-methyl-3-nitropropene.

Although nitro olefins have been prepared as just mentioned, by

AcO CH_2NO_2 (**232**) $\xrightarrow{-AcOH}$ $CHNO_2$ (**233**) $\xrightarrow{+Et_2NH}$ Et_2N CH_2NO_2 (**234**) $\xrightarrow{-Et_2NH}$ CH_2NO_2 (**235**)

(**233**) $-H^+$ ⇅ $+H^+$; $CHNO_2$ (carbanion) ⟷ $^-CH—NO_2$ (**236**) ⟷ $HC{=}NO_2^-$ $\xrightarrow{+H^+}$ (**235**)

(172)

the action of bases upon 2-nitroalkyl esters in hydroxylic solvents, this hardly appears to be a generally applicable, preparative method because of the great sensitivity that nitro olefins exhibit toward nucleophilic reagents. The α-nitro olefins engendered frequently do react further by way of nucleophile addition, so that 2-nitroalkyl esters may in fact serve as convenient substitutes for olefins in many of the reactions discussed in section VI.

Examples relating to the action of ammonia and amines are the preparation of 2-nitroalkylamines described by Heath and Rose[463], and the syntheses of *vic*-nitroamino sugar derivatives reported by Satoh and Kiyomoto[468] and by Baer and coworkers[402,445]. Furthermore, when *trans,trans*-2-nitro-1,3-diacetoxycyclohexane was treated with a mixture of tetrahydrofuran and aqueous ammonia at room temperature, *trans,trans*-2-nitro-1,3-diaminocyclohexane was formed in good yield. Analogous results were obtained with penta-*O*-acetyl-1-nitro-*scyllo*-inositol, in which case the three acetoxyl groups non-vicinal to the nitro group were hydrolyzed as expected, at a rate slower than the elimination–additions that were induced by the nitro group. The nitrodiamines were characterized as their more stable, *N,N'*-diacetyl derivatives[469] (equation 173).

OAc, R, R, AcO, NO_2, R $\xrightarrow{-AcOH}$ OAc, R, R, $-NO_2$, R $\xrightarrow{+NH_3}$ OAc, R, R, H_2N, NO_2, R $\xrightarrow{-AcOH}$

R, R, H_2N, $-NO_2$, R $\xrightarrow{+NH_3}$ NH_2, R, R, H_2N, NO_2, R (173)

R = H or OAc

Lambert and coworkers[422], in the course of their studies on the alkoxylation of α-nitro olefins (see section VI.A), have found that 2-nitroethyl nitrate on refluxing in ethanol for 8 hours gave ethyl 2-nitroethyl ether in 50% yield (equation 174).

$$\begin{aligned} O_2NCH_2CH_2ONO_2 + C_2H_5OH &\longrightarrow O_2NCH{=}CH_2 + C_2H_5ONO_2 + H_2O \\ O_2NCH{=}CH_2 + C_2H_5OH &\longrightarrow O_2NCH_2CH_2OC_2H_5 \end{aligned} \quad (174)$$

Feuer and Markofsky[398] allowed various 2-nitroalkyl acetates to react at low temperatures with alkoxides in alcohols (viz., methyl,

ethyl, *n*-propyl, and *t*-butyl) and obtained good yields (40–78%) of the corresponding alkoxynitroalkanes (equation 175).

$$\mathrm{R\underset{}{\overset{NO_2}{\overset{|}{C}}}HCH_2OAc + {}^-OR' \longrightarrow R\overset{NO_2}{\overset{|}{C}}{=}CH_2 + R'OH + {}^-OAc}$$

$$\mathrm{R\overset{NO_2}{\overset{|}{C}}{=}CH_2 + {}^-OR' \longrightarrow R\overset{NO_2}{\overset{|}{\underset{-}{C}}}CH_2OR' \xrightarrow{H^+} R\overset{NO_2}{\overset{|}{C}}HCH_2OR'} \qquad (175)$$

Baer, Neilson, and Rank[402] heated methyl 2-*O*-acetyl-4,6-*O*-benzylidene-3-deoxy-3-nitro-β-D-glucopyranoside (**211**) for 1 hour in refluxing methanol (or ethanol) in the presence of anhydrous sodium acetate. The corresponding 2-*O*-alkyl derivatives (**237**) were smoothly produced in over 90% yields, no doubt via the nitro olefin **226** (equation 176). It is noteworthy that apparently no stereoisomers

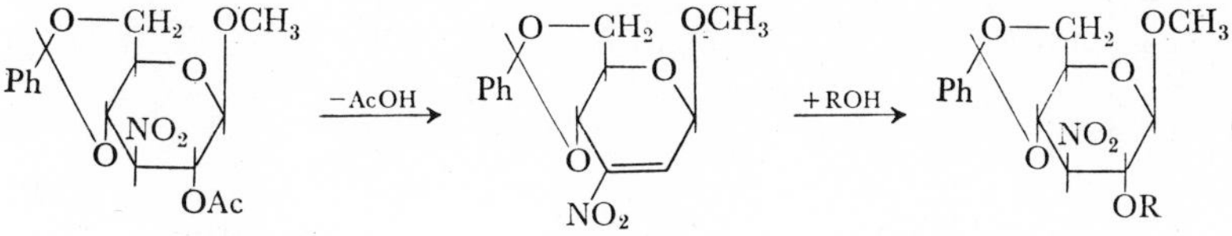

$$(\mathbf{211}) \xrightarrow{-\mathrm{AcOH}} (\mathbf{226}) \xrightarrow{+\mathrm{ROH}} (\mathbf{237}),\ \mathrm{R} = \mathrm{CH_3}\ \text{or}\ \mathrm{C_2H_5} \qquad (176)$$

of **237** were formed in this elimination–addition. The stereoselectivity conforms with that observed in the alkoxide-catalyzed alcohol addition to **226** (see section VI.A). Another point of interest is the stability toward sodium acetate in refluxing alcohol, of the benzylidene linkage adjacent to the nitro group; as discussed earlier (section V.A. 2) this linkage is quite prone to cleavage under somewhat more strongly basic conditions. A mechanism for the action of sodium acetate (and also of sodium alkanenitronates which act similarly) upon nitro esters has been advanced by Feuer and coworkers who examined the use of such esters as substitutes for α-nitro olefins in Michael additions[350,305] (see section III.D) and Diels–Alder reactions[379] (see section IV).

The behavior toward base of a nitroalkyl lactone, namely 3-phthalidylnitromethane (**238**), was studied by Baer and Kienzle[147]. The action of aqueous alkali resulted in an almost instantaneous eliminative lactone opening to give the anion **239** of 2-(2-nitrovinyl)benzoic acid (**240**). This anion, whose formation was revealed by its ultraviolet absorption, was unstable and rapidly added hydroxyl ion to give the dianion **241** of 2-(1-hydroxy-2-nitroethyl)-benzoic acid. The half-life of **239** in 0.01 *N* sodium hydroxide was 2 min, in 1 *M* sodium bicarbonate about 2 hours. Acidification of **241** regenerated **238**, as did acidification of a fresh solution of **239**. In a narrow range around pH 6 an equilibrium in solution between **238** and **240**, in a ratio of about 10:1, was established, but isolation of **240** was not possible (see also section I.B. 6). The action upon **238** of sodium methoxide in methanol followed by acidification led to 2-(1-methoxy-2-nitroethyl)benzoic acid (**242**) (equation 177).

$CH—CH_2NO_2$ (lactone) (**238**) $\xrightarrow[\text{(NaOH or NaOMe)}]{-H^+}$ $CH^-—CHNO_2$ (lactone)

$CH{=}CHNO_2$, CO_2H (**240**)

$+2H^+$ $-H_2O$; H^+ ; $+OH^-$

$CH—CHNO_2^-$, OH, CO_2^- (**241**)

$CH{=}CHNO_2$, CO_2^- (**239**) (177)

$+OCH_3^-$ $+2H^+$

OCH_3, $CH—CH_2NO_2$, CO_2H (**242**)

2-Nitroalkyl esters also behave as potential α-nitro olefins in reactions with thiols[470], sodium hydrogen sulfite[471] or sodium sulfite[472], potassium cyanide[473], and benzyl cyanide[474], which lead, respectively, to 2-nitroalkyl sulfides, 2-nitroalkanesulfonic acids, 2-nitroalkyl cyanides, and 3-nitro-1-phenylalkyl cyanides (see section VI.B).

2-Bromo-2,2-dinitroethyl acetate (**243**) undergoes extremely facile reactions with bases, including relatively weak ones, and good evidence for the intermediate formation of 1,1-dinitroethylene has been adduced[218,475]. Thus, in an attempt to dehalogenate the bromo ester **243** with potassium iodide, the expected potassium salt of 2,2-dinitroethyl acetate (**244**) could not be isolated but underwent, in part, elimination of acetate to give intermediate 1,1-dinitroethylene, which then added surviving **244** to form a Michael adduct, potassium 2,2,4,4-tetranitrobutyl acetate[218] (**245**) (equation 178).

$$2\ BrC(NO_2)_2CH_2OC(=O)CH_3\ (\mathbf{243}) + 4\ KI \longrightarrow 2\ C(=NO_2K)(NO_2)CH_2OC(=O)CH_3\ (\mathbf{244}) + 2\ KBr + 2\ I_2$$

$$\mathbf{244} \xrightarrow{-KOCOCH_3} C(NO_2)_2{=}CH_2 + \mathbf{244} \longrightarrow C(=NO_2K)(NO_2)CH_2{-}C(NO_2)_2{-}CH_2OC(=O)CH_3\ (\mathbf{245}) \qquad (178)$$

When the ester **243** was treated[218,475] with phthalimide sodium (NaA), a mixture of 1,1-dinitro-2-phthalimidoethane sodium salt (**246**) and 1-bromo-1,1-dinitro-2-phthalimidoethane (**247**) was produced. The mechanism shown in equations 179–182 was postulated[475].

$$BrC(NO_2)_2CH_2OCOCH_3\ (\mathbf{243}) + NaA \rightleftharpoons C(=NO_2Na)(NO_2)CH_2OCOCH_3\ (\mathbf{244a}) + BrA \qquad (179)$$

$$\mathbf{244a} \longrightarrow (O_2N)_2C{=}CH_2 + CH_3CO_2Na \qquad (180)$$

$$(O_2N)_2C{=}CH_2 + NaA \longrightarrow NaO_2N{=}C(NO_2)CH_2A\ (\mathbf{246}) \qquad (181)$$

$$\mathbf{246} + BrA \longrightarrow BrC(NO_2)_2CH_2A\ (\mathbf{247}) + NaA \qquad (182)$$

$$A = N(CO)_2C_6H_4 \ \text{(phthalimido)}$$

With absolute methanol at room temperature, 2-bromo-2,2-dinitroethyl acetate formed 1-bromo-1,1-dinitro-2-methoxyethane (**248**) in 72% yield, the same mechanism has been invoked. The ether **248** and the corresponding alcohol **249** were shown to be easily interconvertible by the action of water and methanol, respectively[475] (equation 183).

$$\underset{(\mathbf{248})}{\mathrm{BrC(NO_2)_2CH_2OCH_3}} \underset{+\mathrm{Br^+}}{\overset{-\mathrm{Br^+}}{\rightleftharpoons}} {}^{-}\mathrm{C(NO_2)_2CH_2OCH_3} \underset{+\mathrm{CH_3O^-}}{\overset{-\mathrm{CH_3O^-}}{\rightleftharpoons}} \mathrm{(NO_2)_2C{=}CH_2} \underset{-\mathrm{OH^-}}{\overset{+\mathrm{OH^-}}{\rightleftharpoons}} {}^{-}\mathrm{C(NO_2)_2CH_2OH} \underset{-\mathrm{Br^+}}{\overset{+\mathrm{Br^+}}{\rightleftharpoons}} \underset{(\mathbf{249})}{\mathrm{BrC(NO_2)_2CH_2OH}} \quad (183)$$

In similar investigations it was found that 1,2-dichloro-1,1-dinitroethane, on being dehalogenated with potassium iodide in alcohols, forms 2,2-dinitroethyl alkyl ethers[404] (equation 184).

$$\mathrm{CCl(NO_2)_2CH_2Cl + 2KI \longrightarrow KC(NO_2)_2CH_2Cl + KCl + I_2 \longrightarrow}$$

$$\mathrm{(O_2N)_2C{=}CH_2 + KCl \xrightarrow{ROH} (O_2N)_2CHCH_2OR} \quad (184)$$

Deacylation of β-nitro esters to give the parent nitro alcohol rather than the olefin has been accomplished by p-toluenesulfonic acid-catalyzed transesterification with methanol[54].

VI. ADDITIONS OF NUCLEOPHILES TO NITRO OLEFINS

The electron-withdrawing effect of the nitro group in α-nitroalkenes permits facile nucleophilic additions of alcohols, thiols, amines, and related nitrogenous bases across the olefinic double bond. β-Alkylthio-, β-amino-, and similar β-substituted nitroalkanes are thus accessible and may serve as intermediates in syntheses of amino ethers, amino thioethers, and diamines.

A. *Alkoxylation*

It was noted by early workers that β-nitrostyrene and some of its derivatives including β-bromo derivatives easily add methanol or ethanol in the presence of alkali[140,476–482] (equation 185).

$$\mathrm{PhCH{=}CHNO_2 \xrightarrow{CH_3ONa} PhCH(OCH_3)CH{=}NO_2Na \underset{+NaOH}{\overset{+H^+}{\rightleftharpoons}} PhCH(OCH_3)CH_2NO_2} \quad (185)$$

The alkoxylated product may undergo Michael addition with the starting nitro olefin[479] (equation 186).

The alcohol addition to α,β-disubstituted α-nitroethylenes may give rise to diastereoisomers, as has been demonstrated on the example of α-nitrostilbene[483,484]. The isomers give a common anion, by way of which the thermodynamically less stable isomer may be converted into the more stable one[485].

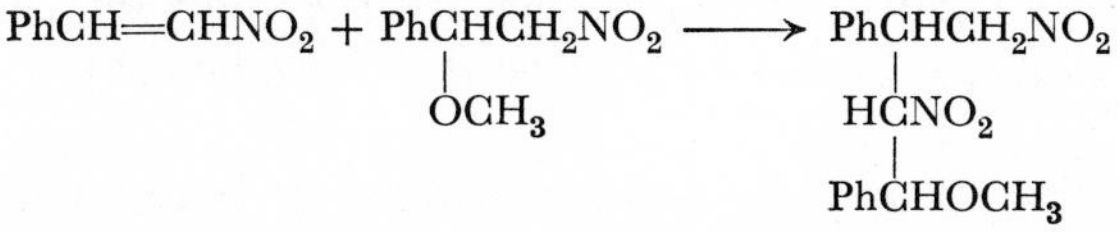

$$PhCH{=}CHNO_2 + PhCH(OCH_3)CH_2NO_2 \longrightarrow PhCHCH_2NO_2{-}HCNO_2{-}PhCHOCH_3 \qquad (186)$$

A large number of β-alkoxynitroalkanes have later been prepared from simple nitro olefins and alcohols[422,486,487].

Addition of alcohols to 3,3,3-trichloro-1-nitropropene occurs by heating alone, even in the absence of a basic catalyst. This has been attributed to the inductive effect of the trichloromethyl group which assists in polarizing the double bond. More than 20 1,1,1-trichloro-2-alkoxy-3-nitropropanes have been obtained in this way[488], but glycolic acid, ethyl glycolate, and glycolonitrile failed to add[489].

Addition of alcohols to certain carbohydrate nitro olefins also occurs with great ease. Thus, methyl 4,6-*O*-benzylidene-2,3-dideoxy-3-nitro-β-D-*erythro*-hex-2-enopyranoside (**250**) in alcoholic solution is alkoxylated rapidly in the cold by catalytic amounts of sodium alkoxide, and the same products **251** arise in the absence of catalyst by short heating[402] (equation 187). The addition to **250** of isopropyl lactate also proceeded well and gave **251** (R = $CH(CH_3)CO_2Pr$-*i*)[490].

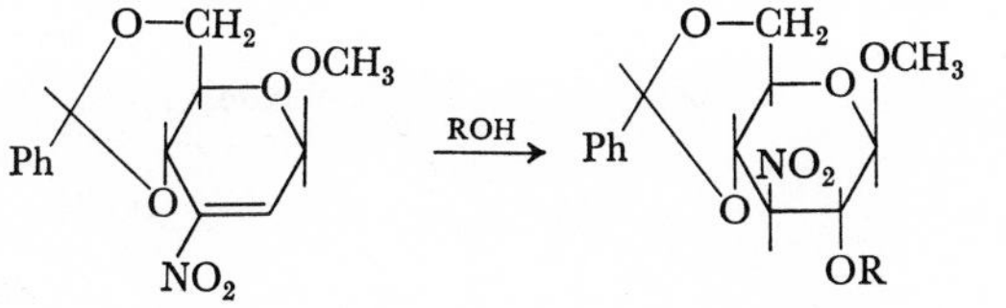

O—CH_2, OCH_3, O, Ph, O, NO_2 **(250)** $\xrightarrow{ROH}$ O—CH_2, OCH_3, O, Ph, NO_2, O, OR **(251)** (187)

R = CH_3, Et, or CH_2Ph

Ph, O, O, O, OCH_3, NO_2

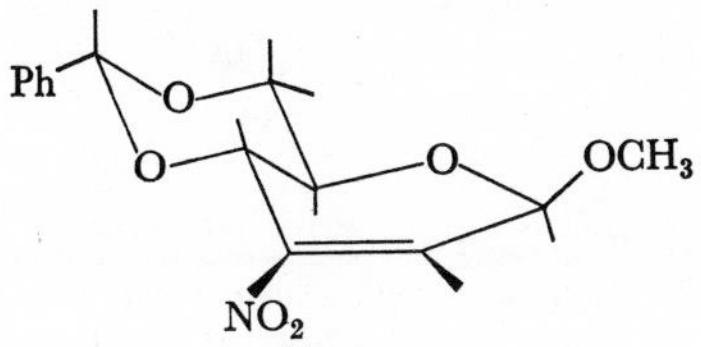

(250a)

The alcohol additions to **250** appear to be exceedingly stereoselective, as yields in excess of 90% have been obtained of the *gluco* derivatives **251**, and no stereoisomers could be isolated or detected. The acetal function represented by the glycosidic center likely is in part responsible, because of its inductive effect, for the great facility of this addition, and the stereochemical course of the reaction appears to be governed by the tendency of the nucleophile to approach, in accordance with Cram's rule, from the less hindered, "lower" side of **250a** and moreover by the tendency of the nitro and alkoxyl groups in **251** to assume the more favorable, equatorial position.

The addition of methoxide ion to sugar derivatives that contain a terminal nitro olefin grouping, has also been investigated. Thus, D-*arabino*-3,4,5,6-tetraacetoxy-1-nitro-1-hexene (**252**) and D-*erythro*-3,4,5-triacetoxy-1-nitro-1-pentene (**253**) gave 1-deoxy-2-*O*-methyl-1-nitro-D-mannitol (**254**) and 1-deoxy-2-*O*-methyl-1-nitro-D-ribitol (**255**), respectively (equations 188 and 189), which could be converted by the Nef reaction into 2-*O*-methyl-D-mannose and 2-*O*-methyl-D-ribose[491]. Again, the preponderant stereoisomers produced were those expected on the basis of Cram's rule as illustrated for the reaction **253** → **255**.

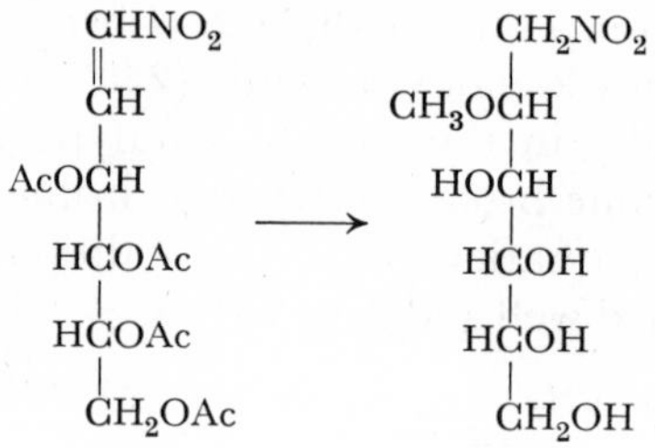

$$\begin{array}{c} CHNO_2 \\ \| \\ CH \\ | \\ AcOCH \\ | \\ HCOAc \\ | \\ HCOAc \\ | \\ CH_2OAc \\ (\mathbf{252}) \end{array} \longrightarrow \begin{array}{c} CH_2NO_2 \\ | \\ CH_3OCH \\ | \\ HOCH \\ | \\ HCOH \\ | \\ HCOH \\ | \\ CH_2OH \\ (\mathbf{254}) \end{array} \qquad (188)$$

$$\begin{array}{c} CHNO_2 \\ \| \\ CH \\ | \\ HCOAc \\ | \\ HCOAc \\ | \\ CH_2OAc \\ (\mathbf{253}) \end{array} \longrightarrow \begin{array}{c} CH_2NO_2 \\ | \\ HCOCH_3 \\ | \\ HCOH \\ | \\ HCOH \\ | \\ CH_2OH \\ (\mathbf{255}) \end{array} \qquad (\mathbf{253}) \longrightarrow (\mathbf{255}) \qquad (189)$$

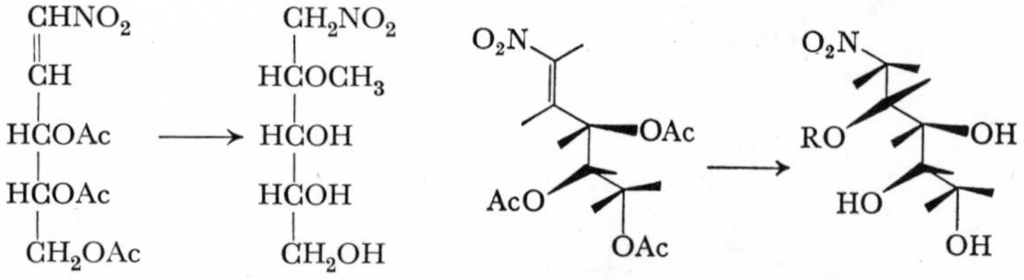

O₂N, OAc, AcO, OAc (**253**) → O₂N, RO, OH, HO, OH (**255**)

At this point should be recalled also the anhydridization of 1-deoxy-1-nitroalditols that was discussed in section I.B. 5a. If this reaction involves, as the authors[88–90] believe, an intermediate dehydration to a polyhydroxy-α-nitroalkene, the ring closure represents an internal alkoxylation of the latter.

Methanol addition in the presence of methoxide to the partially blocked D-glucose derivative **256**, to give **257**, occurred rapidly at room temperature and was followed by the slower loss of the acetyl group at C_3, so that finally a mixture of stereoisomeric 5-*O*-methyl ethers **258** was formed[98] (equation 190).

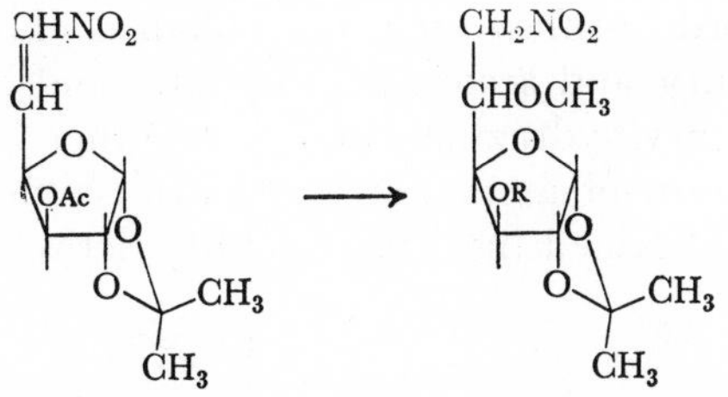

$$\text{(256)}\; CHNO_2{=}CH\text{-furanose(3-OAc, 1,2-}O\text{-isopropylidene)} \longrightarrow CH_2NO_2\text{-}CHOCH_3\text{-furanose(3-OR, 1,2-}O\text{-isopropylidene)} \quad (190)$$

(**256**) (**257**), R = Ac
(**258**), R = H

B. *Addition of Sulfur-Containing Nucleophiles*

Thiols react with α-nitro olefins to give 2-nitroalkyl sulfides. The reaction, which was first studied by Heath and Lambert[470], usually occurs under basic conditions, although instances of addition without catalyst have been observed. The 2-nitroalkyl sulfides can be oxidized with hydrogen peroxide to 2-nitroalkyl sulfones, or reduced with Raney nickel to 2-aminoalkyl sulfides. Oxidation of the latter, as well as reduction of 2-nitroalkyl sulfones, leads to 2-aminoalkyl sulfones. Nitro olefins used included nitroethylene, 1- and 2-nitropropene, and homologs as well as β-nitrostyrene and numerous derivatives; alkanethiols, thiophenols, and thiobenzyl alcohol were employed as addends[470,492–494]. An example in carbohydrate chemistry was reported recently[402].

Hydrogen sulfide adds to α-nitroalkenes without a catalyst. The resulting 2-nitroalkylthiol then may add another molecule of α-nitroalkene giving a bis(2-nitroalkyl)sulfide[470]. Sodium or potassium hydrogen sulfite combines with α-nitroalkenes to yield 2-nitroalkanesulfonates (equation 191), which can be catalytically reduced to 2-aminoalkanesulfonates[471,472].

$$O_2NC(R){=}C(R)_2 \xrightarrow{NaHSO_3} O_2NCH(R)C(R)(R)SO_3Na \quad (191)$$

R = H or alkyl

Arylsulfinic acids also add to nitroalkenes, which provides another route to nitro sulfones[470,493] (equation 192).

$$O_2NCH{=}C(R)_2 + HO_2SPh \longrightarrow O_2NCH_2CR_2SO_2Ph \quad (192)$$

R = H or aryl

C. Addition of Ammonia, Amines, and Other Nitrogenous Bases

Nitrogenous bases readily add to the double bond of α-nitroalkenes to afford β-substituted nitroalkanes. The bases include ammonia, primary and secondary aliphatic and aromatic amines, hydroxylamine, arylhydrazines, and some other hydrazine derivatives. The addition of ammonia to β-nitrostyrene was studied by Worrall[495] who obtained bis(2-nitro-1-phenylethyl)amine (equation 193).

$$2\,PhCH{=}CHNO_2 + NH_3 \longrightarrow PhCH(CH_2NO_2){-}NH{-}CH(CH_2NO_2)Ph \qquad (193)$$

The action of ammonia upon α-nitrostilbenes resulted in the formation of isoxazoline oxides **259** and diaroylarylmethane monoximes **260**, which could be converted into isoxazoles **261**[338] (equation 194).

$$ArCH{=}C(NO_2)Ar \xrightarrow{NH_3} [ArCH(NH_2)CH(NO_2)Ar] \longrightarrow ArCH{=}NH + O_2NCH_2Ar$$

$$ArCH{=}C(NO_2)Ar + O_2NCH_2Ar \longrightarrow ArCHNO_2{-}CH(Ar){-}ArCHNO_2 \xrightarrow{-HNO_2} ArCH{=}C(Ar){-}ArCHNO_2 \rightleftharpoons$$

$$ArCH{=}C(Ar){-}C(Ar){=}N(\rightarrow O)OH \longrightarrow \underset{(\mathbf{259})}{ArCH{-}O{-}N(\rightarrow O){=}C(Ar){-}CH(Ar)}$$

$$\downarrow \qquad\qquad \downarrow \qquad\qquad (194)$$

$$\underset{(\mathbf{260})}{ArC(OH){=}C(Ar){-}C(Ar){=}NOH} \xrightarrow[\text{or acid}]{\text{alkali}} \underset{(\mathbf{261})}{ArC{-}O{-}N{=}C(Ar){-}C(Ar)\ (\text{ring: }ArC{=}CAr)}$$

Later on, Heath and Rose[463] produced 1-nitro-2-aminopropane, 1-nitro-2-amino-2-methylpropane, and 2-nitro-3-aminobutane from 1-nitropropene, 1-nitro-2-methylpropene, and 2-nitro-2-butene, respectively (equation 195). Some α-bromo-α-nitroalkylenes have been found to add ammonia in the same fashion[486].

The reaction has more recently been employed in the synthesis of amino sugars. Thus, *N*-acetyl-D-mannosamine (**264**) can be prepared by the interaction of ammonia with D-*arabino*-3,4,5,6-tetraacetoxy-1-nitro-1-hexene (**262**). The introduction of an amino group at C_2 is accompanied by *N*-acetylation and de-*O*-acetylation. The

$$O_2N{-}\underset{R}{C}{=}\underset{R}{C}{-}CH_3 \xrightarrow{NH_3} O_2N{-}\underset{R}{CH}{-}\overset{NH_2}{\underset{R}{CH}}{-}CH_3 \qquad (195)$$

$$R = H \text{ or } CH_3$$

preponderant product, 2-acetamido-1,2-dideoxy-1-nitro-D-mannitol (**263**) was subsequently subjected to a Nef reaction giving **264** (equation 195)[496–498]. The same reaction sequence when applied to the D-*xylo* isomer of **262** furnished chiefly D-gulosamine (**265**) which was isolated as its hydrochloride[499]. In both cases the configuration of the preponderant product can be predicted by invoking Cram's rule as shown for the addition of methoxide to **253** (equation 189). D-Allosamine and D-altrosamine were synthesized in analogous fashion[499a].

$$\begin{array}{c} HCNO_2 \\ \| \\ CH \\ | \\ AcOCH \\ | \\ HCOAc \\ | \\ HCOAc \\ | \\ CH_2OAc \\ (\mathbf{262}) \end{array} \xrightarrow{NH_3} \begin{array}{c} H_2CNO_2 \\ | \\ AcHNCH \\ | \\ HOCH \\ | \\ HCOH \\ | \\ HCOH \\ | \\ CH_2OH \\ (\mathbf{263}) \end{array} \xrightarrow[2.\ H_2SO_4]{1.\ NaOH} \begin{array}{c} CHO \\ | \\ AcHNCH \\ | \\ HOCH \\ | \\ HCOH \\ | \\ HCOH \\ | \\ CH_2OH \\ (\mathbf{264}) \end{array} \qquad (196)$$

$$\begin{array}{c} CHO \\ | \\ HCNH_2 \\ | \\ HCOH \\ | \\ HOCH \\ | \\ HCOH \\ | \\ CH_2OH \\ (\mathbf{265}) \end{array}$$

Paulsen[459] added ammonia to 3-*O*-acetyl-1,2-*O*-cyclohexylidene-5,6-dideoxy-6-nitro-α-D-*xylo*-hex-5-enofuranose (**266**) in order to synthesize 5-amino-6-nitro and thence 5,6-diamino sugar derivatives.

A synthesis of 2,3-diamino-2,3-dideoxy-D-glucose (**268**) was developed by Baer and Neilson[445] who added ammonia to the

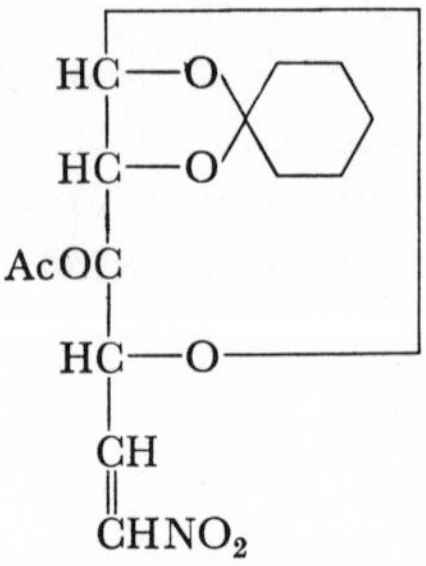

(**266**) 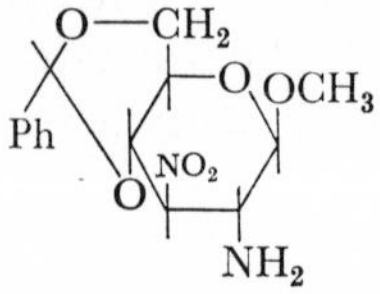 (**267**) 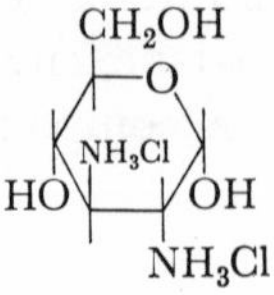 (**268**)

olefin **250**. About 90% of the addition product was methyl 2-amino-4,6-*O*-benzylidene-2,3-dideoxy-3-nitro-β-D-glucopyranoside (**267**) which was subsequently converted into **268** in a number of steps. A minor stereoisomer of **267** that was produced in the ammonia addition was later shown to have the *manno* configuration[8]. The strong preponderance of the *gluco* configuration, in which the nitro and amino groups are equatorially oriented, is in line with the stereochemical course observed in the additions of alcohols to **250** (see section VI.A).

The addition of aromatic amines to β-nitrostyrene and many of its derivatives has been studied in great detail by Worrall[495,500–504]. He found that the capacity of this reaction to take place depends on the structure of the reactants. A nitro group attached to the ring of β-nitrostyrene generally increases the reactivity, and particularly reactive were the 4-chloro-2-nitro and 2-chloro-4-nitro derivatives. On the other hand hydroxyl, methoxyl, and methylenedioxy groups in the ring reduced or abolished the reactivity, and a methyl or phenyl group at the β-carbon atom of the side chain had a similar effect. β-Bromo-β-nitrostyrene and its 2-chloro-4-nitro derivative gave the expected addition products with *p*-toluidine. These were very sensitive, however, and easily decomposed to bromonitromethane and *N*-arylidene-*p*-toluidines (equation 197). Such decomposition was even more pronounced when aniline and some other aromatic amines were used, so that in these cases no pure 2-arylamino-2-aryl-1-bromo-nitroethanes could be isolated.

$$ArCH{=}CBrNO_2 + Ar'NH_2 \longrightarrow ArCH(NHAr')CHBrNO_2 \longrightarrow ArCH{=}NAr' + CH_2BrNO_2 \quad (197)$$

Arylhydrazines also have been found to react with several β-nitrostyrene derivatives. The arylhydrazino adducts formed eliminate a molecule of nitromethane quite readily and give the corresponding arylhydrazones[502,503,505] (equation 198).

$$ArCH{=}CHNO_2 + H_2NNHAr' \longrightarrow ArCH(NHNHAr')CH_2NO_2 \longrightarrow ArCH{=}NNHAr' + CH_3NO_2 \quad (198)$$

Semicarbazide and thiosemicarbazide have been reported to add to β-nitrostyrene to give products of the type $O_2NCH_2CH(C_6H_5)$-$NHNHCONH_2$[495].

The reaction of aliphatic and aromatic amines with α-nitrostilbene was studied by Dornow and Boberg[506]. They obtained stable 1-arylamino-2-nitro-1,2-diphenylethanes with aniline and *p*-toluidine; the same compounds as well as some related ones were produced by the addition of phenylnitromethane to Schiff bases (equation 199).

$$PhC(NO_2){=}CHPh \xrightarrow{H_2NAr} PhCH(NO_2)CH(Ph)(NHAr) \xleftarrow{PhCH_2NO_2} PhCH{=}NAr \quad (199)$$

The analogous adducts of aliphatic amines could not be isolated (with the exception of a rather unstable piperidino derivative) since formation of triphenylisoxazoline oxide (and some triphenylisoxazole) took place, indicating that the amines behaved like ammonia (equation 194).

The reaction between nitroethylene and aniline was first investigated by Wieland and Sakellarios who obtained 1-nitro-2-phenylaminoethane[507]. Later, the reaction was extended to include homologous aliphatic α-nitro olefins as well as different aromatic and aliphatic amines.[463,486,508,509] Reactions were found to proceed rapidly but the yields were variable, which was attributed to the instability of the 1,2-nitroamines formed, particularly when aliphatic amines were employed. Isolation of the nitroamines in the form of hydrochlorides enhances their stability.

Crystalline adducts of *p*-toluidine with several homologous 2-nitro-1-alkenes were obtained in excellent yields[228].

Sowden and coworkers[510] studied the addition of *p*-toluidine, benzylamine, cyclohexylamine, cycloheptylamine, isopropylamine, and ethanolamine to D-*arabino*-3,4,5,6-tetraacetoxy-1-nitro-1-hexene (**262**). Of the two stereoisomeric adducts **269** possible in each case, only one was isolated, in yields ranging from 44 to 75%, and the configurations were not elucidated. Addition of aniline to **262**, however, gave two stereoisomers in yields of 28 and 44%, respectively. No de-*O*-acetylation or O → N acyl migration occurred in these amine additions, in contrast to the reaction with ammonia mentioned above. In attempts to *N*-acetylate the aniline adducts with acetic anhydride in pyridine an unexpected dehydration took place which produced 3,4,5,6-tetra-*O*-acetyl-2-deoxy-2-(*N*-phenylimino)-D-*arabino*-hexononitrile (**270**) from either adduct. Treatment of **270** with aqueous sodium hydroxide led to the displacement of cyanide ion and the formation of D-arabinonic acid anilide **271** (equation 200, for **262** → **269**: R = *p*-$CH_3C_6H_4$, CH_2Ph, C_6H_{11}-*c*, C_7H_{13}-*c*, CH_2CH_2OH, $CH(CH_3)_2$, and Ph; for **269** → **270** → **271**: R = Ph).

$$\mathbf{262}\xrightarrow{RNH_2}\begin{array}{c}CH_2NO_2\\ |\\ CH(NHR)\\ |\\ AcOCH\\ |\\ HCOAc\\ |\\ HCOAc\\ |\\ CH_2OAc\\ (\mathbf{269})\end{array}\xrightarrow[py]{Ac_2O}\begin{array}{c}CN\\ |\\ C{=}NR\\ |\\ AcOCH\\ |\\ HCOAc\\ |\\ HCOAc\\ |\\ CH_2OAc\\ (\mathbf{270})\end{array}\xrightarrow{aq\ NaOH}\begin{array}{c}O{=}CNHR\\ |\\ HOCH\\ |\\ HCOH\\ |\\ HCOH\\ |\\ CH_2OH\\ \\ (\mathbf{271})\end{array}\qquad(200)$$

One of the earliest nucleophilic additions to nitro olefins was that of hydroxylamine to β-nitrostyrene giving *N*-(1-phenyl-2-nitroethyl)hydroxylamine[511]. Analogous adducts were obtained with 1-(2-furyl)-2-nitroethylene, 1-nitro-1-propene, and 1-nitro-1-butene, and it was found that the stability of the products RCH(NHOH)-CH_2NO_2 decreased as R was varied from phenyl to 2-furyl to alkyl[512].

The behavior of α,β-dinitro olefins toward bases has not been investigated extensively. Clapp and coworkers[513] reported that 2,3-dinitro-2-butene and 3,4-dinitro-3-hexene react with ammonia and amines (aniline, *p*-phenylenediamine) under loss of nitrous acid to give nitroimines (equation 201).

$$RC(NO_2){=}C(NO_2)R\xrightarrow[-HNO_2]{+H_2NR'}R{-}\overset{O_2N}{\overset{|}{C}}{=}\overset{NHR'}{\overset{|}{C}}{-}R\longrightarrow R{-}\overset{O_2N}{\overset{|}{C}H}{-}\overset{NR'}{\overset{\|}{C}}{-}R\qquad(201)$$

$$R = CH_3 \text{ or } C_2H_5;\ R' = H,\ Ph,\ \text{or } p\text{-}NH_2C_6H_4$$

N,N'-Dinitroethylenediamine has been found to add two molecules of 1-nitro-1-butene (equation 202) or 1-nitro-1-pentene[514], and similarly, two molecules of methyl vinyl ketone[515] (equation 203).

3,3,5,5-Tetranitropiperidine was added across the olefinic bond in some 1-nitroalkenes and α,β-unsaturated ketones[516].

$$O_2NNHCH_2CH_2NHNO_2 + 2\ O_2NCH{=}CHCH_2CH_3 \longrightarrow CH_3CH_2CH(CH_2NO_2)N(NO_2)CH_2CH_2N(NO_2)CH(CH_2NO_2)CH_2CH_3 \quad (202)$$

$$O_2NNHCH_2CH_2NHNO_2 + 2\ CH_2{=}CHCOCH_3 \longrightarrow CH_3COCH_2CH_2N(NO_2)CH_2CH_2N(NO_2)CH_2CH_2COCH_3 \quad (203)$$

VII. REFERENCES

1. L. Henry, *Compt. Rend.*, **120**, 1265 (1895).
2. L. Henry, *Bull. Soc. Chim. France*, **13**, 999 (1895).
3. H. B. Hass and E. F. Riley, *Chem. Rev.*, **32**, 373 (1943).
4. N. Levy and J. D. Rose, *Quart. Rev.*, **1**, 358 (1947).
5. G. A. Shvekhgeimer, N. F. Piatakov, and S. S. Novikov, *Usp. Khim.*, **28**, 484 (1959). (a) V. V. Perekalin, Nepredel'nye nitrosoedineniya, Goskhimizdat, Leningrad, 1961; English translation by L. Mandel, Unsaturated nitro compounds, Israel Program for Scientific Translations Ltd., Jerusalem, 1964.—Second, revised edition by V. V. Perekalin and A. S. Sopova, Moscow and Leningrad, 1966.
6. J. C. Sowden, *Advan. Carbohydrate Chem.*, **6**, 291 (1951).
7. F. W. Lichtenthaler, *Angew. Chem.*, **76**, 84 (1964).
8. H. H. Baer, *Advan. Carbohydrate Chem.*, **24**, 67 (1969).
9. P. Noble, Jr., F. G. Borgardt, and W. L. Reed, *Chem. Rev.*, **64**, 19 (1964).
10. B. M. Vanderbilt and H. B. Hass, *Ind. Eng. Chem.*, **32**, 34 (1940).
11. H. B. Hass and B. M. Vanderbilt, U.S. Pat. 2,139,120 (1938); *Chem. Abstr.*, **33**, 2149 (1939).
12. H. B. Hass and W. R. McElroy, U.S. Pat. 2,387,019 (1945); *Chem. Abstr.*, **40**, 1171 (1946).
13. F. J. Villani and F. F. Nord, *J. Am. Chem. Soc.*, **69**, 2608 (1947).
14. M. J. Astle and F. P. Abbott, *J. Org. Chem.*, **21**, 1228 (1956).
15. C. J. Schmidle and R. C. Mansfield, *Ind. Eng. Chem.*, **44**, 1388 (1952).
16. L. F. Fieser and M. Gates, *J. Am. Chem. Soc.*, **68**, 2249 (1946).
17. T. Urbański and B. Chylinska, *Rocz. Chem.*, **31**, 695 (1957).
18. C. B. Gairaud and G. R. Lappin, *J. Org. Chem.*, **18**, 1 (1953).
19. S. Byrdy, Z. Eckstein, and J. Plenkiewicz, *Bull. Acad. Pol. Sci., Sér. Sci. Chim.*, **9**, 627 (1961).
20. J. Kamlet, U.S. Pat. 2,151,517 (1939); *Chem. Abstr.*, **33**, 5003 (1939).
21. J. Bourland and H. B. Hass, *J. Org. Chem.*, **12**, 704 (1947).
22. L. Henry, *Compt. Rend.*, **121**, 210 (1895).
23. I. M. Gorskiĭ and S. P. Makarov, *Chem. Ber.*, **67**, 996 (1934); *Zh. Obshch. Khim.*, **4**, 1008 (1934).
24. Tanabe Chemical Industries, Japan Pat. 156,256 (1943); *Chem. Abstr.*, **44**, 2008 (1950).

25. G. Darzens, *Compt. Rend.*, **229**, 1148 (1949).
26. H. P. Otter, *Rec. Trav. Chim.*, **57**, 13 (1938).
27. R. F. B. Cox, U.S. Pat. 2,301,259 (1942); *Chem. Abstr.*, **37**, 2017 (1943).
28. G. Darzens, *Compt. Rend.*, **225**, 942 (1947).
29. K. Namba, S. Iizuka, and M. Yoneno, *J. Chem. Soc. Japan, Ind. Chem. Sect.*, **66**, 1446 (1963); *Chem. Abstr.*, **60**, 11,886 (1964).
30. J. A. Wyler, U.S. Pat. 2,231,403 (1941); *Chem. Abstr.*, **35**, 3265 (1941).
31. E. Schmidt and R. Wilkendorf, *Chem. Ber.*, **52**, 398 (1919).
32. E. C. S. Jones and J. Kenner, *J. Chem. Soc.*, **1930**, 919.
33. E. Schmidt, A. Ascherl, and L. Mayer, *Chem. Ber.*, **58**, 2430 (1925).
34. D. V. Nightingale and J. R. James, *J. Am. Chem. Soc.*, **66**, 352 (1944).
35. L. Bouveault and A. Wahl, *Compt. Rend.*, **134**, 1226 (1902).
36. S. Kanao, *J. Pharm. Soc. Japan*, **50**, 24 (1930); *Chem. Abstr.*, **24**, 2856 (1930).
37. H. Cerf de Mauney, *Bull. Soc. Chim. France*, **7**, 133 (1940).
38. C. A. Sprang and E. F. Degering, *J. Am. Chem. Soc.*, **64**, 1063 (1942).
39. W. Sobótka, Z. Eckstein, and T. Urbański, *Bull. Acad. Pol. Sci., Sér. Sci. Chim. Géol. Géograph.*, **5**, 653 (1957).
40. C. A. Sprang and E. F Degering, *J. Am. Chem. Soc.*, **64**, 1735 (1942).
41. G. D. Buckley and J. L. Charlish, *J. Chem. Soc.*, **1947**, 1472.
42. Z. Eckstein and T. Urbański, *Rocz. Chem.*, **26**, 571 (1952).
43. T. Urbański, Z. Eckstein, and W. Sobótka, *Rocz. Chem.*, **29**, 399 (1955).
44. C. A. Sprang and E. F. Degering, *J. Am. Chem. Soc.*, **65**, 628 (1943).
45. W. Charlton and J. Kenner, *J. Chem. Soc.*, **1932**, 750.
46. O. von Schickh, *Angew. Chem.*, **62**, 547 (1950).
47. J. K. N. Jones, *J. Chem. Soc.*, **1954**, 3643.
48. J. Pauwels, *Bull. Acad. Roy. Belg.*, [3], **34**, 645 (1897).
49. T. Mousset, *Rec. Trav. Chim.*, **21**, 95 (1902).
50. A. Shaw, *Bull. Acad. Roy. Belg.* [3], **34**, 1019 (1897).
51. T. Urbański, H. Dabrowska, B. Lesiowska, and H. Piotrowska, *Rocz. Chem.*, **31**, 687 (1957).
52. Z. Eckstein, E. Grochowski, and T. Urbański, *Rocz. Chem.*, **34**, 931 (1960); *Bull. Acad. Pol. Sci., Sér. Sci. Chim. Géol. Géograph.*, **7**, 289 (1959).
53. Z. Eckstein, A. Sacha, and T. Urbański, *Tetrahedron*, **16**, 30 (1961).
54. Z. Eckstein, E. Grochowski, R. Kowalik, and T. Urbański, *Bull. Acad. Pol. Sci., Sér. Sci. Chim.*, **11**, 687 (1963).
55. H. Plaut, U.S. Pat. 2,616,923 (1952); *Chem. Abstr.*, **49**, 11,701 (1955).
56. S. S. Novikov, I. S. Korsakova, and K. K. Babievskii, *Izv. Akad. Nauk SSSR, Otd. Khim. Nauk*, **1960**, 944; English translation in *Bull. Acad. Sci. USSR, Div. Chem. Sci.*, **1960**, 882.
57. F. I. Carroll, *J. Org. Chem.*, **31**, 366 (1966).
58. F. W. Lichtenthaler and H. O. L. Fischer, *J. Am. Chem. Soc.*, **83**, 2005 (1961).
59. H. O. L. Fischer, E. Baer, and H. Nidecker, *Helv. Chim. Acta*, **18**, 1079 (1935).
60. G. E. McCasland, T. J. Matchett, and M. Hollander, *J. Am. Chem. Soc.*, **74**, 3429 (1952).
61. F. W. Lichtenthaler, *Angew. Chem.*, **73**, 654 (1961).
62. F. W. Lichtenthaler, *Chem. Ber.*, **96**, 845 (1963).
63. F. W. Lichtenthaler, H. Leinert, and H. K. Yahya, *Z. Naturforsch.*, **21b**, 1004 (1966).
64. B. Achmatowicz, M.Sc. Thesis, University of Ottawa, 1963; H. H. Baer, *Tetrahedron*, **20**, Suppl. 1, 263 (1964).
65. A. Pictet and A. Barbier, *Helv. Chim. Acta*, **4**, 924 (1921).
66. J. C. Sowden and H. O. L. Fischer, *J. Am. Chem. Soc.*, **66**, 1312 (1944).

67. J. C. Sowden and H. O. L. Fischer, *J. Am. Chem. Soc.*, **67**, 1713 (1945).
68. J. C. Sowden and H. O. L. Fischer, *J. Am. Chem. Soc.*, **68**, 1511 (1946).
69. J. C. Sowden and H. O. L. Fischer, U.S. Pat. 2,480,785 (1949); *Chem. Abstr.*, **44**, 656 (1950).
70. J. C. Sowden, *J. Am. Chem. Soc.*, **71**, 1897 (1949).
71. J. C. Sowden, *J. Am. Chem. Soc.*, **72**, 808 (1950).
72. J. C. Sowden and H. O. L. Fischer, *J. Am. Chem. Soc.*, **69**, 1963 (1947).
73. J. C. Sowden and R. Schaffer, *J. Am. Chem. Soc.*, **73**, 4662 (1951).
74. J. C. Sowden and R. R. Thompson, *J. Am. Chem. Soc.*, **77**, 3160 (1955).
75. J. C. Sowden and R. R. Thompson, *J. Am. Chem. Soc.*, **80**, 2236 (1958).
76. J. Yoshimura and H. Ando, *J. Chem. Soc. Japan*, **85**, 138 (1964); *Chem. Abstr.*, **61**, 16,140 (1964).
77. J. V. Karabinos and C. S. Hudson, *J. Am. Chem. Soc.*, **75**, 4324 (1953).
78. J. C. Sowden and D. R. Strobach, *J. Am. Chem. Soc.*, **82**, 954 (1960).
79. J. C. Sowden and D. R. Strobach, *J. Am. Chem. Soc.*, **82**, 956 (1960).
80. R. K. Hulyalkar, J. K. N. Jones, and M. B. Perry, *Can. J. Chem.*, **41**, 1490 (1963).
81. J. C. Sowden, *Science*, **109**, 229 (1949); *J. Biol. Chem.*, **180**, 55 (1949).
82. M. Gibbs, *J. Am. Chem. Soc.*, **72**, 3964 (1950).
83. J. O. Lampen, H. Gest, and J. C. Sowden, *J. Bacteriol.*, **61**, 97 (1951).
84. J. C. Sowden and H. O. L. Fischer, *J. Am. Chem. Soc.*, **69**, 1048 (1947).
85. W. W. Zorbach and A. P. Ollapally, *J. Org. Chem.*, **29**, 1790 (1964).
86. J. C. Sowden, *J. Am. Chem. Soc.*, **72**, 3325 (1950).
87. J. Cologne and P. Corbet, *Bull. Soc. Chim. France*, **1960**, 283.
88. J. C. Sowden and M. L. Oftedahl, *J. Org. Chem.*, **26**, 1974 (1961).
89. L. Hough and S. H. Shute, *J. Chem. Soc.*, **1962**, 4633.
90. J. C. Sowden, C. H. Bowers, L. Hough, and S. H. Shute, *Chem. Ind.* (London), **1962**, 1827.
91. J. M. Grosheintz and H. O. L. Fischer, *J. Am. Chem. Soc.*, **70**, 1476 (1948).
92. J. M. Grosheintz and H. O. L. Fischer, *J. Am. Chem. Soc.*, **70**, 1479 (1948).
93. T. Posternak, W. H. Schopfer, and R. Huguenin, *Helv. Chim. Acta*, **40**, 1875 (1957).
94. F. W. Lichtenthaler, *Chem. Ber.*, **94**, 3071 (1961).
95. G. I. Drummond, J. N. Aronson, and L. Anderson, *J. Org. Chem.*, **26**, 1601 (1961).
96. V. Brocca and A. Dansi, *Ann. Chim. (Rome)*, **44**, 120 (1954).
97. F. W. Lichtenthaler, *Angew. Chem.*, **75**, 93 (1963); *Angew. Chem., Intern. Ed. Engl.*, **1**, 662 (1962).
98. H. H. Baer and W. Rank, *Can. J. Chem.*, **43**, 3330 (1965).
99. B. Iselin and H. O. L. Fischer, *J. Am. Chem. Soc.*, **70**, 3946 (1948).
100. H. H. Baer and W. Rank, *Can. J. Chem.*, **43**, 3462 (1965).
101. S. J. Angyal and S. D. Gero, *Aust. J. Chem.*, **18**, 1973 (1965).
102. M. L. Wolfrom, S. M. Olin, and W. J. Polglase, *J. Am. Chem. Soc.*, **72**, 1724 (1950).
    (a) S. S. Novikov, M. S. Burmistrova, and V. P. Gorelik, *Izv. Akad. Nauk SSSR, Otd. Khim. Nauk*, **1960**, 1876; English translation in *Bull. Acad. Sci. USSR, Div. Chem. Sci.*, **1960**, 1748.
    (b) S. S. Novikov, M. S. Burmistrova, V. P. Gorelik, and Y. G. Chkhikvadze, *Izv. Akad. Nauk, SSSR, Otd. Khim. Nauk*, **1961**, 695; English translation in *Bull. Acad. Sci. USSR, Div. Chem. Sci.*, **1961**, 644.
103. H. H. Baer and H. O. L. Fischer, *Proc. Natl. Acad. Sci. U.S.*, **44**, 998 (1958).
104. R. D. Guthrie, *Advan. Carbohydrate Chem.*, **16**, 105 (1961).
105. J. D. Dutcher, *Advan. Carbohydrate Chem.*, **18**, 259 (1963).
106. H. H. Baer and H. O. L. Fischer, *J. Am. Chem. Soc.*, **81**, 5184 (1959).
107. H. H. Baer and A. Ahammad, *Can. J. Chem.*, **41**, 2931 (1963).

107. (a) H. H. Baer and J. Kovář, Unpublished results.
(b) F. Johnson and F. K. Malhotra, *J. Am. Chem. Soc.*, **87,** 5492, 5493 (1965).
(c) R. Caple and W. V. Vaughan, *Tetrahedron Letters*, **1966,** 4067.
108. H. H. Baer and F. Kienzle, *Ann. Chem.*, **695,** 192 (1966).
109. H. H. Baer, *Chem. Ber.*, **93,** 2865 (1960).
110. H. H. Baer, *J. Am. Chem. Soc.*, **83,** 1882 (1961).
111. H. H. Baer, and F. Kienzle, *Can. J. Chem.*, **41,** 1606 (1963).
112. H. H. Baer, and H. O. L. Fischer, *J. Am. Chem. Soc.*, **82,** 3709 (1960).
113. H. H. Baer, *J. Am. Chem. Soc.*, **84,** 83 (1962).
114. A. C. Richardson, *Proc. Chem. Soc.*, **1961,** 255; A. C. Richardson and K. A. McLauchlan, *J. Chem. Soc.*, **1962,** 2499.
115. H. H. Baer and K. Čapek, *Can. J. Chem.*, **47,** 99 (1969).
116. A. C. Richardson, *Proc. Chem. Soc.*, **1961,** 430.
117. A. C. Richardson and H. O. L. Fischer, *J. Am. Chem. Soc.*, **83,** 1132 (1961).
118. H. H. Baer, *J. Org. Chem.*, **28,** 1287 (1963).
119. H. H. Baer, L. D. Hall, and F. Kienzle, *J. Org. Chem.*, **29,** 2014 (1964).
120. G. Baschang, *Ann. Chem.*, **663,** 167 (1963).
121. F. W. Lichtenthaler and H. K. Yahya, *Tetrahedron Letters*, **1965,** 1805.
122. H. H. Baer and A. Ahammad, *Can. J. Chem.*, **44,** 2893 (1966).
123. K. A. Watanabe and J. J. Fox, *Chem. Pharm. Bull.* (Tokyo), **12,** 975 (1964).
124. K. A. Watanabe, J. Beránek, H. A. Friedman, and J. J. Fox, *J. Org. Chem.*, **30,** 2735 (1965).
125. J. Beránek, H. A. Friedman, K. A. Watanabe, and J. J. Fox, *J. Heterocyclic Chem.*, **2,** 188 (1965).
126. F. W. Lichtenthaler, H. P. Albrecht, and G. Olfermann, *Angew. Chem.*, **77,** 131 (1965).
127. F. W. Lichtenthaler and H. P. Albrecht, *Chem. Ber.*, **99,** 575 (1966).
128. H. H. Baer and F. Kienzle, *Can. J. Chem.*, **43,** 3074 (1965).
129. Z. I. Kuznetsova, V. S. Ivanova, and N. N. Shorygina, *Izv. Akad. Nauk SSSR, Otd. Khim. Nauk*, **1962,** 2081; English translation in *Bull. Acad. Sci. USSR, Div. Chem. Sci.*, **1962,** 1995.
130. S. W. Gunner, W. G. Overend, and N. R. Williams, *Chem. Ind.* (London), **1964,** 1523.
131. H. H. Baer and G. V. Rao, *Chem. Ind.* (London), **1965,** 137; *Ann. Chem.*, **686,** 210 (1965).
132. B. Priebs, *Chem. Ber.*, **16,** 2591 (1883); *Ann. Chem.*, **225,** 319 (1884).
133. J. Thiele, *Chem. Ber.*, **32,** 1293 (1899).
134. J. Thiele and S. Haeckel, *Ann. Chem.*, **325,** 1 (1902).
135. M. M. Holleman, *Rec. Trav. Chim.*, **23,** 298 (1904).
136. G. A. Alles, *J. Am. Chem. Soc.*, **54,** 271 (1932).
137. F. W. Hoover and H. B. Hass, *J. Org. Chem.*, **12,** 501 (1947).
138. O. Schales and H. A. Graefe, *J. Am. Chem. Soc.*, **74,** 4486 (1952).
139. Z. Eckstein, T. Kraczkiewicz, A. Sacha, and T. Urbański, *Bull. Acad. Pol. Sci., Sér. Sci. Chim. Géol. Géograph.*, **6,** 313 (1958).
140. K. Rosenmund, *Chem. Ber.*, **46,** 1034 (1913).
141. W. N. Nagai and S. Kanao, *Ann. Chem.*, **470,** 157 (1929).
142. F. W. Hoover and H. B. Hass, *J. Org. Chem.*, **12,** 506 (1947).
143. J. Controulis, M. C. Rebstock, and H. M. Crooks, Jr., *J. Am. Chem. Soc.*, **71,** 2463 (1949).
144. Z. Eckstein, J. Plenkiewicz, and S. Byrdy, *Bull. Acad. Pol. Sci., Sér. Sci. Chim.*, **8,** 623 (1960).

145. F. Heim, *Chem. Ber.*, **44**, 2016 (1911).
146. R. Stewart and L. G. Walker, *Can. J. Chem.*, **35**, 1561 (1957).
147. H. H. Baer and F. Kienzle, *Can. J. Chem.*, **43**, 190 (1965).
148. G. E. Ullyot, J. J. Stehle, C. L. Zirkle, R. L. Shriner, and F. J. Wolf, *J. Org. Chem.*, **10**, 429 (1945).
149. J. Thiele and E. Weitz, *Ann. Chem.*, **377**, 1 (1910).
150. H. H. Baer and B. Achmatowicz, *J. Org. Chem.*, **29**, 3180 (1964).
151. R. D. Campbell and C. L. Pitzer, *J. Org. Chem.*, **24**, 1531 (1959).
152. F. W. Lichtenthaler, *Tetrahedron Letters*, **1963**, 775.
153. L. Henry, *Chem. Ber.*, **30**, 2206 (1897).
154. T. Mousset, *Bull. Acad. Roy. Belg.*, **1901**, 622.
155. J. Maas, *Bull. Acad. Roy. Belg.*, **36**, 294 (1898).
156. R. Wilkendorf and M. Trénel, *Chem. Ber.*, **57**, 306 (1924).
157. E. Schmidt and R. Wilkendorf, *Chem. Ber.*, **55**, 316 (1922).
158. T. Urbański, Z. Eckstein, and H. Wojnowska, *Rocz. Chem.*, **31**, 93 (1957); *Bull. Acad. Pol. Sci., Sér. Sci. Chim. Géol. Géograph.*, **4**, 461 (1956).
159. R. L. Hansche, Brit. Pat. 544,158 (1942); *Chem. Abstr.*, **36**, 6171 (1942); U.S. Pat. 2,298,375 (1942); *Chem. Abstr.*, **37**, 1449 (1943).
160. J. Cologne and G. Lartigan, *Bull. Soc. Chim. France*, **1965**, 738.
161. I. L. Knunyants, L. S. German, and I. N. Rozhkov, *Izv. Akad. Nauk SSSR, Otd. Khim. Nauk*, **1964**, 1946; English translation in *Bull. Acad. Sci. USSR, Div. Chem. Sci.*, **1964**, 1794.
162. F. D. Chattaway and P. Witherington, *J. Chem. Soc.*, **1935**, 1178.
163. S. Malkiel and J. P. Mason, *J. Am. Chem. Soc.*, **64**, 2515 (1942).
164. F. D. Chattaway, J. G. N. Drewitt, and G. D. Parkes, *J. Chem. Soc.*, **1936**, 1294.
165. A. Dornow and G. Wiehler, *Ann. Chem.*, **578**, 113 (1952).
166. D. J. Cook, O. R. Pierce, and E. T. McBee, *J. Am. Chem. Soc.*, **76**, 83 (1954).
167. Z. Eckstein, P. Gluziński, W. Sobótka, and T. Urbański, *J. Chem. Soc.*, **1961**, 1370.
168. G. Fort and A. McLean, *J. Chem. Soc.*, **1948**, 1907.
169. E. F. Degering and C. Sprang, U.S. Pat. 2,332,482 (1943); *Chem. Abstr.*, **38**, 1750 (1944).
170. N. K. Kochetkov and N. V. Dudykina, *Zh. Obshch. Khim.*, **28**, 2399 (1958); English translation in *J. Gen. Chem. USSR*, **28**, 2437 (1958).
171. B. Priebs, *Chem. Ber.*, **18**, 1362 (1885).
172. J. Thiele and H. Lauders, *Ann. Chem.*, **369**, 300 (1909).
173. C. Grundmann and W. Ruske, *Chem. Ber.*, **86**, 939 (1953).
174. S. Kanao, *J. Pharm. Soc. Japan*, **1927**, 1019; *Chem. Abstr.*, **22**, 1588 (1928).
175. W. J. King and F. F. Nord, *J. Org. Chem.*, **14**, 405 (1949).
176. A. M. Simonov and D. D. Dolgatov, *Zh. Obshch. Khim.*, **34**, 3052 (1964); English translation in *J. Gen. Chem. USSR*, **34**, 3088 (1964).
177. A. Dornow and F. Boberg, *Ann. Chem.*, **578**, 101 (1952).
178. C. A. Grob and W. von Tscharner, *Helv. Chim. Acta*, **33**, 1070 (1950).
179. Y. V. Baskov and V. V. Perekalin, *Dokl. Akad. Nauk SSSR*, **136**, 1075 (1961); English translation in *Proc. Acad. Sci, USSR, Chem. Sect.*, **136**, 169 (1961).
180. D. I. Weisblat and D. A. Lyttle, U.S. Pat. 2,570,297 (1951); *Chem. Abstr.*, **46**, 5077 (1952).
181. A. Dornow and A. Frese, *Ann. Chem.*, **581**, 211 (1953).
182. A. Dornow and A. Frese, *Ann. Chem.*, **578**, 122 (1952).
183. K. K. Babievskii, V. M. Belikov, and N. A. Tikhonova, *Dokl. Akad. Nauk SSSR*, **160** 103, (1965); English translation in *Dokl. Chem.*, **160**, 5 (1965).
(a) S. Umezawa and S. Zen, *Bull. Chem. Soc. Japan*, **36**, 1143 (1963).

183. (b) S. Zen, Y. Takeda, A. Yasuda, and S. Umezawa, *Bull. Chem. Soc. Japan*, **40**, 431 (1967).
184. A. Dornow and W. Sassenberg, *Ann. Chem.*, **602**, 14 (1957).
185. H. B. Fraser and G. A. R. Kon, *J. Chem. Soc.*, **1934**, 604.
186. H. J. Dauben, Jr., H. J. Ringold, R. H. Wade, and A. G. Anderson, *J. Am. Chem. Soc.*, **73**, 2359 (1951).
187. T. F. Wood and R. J. Cadorin, *J. Am. Chem. Soc.*, **73**, 5504 (1951).
188. F. F. Blicke, N. J. Doorenbos, and R. H. Cox, *J. Am. Chem. Soc.*, **74**, 2924 (1952).
189. H. B. Hass and J. Bourland, U.S. Pat. 2,343,256 (1944); *Chem. Abstr.*, **38**, 2969 (1944).
190. M. S. Larrison and H. B. Hass, U.S. Pat. 2,383,603 (1945); *Chem. Abstr.*, **40**, 347 (1946).
191. A. Lambert and A. Lowe, *J. Chem. Soc.*, **1947**, 1517.
192. R. A. Smiley and W. A. Pritchett, *J. Chem. Eng. Data*, **11**, 617 (1966).
193. D. V. Nightingale, F. B. Erickson, and N. C. Knight, *J. Org. Chem.*, **15**, 782 (1950).
194. D. V. Nightingale, F. B. Erickson, and J. M. Shackelford, *J. Org. Chem.*, **17**, 1005 (1952).
195. Z. Eckstein, A. Sacha, and T. Urbański, *Bull. Acad. Pol. Sci., Sér. Sci. Chim. Géol. Géograph.*, **5**, 213 (1957).
196. Z. Eckstein, A. Sacha, W. Sobótka, and T. Urbański, *Bull. Acad. Pol. Sci., Sér. Sci. Chim. Géol. Géograph.*, **6**, 621 (1958).
197. Z. Eckstein, A. Sacha, and W. Sobótka, *Bull. Acad. Pol. Sci., Sér. Sci. Chim. Géol. Géograph.*, **7**, 295 (1959); *Rocz. Chem.*, **34**, 1329 (1960).
198. D. V. Nightingale, D. A. Reich, and F. B. Erickson, *J. Org. Chem.*, **23**, 236 (1958).
199. W. E. Noland and R. J. Sundberg, *Tetrahedron Letters*, **1962**, 295; *J. Org. Chem.*, **28**, 3150 (1963).
200. H. O. House and R. W. Magin, *J. Org. Chem.*, **28**, 647 (1963).
201. D. V. Nightingale, S. Miki, D. N. Heintz, and D. A. Reich, *J. Org. Chem.*, **28**, 642 (1963).
202. H. Stetter and J. Mayer, *Angew. Chem.*, **71**, 430 (1959).
203. H. Stetter and P. Tacke, *Chem. Ber.*, **96**, 694 (1963).
204. H. Feuer, G. B. Bachman, and J. P. Kispersky, *J. Am. Chem. Soc.*, **73**, 1360 (1951).
205. K. Klager, J. P. Kispersky, and E. Hamel, *J. Org. Chem.*, **26**, 4368 (1961).
206. K. Klager, *J. Org. Chem.*, **23**, 1519 (1958).
207. M. H. Gold, E. E. Hamel, and K. Klager, *J. Org. Chem.*, **22**, 1665 (1957).
208. E. E. Hamel, Fr. Pat. 1,326,923 (1963); *Chem. Abstr.*, **59**, 13,824 (1963).
209. E. E. Hamel, J. S. Dehn, J. A. Love, J. J. Scigliano, and A. H. Swift, *Ind. Eng. Chem. Res. Develop.*, **1**, 108 (1962).
210. L. W. Kissinger, W. E. McQuistion, M. Schwartz, and L. Goodman, *J. Org. Chem.*, **22**, 1658 (1957).
211. C. O. Parker, W. D. Emmons, A. S. Pagano, H. A. Rolewicz, and K. S. McCallum, *Tetrahedron*, **17**, 89 (1962).
212. C. O. Parker, *Tetrahedron*, **17**, 105 (1962).
213. H. Feuer, A. T. Nielsen, and C. E. Colwell, *Tetrahedron*, **19**, Suppl. 1, 57 (1963).
214. E. S. Lipina, V. V. Perekalin, and Y. S. Bobovich, *Zh. Obshch. Khim.*, **34**, 3635 (1964); English translation in *J. Gen. Chem. USSR*, **34**, 3683 (1964).
215. A. T. Nielsen, *J. Org. Chem.*, **27**, 1993 (1962).
216. A. T. Nielsen, *J. Org. Chem.*, **27**, 2001 (1962).
217. H. Feuer, C. E. Colwell, G. Leston, and A. T. Nielsen, *J. Org. Chem.*, **27**, 3598 (1962).
218. M. B. Frankel, *J. Org. Chem.*, **23**, 813 (1958).

219. E. S. Zonis and V. V. Perekalin, *Zh. Prikl. Khim.*, **33**, 1427 (1960); English translation in *J. Appl. Chem. USSR*, **33**, 1413 (1960).
220. G. Rembarz and M. Schwill, *J. Prakt. Chem.*, **31**, 127 (1966).
221. V. A. Tartakovskii, A. A. Onishchenko, I. E. Chlenov, and S. S. Novikov, *Dokl. Akad. Nauk SSSR*, **167**, 844 (1966); English translation in *Dokl. Chem.* **167**, 406 (1966).
222. M. Hellmann and G. Opitz, *Angew. Chem.*, **68**, 265 (1956).
223. L. Henry, *Bull. Acad. Roy. Belg.*, [3] **32**, 33 (1896).
224. L. Henry, *Chem. Ber.*, **38**, 2027 (1905).
225. H. Cerf de Mauney, *Bull. Soc. Chim. France*, **4**, 1451 (1937); **4**, 1460 (1937).
226. M. Zief and J. P. Mason, *J. Org. Chem.*, **8**, 1 (1943).
227. A. Lambert and J. D. Rose, *J. Chem. Soc.*, **1947**, 1511.
228. A. T. Blomquist and T. H. Shelley, Jr., *J. Am. Chem. Soc.*, **70**, 147 (1948).
229. A. Dornow, A. Müller, and S. Lüpfert, *Ann. Chem.*, **594**, 191 (1955).
230. A. Dornow and W. Sassenberg, *Ann. Chem.*, **606**, 61 (1957).
231. A. Dornow and H. Thies, *Ann. Chem.*, **581**, 219 (1953).
232. H. G. Johnson, *J. Am. Chem. Soc.*, **68**, 12 (1946).
233. G. B. Butler, *J. Am. Chem. Soc.*, **78**, 482 (1956).
234. H. Cerf de Mauney, *Bull. Soc. Chim. France*, **11**, 281 (1944).
235. M. Senkus, *J. Am. Chem. Soc.*, **68**, 10 (1946).
236. M. Senkus, U.S. Pat. 2,421,165 (1947); *Chem. Abstr.*, **41**, 5546 (1947).
237. H. G. Johnson, *J. Am. Chem. Soc.*, **68**, 14 (1946).
238. M. Senkus, *J. Am. Chem. Soc.*, **68**, 1611 (1946).
239. S. Malinovski and T. Urbański, *Rocz. Chem.*, **25**, 183 (1951).
240. T. Urbański and E. Lipska, *Rocz. Chem.*, **26**, 182 (1952).
241. M. Senkus, *J. Am. Chem. Soc.*, **72**, 2967 (1950).
242. T. Urbański and D. Gürne, *Rocz. Chem.*, **28**, 175 (1954).
243. Z. Eckstein and T. Urbański, *Rocz. Chem.*, **30**, 1163 (1956).
244. Z. Eckstein, W. Sobótka, and T. Urbański, *Rocz. Chem.*, **30**, 133 (1956).
245. Z. Eckstein, W. Sobótka, and T. Urbański, *Rocz. Chem.*, **31**, 347 (1957).
246. D. Gürne and T. Urbański, *Rocz. Chem.*, **31**, 855 (1957).
247. D. Gürne and T. Urbański, *Rocz. Chem.*, **31**, 869 (1957).
248. D. Gürne and T. Urbański, *J. Chem. Soc.*, **1959**, 1912.
249. Z. Eckstein, P. Gluziński, W. Hofman, and T. Urbański, *J. Chem. Soc.*, **1961**, 489.
250. T. Urbański, C. Belżecki, and Z. Eckstein, *Rocz. Chem.*, **36**, 879 (1962).
251. Z. Eckstein, P. Gluziński, E. Grochowski, M. Mordarski, and T. Urbański, *Bull. Acad. Pol. Sci., Sér. Sci. Chim.*, **10**, 331 (1962).
252. Z. Eckstein, P. Gluziński, and T. Urbański, *Bull. Acad. Pol. Sci., Sér. Sci. Chim.*, **12**, 623 (1964).
253. Z. Eckstein, P. Gluziński, J. Plenkiewicz, and T. Urbański, *Bull. Acad. Pol. Sci., Sér. Sci. Chim.*, **10**, 487 (1962).
254. Z. Eckstein, P. Gluziński, and J. Plenkiewicz, *Bull. Acad. Pol. Sci., Sér. Sci. Chim.*, **11**, 325 (1963).
255. Z. Eckstein and T. Urbański, *Advan. Heterocyclic Chem.*, **2**, 311 (1963).
256. H. R. Snyder and W. E. Hamlin, *J. Am. Chem. Soc.*, **72**, 5082 (1950).
257. G. B. Bachman and M. T. Atwood, *J. Am. Chem. Soc.*, **78**, 484 (1956).
258. B. Reichert and H. Posemann, *Arch. Pharm.*, **275**, 67 (1937).
259. A. Dornow, O. Hahmann, and R. Oberkobusch, *Ann. Chem.*, **588**, 62 (1954).
260. A. Dornow and A. Müller, *Chem. Ber.*, **89**, 1023 (1956).
261. D. A. Lyttle and D. I. Weisblat, *J. Am. Chem. Soc.*, **69**, 2118 (1947).
262. D. I. Weisblat and D. A. Lyttle, *J. Am. Chem. Soc.*, **71**, 3079 (1949).

263. E. L. Hirst, J. K. N. Jones, S. Minahan, F. W. Ochyński, A. T. Thomas, and T. Urbański, *J. Chem. Soc.*, **1947**, 924.
264. J. K. N. Jones, R. Koliński, H. Piotrowska, and T. Urbański, *Bull. Acad. Pol. Sci., Sér. Sci. Chim. Géol. Géograph.*, **4**, 521 (1956); *Rocz. Chem.*, **31**, 101 (1957).
265. T. Urbański, Z. Biernacki, and E. Lipska, *Rocz. Chem.*, **28**, 169 (1954).
266. T. Urbański and H. Piotrowska, *Rocz. Chem.*, **29**, 379 (1955).
267. T. Urbański and J. Kolesińska, *Rocz. Chem.*, **29**, 392 (1955).
268. T. Urbański and H. Piotrowska, *Rocz. Chem.*, **31**, 553 (1957).
269. W. Tuszko and T. Urbański, *Tetrahedron*, **20**, Suppl. 1, 325 (1964).
270. T. Urbański and R. Koliński, *Rocz. Chem.*, **30**, 201 (1956).
271. T. Urbański, D. Gürne, R. Koliński, H. Piotrowska, A. Jończyk, B. Serafin, M. Szretter-Szmid, and M. Witanowski, *Tetrahedron*, **20**, Suppl. 1, 195 (1964).
272. R. Koliński, H. Piotrowska, and T. Urbański, *J. Chem. Soc.*, **1958**, 2319.
273. H. Piotrowska and T. Urbański, *J. Chem. Soc.*, **1962**, 1942.
274. D. Gürne and T. Urbański, *Rocz. Chem.*, **34**, 881 (1960).
275. D. Gürne, L. Stefaniak, T. Urbański, and M. Witanowski, *Tetrahedron*, **20**, Suppl. 1 211 (1964).
276. D. Gürne, T. Urbański, M. Witanowski, B. Karniewska, and L. Stefaniak, *Tetrahedron*, **20**, 1173 (1964).
277. H. Feuer, G. B. Bachman, and W. May, *J. Am. Chem. Soc.*, **76**, 5124 (1954).
278. E. E. Hamel, *Tetrahedron*, **19**, Suppl. 1, 85 (1963).
279. F. B. Frankel and K. Klager, *J. Am. Chem. Soc.*, **79**, 2953 (1957).
280. F. B. Frankel and K. Klager, *J. Chem. Eng. Data*, **7**, 412 (1962).
281. M. H. Gold, C. R. Vanneman, K. Klager, G. B. Linden, and F. B. Frankel, *J. Org. Chem.*, **26**, 4729 (1961).
282. S. S. Novikov, A. A. Fainzil'berg, S. N. Shvedova, and V. I. Gulevskaya, *Izv. Akad. Nauk SSSR, Otd. Khim. Nauk*, **1960**, 2056; English translation in *Bull. Acad. Sci. USSR, Div. Chem. Sci.*, **1960**, 1908.
283. H. E. Ungnade and L. W. Kissinger, *J. Org. Chem.*, **30**, 354 (1965).
284. E. D. Bergmann, D. Ginsburg, and R. Pappo, *Org. Reactions*, **10**, 179 (1959).
285. F. Arndt, H. Scholz, and E. Frobell, *Ann. Chem.*, **521**, 95 (1936).
286. E. D. Bergmann and R. Corett, *J. Org. Chem.*, **21**, 107 (1956); **23**, 1507 (1958).
287. E. P. Kohler, *J. Am. Chem. Soc.*, **38**, 889 (1916); E. P. Kohler and P. Allen, Jr., *J. Am. Chem. Soc.*, **50**, 884 (1928); numerous papers in the intervening years.
288. M. C. Kloetzel, *J. Am. Chem. Soc.*, **69**, 2271 (1947).
289. L. I. Smith and W. L. Kohlhase, *J. Org. Chem.*, **21**, 816 (1956).
290. S. S. Novikov, I. S. Korsakova, and M. A. Yachovskaya, *Dokl. Akad. Nauk SSSR*, **118**, 954 (1958); English translation in *Proc. Acad. Sci. USSR, Chem. Sect.*, **118**, 151 (1958).
291. S. S. Novikov, I. S. Korsakova, and N. N. Bulatova, *Zh. Obshch. Khim.*, **29**, 3659 (1959); English translation in *J. Gen. Chem. USSR*, **29**, 3618 (1959).
292. V. F. Belyaev, *Khim. Geterotsikl. Soedin. Akad. Nauk Latv. SSR*, **1962**, 215; *Chem. Abstr.*, **63**, 8244 (1965).
293. M. Bourillot and G. Descotes, *Compt. Rend.*, **260**, 3107 (1965).
294. G. Vita and G. Bucher, *Chem. Ber.*, **99**, 3387 (1966).
295. V. I. Isagulyants and E. L. Markosyan, *Dokl. Akad. Nauk Arm. SSR*, **41**, 221 (1965); *Chem. Abstr.*, **64**, 12,542 (1966).
296. V. I. Isagulyants and Z. Poredda, *Zh. Prikl. Khim.*, **37**, 1093 (1964); English translation in *J. Appl. Chem. USSR*, **37**, 1092 (1964).
297. M. C. Kloetzel, *J. Am. Chem. Soc.*, **70**, 3571 (1948).
298. A. Ostaszyński, J. Wielgat, and T. Urbański, *Tetrahedron*, **20**, Suppl. 1, 285 (1964).

299. L. Herzog, M. H. Gold, and R. D. Geckler, *J. Am. Chem. Soc.*, **73,** 749 (1951).
300. H. Shechter and L. Zeldin, *J. Am. Chem. Soc.*, **73,** 1276 (1951).
301. K. Klager, *J. Org. Chem.*, **16,** 161 (1951).
302. A. Solomonovici and S. Blumberg, *Israel J. Chem.*, **3,** 63 (1965).
303. H. Feuer and R. Harmetz, *J. Org. Chem.*, **26,** 1061 (1961).
304. H. Feuer and C. N. Aguilar, *J. Org. Chem.*, **23,** 607 (1958).
305. H. Feuer, G. Leston, R. Miller, and A. T. Nielsen, *J. Org. Chem.*, **28,** 339 (1963). (a) I. S. Ivanova, N. N. Bulatova, and S. S. Novikov, *Izv. Akad. Nauk SSSR, Otd, Khim. Nauk*, **1962,** 1856; English translation in *Bull. Acad. Sci. USSR, Div. Chem. Sci.*, **1962,** 1762.
306. H. B. Hass, *Ind. Eng. Chem.*, **35,** 1146 (1943).
307. V. V. Perekalin and A. S. Sopova, *Zh. Obshch. Khim.*, **24,** 513 (1954); English translation in *J. Gen. Chem. USSR*, **24,** 523 (1954).
308. V. V. Perekalin and A. S. Sopova, *Zh. Obshch. Khim.*, **28,** 675 (1958); English translation in *J. Gen. Chem. USSR*, **28,** 656 (1958).
309. A. S. Sopova and A. A. Temp, *Zh. Obshch. Khim.*, **31,** 1532 (1961); English translation in *J. Gen. Chem. USSR*, **31,** 1420 (1961).
310. V. V. Perekalin and O. M. Lerner, *Zh. Obshch. Khim.*, **28,** 1815 (1958); English translation in *J. Gen. Chem. USSR*, **28,** 1861 (1958).
311. V. V. Perekalin and K. S. Parfenova, *Zh. Prikl. Khim.*, **30,** 1279 (1957); English translation in *J. Appl. Chem. USSR*, **30,** 1353 (1957).
312. V. V. Perekalin and K. S. Parfenova, *Zh. Obshch. Khim.*, **30,** 388 (1960); English translation in *J. Gen. Chem. USSR*, **30,** 412 (1960).
313. Z. Eckstein and J. Plenkiewicz, *Bull. Acad. Pol. Sci., Sér. Sci. Chim.*, **9,** 393 (1961).
314. V. V. Perekalin and O. M. Lerner, *Dokl. Akad. Nauk SSSR*, **129,** 1303 (1959); English translation in *Proc. Acad. Sci. USSR, Chem. Sect.*, **129,** 1143 (1959).
315. E. S. Lipina, V. V. Perekalin, and Y. S. Bobovich, *Zh. Obshch. Khim.*, **34,** 3640 (1964); English translation in *J. Gen. Chem. USSR*, **34,** 3689 (1964).
316. E. S. Zonis, O. M. Lerner, and V. V. Perekalin, *Zh. Prikl. Khim.*, **34,** 711 (1961); English translation in *J. Appl. Chem. USSR*, **34,** 687 (1961).
317. Y. S. Bobovich, E. S. Lipina, and V. V. Perekalin, *Zh. Strukt. Khim.*, **5,** 546 (1964); English translation in *J. Struct. Chem. USSR*, **5,** 504 (1964).
318. E. S. Lipina and V. V. Perekalin, *Zh. Obshch. Khim.*, **34,** 3644 (1964); English translation in *J. Gen. Chem. USSR*, **34,** 3693 (1964).
319. F. Boberg and G. R. Schultze, *Chem. Ber.*, **90,** 1215 (1957).
320. F. Boberg, *Ann. Chem.*, **626,** 71 (1959).
321. V. V. Perekalin and M. M. Zobachova, *Zh. Obshch. Khim.*, **29,** 2905 (1959); English translation in *J. Gen. Chem. USSR*, **29,** 2865 (1959).
322. Y. S. Bobovich, V. V. Perekalin, and A. S. Sopova, *Dokl. Akad. Nauk SSSR*, **134,** 1083 (1960); English translation in *Proc. Acad. Sci. USSR, Chem. Sect.*, **154,** 1125 (1960).
323. A. S. Sopova, V. V. Perekalin, and Y. S. Bobovich, *Zh. Obshch. Khim.*, **31,** 1528 (1961); English translation in *J. Gen. Chem. USSR*, **31,** 1417 (1961).
324. A. S. Sopova, V. V. Perekalin, and V. M. Lebednova, *Zh. Obshch. Khim.*, **33,** 2143 (1963); English translation in *J. Gen. Chem. USSR*, **33,** 2090 (1963).
325. A. S. Sopova, V. V. Perekalin, and O. I. Yurchenko, *Zh. Obshch. Khim.*, **33,** 2140 (1963); English translation in *J. Gen. Chem. USSR*, **33,** 2087 (1963).
326. A. S. Sopova, V. V. Perekalin, and O. I. Yurchenko, *Zh. Obshch. Khim.*, **34,** 1188 (1964); English translation in *J. Gen. Chem. USSR*, **34** 1180 (1964).
327. A. S. Sopova, O. I. Yurchenko, and V. V. Perekalin, *Zh. Org. Khim. SSSR*, **1,** 1707 (1965); English translation in *J. Org. Chem. USSR*, **1,** 1732 (1965).

328. E. P. Kohler and S. F. Darling, *J. Am. Chem. Soc.*, **52**, 1174 (1930).
329. A. S. Sopova, V. V. Perekalin, V. M. Lebednova, and O. I. Yurchenko, *Zh. Obshch. Khim.*, **34**, 1185 (1964); English translation in *J. Gen. Chem. USSR*, **34**, 1177 (1964).
330. V. A. Konkova, G. F. Afanaseva, M. S. Kalyazina, and G. S. Belyaeva, *Zh. Prikl. Khim.*, **37**, 1637 (1964); English translation in *J. Appl. Chem. USSR*, **37**, 1627 (1964).
331. J. P. Freeman and W. D. Emmons, *J. Am. Chem. Soc.*, **78**, 3405 (1956).
332. C. D. Hurd and L. T. Sherwood, Jr., *J. Org. Chem.*, **13**, 471 (1948).
333. A. Dornow and S. Lüpfert. *Ann. Chem.*, **606**, 56 (1957).
334. T. Severin and B. Brück, *Chem. Ber.*, **98**, 3847 (1965).
335. T. Severin, B. Brück, and P. Adhikary, *Chem. Ber.*, **99**, 3097 (1966).
336. E. P. Kohler and G. R. Barrett, *J. Am. Chem. Soc.*, **46**, 2105 (1924).
337. E. P. Kohler and N. K. Richtmeyer, *J. Am. Chem. Soc.*, **50**, 3092 (1928).
338. D. E. Worrall, *J. Am. Chem. Soc.*, **57**, 2299 (1935).
339. A. Lambert and H. A. Piggott, *J. Chem. Soc.*, **1947**, 1489.
340. A. Lambert and H. A. Piggott, Brit. Pat. 584,789 (1947); *Chem. Abstr.*, **41**, 5143 (1947).
341. C. T. Bahner and H. T. Kite, *J. Am. Chem. Soc.*, **71**, 3597 (1949).
342. C. T. Bahner and H. T. Kite, U.S. Pat. 2,477,162 (1949); *Chem. Abstr.*, **44**, 1128 (1950).
343. A. Dornow and H. Menzel, *Ann. Chem.*, **588**, 40 (1954).
344. K. Klager, *J. Org. Chem.*, **20**, 650 (1955).
345. H. Shechter and F. Conrad, *J. Am. Chem. Soc.*, **76**, 2716 (1954).
346. T. M. Khannanov and G. K. Yakomazova, *Izv. Vysshykh Uchebn. Zavedenii, Khim. i Khim. Tekhnol.*, **7**, 237 (1964); *Chem. Abstr*, **61**, 10578 (1964).
347. S. S. Novikov, I. S. Korsakova, and K. K. Babievski, *Izv. Akad. Nauk SSSR, Otd. Khim. Nauk*, **1959**, 1480; English translation in *Bull. Acad. Sci. USSR, Div. Chem. Sci.*, **1959**, 1426.
(a) I. S. Ivanova, Y. V. Konnova, and S. S. Novikov, *Izv. Akad. Nauk SSSR, Otd. Khim. Nauk*, **1962**, 2078; English translation in *Bull. Acad. Sci. USSR, Div. Chem. Sci.*, **1962**, 1985.
348. L. Zeldin and H. Shechter, *J. Am. Chem. Soc.*, **79**, 4708 (1957).
349. K. Klager, *Anal. Chem.*, **23**, 534 (1951).
350. H. Feuer and R. Miller, *J. Org. Chem.*, **26**, 1348 (1961).
351. K. Klager, *Monatsh. Chem.*, **96**, 1 (1965).
352. A. Solomonovici and S. Blumberg, *Tetrahedron*, **22**, 2505 (1966).
353. K. Alder, H. F. Rickert, and E. Windermuth, *Chem. Ber.*, **71**, 2451 (1938).
354. C. F. H. Allen and A. Bell, *J. Am. Chem. Soc.*, **61**, 521 (1939).
355. C. F. H. Allen, A. Bell, and J. W. Gates, Jr., *J. Org. Chem.*, **8**, 373 (1943).
356. S. Sugasawa and K. Kodama, *Chem. Ber.*, **72**, 675 (1939).
357. D. V. Nightingale and V. Tweedie, *J. Am. Chem. Soc.*, **66**, 1968 (1944).
358. H. L. Holmes, *Org. Reactions*, **4**, 60 (1948).
359. W. E. Parham, W. T. Hunter, and R. Hanson, *J. Am. Chem. Soc.*, **73**, 5068 (1951).
360. E. E. van Tamelen and R. J. Thiede, *J. Am. Chem. Soc.*, **74**, 2615 (1952).
361. W. C. Wildman and C. H. Hemminger, *J. Org. Chem.*, **17**, 1641 (1952).
362. D. V. Nightingale, M. Maienthal, and J. A. Gallagher, *J. Am. Chem. Soc.*, **75**, 4852 (1953).
363. J. D. Roberts, C. C. Lee, and W. H. Saunders, Jr., *J. Am. Chem. Soc.*, **76**, 4501 (1954).
364. W. E. Noland and R. E. Bambury, *J. Am. Chem. Soc.* **77**, 6386 (1955).

365. W. E. Noland, R. E. Counsell, and M. H. Fischer, *J. Org. Chem.*, **21**, 911 (1956).
(a) W. E. Noland, B. A. Langager, J. W. Manthey, A. G. Zacchei, D. L. Petrak, and G. L. Eian, *Can. J. Chem.*, **45**, 2969 (1967).
(b) G. I. Poos, J. Kleis, R. R. Wittekind, and J. D. Rosenau, *J. Org. Chem.*, **26**, 4898 (1961).
(c) S. S. Novikov, G. A. Shvekhgeimer, and A. A. Dudinskaya, *Izv. Akad. Nauk SSSR, Otd. Khim. Nauk*, **1961**, 690; English translation in *Bull. Acad. Sci. USSR, Div. Chem. Sci.*, **1961**, 640.
(d) S. S. Novikov, G. A. Shvekhgeimer, and A. A. Dudinskaya, *Izv. Akad. Nauk SSSR, Otd. Khim. Nauk*, **1960**, 1858; English translation in *Bull. Acad. Sci. USSSR, Div. Chem. Sci*, **1960**, 1727.
366. W. C. Wildman and D. R. Saunders, *J. Org. Chem.*, **19**, 381 (1954).
367. N. L. Drake and A. B. Ross, *J. Org. Chem.*, **23**, 717 (1958).
368. N. L. Drake and A. B. Ross, *J. Org. Chem.*, **23**, 794 (1958).
369. A. A. Dudinskaya, G. A. Shvekhgeimer, and S. S. Novikov, *Izv. Akad. Nauk SSSR, Otd. Khim. Nauk*, **1961**, 524; English translation in *Bull. Acad. Sci. USSR, Div. Chem. Sci.*, **1961**, 484.
370. A. A. Dudinskaya, G. A. Shvekhgeimer, and S. S. Novikov, *Izv. Akad. Nauk SSSR, Otd. Khim. Nauk*, **1961**, 522; English translation in *Bull. Acad. Sci. USSR, Div. Chem. Sci.*, **1961**, 482.
371. A. A. Dudinskaya, S. S. Novikov, and G. A. Shvekhgeimer, *Izv. Akad. Nauk SSSR, Otd. Khim. Nauk*, **1965**, 2024; English translation in *Bull. Acad. Sci. USSR, Div. Chem. Sci.*, **1965**, 1988.
372. A. Étienne, A. Spire, and E. Toromanoff, *Bull. Soc. Chim. France*, **1952**, 750.
373. Y. K. Yurev, N. S. Zefirov, and R. A. Ivanova, *Zh. Obshch. Khim. SSSR*, **33**, 3512 (1963); English translation in *J. Gen. Chem. USSR*, **33**, 3444 (1963).
(a) R. R. Fraser, *Can. J. Chem.*, **40**, 78 (1962).
(b) R. J. Ouellette and G. E. Booth, *J. Org. Chem.*, **30**, 423 (1965).
374. R. T. Arnold and P. N. Richardson, *J. Am. Chem. Soc.*, **76**, 3649 (1954).
375. H. E. Zimmerman and T. E. Nevins, *J. Am. Chem. Soc.*, **79**, 6559 (1957).
376. W. E. Noland, H. I. Freeman, and M. S. Baker, *J. Am. Chem. Soc.*, **78**, 188 (1956).
377. H. J. Hudak and J. Meinwald, *J. Org. Chem.*, **26**, 1360 (1961).
378. M. H. Gold and K. Klager, *Tetrahedron*, **19**, Suppl. 1, 77 (1963).
379. H. Feuer, R. Miller, and C. B. Lawyer, *J. Org. Chem.*, **26**, 1357 (1961).
380. M. Senkus, *J. Am. Chem. Soc.*, **63**, 2635 (1941).
381. M. Senkus, *J. Am. Chem. Soc.*, **65**, 1656 (1943).
382. M. Senkus, U.S. Pat. 2,368,071 (1945); *Chem. Abstr.*, **39**, 4098 (1945).
383. M. S. Newman, B. J. Magerlein, and W. B. Wheatley, *J. Am. Chem. Soc.*, **68**, 2112 (1946).
384. G. H. Morey, U.S. Pat. 2,406,504 (1946); *Chem. Abstr.*, **41**, 490 (1947).
385. A. Scattergood and A. L. MacLean, *J. Am. Chem. Soc.*, **71**, 4153 (1949).
386. H. Piotrowska, B. Serafin, and T. Urbański, *Tetrahedron*, **19**, 379 (1963).
387. Z. Eckstein, *Rocz. Chem.*, **27**, 246 (1953).
388. G. B. Linden and M. H. Gold, *J. Org. Chem.*, **21**, 1175 (1956).
389. H. Feuer, A. T. Nielsen, and C. E. Colwell, *Tetrahedron*, **19**, Suppl. 1, 57 (1963).
390. Z. Eckstein and T. Urbański, *Rocz. Chem.*, **30**, 1175 (1956).
391. T. Urbański, Private communication.
392. D. Palut and Z. Eckstein, *Bull. Acad. Pol. Sci., Sér. Sci. Chim.*, **12**, 41 (1964).
393. Z. Eckstein, *Rocz. Chem.*, **30**, 1151 (1956).
394. J. A. Mills, *Advan. Carbohydrate Chem.*, **10**, 1 (1955).
395. H. H. Baer and T. Neilson, *Can. J. Chem.*, **43**, 840 (1965).

396. H. H. Baer, F. Kienzle, and T. Neilson, *Can. J. Chem.*, **43,** 1829 (1965).
397. H. H. Baer and F. Kienzle, *Can. J. Chem.*, **45,** 983 (1967).
398. H. Feuer and S. Markofsky, *J. Org. Chem.*, **29,** 929 (1964).
399. B. Helfrich and M. Hase, *Ann. Chem.*, **554,** 261 (1943).
400. H. O. L. Fischer and H. H. Baer, *Ann. Chem.*, **619,** 53 (1958).
401. H. H. Baer and W. Rank, *Unpublished results.*
402. H. H. Baer, T. Neilson, and W. Rank, *Can. J. Chem.*, **45,** 991 (1967).
403. S. S. Novikov and G. A. Shvekhgeimer, *Izv. Akad. Nauk SSSR, Otd. Khim. Nauk,* **1960,** 307; English translation in *Bull. Acad. Sci. USSR, Div. Chem. Sci.*, **1960,** 279.
404. H. E. Ungnade and L. W. Kissinger, *J. Org. Chem.*, **31,** 369 (1966).
405. Z. Eckstein, T. Urbański, and W. Sobótka, *Bull. Acad. Pol. Sci., Sér. Sci. Chim. Géol. Géograph.*, **5,** 679 (1957).
406. C. H. Hurd and M. E. Nilson, *J. Org. Chem.*, **20,** 927 (1955).
407. A. Dornow and W. Sassenberg, *Ann. Chem.*, **594,** 185 (1955).
408. F. Hofwimmer, *Z. Ges. Schiess–Sprengstoffwesen,* **7,** 43 (1912); *Chem. Zentr.*, **1912,** 1265.
409. F. H. Bergeim, U.S. Pat. 1,691,955 (1928); *Chem. Abstr.*, **23,** 708 (1929).
410. A. McLean, U.S. Pat. 2,399,686 (1946); *Chem. Abstr.*, **40,** 4744 (1946).
411. F. Römer, *Angew. Chem.*, **67,** 157 (1955).
412. A. E. Wilder-Smith, C. W. Scaife, and H. Baldock, Brit. Pat. 586,022 (1947); *Chem. Abstr.*, **41,** 6893 (1947).
413. N. Levy and C. W. Scaife, *J. Chem. Soc.*, **1946,** 1093.
414. N. Levy, C. W. Scaife, and A. E. Wilder-Smith, *J. Chem. Soc.*, **1946,** 1096.
415. N. Levy and C. W. Scaife, *J. Chem. Soc.*, **1946,** 1100.
416. N. Levy, C. W. Scaife, and A. E. Wilder-Smith, *J. Chem. Soc.*, **1948,** 52.
417. H. Baldock, N. Levy, and C. W. Scaife, *J. Chem. Soc.*, **1949,** 2627.
418. H. Wieland and F. Rahn, *Chem. Ber.*, **54,** 1770 (1921).
419. F. G. Bordwell and E. W. Garbisch, Jr., *J. Am. Chem. Soc.*, **82,** 3588 (1960).
420. F. G. Bordwell and E. W. Garbisch, Jr., *J. Org. Chem.*, **27,** 3049 (1962).
421. A. A. Griswold and P. S. Starcher, *J. Org. Chem.*, **31,** 357 (1966).
422. A. Lambert, C. W. Scaife, and A. E. Wilder-Smith, *J. Chem. Soc.*, **1947,** 1474.
423. M. B. Frankel, *J. Org. Chem.*, **27,** 331 (1962).
424. L. W. Kissinger, T. M. Benzinger, H. E. Ungnade, and R. K. Rohwer, *J. Org. Chem.*, **28,** 2491 (1963).
425. W. Steinkopf and M. Kühnel, *Chem. Ber.*, **75,** 1323 (1942).
426. S. P. Lingo, U.S. Pat. 2,471,274 (1949); *Chem. Abstr.*, **43,** 6222 (1949).
427. G. Fort and A. McLean, *J. Chem. Soc.*, **1948,** 1902.
428. S. S. Novikov, V. M. Belikov, and L. V. Epishina, *Izv. Akad. Nauk SSSR, Otd. Khim. Nauk*, **1962,** 1111; English translation in *Bull. Acad. Sci. USSR, Div. Chem. Sci.*, **1962,** 1042.
429. Z. Eckstein and J. Kościelny, *Bull. Acad. Pol. Sci., Sér. Sci. Chim.*, **13,** 11 (1965).
430. E. Czerwińska, Z. Eckstein, J. Kościelny, and R. Kowalik, *Bull. Acad. Pol. Sci., Sér. Sci. Chim.*, **13,** 17 (1965).
431. G. W. Pedlow, Jr., and C. S. Miner, Jr., U.S. Pat. 2,566,365 (1951); *Chem. Abstr.*, **46,** 3068 (1952).
432. A. C. McInnis, Jr., and L. G. Tomkins, *J. Am. Chem. Soc.*, **74,** 2686 (1952).
433. B. M. Vanderbilt, U.S. Pat. 2,177,757 (1939); *Chem. Abstr.*, **34,** 1415 (1940).
434. L. W. Kissinger, M. Schwartz, and W. E. McQuistion, *J. Org. Chem.*, **26,** 5203 (1961).
435. L. Henry, *Bull. Acad. Roy. Belg.*, [3] **34,** 547 (1897).
436. E. Schmidt, G. Rutz, and M. Trénel, *Chem. Ber.*, **61,** 472 (1928).

437. J. B. Tindall, *Ind. Eng. Chem.*, **33**, 65 (1941).
438. C. D. Hurd, S. S. Drake, and O. Fancher, *J. Am. Chem. Soc.*, **68**, 789 (1946).
439. J. Legocki, H. Rodowicz, and J. Hackel, *Przemysl Chem.*, **43**, 148 (1964) *Chem. Abstr.*, **61**, 16166 (1964).
440. H. Feuer and C. Savides, *J. Am. Chem. Soc.*, **81**, 5826 (1959).
441. J. R. Reasenberg and G. B. L. Smith, *J. Am. Chem. Soc.*, **66**, 991 (1944).
442. M. S. Heller and R. A. Smiley, *J. Org. Chem.*, **23**, 771 (1958).
443. H. E. Ungnade, E. D. Loughran, and L. W. Kissinger, *Tetrahedron*, **20**, Suppl. 1, 177 (1964).
444. H. H. Baer, F. Kienzle, and F. Rajabalee, *Can. J. Chem.*, **46**, 80 (1968).
445. H. H. Baer and T. Neilson, *J. Org. Chem.*, **32**, 1068 (1967).
446. H. H. Baer and F. Kienzle, *J. Org. Chem.*, **33**, 1873 (1968).
447. P. K. Bhattacharya, A. C. Ghosh, V. M. Sathe, N. L. Dutta, and Mansa Ram, *Tetrahedron*, **20**, Suppl. 1, 275 (1964).
448. H. Feuer and W. H. Gardner, *J. Am. Chem. Soc.*, **76**, 1375 (1954).
449. S. S. Novikov, G. A. Shvekhgeimer, and N. P. Piatakov, *Izv. Akad. Nauk SSSR, Otd. Khim. Nauk*, **1961**, 375; English translation in *Bull. Acad. Sci. USSR, Div. Chem. Sci.*, **1961**, 351.
450. E. Schmidt and G. Rutz, *Chem. Ber.*, **61**, 2142 (1928).
451. H. Schwarz and J. Nelles, U.S. Pat. 2,257,980 (1941); *Chem. Abstr.*, **36**, 494 (1942).
452. H. B. Hass, A. G. Susie, and R. L. Heider, *J. Org. Chem.*, **15**, 8 (1950).
453. C. Porter and B. Wood, *J. Inst. Petrol.*, **38**, 877 (1952).
454. H. Shechter and J. W. Shepherd, *J. Am. Chem. Soc.*, **76**, 3617 (1954).
455. A. T. Blomquist, W. J. Tapp, and J. R. Johnson, *J. Am. Chem. Soc.*, **67**, 1519 (1945).
456. M. H. Gold, *J. Am. Chem. Soc.*, **68**, 2544 (1946).
457. M. H. Gold, U.S. Pat. 2,414,594; U.S. Pat. 2,414,595 (1947); *Chem. Abstr.*, **41**, 4166 (1947).
458. J. C. Sowden, U.S. Pat. 2,530,342 (1950); *Chem. Abstr.*, **45**, 2971 (1951).
459. H. Paulsen, *Ann. Chem.*, **665**, 166 (1963).
460. Y. V. Baskov, T. Urbański, M. Witanowski, and L. Stefaniak, *Tetrahedron*, **20**, 1519 (1964).
461. H. Irving, *J. Chem. Soc.*, **1936**, 797.
462. H. Irving and H. I. Fuller, *J. Chem. Soc.*, **1948**, 1989.
463. R. L. Heath and J. D. Rose, *J. Chem. Soc.*, **1947**, 1486.
464. F. G. Bordwell and E. W. Garbisch, Jr., *J. Org. Chem.*, **28**, 1765 (1963).
465. L. Bouveault and A. Wahl, *Bull. Soc. Chim. France*, **25**, 808 (1901).
466. Z. Eckstein, T. Urbański, and H. Wojnowska, *Rocz. Chem.*, **31**, 1177 (1957).
467. Z. Eckstein, T. Urbański, and H. Wojnowska, *Bull. Acad. Pol. Sci., Sér. Sci. Chim. Géol. Géograph.*, **5**, 219 (1957).
468. C. Satoh and A. Kiyomoto, *Chem. Pharm. Bull. Japan*, **12**, 615 (1964); *Carbohydrate Res.*, **7**, 138 (1968).
469. H. H. Baer and M. Wang, *Can. J. Chem.*, **46**, 2793 (1968).
470. R. L. Heath and A. Lambert, *J. Chem. Soc.*, **1947**, 1477.
471. R. L. Heath and H. A. Piggott, *J. Chem. Soc.*, **1947**, 1481.
472. M. H. Gold, L. J. Druker, R. Yotter, C. J. B. Thor, and G. Lang, *J. Org. Chem.*, **16**, 1495 (1951).
473. G. D. Buckley, R. L. Heath, and J. D. Rose, *J. Chem. Soc.*, **1947**, 1500.
474. G. D. Buckley, F. G. Hunt, and A. Lowe, *J. Chem. Soc.*, **1947**, 1504.
475. L. J. Winters and W. E. McEwen, *Tetrahedron*, **19**, Suppl. 1, 49 (1963).
476. P. Friedländer and J. Möhly, *Ann. Chem.*, **229**, 210 (1885).
477. J. Thiele and S. Haeckel, *Ann. Chem.*, **325**, 1 (1902).

478. B. Flürsheim, *J. Prakt. Chem.*, **66**, 16 (1902).
479. J. Meisenheimer and F. Heim, *Chem. Ber.*, **38**, 466 (1905).
480. J. Meisenheimer and F. Heim, *Ann. Chem.*, **355**, 260 (1907).
481. B. Flürsheim and E. L. Holmes, *J. Chem. Soc.*, **1932**, 1458.
482. B. Reichert and W. Koch, *Arch. Pharm.*, **273**, 265 (1935).
483. J. Meisenheimer and F. Heim, *Ann. Chem.*, **355**, 269 (1907).
484. F. Heim, *Chem. Ber.*, **44**, 2013 (1911).
485. A. Dornow and F. Boberg, *Chem. Ber.*, **83**, 261 (1950).
486. J. Loevenich, J. Koch, and U. Pucknat, *Chem. Ber.*, **63**, 636 (1930).
487. W. J. Seagers and P. J. Elving, *J. Am. Chem. Soc.*, **71**, 2947, (1949).
488. I. Thompson, S. Louloudes, R. Fulmer, F. Evans, and H. Burkett, *J. Am. Chem. Soc.*, **75**, 5006 (1953).
489. H. Burkett, G. Nelson, and W. Wright, *J. Am. Chem. Soc.*, **80**, 5812 (1958).
490. H. H. Baer and F. Kienzle, *J. Org. Chem.*, **32**, 3169 (1967).
491. J. C. Sowden, M. L. Oftedahl, and A. Kirkland, *J. Org. Chem.*, **27**, 1791 (1962).
492. R. Trave, *Gazz. Chim. Ital.*, **79**, 233 (1949).
493. L. F. Cason and C. C. Wanser, *J. Am. Chem. Soc.*, **73**, 142 (1951).
494. A. Mustafa, A. Hamid, E. Harnash, and M. Kamel, *J. Am. Chem. Soc.*, **77**, 3860 (1955).
495. D. E. Worall, *J. Am. Chem. Soc.*, **49**, 1598 (1927).
496. A. N. O'Neill, *Can. J. Chem.*, **37**, 1747 (1959).
497. J. C. Sowden and M. L. Oftedahl, *J. Am. Chem. Soc.*, **82**, 2303 (1960).
498. S. D. Gero and J. Defaye, *Compt. Rend.*, **261**, 1555 (1965).
499. J. C. Sowden and M. L. Oftedahl, *J. Org. Chem.*, **26**, 2153 (1961).
(a) M. B. Perry and J. Furdová, *Can. J. Chem.*, **46**, 2859 (1968).
500. D. E. Worall, *J. Am. Chem. Soc.*, **60**, 2841 (1938).
501. D. E. Worall and F. Benington, *J. Am. Chem. Soc.*, **60**, 2844 (1938).
502. D. E. Worall, *J. Am. Chem. Soc.*, **60**, 2845 (1938).
503. D. E. Worall, *J. Am. Chem. Soc.*, **43**, 919 (1921).
504. D. E. Worall, *J. Am. Chem. Soc.*, **62**, 3253 (1940).
505. C. Musante, *Gazz. Chim. Ital.*, **67**, 579 (1937).
506. A. Dornow and F. Boberg, *Ann. Chem.*, **578**, 94 (1952).
507. H. Wieland and E. Sakellarios, *Chem. Ber.*, **52**, 898 (1919).
508. J. Loevenich and H. Gerber, *Chem. Ber.*, **63**, 1707 (1930).
509. G. B. Bachman and D. E. Welton, *J. Org. Chem.*, **12**, 208 (1947).
510. J. C. Sowden, A. Kirkland, and K. O. Lloyd, *J. Org. Chem.*, **28**, 3516 (1963).
511. T. Posner and O. Unverdorben, *Ann. Chem.*, **389**, 114 (1912).
512. C. D. Hurd and J. Patterson, *J. Am. Chem. Soc.*, **75**, 285 (1953).
513. L. B. Clapp, J. F. Brown, Jr., and L. Zeftel, *J. Org. Chem.*, **15**, 1043 (1950).
514. I. S. Ivanova, Y. V. Konnova, and S. S. Novikov, *Izv. Akad. Nauk SSSR, Otd. Khim. Nauk*, **1962**, 920; English translation in *Bull. Acad. Sci. USSR, Div. Chem. Sci.*, **1962**, 858.
515. I. S. Ivanova, N. N. Bulatova, and S. S. Novikov, *Izv. Akad. Nauk SSSR, Otd. Khim. Nauk*, **1962**, 1858; English translation in *Bull. Acad. Sci. USSR, Div. Chem. Sci.*, **1962**, 1765.
516. I. S. Ivanova, Y. V. Konnova, N. N. Bulatova, and S. S. Novikov, *Izv. Akad. Nauk SSSR, Otd. Khim. Nauk*, **1962**, 1686. English translation in *Bull. Acad. Sci. USSR, Div. Chem. Sci.*, **1962**, 1603.

CHAPTER 4

Biochemistry and Pharmacology of the Nitro and Nitroso Groups

JAN VENULET

Drug Research Institute,
Warsaw, Poland

and

ROBERT L. VANETTEN*

Department of Chemistry,
Purdue University,
Lafayette, Indiana

* Recipient of N.I.H. Research Career Development Award No. K4 GM 17,620 from the National Institute of General Medical Sciences.

I. INTRODUCTION

Biochemistry and pharmacology of compounds possessing the nitro and nitroso groups include some of the most interesting areas of current research. This is both an asset and a liability. It is difficult to discuss topics which encompass areas of active research since they constantly change in detail as the research progresses. On the other hand, the intensive research currently in progress involving the biochemical reactions of nitrogen compounds is a reflection of the great importance of the field. It is beyond the scope of the present chapter to attempt a treatment of the details of the interdependence of chemical structure and biological activity even in a limited area such as nitro and nitroso compounds. Indeed, such a treatment is not yet possible. However, by considering just some of the most important examples, ample opportunity is provided for describing the major outlines of metabolic reactions of nitro and nitroso compounds and for discussing the pharmacology and toxicology of some of the important representatives of these classes of compounds.

It is also difficult to sharply limit the discussion to reactions or effects characteristic of the nitro and nitroso groups since biological effects are the result of an interaction between some biological receptor unit and an entire molecule, not just a particular side chain or substituent. This is perhaps most effectively illustrated by an example from the work of Landsteiner and Jacobs, who found that while 1,2,4-trinitrobenzene has potent allergenic sensitizing properties the 1,3,5 isomer does not elicit such effects. Thus a discussion of the biochemistry and pharmacology of the nitro and nitroso groups must necessarily be concerned with molecular systems having a variety of potential functional groupings and capable of eliciting a variety of physiological and biochemical effects. The nitro and nitroso groups participate to varying extents in the chemical reactions involving highly important biochemical and physiological responses and the scientist who is not particularly familiar with biochemistry should find these discussions interesting. In addition, it is felt that an important aim of the present chapter is to call attention to potential toxicological hazards and other threats to health.

The biochemical reactions of the nitro and nitroso groups are very much interrelated with the biochemistry of the amino group, and the reader is referred to an excellent chapter on this topic in another volume of this series[1]. Because of the involvement of nitro and nitroso compounds in some of the most fundamental biochemical processes a short discussion of some appropriate areas of biochemistry precedes the detailed discussion of biochemical and pharmacological effects of these compounds.

II. BIOLOGICAL OXIDATION–REDUCTION PROCESSES AND OXIDATIVE PHOSPHORYLATION

The metabolic effects of nitro and nitroso compounds are very much involved with some of the most important biochemical reactions occurring in living systems. Aerobic organisms derive their energy from the oxidation by oxygen of a variety of foodstuffs. Coincidentally with these oxidative processes there are numerous biosynthetic reactions involving oxidative and reductive steps which result in the formation of important metabolites. Oxidation–reduction reactions and the reactions of oxidative phosphorylation may be affected by the presence of nitro and nitroso compounds. For these reasons it seems appropriate to describe some of the major details of these processes. For more extended discussions the reader may wish to refer to some recent biochemistry textbooks[2,3].

A. Some Biochemical Oxidation–Reduction Reactions

The common oxidative enzymes employ members of either of two groups of coenzymes as one of the reactants. These coenzymes are the pyridine and the flavin nucleotides. The pyridine nucleotides include nicotinamide adenine dinucleotide or NAD^+ (**1a**), also less informatively called diphosphopyridine nucleotide (DPN^+), and nicotinamide adenine dinucleotide phosphate or $NADP^+$ (**1b**), also called TPN. The nicotinamide coenzymes are reduced in the various reversible enzyme-catalyzed reactions such as shown in reaction 1 involving the transformation of a reduced substrate SH_2 into some oxidized substrate S. The typical reduced substrate might be, for example, an alcohol which is oxidized to an aldehyde or ketone. The other products of such a transformation are hydrogen ion and the reduced coenzymes NADH (**2a**) or NADPH (**2b**).

Another major group of oxidation–reduction coenzymes are the flavin coenzymes flavin mononucleotide or FMN (**3a**) and flavin adenine dinucleotide or FAD (**3b**). These coenzymes also participate as hydrogen acceptors or donors in reversible enzyme-catalyzed oxidation–reduction reactions to form the reduced coenzymes $FMNH_2$ (**4a**) or $FADH_2$ (**4b**) (equation 2).

The formation of reduced coenzymes such as NADH or $FADH_2$ by oxidation of some energy-rich foodstuff merely represents the transfer of chemical energy from one molecular species to another. The advantage of the coenzymes as a system is that they represent

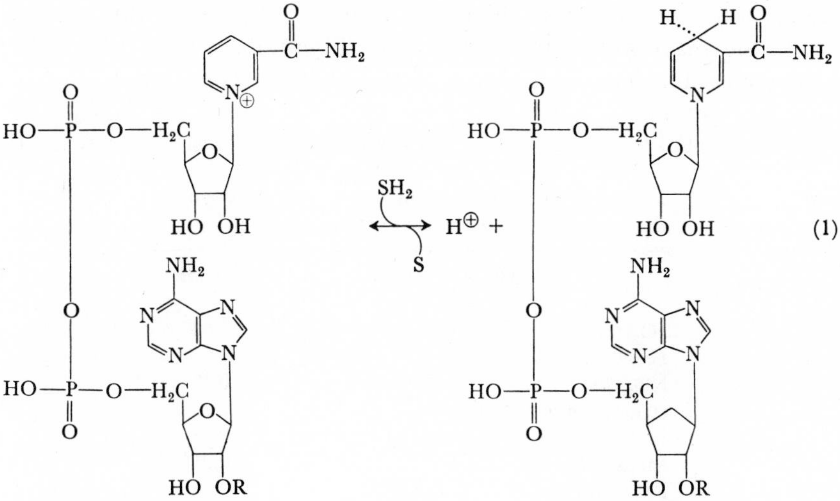

(1)

(**1a**), R = H
(**1b**), R = PO_3H_2

(**2a**), R = H
(**2b**), R = PO_3H_2

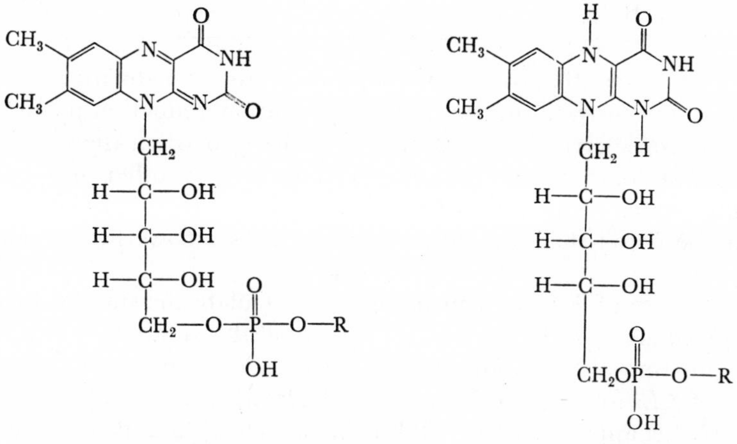

(2)

(**3a**), R = H
(**3b**), R = AMP

(**4a**), R = H
(**4b**), R = AMP

where AMP =

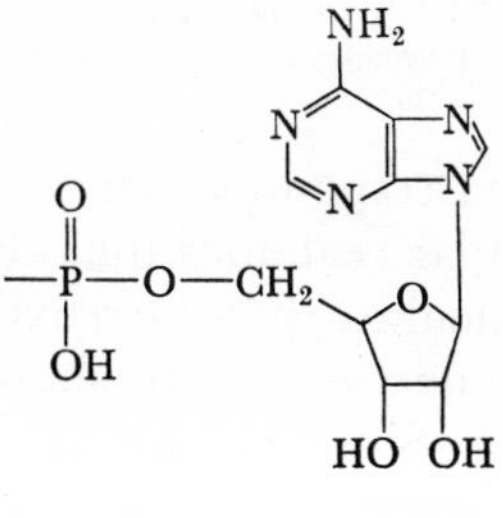

common intermediates which can then be utilized in the energy-yielding reactions common to widely different kinds of tissues and enzymes. The energy content of the reduced coenzymes may be readily estimated. Standard electrode potentials at pH 7 (E_0') are given in Table 1 for a few biologically important reactants. It

TABLE 1. Reduction potentials of some biochemical systems at pH 7.

Half-cell	E_0', volts
$\frac{1}{2}O_2/H_2O$	0.82
Fe^{3+}/Fe^{2+}	0.77
NO_3^-/NO_2^-	0.42
Cytochrome *c* Fe^{3+}/Fe^{2+}	0.22
Methemoglobin/hemoglobin (Fe^{3+}/Fe^{2+})	0.17
Methylene blue, ox/red	0.01
$FMN/FMNH_2$	−0.20
Acetoacetic acid/3-hydroxybutyric acid	−0.27
$NAD^+/NADH$	−0.32
Ferredoxin Fe^{3+}/Fe^{2+}	−0.42
$H^+/\frac{1}{2}H_2$	−0.42

might be noted that the iron-containing protein cytochrome *c* is included to illustrate the fact that the oxidation–reduction potential of the ferrous–ferric couple is markedly changed when the metal is chelated as part of the protein. Ferredoxin is another iron-containing protein which will be discussed in more detail later. It might be noted that the reduced ferredoxin is a powerful reducing agent[4,5].

The values of Table 1 can be used to calculate the standard free energy change at pH 7 ($\Delta G^{\circ\prime}$) for a typical oxidation–reduction reaction from the relationship $\Delta G^{\circ\prime} = -nFE_0'$. For example, the value of $\Delta E'$ for reaction 3 can be calculated to be +0.05 volts so that $\Delta G^{\circ\prime}$ is approximately −2 kcal/mole. The sign and magnitude

$$\text{HO—CH(CH}_2\text{CO}_2\text{H)(CH}_3) + \text{NAD}^{\oplus} \rightleftharpoons \text{O=C(CH}_2\text{CO}_2\text{H)(CH}_3) + \text{NADH} + \text{H}^{\oplus} \qquad (3)$$

of this quantity correspond closely to the expectation we have from knowledge of similar reactions that occur in living cells. In a similar fashion we can calculate that the reoxidation of the reduced coenzyme NADH by molecular oxygen (reaction 4) would have a $\Delta G^{\circ\prime}$ value of −52 kcal/mole. Similarly we can estimate that the reoxidation of the reduced coenzyme $FMNH_2$ by molecular oxygen (reaction 5)

$$NADH + \frac{1}{2}O_2 + H^+ \longrightarrow NAD^+ + H_2O \quad (4)$$

has a $\Delta G^{\circ\prime}$ value of -47 kcal/mole.

$$FMNH_2 + O_2 \longrightarrow FMN + H_2O + \frac{1}{2}O_2 \quad (5)$$

B. Oxidative Phosphorylation and the Electron Transport Sequence

The relatively large negative free energy changes for the reoxidation by molecular oxygen of the reduced flavin and nicotinamide coenzymes is consistent with the fact that these reoxidations are major sources of free energy in metabolic reactions. In particular, a substantial amount of the free energy of these reactions may be trapped in biochemically useful form as the energy-rich pyrophosphate derivatives adenosine diphosphate (ADP) and adenosine triphosphate, or ATP (**5**). The structures of ADP and AMP (adenosine monophosphate) should be obvious on consideration of structure **5**. The free energy change accompanying the hydrolysis

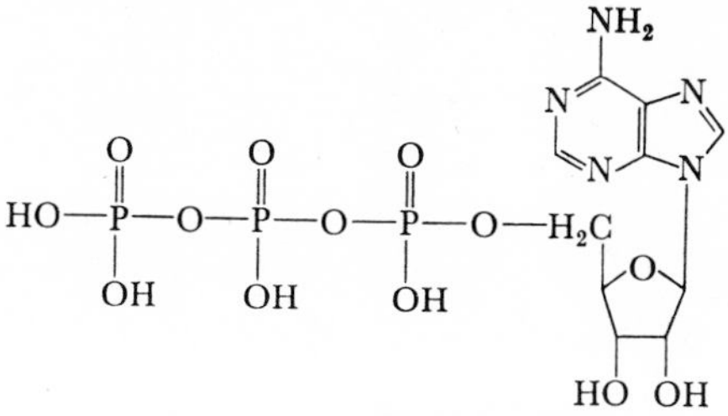

HO—P(=O)(OH)—O—P(=O)(OH)—O—P(=O)(OH)—O—H_2C—[ribose (HO, OH)]—adenine (NH_2)

(**5**)

of either of the pyrophosphate bonds of ATP is estimated to be -8 to -10 kcal under physiological conditions. The energy content of this group is utilized in a large number of coupled reactions so as to result in an energetically favorable overall reaction. The amount of ATP available to the cell is an exceedingly important quantity which affects sensitive control mechanisms and can thus greatly affect the metabolic pattern of an entire organism.

The major source of ATP in aerobic organisms are the processes of oxidative phosphorylation in which reduced nicotinamide or flavin nucleotides are reoxidized using molecular oxygen. This oxidation is coupled to a series of electron-transfer reactions which lead to a generation of adenosine triphosphate from adenosine diphosphate and inorganic phosphate. A schematic representation of part of the presently accepted electron-transport sequence is shown in Scheme 1. In this diagram SH_2 and S are reduced and

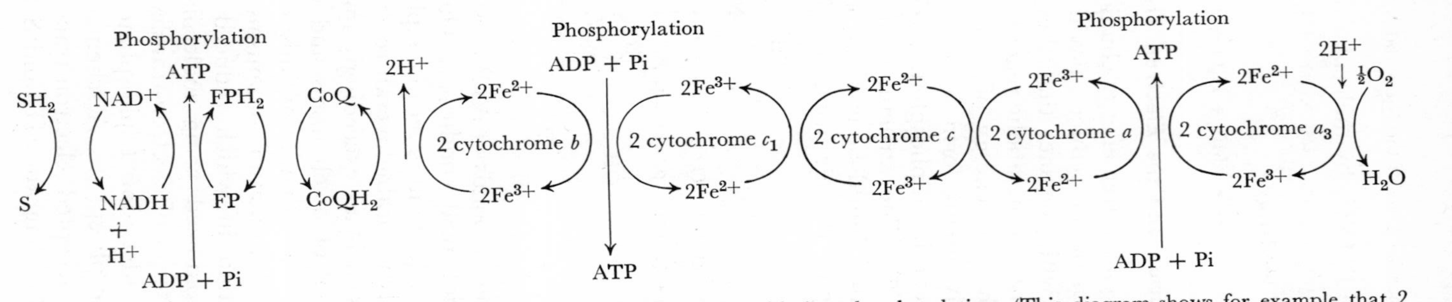

SCHEME 1. The electron transport sequence and probable sites of coupled oxidative phosphorylation. (This diagram shows for example that 2 molecules of reduced cytochrome *b* react with 2 molecules of oxidized cytochrome c_1, and this oxidation–reduction process is coupled with the transformation of ADP to ATP.)

oxidized metabolites, FPH_2 and FP are reduced and oxidized flavin derivatives (cf. **4a** and **3a**) which are bound to certain proteins, $CoQH_2$ and CoQ are hydroquinone and quinone derivatives of the type of coenzyme Q (**6**), and the various cytochromes (b, c_1, c, etc.) are proteins containing chelated iron atoms and having distinctive oxidation–reduction potentials.

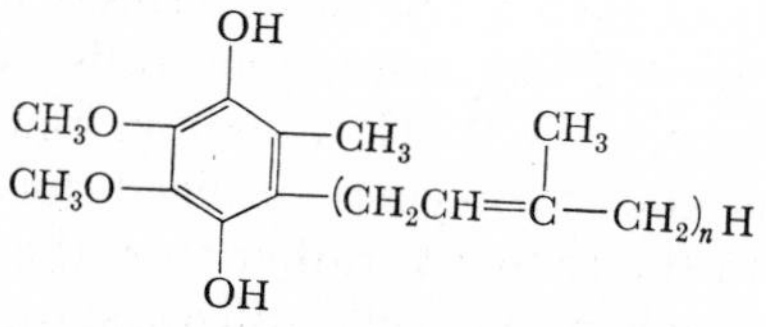

OH

CH_3O — — CH_3 CH_3

CH_3O — — $(CH_2CH{=}\overset{|}{C}{-}CH_2)_n H$

OH

(**6**, where n = integers up to 10)

Scheme 1 represents the reoxidation of nicotinamide coenzymes. In addition, the other major group of coenzymes, the flavins (**4a** or **4b**) also may be reoxidized via the electron-transport sequence but they enter the reaction sequence at the point of CoQ. It may be considered that there are two distinct branches of the electron-transport sequence, only one of which is directly represented in Scheme 1. This point is made because the second branch, entering as it does after one of the sites of oxidative phosphorylation, means that metabolic reactions leading to the formation of reduced flavin coenzymes potentially yield less chemically useful energy than reactions which result in formation of reduced nicotinamide coenzymes. The energy yield may be expressed in terms of the P/O ratio, by which is meant the ratio of (terminal) pyrophosphate linkages of ATP formed to the number of oxygens converted to water. The reoxidation of the nicotinamide coenzymes via the electron-transport sequence proceeds with a P/O ratio approaching 3, while for reoxidation of the flavin coenzymes the P/O ratio approaches 2 consistent with coupling of this reoxidation at a point past the flavoproteins in Scheme 1.

An important point regarding oxidative phosphorylation is that, in normal intact cells, phosphorylation is obligatorily coupled to the reoxidation of the coenzymes; in brief, oxidation is coupled to phosphorylation. The important implication of this statement can quickly be made apparent. The metabolic utilization of foodstuffs depends on the availability of oxidized nicotinamide and flavin coenzymes and also other factors such as the presence in the cell of sufficient ATP to result in the rapid formation of the phosphorylated derivatives which are the actual substrates of many of the metabolic transformations. In turn, the availability of oxidized coenzymes depends on the activity of the electron-transport sequence or on various biosynthetic steps which require the reduced coenzymes. Since under physiological conditions oxidation is obligatorily coupled to phosphorylation, this means that in normal cells the reoxidation of coenzymes via the electron-transport sequence cannot occur unless ADP, inorganic phosphate, and oxygen are present. The utility of such a system in terms of metabolic control is apparent. When the cell has ample foodstuffs available then ATP will be in abundance while AMP and/or inorganic phosphate will be limiting. Thus the coenzymes utilized for the oxidation of foodstuffs will be in the reduced, rather than the necessary oxidized forms. Conversely, when the cell utilizes substantial amounts of ATP for biosynthetic reactions, muscular work, etc., then ADP and inorganic phosphate will become available, oxidative phosphorylation will occur, and reduced coenzymes will be transformed into the oxidized forms. In turn, the availability of oxidized coenzymes will permit further oxidative metabolism of foodstuffs.

It should be apparent from the foregoing discussion that any substance which can interfere with the availability of reduced or oxidized coenzymes or which can interfere with the coupling of oxidation to phosphorylation will have the potential for disrupting the most fundamental life processes. We shall see that some nitro and nitroso compounds have these potentials.

C. *Biochemical Oxidation–Reduction Processes and the Nitro and Nitroso Groups*

A great variety of examples of the reduction of nitrogen compounds might be cited. For example, the work of Mortenson and his colleagues on the reduction of nitrogen to ammonia by cell-free extracts could be discussed[6–9], or enzymatic systems which employ NADH to reduce other nitrogen heterocycles[10,11]. However, the most appropriate reactions with which to introduce the subject are the various oxidation–reduction processes involving nitrite or nitrate compounds. Some recent examples can be cited[12–21].

One of these examples[16] concerns the reduction of aromatic nitro and nitroso compounds by the protein ferredoxin. Ferredoxin is an iron-containing protein which can act as a powerful reducing agent[4,5]. It has been isolated from a variety of plant and bacterial sources, but the possibility of a very similar reductive metabolism of nitro and nitroso compounds applies just as well to animals as to these organisms. Wessels described the reduction of nitrosophenol,

2,4-dinitrophenol, nitrobenzene, and *m*-dinitrobenzene by chemically reduced ferredoxin[16]. The latter reaction is illustrated in equation 6 where $Fd(Fe^{2+})$ represents the reduced form of the

$$m\text{-}C_6H_4(NO_2)_2 + 6Fd(Fe^{2+}) + 6H^+ \longrightarrow m\text{-}NO_2C_6H_4NH_2 + 6\ Fd(Fe^{3+}) + 2H_2O \tag{6}$$

metalloprotein ferredoxin. In this reaction, ferredoxin acts as a one-electron donor[5] in the reduction of *m*-dinitrobenzene to *m*-nitroaniline. Similarly, 2,4-dinitrophenol is reduced to 2-amino-4-nitrophenol. Reduced ferredoxin could be produced in turn from the oxidized form of the metalloprotein and excess reduced nicotinamide adenine dinucleotide phosphate (**2b**) in the presence of appropriate enzymes[16].

In a similar manner it has been shown that the flavin coenzymes can be utilized to reduce nitro compounds. When the reduced flavin coenzyme (**4a**) is mixed with 2,4-dinitrophenol the phenol is reduced to 2-amino-4-nitrophenol[15]. This result was part of a study dealing with a bacterial nitrite reductase which reduced nitrite ion to nitric oxide and nitrous oxide. There are related enzymes which catalyze the reduction of nitrite to ammonium ion[20]. A number of coenzymes can serve as the ultimate reducing agents in such reactions. No attempt is made here to describe current research on aspects of the oxidation–reduction reactions of nitrogen compounds, but a few examples have been chosen so as to illustrate possible reaction mechanisms discussed in the chapter.

D. The Uncoupling of Oxidation from Phosphorylation

One further concept of great importance remains to be discussed before we begin a detailed discussion of the biochemistry and pharmacology of the nitro and nitroso groups. This is the concept of the uncoupling agent. In section II.B the process of oxidative phosphorylation was discussed and it was pointed out that in normal cells the reoxidation of coenzymes is obligatorily coupled to phosphorylation. Phrased differently, the utilization of molecular oxygen in the electron-transport sequence requires that ATP be generated ismultaneously (see Scheme 1). In the normal cell this is an important mechanism for physiological control of energy balance and food utilization. It is obvious that if the coenzymes were simply reoxidized directly by molecular oxygen rather than via the electron-transport sequence there would be no ATP produced.

Since ATP is the most important reservoir of chemically useful energy any substance which can interfere with oxidative phosphorylation interferes with a majority of the important degradative and synthetic reactions of aerobic organisms. Certain types of aromatic nitro compounds can act as powerful uncoupling agents. One particularly potent uncoupling agent is 2,4-dinitrophenol, and the literature of even a few years is replete with examples of this effect of the compound[22–266]. The uncoupling action of the phenol results in a reoxidation of the reduced coenzymes but without a concomittant formation of biochemically useful ATP. Because the coenzymes are converted back to their oxidized forms they can again be utilized for the oxidative metabolism of additional foodstuffs. The lack of ATP causes the organism to seemingly starve in the midst of plenty.

III. BIOCHEMISTRY AND PHARMACOLOGY OF NATURALLY OCCURRING COMPOUNDS CONTAINING THE NITRO OR NITROSO GROUP

Although the number of naturally occurring nitro and nitroso compounds is relatively small, the group is still important because it contains compounds such as antibiotics and carcinogens. A thorough discussion of any naturally occurring compound would require detailed knowledge of the biosynthesis as well as biochemical functioning of the compound. In most cases, our knowledge of the biochemical transformations of nitro and nitroso compounds is too limited to permit such detailed discussions. However, it is possible to suggest certain analogies between well understood and poorly understood transformations. This is, of course, a reasonable approach in view of the economy of nature.

A. Antibiotics

1. Chloramphenicol

Among naturally occurring nitro compounds chloramphenicol is most important from a practical point of view. It was also the first natural product recognized to possess an aromatic nitro group.

a. Origin and structure. Chloramphenicol was obtained in crystalline form from cultures of *Streptomyces venezuelae*, a mold first isolated from soil obtained near Caracas, Venezuela[267]. The chemical structure of chloramphenicol was established by Rebstock and coworkers to be D-(1)-*threo*-1-*p*-nitrophenyl-2-dichloroacetamido-1,3-propanediol[268] (**7**). Because of the presence of two asymmetric

$$O_2N-C_6H_4-CH(OH)-CH(NHCOCHCl_2)-CH_2OH$$

(**7**)

carbon atoms in the molecule there are D- and L-*threo* and D- and L-*erythro* isomers; of these, only the D-*threo* isomer exhibits the full antibiotic activity.

The biosynthesis of chloramphenicol (**7**) has been the subject of numerous practical and theoretical investigations. The practical importance of such a study is obvious, for it led to efficient large-scale production of the antibiotic at a time when efficient chemical methods of synthesis were not available. Because **7** is a fermentation product it is readily possible to manipulate culture conditions and select mutant strains so as to maximize the yield of the desired metabolite. One important culture condition which can be easily varied is the presence of specific substrates which can act as particularly efficient precursors of the desired metabolite. This also provides the basis of a powerful technique for the investigation of biosynthetic pathways, since isotopically labeled precursors may be employed to measure overall rates of incorporation and determine the origins of particular atoms in the final metabolic product.

The formation of ^{14}C-labeled **7** has been compared with the formation of labeled aromatic amino acids such as phenylalanine (**8**) and tyrosine (**9**) when various ^{14}C-labeled glucose isomers were

$$C_6H_5-CH_2-CH(NH_2)-CO_2H$$

(**8**)

$$HO-C_6H_4-CH_2-CH(NH_2)-CO_2H$$

(**9**)

utilized by a chloramphenicol-producing *Streptomyces* species[269]. The resulting labeling patterns of the C-6–C-3 phenylpropanoid skeleton of **7** and of the aromatic amino acid **8** isolated from the protein of the mold were found to be similar, suggesting a common pathway for the two substances. On the other hand, an earlier study[270] established that the phenylpropanoid amino acids were not converted to **7** without prior degradation. This suggests that although there may be a common intermediate for both **7** and **8** there must be some irreversible branch point that precludes facile interconversions. Based on current knowledge of the pathway for formation of the aromatic amino acids it has been suggested that shikimic acid (**10**) may serve as this common intermediate[271]. Although **10**

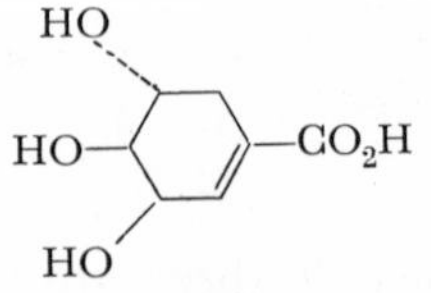

HO

HO— —CO_2H

HO

(**10**)

is only utilized to a limited extent by *Streptomyces* species it was incorporated into the antibiotic as well as into **8** and **9** of proteins. More extended studies of the incorporation of labeled potential precursors have led one group to postulate the relationships shown in Scheme 2[271] (shown on page 215).

Additional support for the biosynthetic relationships shown in Scheme 2 is provided by the fact that α-*N*-dichloracetyl-L-*p*-aminophenylserinol (**11**) has been isolated from chloramphenicol-producing cultures of *Streptomyces venezuelae*[272]. This, together with the fact that DL-*threo*-*p*-nitrophenylserine and *p*-nitrophenylserinol[273] (**12**) are not

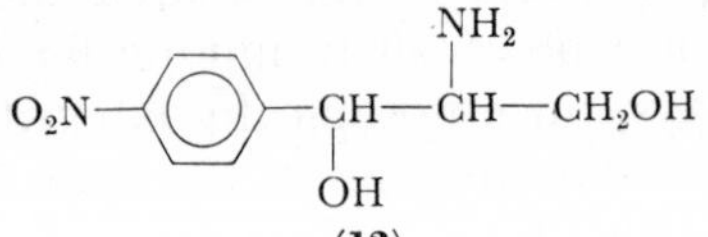

NH_2

O_2N— —CH—CH—CH_2OH

OH

(**12**)

utilized for the formation of **7** when added to the culture media suggest that the formation of the nitro group is one of the last reactions in the biosynthetic pathway.

Under natural conditions the mold mycelia produce the nitro group from inorganic precursors. The presence of ammonium sulfate or of nitrate or nitrite salts permits the formation of **7**[274]. Ammonium ion appears to be more stimulatory than nitrate salts, much as in the case of the biosynthesis of β-nitropropionic acid[275]. Significantly, ^{15}N-labeled nitrate in the culture medium is not efficiently incorporated into the nitro group of **7**. These results are consistent in that in order for nitrate ion to serve as a biological precursor of the aromatic nitro group of **7** it must first be reduced to the level of ammonia which is then utilized in reactions which lead to the aromatic amine derivative **11**. In this regard it is probably significant that although **7** is known to inhibit[276] the formation of

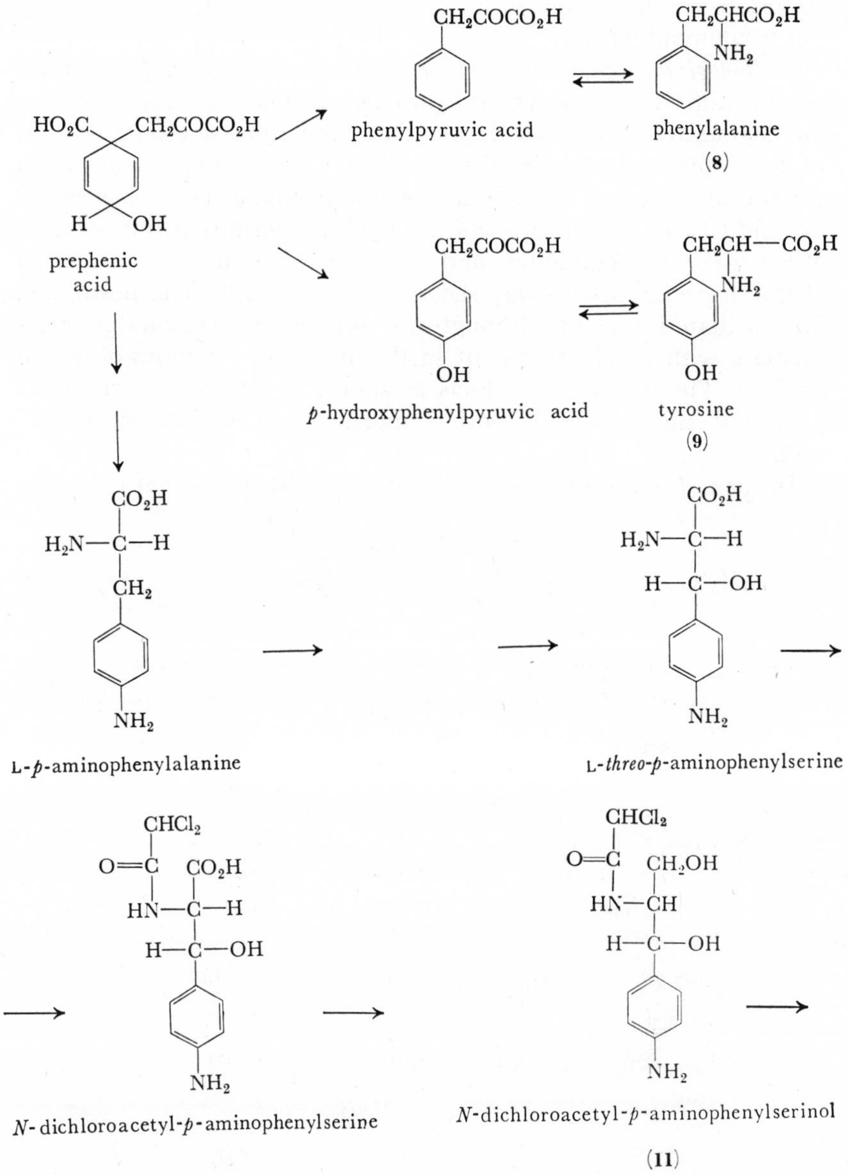

CH_2COCO_2H

CH_2CHCO_2H

NH_2

HO_2C CH_2COCO_2H

phenylpyruvic acid

phenylalanine

(8)

H OH

prephenic acid

CH_2COCO_2H

$CH_2CH—CO_2H$

NH_2

OH

OH

p-hydroxyphenylpyruvic acid

tyrosine

(9)

CO_2H

$H_2N—C—H$

CH_2

NH_2

L-*p*-aminophenylalanine

CO_2H

$H_2N—C—H$

H—C—OH

NH_2

L-*threo*-*p*-aminophenylserine

$CHCl_2$

O=C CO_2H

HN—C—H

H—C—OH

NH_2

N-dichloroacetyl-*p*-aminophenylserine

$CHCl_2$

O=C CH_2OH

HN—CH

H—C—OH

NH_2

N-dichloroacetyl-*p*-aminophenylserinol

(11)

SCHEME 2. Proposed pathway for biosynthesis of chloramphenicol[271].

certain types[277] of proteins it is in fact found to be stimulatory with regard to the formation of an enzyme capable of reducing nitrate ion to ammonium ion[278,279].

b. Biological degradation. Many important biological reactions have distinctive pathways for synthetic, or anabolic reactions, and for degradative, or catabolic reactions. It is thus not surprising to find a similar situation for the case of **7**. For example, a bacterium capable of utilizing **7** as a sole carbon source has been described[280,281]. Careful examination of the culture medium permitted the isolation of a variety of metabolites, and this in turn led to the postulation of the degradation pathway shown in Scheme 3[281]. The postulated intermediate imine **13**, although not isolated, is a reasonable intermediate with much precedent in the metabolic reactions of amino acids[282]. The eventual products as shown in Scheme 3 are acetic acid and succinic acid, both important metabolites in normal metabolism.

In man, **7** is rapidly absorbed from the gastrointestinal tract and

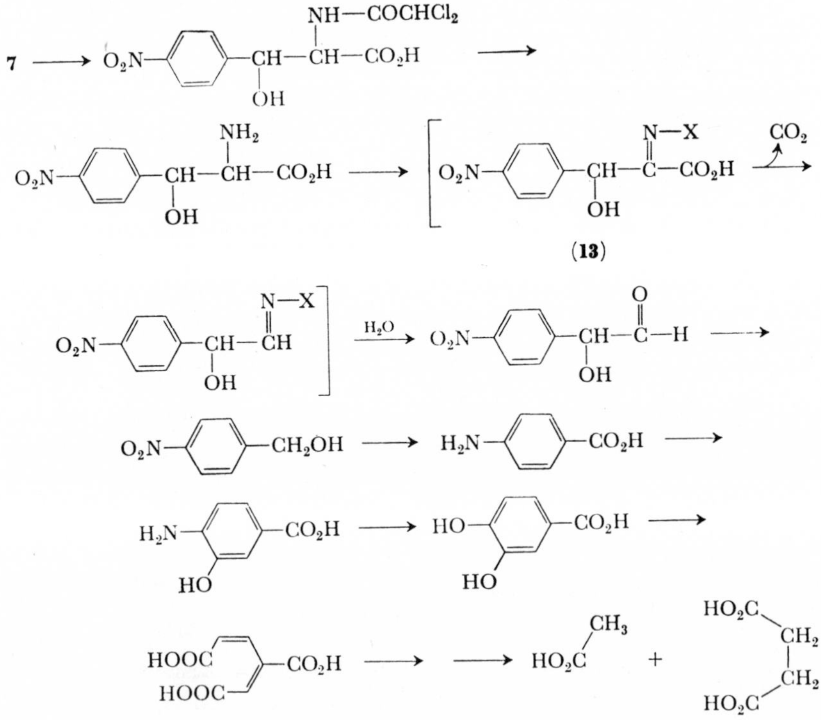

7 → $O_2N-C_6H_4-CH(OH)-CH(NH-COCHCl_2)-CO_2H$ →

$O_2N-C_6H_4-CH(OH)-CH(NH_2)-CO_2H$ → [$O_2N-C_6H_4-CH(OH)-C(=N-X)-CO_2H$] **(13)** $\xrightarrow{-CO_2}$

[$O_2N-C_6H_4-CH(OH)-CH=N-X$] $\xrightarrow{H_2O}$ $O_2N-C_6H_4-CH(OH)-C(=O)-H$ →

$O_2N-C_6H_4-CH_2OH$ → $H_2N-C_6H_4-CO_2H$ →

$H_2N-C_6H_3(HO)-CO_2H$ → $HO-C_6H_3(HO)-CO_2H$ →

$(HOOC)(HOOC)C_4H_2-CO_2H$ → → HO_2C-CH_3 + $HO_2C-CH_2-CH_2-CO_2H$

SCHEME 3. Hypothetical pathway for microbial catabolism of chloramphenicol[281].

the maximal level of **7** in the blood is reached in about 2 hours followed by a gradual decrease over a 12-hour period. Some 60% of the drug is bound to the blood serum albumins. High concentrations of **7** are found in the liver and kidneys, whereas only small amounts are found in the brain and cerebrospinal fluid[283]. Less than 10% of the administered dose is excreted unchanged, the remainder undergoing metabolic transformations in part influenced by the composition of the intestinal flora. As much as 90% of the administered dose appears in the daily urine, but this material possesses less than 10% of the biological activity of free **7**. This is because most of the aromatic compounds derived from **7** are present instead as hydrolysis products such as **12** or glucuronic acid conjugates such as **14**. This is a relatively common means of detoxification or

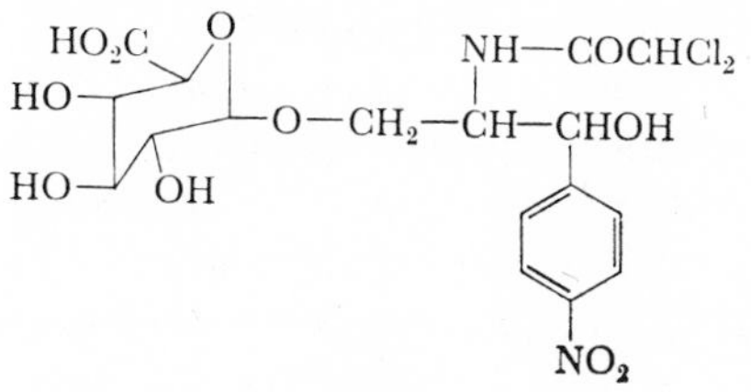

HO_2C O NH—$COCHCl_2$
HO O—CH_2—CH—CHOH
HO OH
NO_2

(**14**)

elimination of a potentially toxic material as inactive derivatives.

The metabolism of **7** by rats is known to differ significantly from that by man. Small amounts are excreted into urine as the glucuronic acid conjugate, but large amounts of the conjugated **7** are excreted with bile into the intestines. Particularly in the cecum the intestinal flora cause a number of transformations of **7**, including hydrolysis and reduction reactions. The arylamines resulting from bacterial reduction are partially absorbed into the blood stream and are then typically acetylated, conjugated with glucuronic acid, and then excreted in the urine[284,286]. Significantly, less of the administered dose in rats is excreted as compounds having intact nitro groups; reduction products such as *p*-aminobenzoic acid are observed[286,287]. The different products are partly a function of the particular intestinal flora. For example, it has been shown that *Bacillus mycoides* and *B. subtilis* produce mainly inactive nitro compounds, whereas *Escherichia coli* produces inactive arylamine derivatives[288]. Among the specific compounds which have been identified are: **12**, α-amino-β-hydroxy-*p*-nitropropiophenone, *p*-nitrobenzaldehyde, *p*-nitrophenylserine, and *p*-nitrobenzoic acid.

Although the predominant action of rat and guinea pig liver and kidney is to cause hydrolysis of **7** to **12** they also cause reduction to arylamines. The reduction of **7** is enzyme catalyzed, has a pH optimum of 7.8, and is activated by the coenzymes NADH (**2a**) and $FMNH_2$ (**4a**) and also by ethyl alcohol. Because the reduced coenzymes **2a** and **4a** result from enzymatic oxidation of ethyl alcohol to acetaldehyde, at least in the presence of the enzyme alcohol dehydrogenase, the role of the alcohol may be to cause formation of the reduced coenzymes. Enzymatic activity is found in the cytoplasm of the cell as well as in the microsomal particles found in cells[289], and both of these locations contain a number of reversible oxidation–reduction enzyme systems which employ the nicotinamide and flavin coenzymes. Significantly, known inhibitors of nitrate reductases including dipyridyl, silver nitrate, oxygen, sodium salicylate, and ammonium sulfate all have an inhibitory effect on the course of this reduction reaction[290,291].

c. Mechanism of action on microorganisms. Chloramphenicol (**7**) possesses a broad antibacterial spectrum. Its action is mainly bacteriostatic, arresting the growth of microorganisms and thus permitting the natural defenses of the body to cope with the foreign organisms. Against certain strains, however, **7** has a definite bactericidal action, actively killing the microorganisms. Some strains are resistant to the action of **7** and mutant strains can be selected which show greater and greater resistance. It should be remembered that the processes of evolutionary selection are much more readily apparent with microorganisms because of the large numbers of individual cells and the short generation times. Several mechanisms for resistance to **7** have been recognized. Some resistant microorganisms are found to have a cellular membrane which is impermeable to the antibiotic[292] but in other instances, the organism may be capable of degrading the antibiotic as discussed in the preceding part of this chapter.

In the cell, **7** acts to inhibit the biosynthesis of proteins[293–295] as illustrated by some recent examples[296–319]. The effect of **7** may be effectively illustrated by some data of Kroon[320], who measured the relative incorporation of ^{14}C-labeled leucine into proteins being synthesized in controlled incubation mixtures (Table 2). It may be seen that the effects of the two added compounds are quite marked. The effect of 2,4-dinitrophenol is probably explained by its interference with the metabolic processes which generate the energy-rich phosphate compound ATP needed in the cell for peptide bond formation. However, the usual cellular extracts or homogenates have some residual or endogenous ATP so that there is still a small

TABLE 2. Effect of inhibitors on leucine-^{14}C incorporation into beef heart mitrochondrial protein[320].

Expt.	Added inhibitor	Specific activity of isolated protein, counts/min/mg
1	None	180
	5×10^{-4} *M* 2,4-dinitrophenol	64
2	None	217
	1.5×10^{-4} *M* chloramphenicol	7
3	None	300
	5×10^{-4} *M* 2,4-dinitrophenol	29
	1.5×10^{-4} *M* chloramphenicol	12

net incorporation of ^{14}C-leucine into the protein even in the presence of the uncoupling agent.

The effect of **7** is dramatic as it causes an almost complete halt in the synthesis of proteins. It appears that **7** acts to inhibit the formation of a complex of ribonucleic acids and high molecular weight nucleoproteins which is a prerequisite for protein synthesis[321–328]. Cellular mechanisms recognize that protein formation is not occurring and attempt to compensate for this by producing more of the particular types of ribonucleic acid involved in protein synthesis. The result is that when protein synthesis is inhibited by **7** there is an accumulation of ribonucleic acids in the cell. The accumulated ribonucleic acid can be hydrolyzed back to nucleotides upon removal of **7**[329]. It is interesting to note that the ability of **7** to inhibit protein synthesis has been observed with cell free experimental systems obtained from microorganisms which are resistant to the bacterostatic action of **7** by virtue of the impermeability of their cellular membranes[330]. It is also significant that the L-*threo* and L-*erythro* isomers of **7** have little effect on protein synthesis[322]. Thus, the biological effects of **7** are presently well explained at the subcellular level and it remains only to specify more exactly the nature of the nucleoprotein receptor site; some progress along these lines is being made[328].

d. Tolerance and toxicity. Dangerous physiological reactions sometimes result from the administration of **7**. For example, the blood cell forming capabilities of bone marrow may be interfered with and sporadic skin eruptions may occur. These complications are probably associated with the impaired ability to synthesize proteins[331]. Other side effects include gastrointestinal disorders such as nausea, vomiting, and diarrhea. A particularly strong reaction is seen in newborn infants, probably because they have not yet developed efficient mechanisms for detoxification and excretion.

e. Related substances. Not surprisingly, the discovery that choramphenicol (**7**) is a potent antibiotic and yet possesses a relatively simple chemical structure has stimulated much research aimed toward the discovery of related biologically active compounds. It appears, however, that even slight changes in **7** can cause the loss of biological activity. This is of course apparent first of all from consideration of the results of studies using stereoisomers of **7**, since only the D-(−)-*threo* isomer exhibits biological activity. Compounds in which the nitro group was placed in the *ortho* or *meta* position were all inactive, as were halogen, methoxy, phenoxy, as well as nitrophenyl derivatives[332–336]. However, the azidoacetamide derivative **15** has been introduced into chemotherapeutics under the name 'Leukomycin N.' Phillips has described the synthesis of compounds with potential antiviral activity, including *N*-(2,5-dimethoxy-4-nitrophenethyl)dichloracetamide (**16**) which is found to possess biological activity[337,338].

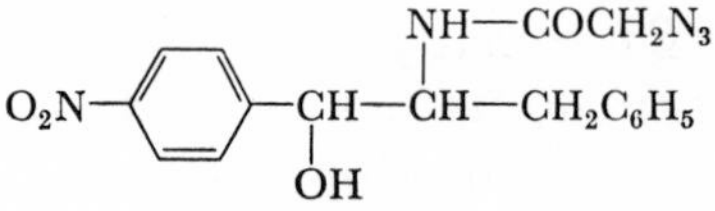

$NH-COCH_2N_3$
$O_2N-C_6H_4-CH(OH)-CH-CH_2C_6H_5$

(**15**)

H_3CO
$O_2N-C_6H_2-CH_2-CH_2NH-COCHCl_2$
OCH_3

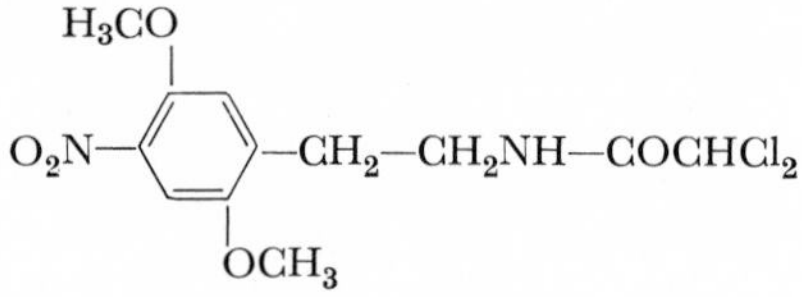

(**16**)

A commonly employed derivative of **7** is the palmitate ester. Esterification with a long saturated fatty acid imparts several advantages from a therapeutic point of view as the ester is less toxic and nearly tasteless. The ester is slowly hydrolyzed in the gastrointestinal tract, freeing **7** which can then be absorbed into the cells. In this way more even and sustained levels of the drug are present in the organism.

2. Other antibiotics

a. 2-Nitroimidazole. Azomycin or 2-nitroimidazole (**17**), was isolated from *Nocardia mesenterica*. It has a relatively low toxicity and exhibits a broad antibiotic activity even against protozoa[340].

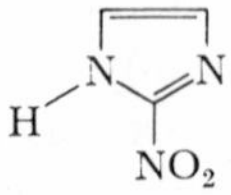

H N N
NO_2

(**17**)

b. p-Nitrobenzylpenicillin. The addition of *p*-nitrophenylacetic acid to culture media inoculated with *Penicillium chrysogenum* results in the formation[341] of *p*-nitrobenzylpenicillin (**18**). There are also

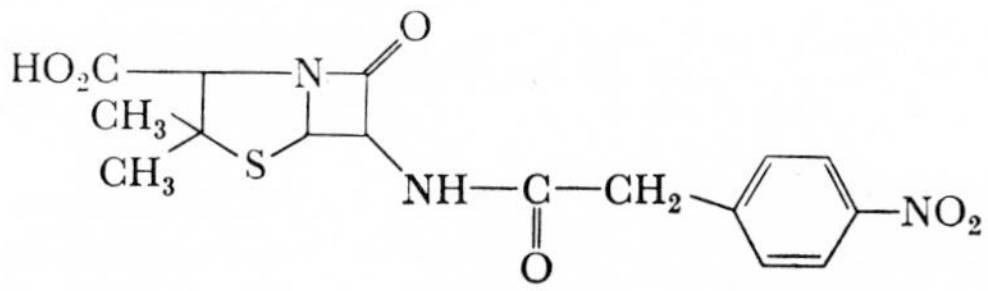

(**18**)

reports of the isolation of 2-hydroxy-3-nitrophenylacetic acid from penicillin cultures thus suggesting that these microorganisms can bring about biochemical nitration reactions[342].

The degradation, detoxification, and excretion of **18** *in vivo* may be predicted to follow a course similar to that of the corresponding benzyl derivative, with formation of penicilloic acid (**19**) and phenaceturic acid (**20**) as shown in reaction 7[343]. In forming **20** it may be seen that the nitro-substituted carboxylic acid has been reacted with the amino acid glycine. This is a relatively common means of detoxification[344].

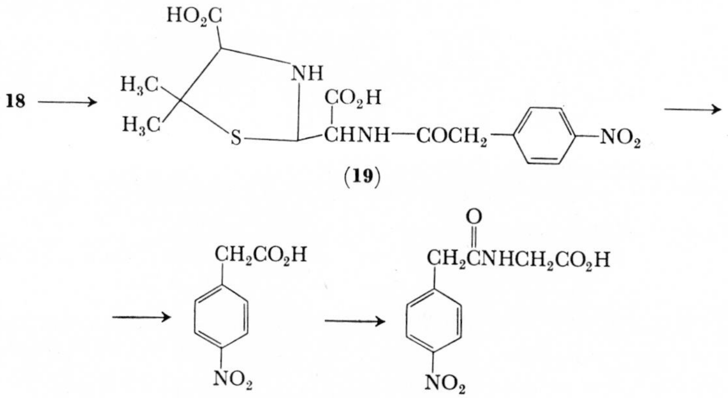

18 ⟶ (**19**) ⟶ ⟶ (**20**) (7)

c. Aristolochia acids. The *Aristolochiaceae* plant family includes over 200 representatives, many of them growing in Mediterranean regions. Medicinal preparations using various parts of these plants have been employed since ancient times. The structure of Aristolochia acid I, isolated from *Aristolochia clematitis* L., has been established by Pailer and his colleagues to be 3,4-methylenedioxy-8-methoxy-10-nitrophenanthrene carboxylic acid (**21**)[345,346]. Aristolochia acid

II differs only by the absence of the 8-methoxy function. These acids exhibit weak antitumor and bacteriostatic activities[347,348]. However, they have a strong potentiating effect on phagocytosis, the ingestion of foreign microorganisms by the white blood cells. Thus, the

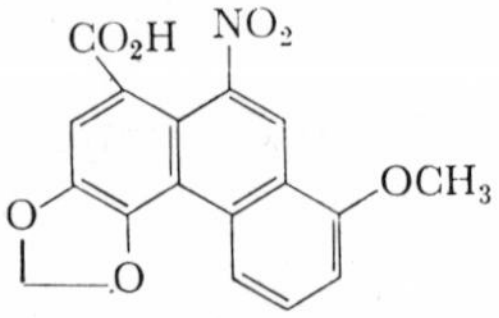

(**21**)

Aristolochia acids have potential therapeutic value in infections where phagocytosis represents the essential mechanisms of defense[348]. The isolation of this material represents one example of the repeated instances where a therapeutically active substance was isolated from an ancient nostrum.

d. 3-Nitropropionic acid. This compound was first isolated in 1920 as a component of the glycoside hiptagen from the root of the Javanese tree *Hiptaga madablota*. The structure of the acid component was finally identified as 3-nitropropionic acid (**22**)[349]. It is fairly

$$O_2N—CH_2CH_2CO_2H$$

(**22**)

widely distributed in the plant kingdom, having been found in the roots of other plants such as *Viola odorata*[350], numerous fungi or molds such as *Aspergillus flavus*, *Penicillium atrovenetum*, and *Aspergillus oryzae*[350–352], and in the legume *Indigofera endecaphylla*[353]. The latter plant had been introduced into Hawaii as a forage and cover crop, but when it served as fodder for cattle and sheep there occurred severe symptoms of intoxication that were ascribed to **22**. The widespread occurrence of **22** suggests that it may play some important biochemical function. The toxicity to animals is reminiscent of the ambiguous biochemical role played by the alkaloids. Despite some early uncertainty[350] however, a definite antibiotic activity of **22** against certain *Bacillus* species has been recognized[351].

A number of interesting studies on the biosynthesis of **22** have been conducted. Growing cultures of *Penicillium atrovenetum* bring about the formation of **22** when the medium contains ammonium ions and four-carbon dicarboxylic acids[352,354,355]. A particularly efficient incorporation of carbons from L-aspartic acid occurred in the biosynthesis of **22**, and it was established that carbons 2, 3, and 4 of aspartic acid become carbons 3, 2, and 1 of **22**, respectively[356–359].

In the growth of the mold mycelium the production of **22** is particularly great during the early stages of growth and the total amount in the medium decreases when the growth reaches the end of the logarithmic phase[352,355]. More than 60% of the ammonium nitrogen metabolized by the mycelia to compounds other than protein can be isolated as **22.** In contrast to the facile incorporation of ammonium nitrogen it was found[359] that nitrate nitrogen was not utilized for the formation of **22** and, indeed, caused a decrease in its formation[355]. The amino group of aspartic acid was utilized in preference to ammonium nitrogen in the biosynthesis of **22**[358,359], and it appears that the stimulatory effect of ammonium ion is simply related to an increase in the relative amounts of aspartic acid (**23**) by means of a pair of coupled enzyme-catalyzed reactions (7a and 7b)

$$\mathrm{NADH} + \mathrm{NH_4^{\oplus}} + \mathrm{HO_2C{-}C({=}O){-}CH_2{-}CH_2{-}CO_2H} \underset{\text{dehydrogenase}}{\overset{\text{glutamic}}{\rightleftharpoons}} \mathrm{NAD^{\oplus}} + \mathrm{H_2N{-}CH(CO_2H){-}CH_2{-}CH_2{-}CO_2H} \qquad (7a)$$

$$\mathrm{H_2N{-}CH(CO_2H){-}CH_2{-}CH_2{-}CO_2H} + \mathrm{HO_2C{-}C({=}O){-}CH_2{-}CO_2H} \underset{\text{transaminase}}{\overset{\text{glutamic–oxalacetic}}{\rightleftharpoons}} \mathrm{HO_2C{-}C({=}O){-}CH_2{-}CH_2{-}CO_2H} + \underset{(\mathbf{23})}{\mathrm{H_2N{-}CH(CO_2H){-}CH_2{-}CO_2H}} \qquad (7b)$$

involving α-ketoglutaric acid, glutamic acid, oxalacetic acid, and the reducing agent NADH (**3a**). The same reaction is possible using the coenzyme NADPH, and these reactions are known to represent a major mechanism for the reversible incorporation or removal of amino nitrogen into amino acids and proteins. In view of the possibility of reduction of nitrite ion into ammonia and its subsequent reaction with oxalacetic acid it is not surprising that both nitrite ion and oxalacetate can be used in the biosynthesis of **22** although they are much less efficient than aspartic acid as precursors.

An enzyme termed 'β-nitroacrylic acid reductase' was obtained from extracts of *Penicillium atrovenetum*[360]. The enzyme catalyzed the reduction of 3-nitroacrylic acid (**24**) as shown in reaction 7c. The activity appeared to be due to a single enzyme having a pH optimum of ~5 and was moreover highly specific; a number of unsaturated compounds, including acrylic acid, crotonic acid, fumaric acid, and cinnamic acid were not reduced by excess NADPH in the presence of this enzyme. This specificity together with the fact that reaction

$$\underset{(\mathbf{24})}{O_2N{-}CH{=}CH{-}CO_2H} + NADPH + H^{\oplus} \longrightarrow O_2N{-}CH_2{-}CH_2{-}CO_2H + NADP^{\oplus} \quad (7c)$$

7c was the first demonstration of the enzymatic formation of **22** *in vitro*, suggested that **24** might be on the normal pathway for biosynthesis of the saturated nitro compound[360]. Further support for this hypothesis was obtained by measuring the amount of ^{14}C-labeled aspartic acid which was transformed into **22** in the presence of varied concentrations of **24**. Indeed, when the concentration of added **24** in the culture medium reached a level of approximately 2 μmole/ml the specific activity of resulting **22** decreased almost to zero, indicating that **24** was being used to produce **22.** A similar experiment employing cold aspartic acid and ^{14}C-labeled **24** confirmed that the carbons of **23** were utilized to produce **22** under these conditions[360]. It should be noted that these results do not prove that **24** is on the natural pathway for biosynthesis of **22** although they certainly are consistent with such a possibility.

B. Other Compounds of Natural Origin

1. Aureothin

This antibiotic is produced by *Streptomyces netropsis* and *S. thiolutens* and is reported to have structure **25**[361,362].

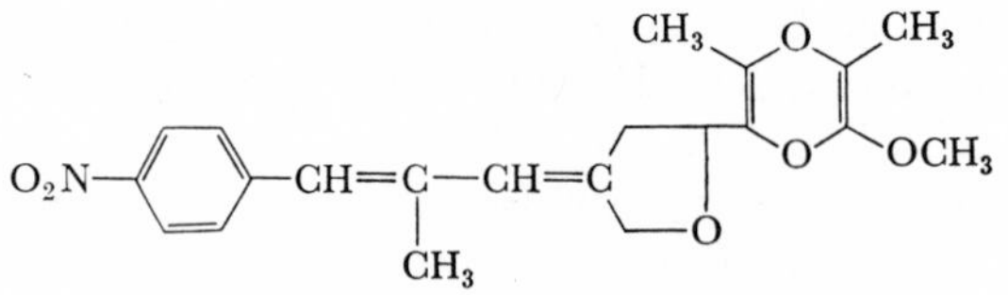

(**25**)

2. 1-Phenyl-2-nitroethane

Compound **26** has been isolated from the wood and bark of various plants such as *Aniba canellila* and *Octotea pretiosa*. It is a

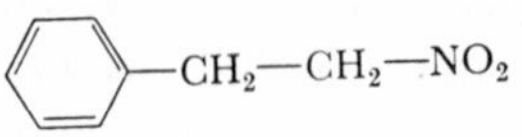

(**26**)

constituent of certain essential oils and possesses a quite characteristic odor[363,364].

3. *p*-Methylnitrosaminobenzaldehyde

This interesting *N*-nitroso compound **27** has been isolated from the basidiomycete *Clitocybe suaveolens*[365]. In view of the discussion of

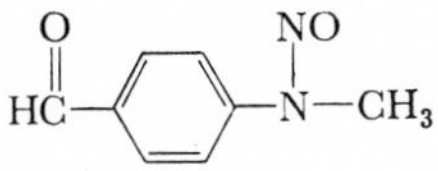

$$OHC-C_6H_4-N(NO)-CH_3$$

(**27**)

N-nitroso compounds which is given in the next section (III.B. 4) and in section IV.A. 5, it seems likely that **27** can exhibit carcinogenic properties.

4. Methylazoxymethanol glycosides

A naturally occurring compound related to the *N*-nitroso compounds is the glycoside cycasin (**28**) isolated from several tropical cycad plants[366,367]. The discovery of this material is an interesting

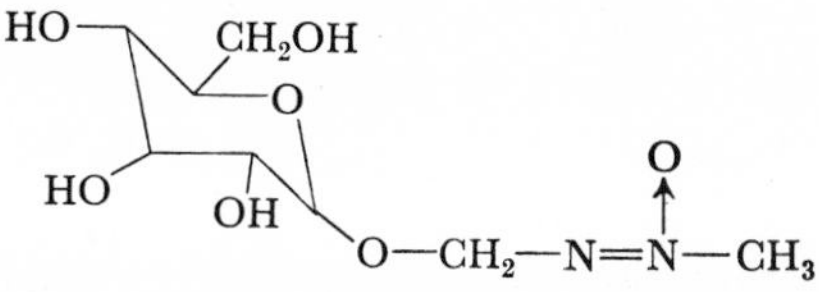

$$\text{Glucosyl}-O-CH_2-N{=}\overset{O}{\overset{\uparrow}{N}}-CH_3$$

(**28**)

tale. Ground meal from the seeds of the cycad plants, *Cycas circinalis L.*, were used as food by tribes on Guam and other islands in the Mariana chain. It was observed that these peoples also suffered a high incidence of certain diseases, including carcinoma of the liver[368]. When cycad seed meal was fed to rats there resulted a number of liver and kidney tumors and other carcinomas[369,370]. The toxic principle is now recognized to be **28**.

Interestingly, when **28** is fed to germ-free rats or when it is injected it is not carcinogenic. On the other hand, the aglycone methylazoxymethanol is a potent carcinogen. Apparently, the intestinal bacteria act to hydrolyze **28**, thus freeing methylazoxymethanol which can then be absorbed and act as a carcinogen[371]. It might also be noted that when **28** was fed to pregnant rats from the 17th to the 19th day of pregnancy, tumors resulted in the surviving offspring[372]. This means that compounds related to the *N*-nitroso compounds may present considerable health hazards which may not be immediately apparent. This topic will be dealt with in more detail in section IV.A. 5.

Interesting studies relating to the biochemical mode of action of **28** have been conducted. Riggs has observed that **28** can be employed under certain conditions to cause methylation of phenol to form anisole[367]. This suggests that **28** may function *in vivo* as an alkylating agent. Indeed methylazoxymethyl acetate (**29**) has been shown to alkylate the guanine residues of cellular ribonucleic acid (RNA) and deoxyribonucleic acid (DNA) with the formation of 7-methylguanine (**30**)[373]. The alkylation of rat RNA and DNA,

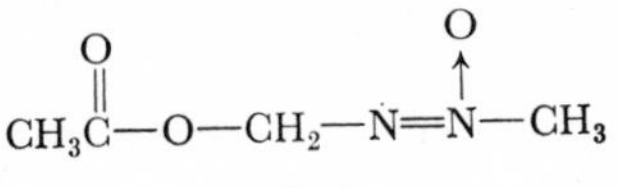

$$CH_3\overset{\overset{O}{\|}}{C}-O-CH_2-N=\overset{\overset{O}{\uparrow}}{N}-CH_3$$

(**29**)

(**30**) [structure: 7-methylguanine, with OH, CH_3 on N-7, and H_2N]

following the administration of cycasin, has been described[374]. Formation of **30** was greater in liver RNA than in kidney or small intestine RNA, and in turn DNA was more highly alkylated than RNA. Possibly these effects are closely related to the inhibition of protein synthesis in liver following administration of **28**. Because of the similarity in biological effects of **28** and diethyl nitrosamine (section IV.A. 5) as well as other *N*-nitroso compounds it has been suggested[374] that they act via a common intermediate, perhaps diazomethane, which is obviously a powerful alkylating agent and possesses known carcinogenic properties[375].

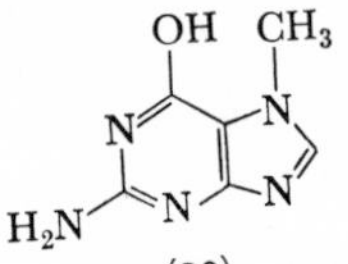

The investigation of the biological site and mechanism of action is expected to be greatly facilitated by the availability of isotopically labeled **29**. Both ^{14}C- and ^{3}H-labeled compounds were obtained by means of reaction sequence 8[376].

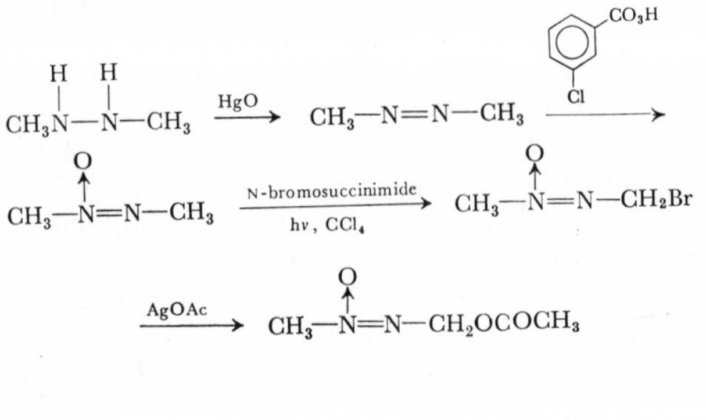

$$CH_3\overset{H}{\overset{|}{N}}-\overset{H}{\overset{|}{N}}-CH_3 \xrightarrow{HgO} CH_3-N=N-CH_3 \xrightarrow{m\text{-}ClC_6H_4CO_3H}$$

$$CH_3-\overset{\overset{O}{\uparrow}}{N}=N-CH_3 \xrightarrow[h\nu,\ CCl_4]{N\text{-bromosuccinimide}} CH_3-\overset{\overset{O}{\uparrow}}{N}=N-CH_2Br \qquad (8)$$

$$\xrightarrow{AgOAc} CH_3-\overset{\overset{O}{\uparrow}}{N}=N-CH_2OCOCH_3$$

C. The Biological Role of Natural Nitro Compounds

When dealing with a natural product, there are two points of particular interest—the biosynthetic pathway and the role played by the compound *in vivo*. We have seen examples where some progress has been made in elucidating the pathway of biosynthesis, but our understanding of the biological role played by nitro and nitroso compounds is exceedingly limited. It may be that the natural role played by antibiotics is indeed to protect the host organisms from bacterial or fungal invasion and its consequences. Other possible roles may be those of insect attractants or repellants. For example, 1-phenyl-2-nitroethane (**26**) might serve as a repellent of parasitic insects. The glycoside cycasin (**28**) and also 3-nitropropionic acid (**22**) cause severe near-toxic symptoms on ingestion of the plant sources, and this may serve to protect the plants. At the present time these remarks are largely speculation. It is clear that this is a problem which should be carefully examined in the natural habitat by ecologists.

IV. BIOCHEMISTRY AND PHARMACOLOGY OF SYNTHETIC COMPOUNDS CONTAINING THE NITRO OR NITROSO GROUP

Man purposefully brings a large number of synthetic compounds into contact with living matter. These actions may be roughly divided up into those directed at toxic effects and those directed at therapeutic effects. From a biological point of view there may be no essential difference between these two uses since in each instance we are interested in affecting living matter. When we introduce a given substance into the body for the purpose of destroying microorganisms we talk about the therapeutic effect of the drug, whereas when dealing with an insecticide we are concerned with its toxic properties against insects. Obviously, from a practical point of view these cases are vastly different in that very distinctive specificity requirements are set before each group of toxic agents. It is again appropriate to organize the discussion around the various compounds because our knowledge of the metabolic role of the nitro group is too fragmentary to permit an integrated discussion from that point of view. Moreover, the role of the nitro group in many instances can be recognized to be relatively minor but the types of compounds involved are of such substantial interest that the organic chemist will still consider the discussion to be entirely appropriate.

A. Substances Utilized for their Toxic Properties

I. Insecticides

a. Esters of phosphoric acid. Among the large group of toxic esters of phosphoric acid there are a number of compounds with nitro substituents. Many of these substances may be represented by the general formula **31**, and possess substituents such as those shown in Table 3. Many of the toxic phosphorus esters[377] can be represented by the general formula **31** where X may be oxygen or sulfur, R_1 and R_2 are alkyl or aryl groups, and R_3 is a nitroaryl or, rarely, a nitroalkyl group.

TABLE 3. Examples of toxic phosphorus esters having nitro substituents.

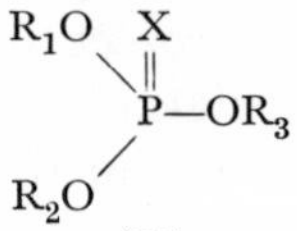

$$(R_1O)(R_2O)P(=X)-OR_3$$

(**31**)

Compound	R_1; R_2	X	R_3
Parathion	C_2H_5; C_2H_5	S	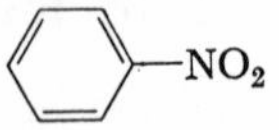 $-NO_2$
Paraoxon	C_2H_5; C_2H_5	O	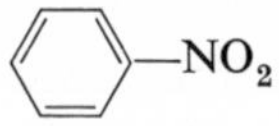 $-NO_2$
Methylparathion	CH_3; CH_3	S	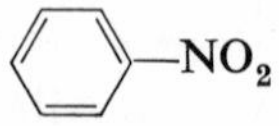 $-NO_2$
Chlorthion	CH_3; CH_3	S	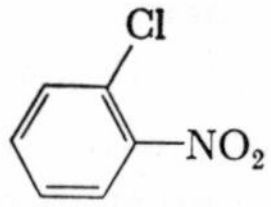 Cl; $-NO_2$
Dicapthon	CH_3; CH_3	S	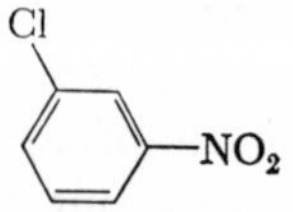 Cl; $-NO_2$
Metathion	CH_3; CH_3	S	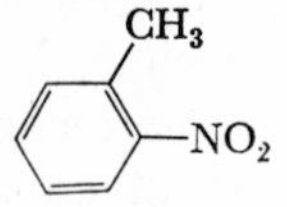 CH_3; $-NO_2$

The toxic effects of these esters occur as the result of an inhibition of acetylcholinesterase, an enzyme required for the continued transmission of nerve impulses from one nerve cell to another. Transmission of an electrical impulse across the synapses (junctions between individual nerve cells), and to muscle cells does not occur directly, but occurs instead via chemical mediators. Synaptic junctions thus act as transducers, transforming electrical impulses into chemical energy. One chemical which is released by the nerve action potential during the transmission of nerve impulses is acetylcholine (**32**), and before another nerve impulse can be transmitted by these cells the acetylcholine must be hydrolyzed by the enzyme acetylcholinesterase (reaction 9). If this hydrolysis is not carried out

$$\underset{(\mathbf{32})}{(CH_3)_3\overset{\oplus}{N}{-}CH_2CH_2O{-}\overset{O}{\overset{\|}{C}}CH_3} \xrightarrow[H_2O]{\text{acetylcholinesterase}} (CH_3)_3\overset{\oplus}{N}{-}CH_2CH_2OH + CH_3COO^{\ominus} + H^{\oplus} \quad (9)$$

the repolarization of the nerve endings is impossible and the transmission of subsequent nerve stimuli is impaired. Clearly any compound capable of interfering with the transmission of nerve impulses has the potential for interfering with the most fundamental physiological processes.

The toxic effects of such nitrophenyl phosphate esters in animals are almost entirely a consequence of the phosphorylation of acetylcholinesterase. Although nitrophenols are produced on hydrolysis which have distinct toxic effects, the amounts of these materials are generally relatively small. Rather, poisoning occurs as the result of an accumulation of abnormally large quantities of acetylcholine at the nerve endings. The first symptoms of poisoning appear after about 20% of the acetylcholinesterase has been blocked The initial manifestations of poisoning are nausea, sweating, excessive flow of saliva, painful intestinal spasms, defecation, enuresis, lacrimation, and lack of appetite. These are followed by muscular twitching, spasms, and then muscular weakness and paralysis. Symptoms caused by effects on the central nervous system (particularly by less than acute doses) include giddiness, apprehension, restlessness, insomnia, and nightmare dreams. Death follows motor incoordination, coma, irregular breathing (Cheyne–Stokes respiration), and finally paralysis.

It is appropriate at this point to consider in more detail the enzymatic hydrolysis reactions catalyzed by acetylcholinesterase. They follow the general sequence of reaction 10. In this sequence the enzyme E and substrate S (such as **32**) undergo a rapid reversible complex formation to form E · S. This complex can react to form a covalent intermediate ES′ accompanied by a release of part of the substrate P_1. A subsequent step described by rate constant k_3

$$E + S \underset{k_{-1}}{\overset{k_1}{\rightleftharpoons}} (E{\cdot}S) \xrightarrow[P_1]{k_2} ES' \xrightarrow[P_2]{k_3} E \qquad (10)$$

involves the decomposition of ES′ to regenerate the enzyme and release product P_2. For the particular case of **32** and acetylcholinesterase these transformations may be schematically represented as shown in Scheme 4. In this scheme the enzyme is represented as

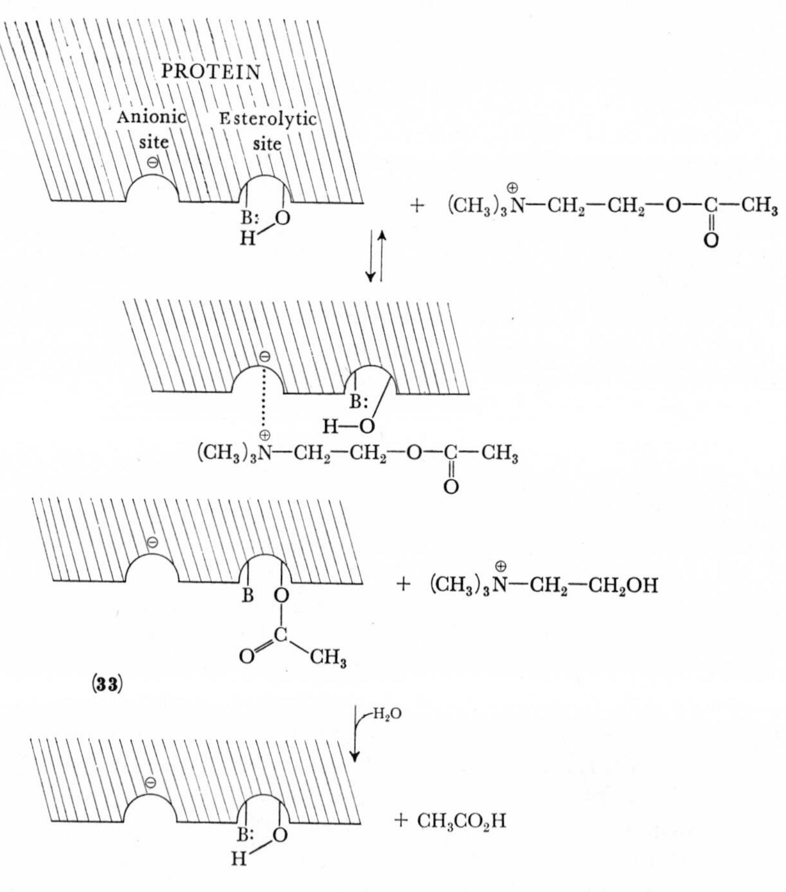

Scheme 4. Representation of the hydrolytic action of acetylcholinesterase. (After Greenberg and Nachmansohn, cf. reference 378.)

having an esteratic site containing a hydroxyl group and a general-base catalyst B. These groups are in fact the side-chain hydroxyl group of the amino acid serine and the side-chain imidazole group of the amino acid histidine. In addition the enzyme has an anionic binding site responsible for much of the enzymatic specificity. From a variety of studies[378] it may be estimated that the distance between the anionic and esteratic sites is 2.5–6 Å as would seem reasonable from the size of the usual substrate **32**[378–380]. Following initial formation of the E · S complex there occurs a general-base-catalyzed transesterification reaction which results in formation of an acyl–enzyme intermediate **33** and free choline. The concept of the covalent complex ES′ as an acyl enzyme began with the work of Balls[381] and of Hartley and Kilby[382], and has been brilliantly extended by the work of Bender[383] and others. Finally, the acyl enzyme **33** undergoes a general-base-catalyzed hydrolysis to regenerate the enzyme and free the carboxylic acid portion of the original substrate molecule.

The toxic properties of phosphorus esters depend upon changes in the relative magnitudes of the acylation rate constant, k_2, and the deacylation rate constant, k_3 (reaction 10). In the course of the normal hydrolysis reaction these rate constants are both large and the enzyme of the nerve cells can thus rapidly destroy **32** which is freed at the time of nerve impulse transmission. The small amount of enzyme which is present in these tissues can catalyze the hydrolysis of large amounts of **32**. However, toxic phosphorus esters result in the formation of a much more stable phosphoryl enzyme intermediate[382,384–386] for which k_3 is very small. This was first recognized for the case of the nerve gas DFP, diisopropyl fluorophosphate (**34**)[382]. As might be expected it is possible to determine that the

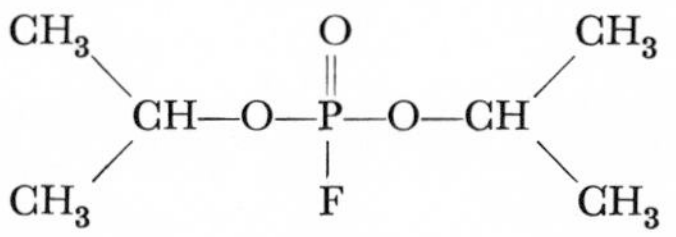

$$(CH_3)_2CH{-}O{-}P(=O)(F){-}O{-}CH(CH_3)_2 \quad \textbf{(34)}$$

$$(C_3H_7O)_2P(=O){-}O{-}CH_2{-}CH(NH_2)CO_2H \quad \textbf{(35)}$$

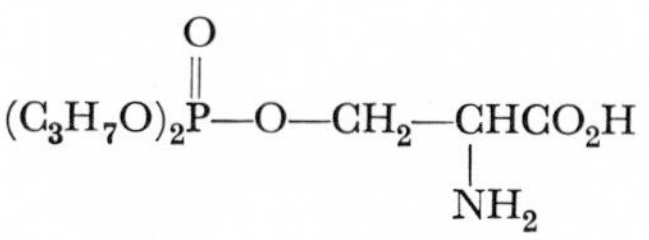

phosphate group becomes covalently bonded to a serine group at the active site of the enzyme[387] and following complete hydrolysis of the protein it is possible to isolate the phosphorylated serine derivative **35**.

It is not entirely correct to say that acetylcholinesterase is irreversibly inhibited by toxic phosphorus esters although in the normal physiological state of the organism this is effectively the result. There are, however, certain compounds which can reverse the toxic effect by phosphorylation of the enzyme. Following the leads suggested by Wilson[388,389] a number of compounds were developed which included powerful nucleophilic groups located a certain distance from a cationic group, thus ensuring that the potential reactivating group is bound to the enzyme in the proper steric orientation as to permit reaction with the phosphate group on the serine at the active site of the enzyme[390–394]. One of the most successful of these compounds is PAM, 2-pyridinealdoxime methiodide (**36**).

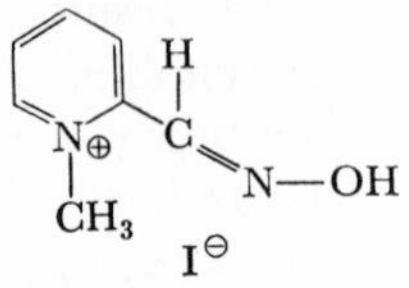

$$\text{(pyridinium ring, } N^{\oplus}\text{–}CH_3\text{)–C(H)=N–OH} \quad I^{\ominus}$$

(**36**)

In view of the fact that **34** is a potent irreversible inhibitor of acetylcholinesterase it is clear that the presence of a nitroalkyl or nitroaryl substituent is not an indispensible requirement for biological activity. Rather, the nitro group is important because it affects reactivity and at the same time imparts some desirable characteristics of specificity and volatility. The use of the nitro group in affecting reactivity can be illustrated by data of Markov and his coworkers[395]. They report data for the reaction of a number of *m*-nitrophenyl and *p*-nitrophenyl phosphates and phosphonates with acetylcholinesterase, and data for the *meta* and *para* isomers of **37** may be cited. The value of k_2 (see reaction 10) was 1.4×10^8 l./(mole sec) for the *para* isomer of **37**, and 1.6×10^7 l/(mole sec) for the *meta* isomer. Interestingly, the alkaline hydrolysis rates of these two compounds are very similar[396]. Another obvious benefit of compounds such as **37**, which have nitrophenyl substituents, is

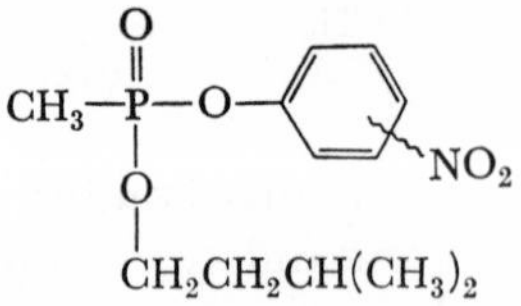

$$CH_3\text{–}P(=O)(\text{–}O\text{–}C_6H_4NO_2)\text{–}O\text{–}CH_2CH_2CH(CH_3)_2$$

(**37**)

that they have only a limited volatility, clearly a necessary property if relatively non-selective poisons are to be utilized in a specific area.

The product P_1 (reaction 10), which is released on phosphorylation of acetylcholinesterase by compounds such as **37** is, of course, a nitrophenol. For the case of the *para* isomer the *p*-nitrophenol will be substantially ionized at physiological pH values because its pK_a is 7. Indeed, this is why such compounds are excellent phosphorylating reagents *in vivo*; one of the potential leaving groups in a nucleophilic displacement reaction is the nitrophenoxide anion, a relatively good leaving group. From the particular standpoint of our interest in the metabolism of nitro compounds *in vivo*, it is clear that a major concern in the metabolism of insecticides such as those in Table 3 will be the fate of the nitrophenyl group.

The metabolism of parathion in cattle can take a number of courses but the major ones as shown in Scheme 5 involve reduction,

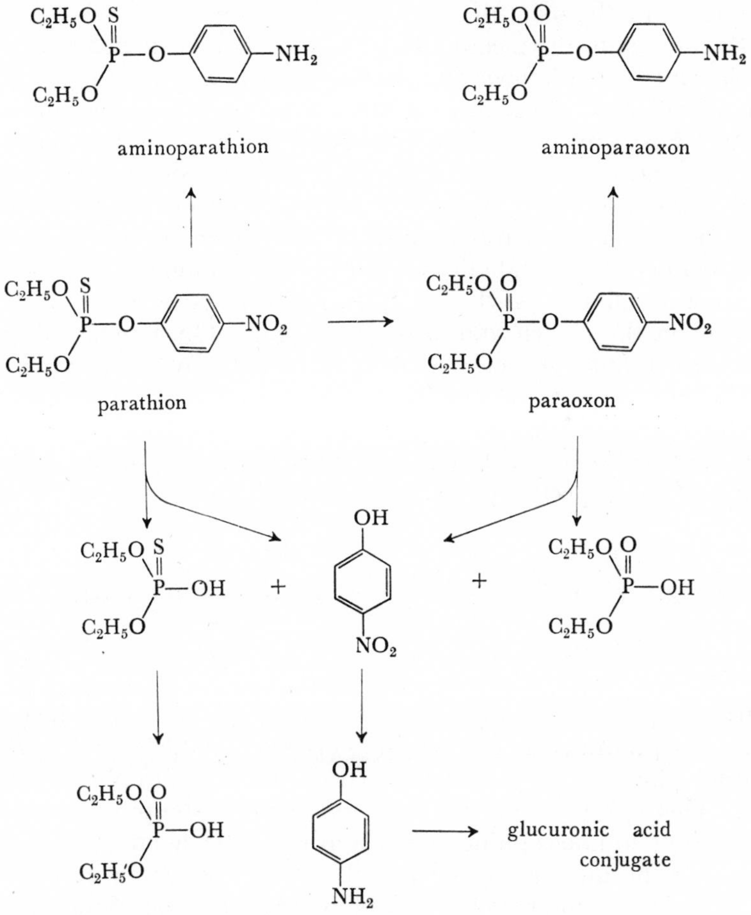

aminoparathion

aminoparaoxon

parathion

paraoxon

glucuronic acid conjugate

SCHEME 5. Metabolism of parathion in cattle[377].

hydrolysis, and/or replacement of sulfur by oxygen[377]. The replacement of sulfur by oxygen as exemplified by the transformation of parathion to paraoxon is an interesting reaction which is required for activation *in vivo*[397,398]. The activation process is promoted by the reduced coenzymes NADH and NADPH[399]. It is also of interest that this activation process is enzyme catalyzed and in some cases the rates differ with the sexes. For example, the activation of parathion is some ten times more rapid in female rats than in male rats[400] leading to a selective toxicity of parathion toward the female[401]. Similar activation reactions (though not necessarily sex linked) occur in insects[402,403]. Not surprisingly, a number of metabolic poisons, such as iodoacetate, cyanide, chloropicrin, and azide ion act as inhibitors of the activation process[403].

In contrast to the metabolism of parathion by cattle as described in Scheme 5, it is known that the metabolism by rats[404] or by dogs takes a different course because only a small part of the administered dose[405] is excreted in the form of amino derivatives. In the ruminant, large quantities of the aminophenols have been found in the blood and in the urine[406,407]. Apparently they arise from the action of the large numbers of microorganisms in the rumen. It is well known that many of the metabolic reactions of the ruminants are changed relative to animals which lack the unusually active microbial flora present in the rumen and associated organs. On a practical level this means that insecticides such as those noted in Table 3 are not particularly toxic to ruminants.

Differences in the effects of reduction of the nitro group upon the insecticidal action of aryl and alkyl nitro compounds, such as **37a** and **38**, respectively, are quite marked. In the case of the aryl

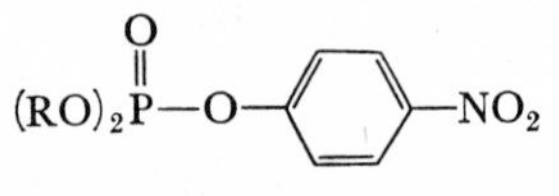

$(RO)_2P(=O)-O-C_6H_4-NO_2$

(**37a**)

$(RO)_2P(=O)-O-CH_2CH_2NO_2$

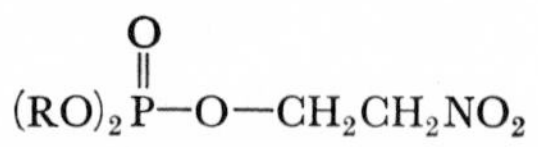

(**38**)

nitro compound it is clear that reduction will sharply reduce the electron-withdrawing characteristics of the aromatic ring; the pK_a of *p*-aminophenol is 9.4[409] as compared to that of *p*-nitrophenol which is 7. This will have the effect of making the phosphate ester less susceptible to nucleophilic attack. On the other hand, the effect of reducing the nitro group of **38** will be to produce an alkylamine which will be protonated in the physiological range of pH values;

this will result in the phosphate group being even more activated toward nucleophilic displacement reactions. This is similar to the situation observed for esters of α-amino acids, which are exceedingly rapidly hydrolyzed in relatively dilute alkaline solutions.

Recession of the sublethal toxic effects and spontaneous reactivation of cholinesterase takes place much faster in insects than in mammals[408]. Of course, these processes occur via hydrolytic reactions. The relative proportions of hydrolysis products isolated, following administration of various insecticides to the cockroach, are shown in Table 4[410]. In each case, the nitrophenoxy group is most readily removed by hydrolysis, although in the case of methyl parathion it may be seen that significant amounts of phosphodiester having an intact nitrophenyl group are formed. This may be a consequence of the lessened steric hindrance offered by the methyl groups as opposed to the ethyl groups of parathion itself. This effect is offset by introduction of substituents, such as a chloro substituent on the aromatic ring as evidenced by the results obtained for the last two compounds in Table 4. In this regard it is undoubtedly significant that the pK_a of, for example, 3-chloro-4-nitrophenol is approximately one pK unit smaller than that of *p*-nitrophenol[411,412]. Clearly, a delicate balance must be struck between hydrolytic instability, volatility, solubility, and toxicity to the correct organism.

b. Chloropicrin. Trichloronitromethane or chloropicrin was used for some time as a fumigant for agricultural uses. The compound, being a potent poison, destroys insects, fungi, weeds, and soil-inhabiting nematode worms. It has a sharp odor which can be noticed even at concentrations as low as 8–5 μg/l.; this sharp odor acts as a warning and indeed, chloropicrin may be added to less odorous fumigants to provide warning of their presence. Chloropicrin also has a strong irritating effect on the skin and mucous membranes and was used as a vomiting agent during World War I. These properties are particularly associated with the presence of chlorine in the molecule. Chloropicrin also acts as an oxidizing agent to cause the formation of disulfide bridges between cysteine residues of reduced hemoglobin[413] (reaction 11), while the presence of the nitro group may contribute to methemaglobinemia.

$$2R{-}SH + 2ClR' \longrightarrow 2HCl + R{-}S{-}S{-}R + R_2' \qquad (11)$$

c. Nitro analogs of DDT. Derivatives which exhibit marked biological activity are those represented by structure **39** where R = methyl or ethyl. These substances and related ones such as **40**

TABLE 4. Hydrolysis products of some insecticides in the American cockroach[409].

Compound	Structure	Relative percentage of hydrolysis products			
		$(RO)_2P(O)OH$	$(RO)_2P(S)OH$	$(RO)(HO)P(O)OX$	$(RO)(HO)P(S)OX$
Parathion	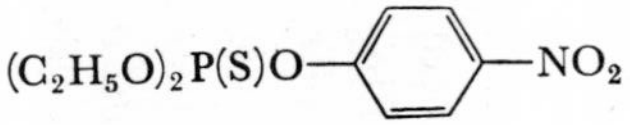 $(C_2H_5O)_2P(S)O$—⟨ring⟩—NO_2	28	67	3	2
Methylparathion	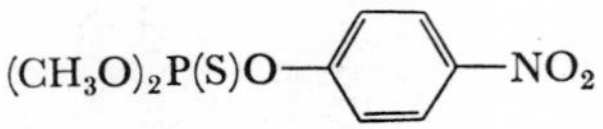 $(CH_3O)_2P(S)O$—⟨ring⟩—NO_2	27	47	3	23
Chlorthion	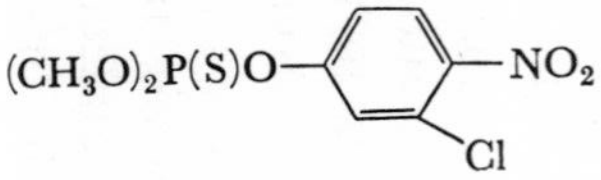 $(CH_3O)_2P(S)O$—⟨ring, Cl⟩—NO_2	2	97	Trace	1
Dicapthon	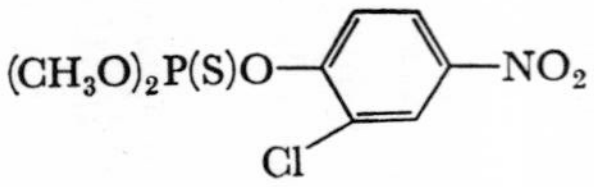 $(CH_3O)_2P(S)O$—⟨ring, Cl⟩—NO_2	3	93	Trace	4

$(Cl-C_6H_4)_2-CH_2-CH(R)-NO_2$

(**39**)

$(Cl-C_6H_4)_2-CH-CCl_3$

(**40**)

exhibit marked insecticidal action although the effects may also be lessened by the development of resistant strains[414].

d. Other substances. Products such as **41** and **42** obtained by

$ClCH_2SO_2-C_6H_4-NH-N{=}C(R)-NO_2$

(**41**)

$ClCH_2SO_2-C_6H_4-N{=}N-C(R)_2-NO_2$

(**42**)

condensation of *p*-chloromethylsulfonylbenzenediazonium chloride with nitroparaffins possess insecticidal activity[415]. They differ significantly in their effects on different insect species. Nitro derivatives of phenoxyacetic acids, widely used as herbicides, also possess activity against ticks and mites (acaricidal activity). Thus compounds such as **43** and **44** may be useful[416,417] in combating insect

$(O_2N)(Cl)_2C_6H_2-OCH_2OC(=O)-CH_2O-C_6H_2(Cl)_2(NO_2)$

(**43**)

$Cl_2C_6H_3-S-CH_2CH_2OC(=O)CH_2O-C_6H_2(Cl)_2-NO_2$

(**44**)

pests known to transmit a number of dangerous diseases.

2. Molluscicides

Certain species of mollusks, particularly the fresh water snails, may act as carriers of disease or as garden pests. Biological activity toward mollusk species is exhibited by *N*-(2-chloro-4-nitrophenyl)-5-chlorosalicylanilide (**45**) even at concentrations as low as 0.3 ppm.

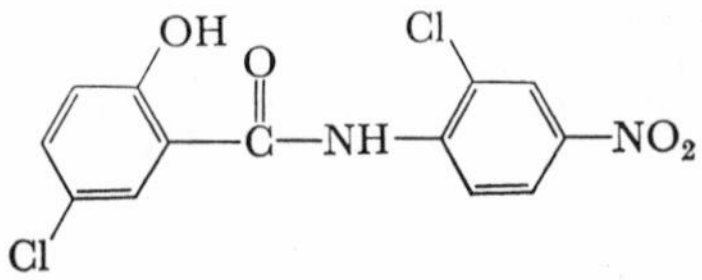

(**45**)

The position of the nitro group is of considerable importance[418]. The 2-nitro compound exhibits only $\frac{1}{3}$ the biological activity of the 4 compound, and the 3 isomer exhibits only $\frac{1}{300}$ of the activity. Reduction of the nitro group to an amino group results in a much less active compound. Compound **45** has a very low toxicity toward higher animals. The LD_{50} for oral administration of **45** to rats exceeds 5 g/kg. Because **45** also shows activity against tapeworms it has been introduced into therapeutic use and will be discussed in more detail in IV.B. 1c.

3. Fungicides

Various fungi cause substantial economic losses as the result of their growth on and destruction of plants, or as the result of damage caused during storage of various products. The introduction of a nitro group into various aliphatic or aromatic hydrocarbons often results in substances with fungicidal activity[419]. Simultaneous halogenation of the compounds produces an even further rise in effectiveness. *p*-Nitrophenol exhibits fungicidal properties and is used for preserving leather products. The nitrosopyrazole **46** exhibits

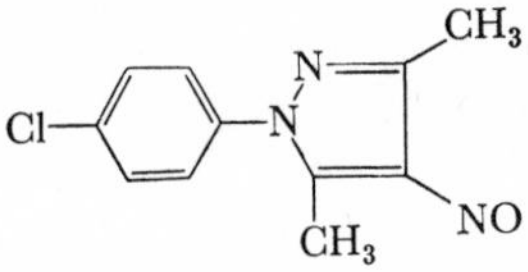

(**46**)

fungicidal activity although this compound has not been widely used[420].

A number of 2-nitro-1,3-propanediol derivatives of the general formula **47** exhibit fungicidal activity including activity against

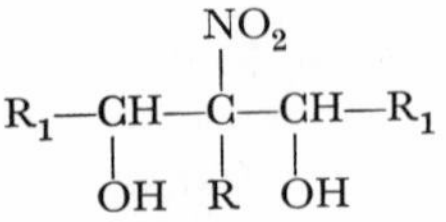

$$R_1\text{—}\underset{\text{OH}}{\underset{|}{\text{CH}}}\text{—}\overset{\text{NO}_2}{\overset{|}{\underset{\text{R}}{\underset{|}{\text{C}}}}}\text{—}\underset{\text{OH}}{\underset{|}{\text{CH}}}\text{—}R_1$$

(**47**)

important plant pathogens[421,422]. Specific examples of the types of substitution include the compound with R_1 = methyl or isobutyl and R = bromine. Related nitro compounds include **48–50**[423–425].

R_1—CH—O / R, O_2N—C / SO / CH—O—R_1

(**48**)

O_2N—CH(CH_2OH)—CHOH—C≡C—$C_{13}H_{28}$

(**49**)

O_2N—(furan)—CH(OH)—C(R_1)(R)(NO_2)

(**50**)

Aryl nitroparaffins such as **51** and **52** are potent fungicides. Compound **51** exhibits particularly marked activity when the nitro

O_2N—C_6H_4—CH(R_1)—C(R_2)(Br)—NO_2

(**51**)

C_6H_5—Y—CH(R_1)—CH_2NO_2

(**52**)

group is in the *para* position and R_1 is bromine or methoxy[426]. Compounds of the type of **52**, where Y is sulfur or selenium, are active as fungicides and acaricides[427,428].

The last group of fungicides to be considered are nitro olefins related to 2-nitrostyrene. 2-Nitrostyrene itself (**53**, X = H) is a

X—C_6H_4—CH=C(R)—NO_2

(**53**)

fungicide useful in medicine and agriculture[426,429], while the *m*- or *p*-fluoro derivatives of **53** are potent insecticides[430,431]. Compounds possessing structure **52**, with Y being S or Se and R_1 being phenyl, have both fungicidal and acaricidal activity. The fact that a number of saturated compounds such as **52** exhibit fungicidal activity tends to diminish the attractiveness of one hypothesis which might be advanced to explain the biological working of **53**. Huitric and his colleagues examined the properties of a number of nitrostyrene derivatives[432]. They noted that the nitro group is strongly electron withdrawing, and this effect may be enhanced in some instances by halogen substituents. Because the carbon–carbon double bond of **53** may be highly activated toward nucleophilic addition reactions, it seems possible that the biological activity may be involved in reactions with a variety of biologically important nucleophiles[433] such as the sulfhydryl group. While this is a possibility, it would seem to require that either the saturated derivatives first undergo a reaction to form the unsaturated compounds as a prelude to exerting a biological effect, or alternatively, that the biological effects of the saturated compounds such as **51** or **52** are brought about by some completely different mechanism.

4. Herbicides

The problem of developing effective herbicides has two aspects. One may seek a material which is toxic to all types of plants or, more commonly, one seeks a material which exhibits selective toxicity. The presence or absence of a nitro group is not particularly related to the matter of specificity so that little attempt will be made here to discuss the detailed spectrum of activity of various herbicides.

a. Derivatives of dinitrophenol. A variety of dinitrophenol derivatives has been shown to possess herbicidal activity. Some of the most effective of these may be represented by the general formula **54**[434–439]. Pianka and Browne recently discussed the activity of a

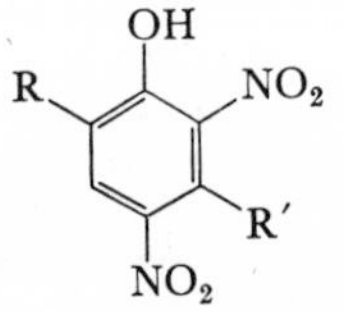

OH
R NO$_2$
R′
NO$_2$

(**54**)

number of these materials and of their esters, carbonates, and ethers[440]. They reached a number of conclusions regarding the effectiveness of such compounds. Aromatic esters, carbonates, or ethers derived from the compounds in Table 5 were substantially less active than the free phenols. On the other hand, acetylation of the phenols caused an enhancement of the activity although longer chain esters had a variable effect. This may be the result of improved penetration of the material into the biological site of action whereupon the esters are hydrolyzed to the free phenols. The most active

TABLE 5. Dinitrophenol derivatives with herbicidal activity[440].

Compd	Trivial name	Substituents on formula **54**[a]	
		R	R′
55	DNOC	CH_3	H
56	Dinoseb	$CH(CH_3)C_2H_5$	H
57	Dinoterb	$C(CH_3)_3$	H
58	Dinosam	$CH(CH_3)C_3H_7$	H
59	—	$CH(CH_3)C_2H_5$	CH_3
60	Medinoterb	$C(CH_3)_3$	CH_3

[a] Proper nomenclature would seem to require that for example, **60** be named 2,4-dinitro-3-methyl-6-*t*-butylphenol, but it is often seen as 2-*t*-butyl-4,6-dinitro-5-methylphenol.

substances were **57** and **60**. This activity was attributed[441] to the ability of these phenols to form relatively stable free radicals. The free radicals could then 'interfere with the vital processes in the plant[440].' A similar action was advanced[437,438,442] to explain the acaricidal activity of these phenols. Based on a large number of observations Kirby and his coworkers concluded that either 2,4- or 2,6-dinitrophenols exhibit maximal herbicidal and fungicidal activities[443]. Comparisons of mono-, di-, and trinitrophenols showed the dinitrophenols to possess maximal biological activity[444,445]. The mechanism by which these phenols bring about their toxic action remains uncertain. While the alkylnitrophenols may be expected to form relatively reactive free radicals, it does not seem possible to eliminate an explanation based on the uncoupling of the fundamental processes of oxidative phosphorylation. For example, both 2,4-dinitrophenol and 2,6-dinitrophenol act as potent uncouplers of oxidative phosphorylation[446]. The mechanism of action of these herbicides remains an interesting problem.

b. Other substances. The selective weed-killing properties of phenoxyacetic acids encouraged the synthesis and evaluation of their nitro derivatives. According to Sexton[447] an essential role in the action of these herbicides is played by the steric relationship involving the aromatic ring and the ester grouping. Thus, *cis* and *trans* cinnamic acid esters differ in activity. A selective herbicidal activity is exhibited by compounds such as **61** and **62**. But since they may be considered to be derivatives of chlorophenoxyacetic acid it is

$$Cl_3C(O_2NCH_2)CH{-}O\overset{O}{\overset{\|}{C}}{-}CH_2O{-}C_6H_3Cl_2$$

(61)

$$Cl{-}C_6H_3(CH_3){-}O{-}CH_2\overset{O}{\overset{\|}{C}}OCH_2{-}C(O_2N)(CH_2{-}O)(CH_2{-}O)C(CH_3)_2$$

(62)

not clear to what extent the nitro group contributes to the biological activity[448,449].

5. Carcinogenic *N*-nitroso compounds

N-Nitroso compounds such as dimethylnitrosamine (**63**) and *N*-nitroso-*N*-methylurethane (**64**) are of considerable chemical

$$CH_3{-}N(CH_3){-}NO \qquad CH_3{-}N(C({=}O)OC_2H_5){-}NO$$

(63) **(64)**

interest but it must be noted that such *N*-nitroso compounds are exceedingly dangerous. Because of reports[450] from industrial laboratories on the toxicity of **63**, Barnes and Magee investigated the effects of that substance on rats. Doses less than 25 mg of **63** per kilogram of body weight produced profound lesions of the liver as well as hemorrhages into the liver and lungs and death[451]. Similar toxicity was noted with a variety of other animals including rabbits and dogs[450,451]. This corresponds to observations made in the case of chemical workers[452–454], and applies to a variety of *N*-nitroso compounds[455]. The detailed pathological changes brought about in various tissues by *N*-nitroso compounds have been nicely reviewed[456].

A toxic property of the *N*-nitroso compounds which is of considerable biological interest and which should be of great concern to chemists working with such compounds is the potent carcinogenic character of *N*-nitroso compounds[456]. The result was first observed by Magee and Barnes[457] as an extension of the work on the toxicity of **63**[451].

TABLE 6. Acute toxicity of *N*-nitroso compounds[455,458,459].

Compound	Acute oral toxic dose in rats (LD_{50}), mg/kg
N-Nitrosodimethylamine	27–41
N-Nitrosodiethylamine	216
N-Nitroso-*n*-butylmethylamine	130
N-Nitroso-*t*-butylmethylamine	700
N-Nitroso-*t*-butylethylamine	1600
N-Nitrosomethylphenylamine	200
N-Methyl-*N*-nitrosourethane	240

a. Toxic and carcinogenic properties. A number of the *N*-nitroso compounds are highly toxic, producing liver and lung damage, bleeding, convulsions, and coma. In many instances the toxic doses are quite small (Table 6). The toxicity[460,461] of **64** and more particularly its carcinogenic properties have led several groups to suggest that *p*-tosylmethylnitrosamine be used as a precursor for diazomethane because it is apparently much less dangerous[462,463].

Magee and Barnes state that 'there is no correlation between the acute toxic effects and carcinogenic activity[456].' Since their initial report[457] of the development of malignant liver tumors in rats following administration of **63**, there has developed an extensive literature dealing with the carcinogenic character of *N*-nitroso compounds. Much of these data are carefully reviewed in an excellent article by Magee and Barnes[456]. Unquestionably, the most significant result of such studies is that single doses are sufficient to cause the development of malignant tumors in different organs. For example, single oral doses of **64** were sufficient to induce tumors of the stomach and esophagus in rats[464], while single inhalations of *N*-nitroso-*N*-methylvinylamine (**65**) are reported to induce the development of cancers in the nose of rats[456]. The oral administration of a few doses of *N*-nitrosomorpholine (**66**) leads to the formation

$$CH_2{=}CH{-}N(CH_3){-}NO$$

(**65**)

$$O\langle {}^{CH_2-CH_2}_{CH_2-CH_2} \rangle N{-}NO$$

(**66**)

of kidney tumors[466], while the repeated local administration of *N*-nitroso-*N*-methylurea (**67**) to the skin of mice and hamsters leads

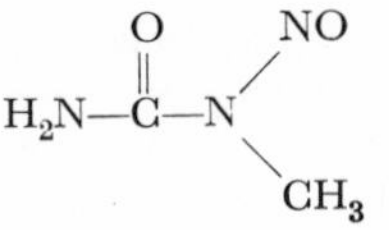

$$H_2N{-}\overset{\displaystyle O}{\overset{\|}{C}}{-}N(NO)(CH_3)$$

(**67**)

$$\text{N-nitrosoazetidine: ring } CH_2{-}CH_2{-}N(NO){-}CH_2$$

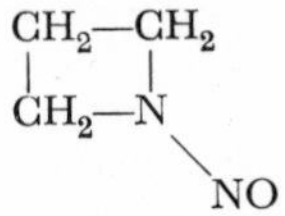

(**68**)

to development of skin cancer[467]. Similar data could be cited for a wide variety of *N*-nitroso compounds administered in various ways over varying periods of time and resulting in the development of a variety of cancers in different organs in a number of experimental animals[456].

The difference between acute toxicity and carcinogenicity of *N*-nitroso compounds is nicely illustrated by a report concerning the properties of *N*-nitrosoazetidine (**68**). This substance was found to possess a low toxicity (the LD_{50} for rats was not fully established but, with doses of 1.6 g/kg of body weight, only 2 of 5 rats died[468]). Compound **68** (5 mg/day) was administered to 20 rats for 46 days. In the 16 rats surviving after 80 weeks it was found that a number of tumors had developed in 15 of the animals[468]. These tumors included lung, kidney, liver, adrenal, intestinal, uterine, mammary, and other tumors. Similar but less pronounced effects were observed with mice. Indeed, the results of such studies on the carcinogenicity of *N*-nitroso compounds should give pause to chemists and technicians handling such materials.

b. Mutagenic properties. An important property of the *N*-nitroso compounds is their ability to function as mutagenic agents, a property which has been related by one author to the ability of these substances to act as chromosome-breaking agents[469]. Substances such as *N*-methyl-*N*-nitroso-*N'*-nitroguanidine (**69**) are exceedingly

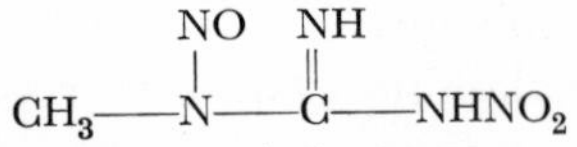

$$CH_3{-}N(NO){-}C({=}NH){-}NHNO_2$$

(**69**)

powerful mutagenic agents, for example causing the development of large numbers of mutant bacteria. Possibly related to the mutagenic properties of compounds such as **69** is the observation that it causes growth abnormalities in rat fetuses when administered on the 15th or 16th day of pregnancy[470]. Mention has also been made of a similar teratogenic effect of the methylazoxymethanol glycoside cycasin[372].

There appears to be a rough parallel between carcinogenic effects and mutagenic effects. For example, **69** is a potent mutagenic agent and is also a powerful carcinogen, and even single doses can produce malignant stomach tumors in the rat[471]. Conversely, *N*-ethyl-*N*-*t*-butylnitrosamine is neither carcinogenic[472] or mutagenic[473] at least in the particular test organisms employed. A more extended discussion of this complicated relationship is available[456].

c. Biochemical reactions. It appears that *N*-nitroso compounds are relatively rapidly metabolized in whole animals. For example, in the rat or mouse, administration of ^{14}C-labeled **63** leads to the majority of the isotope appearing as $^{14}CO_2$ after 6–10 hours[474]. Similar experiments with ^{15}N-labeled **63** resulted in the production of isotopically labeled urea and proteins, indicating that the nitroso group as well as the amino group is transformed metabolically into ammonia, which is in equilibrium with the endogenous ammonia 'pool' of the organism[475]. Thus, any interpretation of the mechanism of action of carcinogenic *N*-nitroso compounds will be made difficult by the fact that such substances can be metabolized to a large extent by pathways which are of importance to the normal animal.

The difficulty of distinguishing between normal metabolic interactions and those due to the distinctive receptor sites active in carcinogenesis, of course, contributes to the difficulty of accounting for the potent carcinogenic properties of the *N*-nitroso compounds. However, extensive work in this area is being carried out, and it is of interest to discuss at least one important hypothesis advanced to explain the biochemical mode of action of these substances. Alkyl derivatives of *N*-nitroso compounds are convenient sources of diazomethane. It has been proposed that compounds such as **63** may be degraded metabolically to diazoalkanes such as diazomethane and that these may act as alkylating agents which in some way are responsible for the induction of cancers[476]. Compounds such as *N*-methyl-*N'*-nitro-N-nitrosoguanidine (**69**) may decompose under acidic conditions to nitrous acid and under alkaline conditions to diazomethane either of which is exceedingly toxic to vital cellular constituents[477,478]. Heath designed an experiment to test this interesting hypothesis and to measure the importance of diazoalkanes as toxic metabolites[476]. He examined the physiological effects brought about by two isomeric nitrosamines, *n*-butylmethylnitrosamine (**70**) and *t*-butylmethylnitrosamine (**71**)[476]. Note that the unbranched carbon chain isomer can be degraded in a manner

$$CH_3CH_2CH_2CH_2—N(CH_3)—NO \qquad (CH_3)_3C—N(CH_3)—NO$$

(**70**) (**71**)

similar to that proposed for dimethylnitrosamine to give either diazomethane or methyldiazonium ion (reaction 12)[477,478]. However, the *t*-butyl compound, lacking as it does an α-hydrogen, cannot be degraded to a diazoalkane. In fact, it was found that marked pathological changes occurred in the liver of rats dosed with **70**, but little

$$CH_3{-}N(CH_3){-}N{=}O \xrightarrow[O_2]{NADPH} HOCH_2{-}N(CH_3){-}N{=}O \rightarrow HCHO + H{-}N(CH_3){-}N{=}O$$

$$H{-}N(CH_3){-}N{=}O \rightarrow CH_3{-}N{=}N{-}OH$$

$$CH_3{-}N{=}N{-}OH \rightarrow {}^{\oplus}N_2{-}CH_2^{\ominus} \rightarrow ?$$

$$CH_3{-}N{=}N{-}OH \rightarrow {}^{\oplus}N_2{-}CH_3 \rightarrow N_2 + CH_3^{\oplus}$$

change was observed following the administration of large amounts of **71**. These results were extended by examining the metabolism of ^{14}C-labeled **70** and **71** in the rat[479]. Labeled carbon dioxide was rapidly produced from the metabolism of **70** and again represented a major pathway. However, only a few percent of the isotopic carbon in **71** was transformed into carbon dioxide. These marked differences in the extent of metabolism of **71** and compound such as **70** parallel the marked difference in carcinogenic and toxic effects.

Thus, it would seem that a requirement for the carcinogenic character of such *N*-nitroso compounds is that they can be metabolized to monoalkylnitrosamines or to diazoalkanes or diazonium compounds[479]. Confronted with such evidence, the chemist might suggest the possibility of an alkylation of some cellular constituents as an important requirement for carcinogenesis by *N*-nitroso derivatives. Indeed, the study of such an alkylation of proteins[480] and nucleic acids[481–491] is an exceedingly important area of research at the present time. One reaction which has been found to occur to a significant extent is the methylation[481,484] or ethylation[485] of the nitrogen bases present in the cellular ribonucleic and deoxyribonucleic acids. With **63** the predominant product *in vivo* is 7-methylguanine (**30**), which was discussed in connection with the biochemical effect of the related carcinogenic *N*-nitroso compound cycasin (III.B. 4). Indeed, the same methylated purine can be found in hydrolysates of rat liver ribonucleic acids which were obtained from rats treated with cyclic *N*-nitroso compounds such as *N*-nitropyrrolidine[486]. Compound **30** was isolated from both the ribonucleic acids and deoxyribonucleic acids of rats treated with either **63** or cycasin (**28**)[374]. In view of the mutagenic[469] character of the *N*-nitroso compounds[492–495], as well as the involvement of carcinogenesis with the genetic material, it would appear that the alkylating ability of derivatives of *N*-nitroso compounds is connected with the induction of cancers. This hypothesis is currently the subject of intensive research[456]. However, a recent report suggests that the methylation of guanine or cytosine may not be the actual mutagenic event caused by compounds such as nitrosoguanidines[491]. Thus, at the present time it cannot be said that alkylation of the genetic material of the cell or, indeed, alkylation of any cellular substance is causally related to carcinogenesis. In any event, it is entirely clear that the *N*-nitroso compounds represent exceedingly dangerous substances, particularly because their effects on health may not be evidenced for a long time after exposure. Even more frightening is the conclusion which may be drawn from studies on small animals that a single exposure to certain *N*-nitroso compounds may lead to the development of malignant tumors.

B. *Substances Utilized for their Therapeutic Properties*

I. Drugs used for combating live organisms in the body of the host

a. Antibacterial drugs. Amongst the most important of the antibacterial drugs are the nitrofurans[496]. When investigating the structure–activity relationships applicable to various furan derivatives, Dodd and Stillman observed that compounds possessing a nitro group in the 5 position exhibited an enhanced antibacterial activity over derivatives having, for example, simply a side chain in the 2 position[497]. Extensions of such investigations have led to compounds such as 5-nitro-2-furfuraldehyde semicarbazone (**72**; nitrofurazon), *N*-(5-nitro-2-furfurylidine)-1-aminohydantoin (**73**, nitrofurantoin), and 3-(5-nitrofurfurylideneamino)-2-oxazolidinone (**74**; Furazolidone). These substances are potent, broad-spectrum antibacterials.

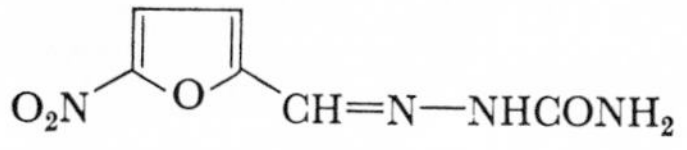

O_2N—(furan)—CH=N—NHCONH$_2$

(**72**)

O_2N—(furan)—CH=N—N(hydantoin)NH

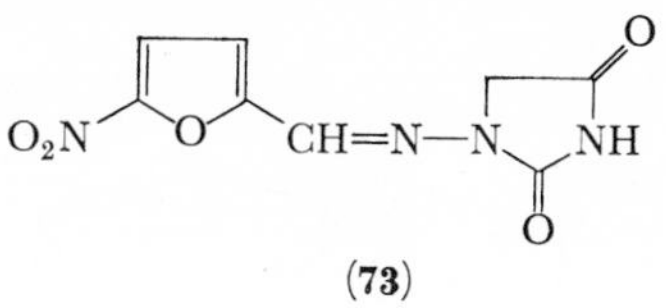

(**73**)

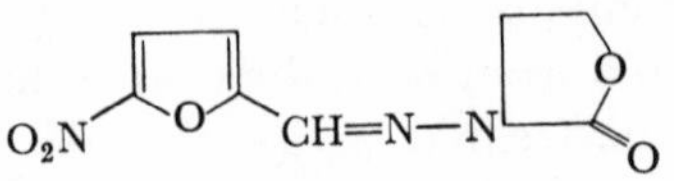

(74)

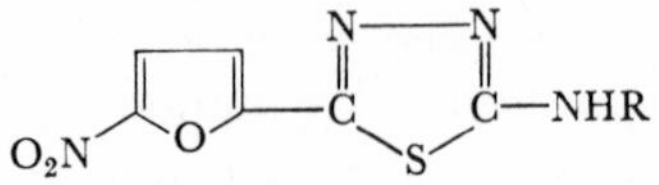

(75)

Compound **73** is used orally in the treatment of urinary tract infections. Others exhibit antifungal or antiprotozoal activity. Related derivatives include compounds having thiadiazolinone rings in addition to the nitrofuran ring and have the structures such as **75–77**, where R and R′ are alkyl or aryl groups and X is halogen

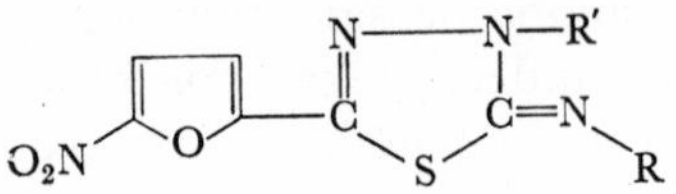

(76)

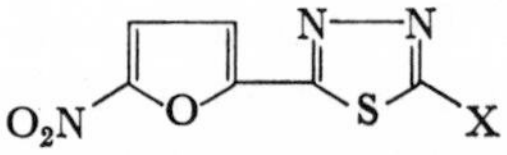

(77)

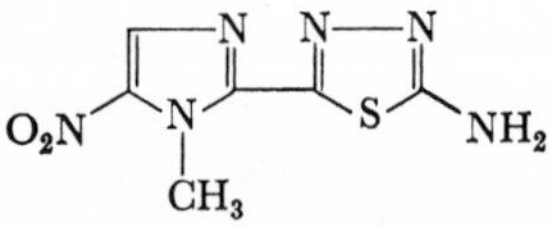

(78)

or phenoxy. Recently compound **78** has been shown to be a powerful antimicrobial and antiparasitic agent of broad specificity[498].

All of the examples cited have a constant structural factor, and removal of the 5-nitro group very greatly reduces the antibacterial properties of such substances. While the hydrazine-like substituent in the side chain may be of some importance[499,500] it appears that it is more involved in influencing the penetration of the drug into the proper site and/or decreasing the amount of hydrolytic decomposition[499]. Not surprisingly, one mechanism of metabolism and detoxification of drugs such as **72** is reduction of the nitro group[501,502]. The production of aminofurans from compounds having the general formula of **79**, where R_2 = H has been described[503–505]. However,

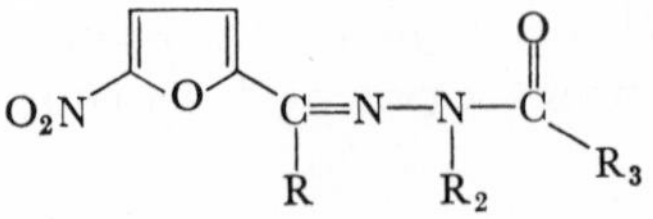

(79)

it appears that a different degradative pathway is followed when R_2 and R_3 in **79** are part of a cyclic system or are alkyl groups[506].

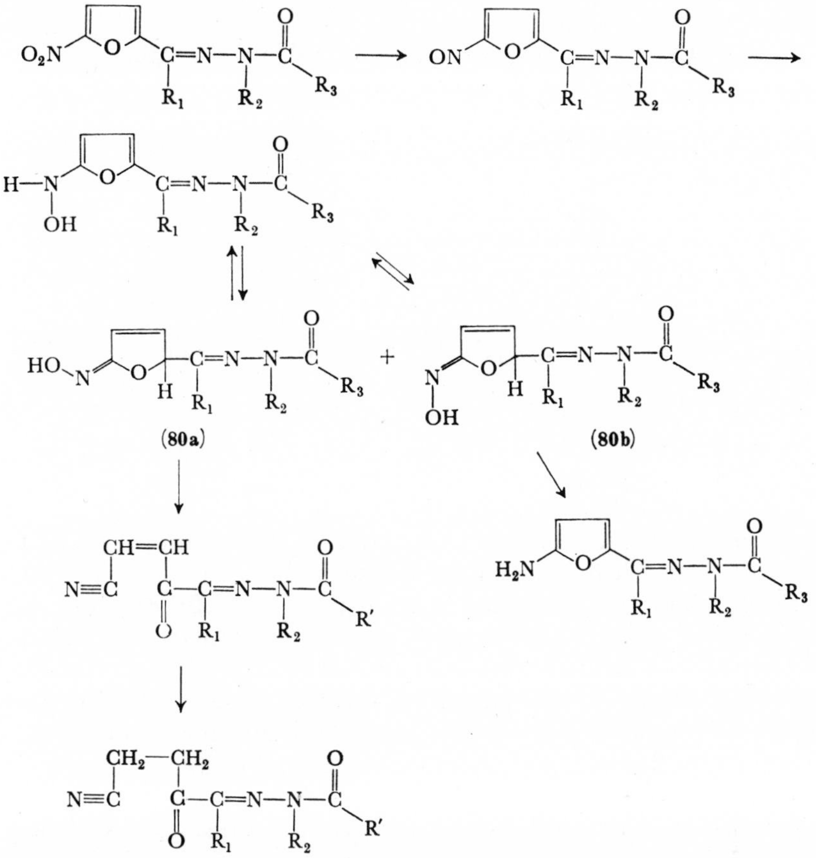

SCHEME 6. Postulated pathway for the aerobic microbial degradation of nitrofurans[506].

These conclusions have been summarized in the form of a postulated pathway for the biological degradation of the nitrofurans[506] (Scheme 6). It may be seen that a critical point of this postulated scheme is the fate of the (hypothetical) oxime **80**, since isomer **80a** can undergo a *trans* elimination of water to yield the open-chain nitrile. Such nitrile derivatives have been demonstrated to be the major products of bacterial degradation of certain nitrofuran compounds[506]. In both pathways, of course, the nitrofuran ring is destroyed, and it is well established from these and other studies that the integrity of the nitrofuran ring is essential for bactericidal activity[507–509].

Hirano and his colleagues have carried out an interesting study of the antibacterial activity of nitrofurans **72**, **74**, and **81–84** as a

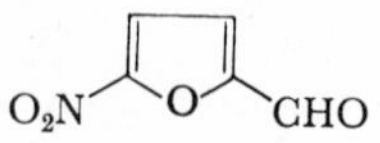

(81)

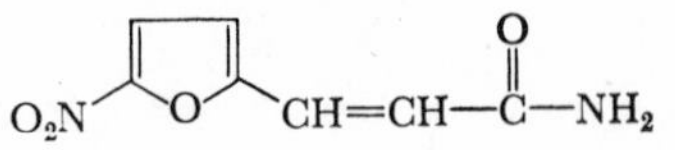

(82)

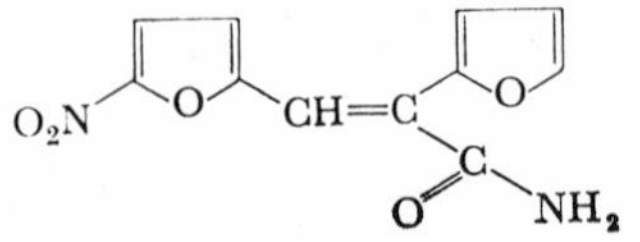

(83)

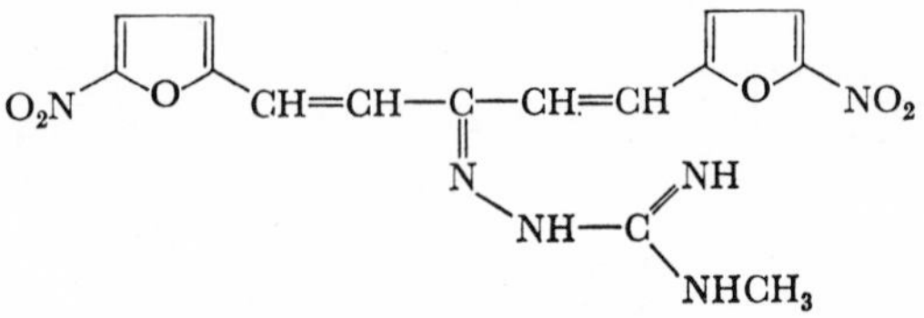

(84)

function of certain electronic parameters[510]. They employed LCAO–MO methods to calculate parameters such as the nucleophilic superdelocalizability of the carbon atom at position 5 of the furan ring, since other workers had found[511] that this parameter was closely related to the antibacterial activity exhibited by 4-nitroquinoline *N*-oxides. However, this parameter was found not to be related to the antibacterial activity in the case of the nitrofurans (Table 7). Included in Table 7 are values for the energy of the lowest empty molecular orbital as well as the polarographic half-wave reduction potential and the antibacterial activity of the substances toward

TABLE 7. Electronic parameters and antibacterial activity of some nitrofuran compounds[510].

Compound	Nucleophilic reactivity index	Energy of lowest unoccupied orbital	Polarographic reduction potential	Relative antibacterial activity
Furan	0.72	−0.91		1000
Furfural	1.14	−0.36	−1.09	1000
2-Nitrofuran	0.12	−0.32	−0.43	100
81	1.91	−0.18	−0.08	25
72	1.87	−0.27	−0.18	7
82	1.61	−0.21	−0.15	6.2
74	1.87	−0.27	−0.11	5
83	1.54	−0.20	−0.16	1.5
84	0.78	−0.23	−0.14	1

Staphyloccoccus aureus expressed in terms of the dilution (microliters/milliliter) needed for antibacterial action. There is no correlation apparent between the antibacterial activity and the nucleophilic reactivity index. If other factors such as permeability, diffusibility, and solubility of these substances in the cells and tissues are the same, then this lack of correlation suggests that the antibacterial activity of the nitrofurans does not depend on a nucleophilic displacement on the 5 position of the furan ring. Such a displacement might conceivably have occurred involving biologically important thiol compounds of the microbe[511,512].

There is, however, a rough correlation of antibacterial activity with the energy of the lowest vacant orbital or with the closely related reduction potential (Table 7). Thus, the authors suggest that the reduction of the nitrofuran derivatives is the most important process in the bactericidal action of these substances[510]. The authors suggest that NADH (**2a**) probably acts as the coenzyme in such a process. Interestingly, they established that complex formation occurs between **82** and NADH and by use of a Benesi–Hildebrand type of plot and difference spectrophotometry they determined the association constant for complex formation between NADH and **82** to be 4×10^3 l./mole and probably involved a 1:1 complex. This suggests that the key reaction in the first stage of bactericidal action of the nitrofurans is the formation of a molecular complex probably involving the intermediates of the electron-transport sequence (Scheme 1). One particular mode of action of the nitrofurans thus might be to act as an irreversibly reduced acceptor following NADH but before the flavin coenzymes (flavoproteins) in Scheme 1. However, a number of substituted nitrofurans have recently been shown to be effective competitive inhibitors of the soluble cellular nitroreductase enzymes so that it is possible that no chemical transformation need occur for these drugs to exert their biochemical effects[513].

If the nitrofurans do exert their biochemical effects by interfering with the electron-transport sequence of the invading organism it might also be suspected that the nitrofurans have the potential for causing toxic reactions in the host organism. Although the precise biochemical cause may be different there are indeed significant adverse physiological reactions which may occur during medical use of the nitrofurans[513]. This may take the form of jaundice and apparent liver disease[514–517] or a pulmonary edema[518–526] which is frequently mistaken as a symptom of cardiac failure. These effects may be the result of an allergic reaction[525,526]. Hemolytic anemia during treatment with drugs such as nitrofurantoin has also been observed[513] and is more common among a small percentage of Negroes and ethnic groups of Mediterranean and Near East origin. This is attributed to a genetic abnormality[527] of the red blood cells in that the erythrocytes of these individuals are deficient in the enzyme glucose-6-phosphate dehydrogenase. In turn this results in an abnormally rapid destruction of the red blood cells and thus anemia develops. Other side reactions observed during medical use of nitrofuran derivatives are polyneuropathy involving both the central and peripheral nervous systems[528–531].

The effectiveness as antituberculosis drugs of a variety of aliphatic aromatic and heterocyclic compounds containing the nitro group has been investigated by Urbański and coworkers[532–534]. In view of the known tuberculostatic properties of the hydroxamic acids[535–537], it appeared possible that various nitro compounds might exhibit similar properties since in the course of reductive metabolism of the nitro group there may be produced hydroxamic acids, at least as transient intermediates. A number of the materials examined by Urbański, *et al.*, showed good tuberculostatic activity *in vitro*, including 5-nitro-5-ethyltetrahydro-1,3-oxazine (**85**), diethyl (α-nitromethylbenzyl)malonate (**86**), *p*-chloro-*N*-(3-methylamino-2-nitropropylidene)aniline (**87**), and 2-bromo-2-nitro-5-methyl-1,3-hexanediol (**88**). Related to these substances are aromatic *N*-oxides

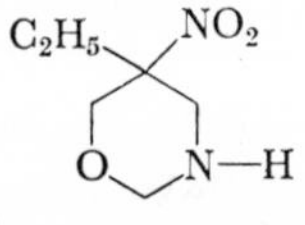

C_2H_5, NO_2; O, N—H

(**85**)

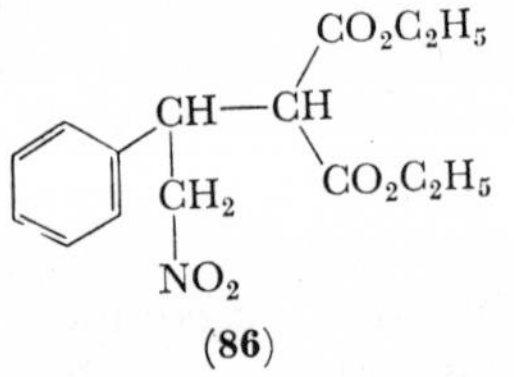

$CO_2C_2H_5$; CH—CH; CH_2; NO_2; $CO_2C_2H_5$

(**86**)

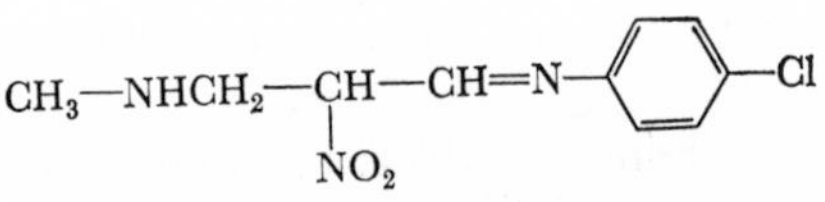

CH_3—$NHCH_2$—CH(NO_2)—CH=N—C_6H_4—Cl

(**87**)

$(CH_3)_2CH$—CH_2—CH(OH)—C(Br)(NO_2)—CH_2OH

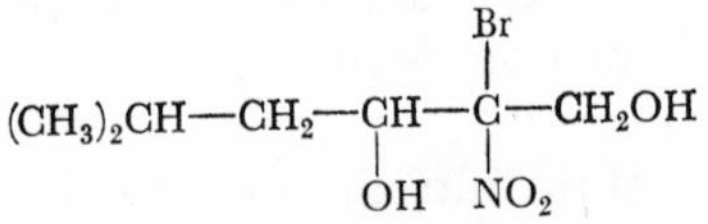

(**88**)

which may be considered to be tautomers of cyclic hydroxamic acids. A number of substituted hydrazides are well recognized as active antituberculosis agents; 4-nitrobenzalisonicotinic acid hydrazide (**89**) might be cited as one example. Significant antibacterial and antifungal activities were exhibited by compounds such as **90**. The nitro group potentiates the biological activity shown by these compounds[538].

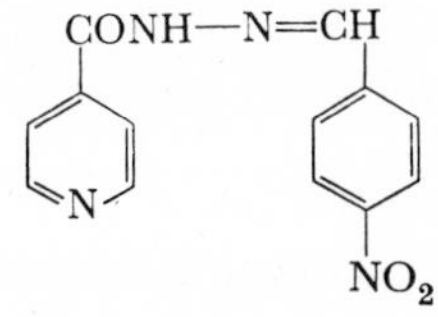

CONH—N=CH

NO_2

N

NO_2

N

O

(**89**) 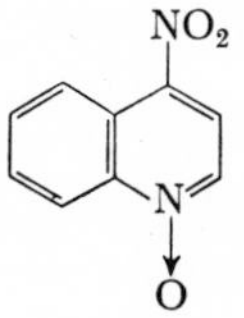 (**90**)

b. Antiprotozoal drugs. A number of derivatives of the general formula **91** have been found useful as antiprotozoal drugs[539,540]. One of the more effective antiamebic drugs is chlorophenoxamide (**91**, X = O, R_1 = CH_2CH_2OH, and R_2 = $COCHCl_2$). Chlorophenoxamide is of low oral toxicity to the host organism, probably

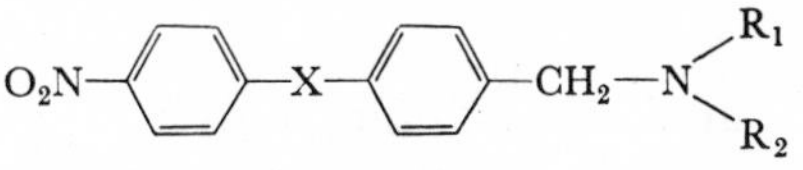

O_2N—⟨⟩—X—⟨⟩—CH_2—N(R_1)(R_2)

(**91**)

because it is not well absorbed from the gastrointestinal tract. After a single oral dose **91** could be detected in the feces for 48 hours[541]. Only low levels of nitro or amino derivatives of the drug could be detected in the blood, while both free and combined amino derivatives were found in the urine; about 8% of the dose was excreted in the urine as nitro compounds. It may be presumed that the intestinal flora can reduce the nitro group of such compounds to amino derivatives and this may represent one means of detoxifying the substance.

The nitroimidazole derivative **92** has been found to be effective

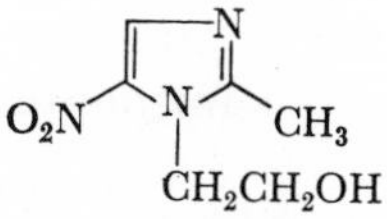

N

O_2N N CH_3

CH_2CH_2OH

(**92**)

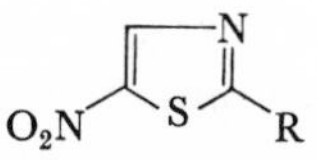

N

O_2N S R

(**93**)

R=$NHCOCH_3$, NHCO—(thienyl), or N⟨⟩$NCOCH_3$

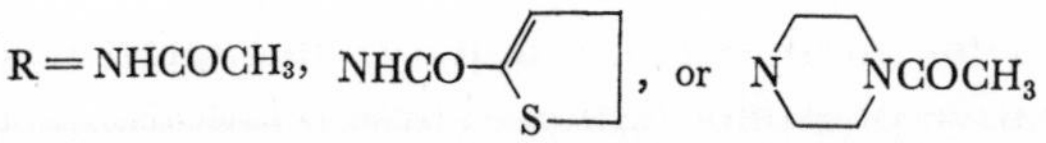

against *Trichomonas vaginalis* both by local and by oral administration[542]. Since the drug is effective even when administered orally it follows that it is well absorbed from the gastrointestinal tract. Although well tolerated in therapeutic doses, it was found that higher doses caused serious side effects. These included disorders of the central nervous system such as tremor, atoxia, and muscular weakness. There were also reports of gastrointestinal disturbances similar to those observed during nitrofuran administration, as well as dermal and hematological changes probably caused by an allergic reaction. A group of related nitrothiazole compounds possessing primarily antitrichomonal activity may be represented as **93.** It may be seen that these substances resemble somewhat the nitroimidazole antibiotic azomycin (III.A. 2a).

It is perhaps appropriate at this point to mention the drug 1-(*p*-nitrophenyl)-2-amidineurea (**94**) which has been tested as an

$$O_2N-C_6H_4-NHCONHC(=NH)-NH_2 \cdot HCl$$

(**94**)

antimalarial[543,544], and also exhibits activity *in vitro* against *Mycobacterium tuberculosis*[532]. The nitro group is not required for activity but causes the material to be relatively less toxic than, for example, the chloro analog.

c. Antihelminthic drugs. An important property of therapeutically useful drugs employed to combat intestinal worms is that they are not readily absorbed through the wall of the intestine following oral administration, since otherwise they might produce a general toxic effect. One nitro compound of wide therapeutic use is compound **45**[545–547]. Only a small part of the drug is absorbed unchanged from the intestine[548], and because it is poorly absorbed this drug is effective only with intestinal worms. The nitro group is reduced to the corresponding amine by intestinal bacteria resulting in a loss of effectiveness. The therapeutic effectiveness of this and related compounds may be attributed to the balancing of absorption from the intestine with a general toxic effect based on the uncoupling of oxidation from phosphorylation.

Compounds such as **45** have been intensively investigated with regard to the relationship between antihelminthic activity *in vivo* and *in vitro*, molluscicidal activity, and effectiveness as uncouplers of oxidative phosphorylation. These relationships may be illustrated by the data shown in Table 8[418,549]. The introduction merely of an increasing number of halogen atoms potentiates the antihelminthic effect but this is accompanied by increasing toxicity to the host. Such compounds lack uncoupling activity. The addition of a nitro group in the 4′ position results in a compound with a good therapeutic index. It should also be noted that compounds with

TABLE 8. Some structure-activity effects observed with phenylsalicylanilide derivatives[418–549].

R	2′	3′	4′	5′	Antihelminthic activity in vivo	Antihelminthic activity in vitro[a]	Molluscicidal activity[a]	Uncoupling effect[b]
Cl					None	>10		0
Cl		Cl			Transient	10	10	0
Cl		NO_2			None	2	3	28
Cl	NO_2		NO_2		Transient	3		16
Cl	Cl		NO_2		Cure	0.1	0.3	100
Cl	Cl		NO_2	Cl	Transient	1	0.3	36
Cl		NO_2	Cl		Trace	3	100	12
Cl	NO_2		Cl		Transient	1	1	22
Cl	NO_2			Cl	None	3		0
Cl	Cl		NH_2		None	>10		0
I	I		NO_2		Transient	0.3		30
Br	Br		NO_2		Cure	0.3		100

[a] In parts per million needed to produce a given effect in standardized tests.
[b] In percent uncoupling of oxidative phosphorylation caused by the addition of 10^{-5} M of the test compound to a standard assay system.

therapeutic effectiveness are also compounds which act as strong uncouplers of oxidative phosphorylation, and the two compounds cited in Table 8 which bring about a cure are also the two most potent uncoupling agents.

It is significant that replacement of the nitro group by the amino group abolishes all biological effects observed in these tests (Table 8). It is probable that such a reduction reaction occurs in the alimentary tract under the influence of the intestinal flora. This would be consistent with the lessened effectiveness of the nitro-substituted drugs such as **45** when administered in cases of infection by young parasites[545]. During this stage of their life cycle the parasites locate themselves in the lower portion of the small intestine where a greater proportion of the drug can be found as the amino reduction product. The mature parasites migrate to the upper portion of the small intestine; here the intact nitro form of the drug is still present in significant amounts and acts to poison the parasitic worms. Again, the mechanism of action of these substances can be assumed to be their interference with the fundamental energy yielding processes of oxidative phosphorylation. Their selective toxicity toward the parasites rather than the host is due to the limited absorption of the drug from the intestine.

The detoxification of these aromatic nitro derivatives by reduction to the amino compound is consistent with numerous studies including those on the nitroguanidine derivative **94** which also is known to have antihelminthic activity and undergoes reduction, probably by bacterial nitroreductases[550–552].

2. Substances acting directly on macroorganisms

a. Cytostatic compounds. Chemicals which stop the growth of certain cells are potentially very valuable if, for example, they possess a selective action against microorganisms (antibiotics) or against cancers (antineoplastic agents). The antineoplastic activities of 1,3-oxazine derivatives such as **95** have been described[533].

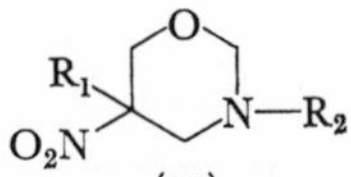

(95)

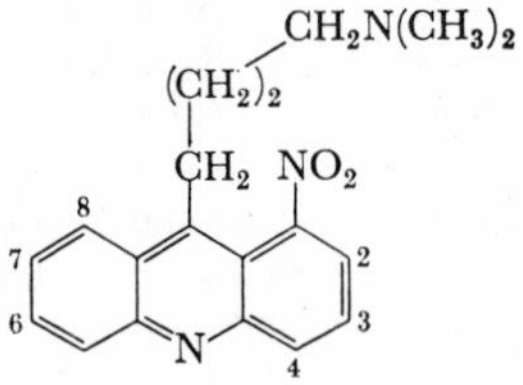

(96)

Compounds where $R_1 = R_2 = CH_3$ or $R_1 = C_3H_7$ and $R_2 = C_6H_{11}$ were active *in vivo*, causing an inhibition in the growth of certain tumors of mice[554–556]. The importance of the nitro substituent as well as effects of substitution at position 2 in **95** have been discussed[556].

Antineoplastic activity is also exhibited by a number of acridine derivatives. For example, 1-nitro-9-(4-dimethylaminobutyl)acridine (**96**) exhibits activity against Ehrlich ascites sarcoma and in other tests[557–559]. Of the many compounds examined the most active derivatives possessed a strongly electrophilic substituent such as a nitro group in the 1 position of the substituted acridine ring. These materials are active *in vivo* and the therapeutic utility of these materials has been explored[560,561]. The biological effects of such nitroacridine derivatives are diverse, including changes in the levels of such hydrolytic enzymes as acid and alkaline phosphatases, structural and functional disturbances in the intestinal epithelia, water and mineral metabolism, intestinal disorders, and interference with spermatogenesis[562–564]. The LD_{50} for oral administration of **96** to rats is only 21 mg/kg; for intravenous administration it is some 20 times less, indicating that either the drug is reduced and detoxified by the intestinal microbial flora, or that it is poorly absorbed through the intestinal walls.

The biochemical mode of action of nitroacridines such as **96** remains uncertain although it is reasonable to assume that the nitro compounds act by the same mechanism as a variety of other acridines. Acridine derivatives have been shown to interfere with the replication of deoxyribonucleic acids, a process which is required for cell division and tissue (or tumor) growth[565,566]. The molecular basis for this inhibition appears to depend on the specific interaction of the planar acridine ring system with the nucleic acid helices. The acridine ring may slip between the planar stacked purine and pyrimidine bases (intercalation)[567] and assume a form stabilized by hydrophobic or charge-transfer interactions. If this is the case, then the nitro compound itself may be responsible for the cytostatic activity. Alternatively, the nitro compound may be absorbed to the site of action and then reduced to an amino derivative. The resulting polycationic acridine molecule might then interact electrostatically with the polyanionic deoxyribonucleic acid chain. Although it has been suggested that the antitumor activity of the nitroacridine derivatives is unrelated to deoxyribonucleic synthesis[561], such studies were directed toward determining the percent deoxyribonucleic acids in tumors and normal liver; such percentage values might not be expected to undergo much change.

Antitumor activity is also exhibited by compounds such as **97** and **98**[568,569]. Compound **98** is not particularly toxic and yet it exhibits moderate activity against sarcoma 180 in mice[570].

An additional type of alkyl nitro compound which exhibits activity against certain types of neoplasms is **99**[571]. One hypothesis

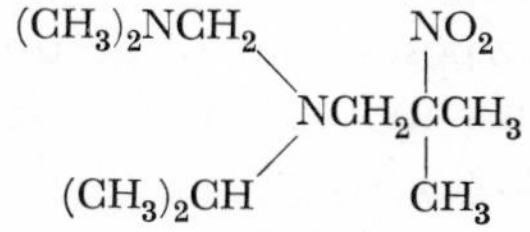

$(CH_3)_2NCH_2$—N(—$CH(CH_3)_2$)—$CH_2C(NO_2)(CH_3)CH_3$

(**97**)

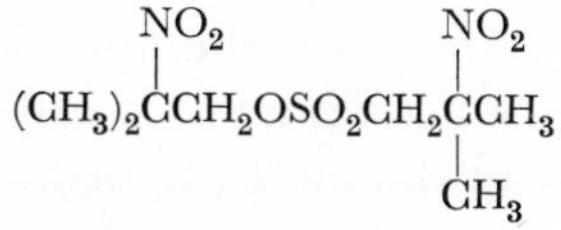

$(CH_3)_2C(NO_2)CH_2OSO_2CH_2C(NO_2)(CH_3)CH_3$

(**98**)

CH_3—$C(CH_2Cl)_2$—NO_2

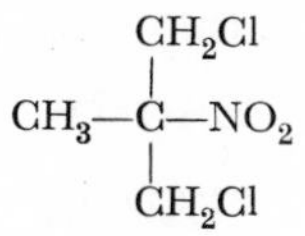

(**99**)

which can be advanced to explain the biological activity of compounds such as **99** is that in the cell they are reduced to an amine which, in turn, can undergo cyclization to a substituted aziridinium ion (reaction 13). This quaternary ammonium compound **100** can

$$\mathbf{99} \longrightarrow \longrightarrow CH_3\text{—}\underset{\underset{CH_2Cl}{|}}{\overset{\overset{CH_2Cl}{|}}{C}}\text{—}NH_2 \rightleftharpoons CH_3\text{—}\overset{\overset{CH_2Cl}{|}}{C}\underset{CH_2}{\diagdown\diagup}\overset{+}{N}H_2\ Cl^{-} \qquad (13)$$

(**100**)

act as a powerful alkylating agent[572] capable of reaction with such vital cellular constituents as the nucleic acids or sulfhydryl compounds. The initial product could undergo recyclization to a second aziridinium derivative **101** which could in turn react with a second nucleophile HNuc′ (reaction 14). Thus, the nitro compound **99**

$$\mathbf{100} + HNuc \longrightarrow CH_3\text{—}\underset{\underset{Nuc}{|}}{\underset{CH_2}{|}}{\overset{\overset{CH_2Cl}{|}}{C}}\text{—}NH_2 \rightleftharpoons CH_3\text{—}\underset{\underset{Nuc}{|}}{\underset{CH_2}{|}}{\overset{\overset{CH_2}{|\diagdown}}{C}}\text{—}\overset{+}{N}H_2\ Cl^{-}$$

(**101**)

$$\mathbf{101} + HNuc' \longrightarrow CH_3\text{—}\underset{\underset{Nuc}{|}}{\underset{CH_2}{|}}{\overset{\overset{Nuc'}{|}}{\overset{CH_2}{|}}{C}}\text{—}NH_2 \qquad (14)$$

probably exhibits cytostatic activity by virtue of the ease with which it may be transformed in the cell into a bifunctional alkylating agent. These agents have attracted considerable interest in cancer chemotherapy[572].

b. Spasmolytic compounds. A number of nitric and nitrous acid esters are utilized for therapeutic purposes, and while they are not properly nitro compounds it still seems appropriate to briefly discuss them. One of the most important alkyl nitrates is probably glyceryl trinitrate (nitroglycerin, **102**) which has been widely used for the

$$\begin{array}{c} CH_2ONO_2 \\ | \\ CHONO_2 \\ | \\ CH_2ONO_2 \end{array}$$

(**102**)

treatment of spasms of coronary vessels (angina). There are a number of other alkyl nitrate derivatives such as amyl nitrate, erythrityl tetranitrate, and mannityl hexanitrate which are similarly employed.

For some time it was thought that organic nitrates act only after the nitrate ion is split off in the cell and reduced to the nitrite ion. It is true that nitrite ions are observed in the blood following administration of **102** or other nitrate esters[573,574]. However, these are probably not the agents responsible for the vasodilating effect which leads to the lowering of the blood pressure on administration of large amounts of **102**, since the amount of nitrite ions appears to be too small to account for the intensity of the phenomenon. Other evidence suggests that the nitrate esters produce vasodilation without previous hydrolysis and without the appearance of nitrites[575,576]. However, it does seem possible that the alcoholic part of the esters acts as a carrier which allows the nitrate precursor to enter the target cells and tissues much in the same manner as has been seen for other drugs.

The fate of nitrate esters in the organism is an interesting problem. The formation of nitrite ions *in vitro* during incubation of alkyl nitrates with tissue homogenates has been demonstrated[577]. In liver mitochondria, an intracellular particle which is the site of a considerable proportion of cellular oxidative metabolism and oxidative phosphorylation, there is an enzyme called glutathione–organic nitrate reductase which catalyzes the reduction of the nitrate group simultaneously with the oxidation of the sulfhydryl-containing tripeptide glutathione (GSH) as diagrammed in reaction 15[578].

$$2GSH + RONO_2 \longrightarrow GSSG + RONO + H_2O \qquad (15)$$

The alkyl nitrite which results from the reduction reaction can undergo hydrolysis to form free nitrite ions[579]. The other products following hydrolysis are the 1,2- and 1,3-dinitroglyceryl esters[580,581]. These denitration products possess much lower biological activities than the fully esterified derivatives[581]. The reductive metabolism of the nitrate compounds and hydrolytic metabolism of the nitrites is consistent with the fact that methemoglobinaemia (section V) occurs to only a limited extent following administration of large doses of nitrates.

The physiological effects caused by the alkyl nitrate esters consist primarily of the relaxation of all plain (smooth) muscles including those of the blood vessels. Large doses produce a marked dilation of the blood vessels and a resulting fall in arterial blood pressure.

In turn, this may result in an insufficient supply of oxygen and fainting. Nitrate therapy may also be accompanied by headaches due to effects on blood vessels of the brain.

c. Miscellaneous compounds. Asymmetric alkyl 2,4-dinitrophenyl disulfides undergo a facile reaction with the sulfhydryl group of the amino acid cysteine, giving rise to 2,4-dinitrothiophenols and the corresponding *S*-alkyl cysteine (reaction 16)[582,583]. It was also

R—S—S(2,4-$(NO_2)_2C_6H_3$) + H_3N^+—CH—COO^- (CH_2—CH_2—SH) ⟶ HS(2,4-$(NO_2)_2C_6H_3$) + H_3N^+—CH—COO^- (CH_2—CH_2—S—S—R)

(103) (16)

observed that compounds such as the *S*-hydroxyethyl derivative **103**, R = CH_2CH_2OH, are potent poisons of muscular action[584]. Reaction with the sulfhydryl groups of the contractile proteins of muscle could readily explain the loss of muscle function following treatment with a drug of structure **103**. When the *S*-hydroxyethyl compound was administered intraperitoneally to mice in a dose of 28 mg/kg of body weight there resulted a reversible paralysis of the dorsal muscles. Investigations with the worm *Enchytraeus albidus* revealed the expected presence of 2,4-dinitrothiophenol following the administration of **103**.

The facile reaction of the nitro-substituted disulfides is the basis of the Ellman method[585] for the determination of free sulfhydryl groups in proteins using 5,5′-dithiobis(2-nitrobenzoic acid) (**104**).

O_2N, HO_2C —S—S— NO_2, CO_2H **(104)**   —S—S— O_2N, NO_2 **(105)**

Because the resulting *p*-nitrothiophenolate anion absorbs strongly even above 400 mμ (where proteins do not absorb significantly) the extent of the disulfide interchange reaction analogous to reaction 16 can be determined by simple spectrophotometric techniques. The major function of the nitro substituents is to shift the pK_a of the thiophenol and cause the ionized phenol to absorb in a convenient region of the visible spectrum. In addition the Ellman reagent possesses a carboxyl group so as to increase the solubility of the reagent in aqueous solution. Sulfhydryl–disulfide interchange reactions are well-recognized phenomena in chemical and biochemical systems[586]. It is of some interest that *m*-dinitrophenyl disulfide (**105**) has achieved some use in veterinary medicine as a coccidiostat.

Aromatic nitro compounds exhibit a number of other interesting activities. For example, compounds **106** and **107** are powerful

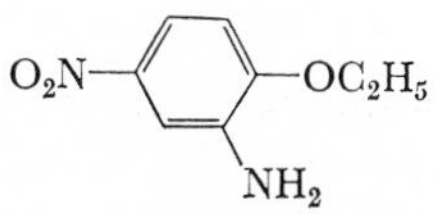

$O_2N-C_6H_3(NH_2)-OC_2H_5$

(**106**)

$O_2N-C_6H_4-NHC(=O)NHCH_2CH_2CO_2H$

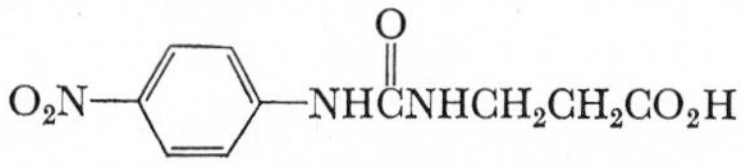

(**107**)

sweetening agents. In addition to these compounds, it is possible to cite a variety of other nitro derivatives which exhibit biological activities as analgetics, antitussive agents, hypertensives, hypnotics, plant hormones, pituitary inhibitors, and further antibacterial antifungal or antitrichomal drugs. For most of these compounds, only a small or peripheral role is played by the nitro group in causing biological activity.

3. The importance of the nitro group in drugs

A variety of preparations containing nitro groups are of therapeutic utility in the treatment of infectious diseases. The utility of the drug is not, in general, due to its effect on the host organism, but rather is due to the fact that the drug exerts a selective effect on the invading parasite. The toxicity of the drugs which have been discussed is a relative property; there may be a fairly general toxicity toward the metabolic systems which are, after all, similar in even seemingly widely different organisms. The deciding factors which affect the therapeutic utility of a drug are chemical and physical properties which may be largely unrelated to the presence of nitro or nitroso groups. Frequently, properties which govern permeability through membranes such as dissociation constants, polarity, hydrophilic–hydrophobic character, etc., may be exceedingly important. These properties might for example dictate that, at a given pH, a drug is absorbed by intestinal parasites but not through the intestinal wall of the host.

Another complicating factor in affecting therapeutic utility is the balance between medicinal action and detoxification. Drugs which are rapidly detoxified may be of limited medical use. However, it is easy to imagine situations where the detoxification route might be selectively and purposely blocked so as to enhance the therapeutic effect of a drug. For example, the reduction of nitro substituents to amino groups (perhaps followed by *N*-acylation) is a relatively common route for the detoxification of drugs bearing nitro substituents. If, for example, this reduction were carried out by the microbial flora of the intestines, then it should be possible to enhance the effect of certain nitro-substituted drugs by the simultaneous administration of drugs which diminish or abolish the activity of the intestinal flora.

Because of the frequently observed correlation between therapeutic effect and activity in the uncoupling of oxidative phosphorylation, it may be concluded that many nitro-substituted drugs act by causing severe disorders in the fundamental metabolic processes of the target organism. Because oxidative phosphorylation occurs in a very nearly identical fashion in all of the higher organisms, it is clear that drug toxicity toward the host as well as the parasite is a real possibility. Thus, the therapeutic utility of a drug is highly dependent on the phenomena associated with selective permeability and drug absorption.

C. Technically Important Compounds

There are obviously a very large number of nitro and nitroso compounds produced for various technical purposes, and it is impossible to enumerate all of them. However, some of these compounds are potentially highly toxic. One of the most important toxic symptoms, namely methemoglobinemia will be discussed at some length in section V.

I. Aliphatic nitro compounds

Nitroparaffins are of great industrial importance as explosives, solvents, reactants, etc. From a structural point of view, these substances may be primary, secondary, or tertiary nitro compounds, and mono- or polynitro compounds. From a physical point of view, they are often oily liquids which may be moderately volatile or which may be readily absorbed through the skin. The toxic symptoms evoked by these substances possess some common features but are frequently distinctive and vary with the route of administration[587]. When the nitroparaffins penetrate the organism by the respiratory tract they cause irritation of the upper respiratory passages and mucous membranes of the mouth and eyes, with consequent coughing, salivation, and lacrimation. The intensity of these symptoms increases with the increasing chain length of the nitroparaffin. In addition, polynitroparaffins exert a stronger irritating effect than do the corresponding monosubstituted analogs. Unsaturated or chlorinated nitro compounds are still more irritating.

In addition to local symptoms there may result a general malaise and depression of the central nervous system. Nitroparaffins are also fairly general cellular poisons, particularly against organs such as the liver[588] but to a lesser extent causing damage to the heart and kidneys. In the case of acute toxic exposures to nitroparaffins death usually comes as the result of paralysis of the respiratory system, while in the case of chronic exposure to poisonous doses it is found that liver damage is the important symptom. Lethal doses of homologous nitroparaffins from nitromethane to nitrobutane are very similar; when administered orally to rabbits the LD_{50} ranges from 0.25 to 1.0 g/kg[587]. On the other hand, the chlorinated nitroparaffins are 5–10 times more toxic. Any of the nitroparaffins may be quite irritating to the skin. Oral ingestion results in irritation of the alimentary tract, pain, colic, diarrhoea, and bleeding from the intestinal mucosa. Damage to the blood vessels may result in certain organs following the administration of acute doses of nitroparaffins but there are only indirect effects on arterial blood pressure and respiration[589]. Changes in the lungs are particularly intense following exposure to chloropicrin. Methemoglobinemia is not an important result of exposure to nitroparaffins but was, for example, detected following the administration of 2-nitropropane to cats by inhalation[588].

The nitroparaffins are quickly eliminated from the blood, partly by respiration and partly by metabolic transformations resulting in cleavage of the carbon–nitrogen bond[590,591]. The general course of this transformation is represented in reaction 17[587].

$$RCH_2NO_2 \xrightarrow{[o]} \longrightarrow RCHO + H^{\oplus} + NO_2^{\ominus} \qquad (17)$$

The nitro olefins possess a strong local irritant action and are relatively toxic substances. Manifestations of poisoning by these materials are hyperemia, increased mucosal secretion, lacrimation, and coughing. With prolonged exposure there occurs cyanosis, breathlessness, lowering of the arterial blood pressure, hyperexcitability and convulsions, and finally a deep depression of the central nervous system, coma, and death. This sequence of events resembles, to a considerable degree, the effects due to inhalation of narcotics.

2. Aromatic nitro compounds including nitrophenols

a. Toxic properties. Aromatic nitro compounds are of considerable technical importance as solvents and as synthetic intermediates for dyes and explosives. Nitrobenzene is quite toxic and is readily absorbed through the skin. In acute poisoning there occur changes in the blood such as lowering of the hemoglobin level and metahemoglobinemia, and also symptoms of intoxication such as vomiting, colic, headaches, breathlessness, and cyanosis. Additional effects on the central nervous system result in feelings of anxiety and convulsions. A variety of nitrophenols affect the liberation of acetylcholine at the nerve endings to the small intestines and certain muscles[592]. Individual aromatic nitro compounds may cause specific symptoms of intoxication in addition to the general ones just described. For example, *m*-dinitrobenzene can accumulate in the lipid-rich adipose tissues where it can remain for extended periods. On ingestion of alcohol, the residues of *m*-dinitrobenzene can be washed out of the adipose tissues and severe toxic symptoms including cyanosis, headaches, and vomiting can result. Chronic intoxication with nitroaromatics is observed to result in cirrhosis or acute atrophy of the liver. Yellow coloring of the hair and nails occurs in poisoning with *m*-dinitrobenzene. A similar observation has been made for the case of *p*-nitrobenzoic acid[593].

A different toxic property is possessed by 1-chloro-2,4-dinitrobenzene, for it is known to cause strong allergic reactions particularly characterized by skin eruptions and other dermal disorders[594–596]. Such aromatic dinitro compounds are known to be activated with respect to nucleophilic substitution reactions in contrast, for example, to chlorobenzene itself. As a result, 1-chloro- or 1-fluoro-2,4-dinitrobenzene undergoes a facile reaction with nucleophiles such as amino groups which may be part of a protein. Such a reaction is utilized to introduce the 2,4-dinitrophenyl group into a variety of polypeptide and protein molecules for the purpose of synthesizing 'unnatural' proteins which will in turn elicit antibody formation[597–607]. The reaction of amino groups with 1-fluoro-2,4-dinitrobenzene is, of course, the basis for the Sanger method for the determination of the *N*-terminal amino acid of a peptide chain[608]. In addition to the reaction with *N*-terminal amino functions, the halodinitrobenzenes also react with other groups found in proteins including the side-chain primary amino group of the amino acid lysine. If, for example, the low molecular weight polypeptide hormone insulin is treated with 1-fluoro-2,4-dinitrobenzene, it is possible to introduce one dinitrophenyl group corresponding to the presence of one lysine residue. The dinitrophenyl group together with a restricted region of adjacent amino acids acts as a determining site in antigen–antibody reactions[609]. Thus the allergenic properties of 1-chloro-2,4-dinitrobenzene are probably due to an antigen–antibody reaction which results after the dinitrophenyl group is covalently attached to natural proteins.

b. Metabolic transformations. In the course of the discussions presented in this chapter there have been a number of examples of metabolic transformations. It will be useful to discuss briefly the metabolic transformations undergone by some of the technically important aromatic nitro compounds. The metabolism of nitrobenzene by the rabbit has been investigated in some detail[610,611]. As might be anticipated, reduction to the amino group represented an important metabolic reaction, but the amount of aniline which is found in the metabolites is less than 1% of the products. Rather, *p*-aminophenol and its conjugates represented over 30% of the original dose isolated and identified over a 5-day period. The conjugates were largely the glucuronic acid derivative or the *N*-acetyl derivative of the glucuronide[610]. In addition, conjugates of *m*- and *p*-nitrophenols were isolated in significant amounts from the urine. The enzymatic basis for formation of glucuronic acid conjugates of *p*-nitrophenol as a means of detoxification has been studied[612,613]. A significant amount of the administered dose of nitrobenzene remained in the tissues for extended periods, presumably in the form of various partially metabolized derivatives.

The metabolism of *m*-dinitrobenzene by rabbits also involves reduction to the amino derivatives, and the formation of aminophenols (or conjugates)[611]. Isolated *m*-nitroaniline and *m*-phenylenediamine represented about one-third of the administered dose, while 2-amino-4-nitrophenol and 2,4-diaminophenol isolated from the urine, together represented as much as 50% of the administered dose. In addition, small amounts of 2-nitro-4-aminophenol were isolated.

The metabolism of nitrotoluenes by the higher animals follows an interesting course in that the methyl group may be oxidized. For example, *o*-nitrotoluene on administration to dogs was found to lead to production of *o*-nitrobenzyl alcohol and *o*-nitrobenzoic acid[614]. The alcohol was isolated as the glucuronic acid conjugate. The metabolism of *p*-nitrotoluene similarly results in formation of the benzoic acid derivatives. Much of the acid was excreted as a conjugate[615].

A relatively unusual pathway for metabolism of an aromatic nitro compound has been reported for the case of the polychloro derivatives. For example, when 2,3,5,6-tetrachloronitrobenzene is administered to rabbits[616] the major portion of the dose is excreted unchanged in the faeces. However, in accord with the metabolism of certain of the nitrobenzenes already discussed there was isolated ~10% of the tetrachloroamino compound and ~15% of 2,3,5,6-tetrachloro-4-aminophenol. In addition, about 15% of a mercapturic acid derivative **108** was isolated. This compound is a conjugate

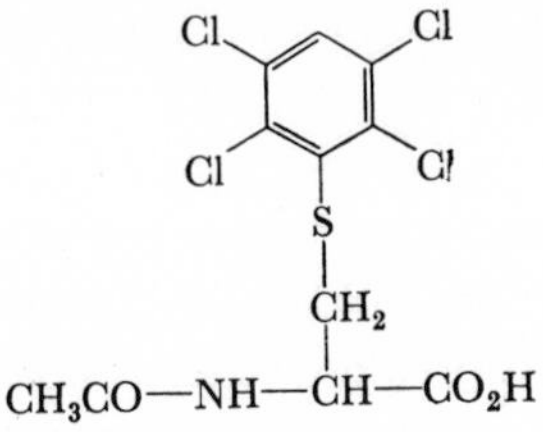

Cl Cl Cl Cl (tetrachlorophenyl)—S—CH_2

CH_3CO—NH—CH—CO_2H

(**108**)

of the aromatic ring with the *N*-acetyl derivative of the amino acid cysteine.

As might be anticipated from the foregoing discussion, the metabolism of the nitrophenols involves reduction and conjugation as major routes. For example, the relatively toxic *p*-nitrophenol is largely excreted by rabbits as the glucuronic acid conjugate. Smaller amounts may be excreted as the sulfate (sulfuric acid conjugate). Reduction to the amino compound also occurs[617]. Similar processes of reduction and conjugation can be cited for the dinitrophenols such as 2,4-dinitrophenol.

c. Effects on oxidative phosphorylation. One important toxic property of certain technically useful aromatic nitro compounds is the effect of nitrophenols on the energy-yielding processes of oxidative phosphorylation. This has, of course, been a recurring theme throughout this chapter. The mechanism by which dinitrophenols exert an effect on the sequence of oxidative phosphorylation remains uncertain. Differences in the uncoupling activity of isomeric dinitrophenols have been considered, and it was found[618] that the relative potencies of the dinitrophenols could be ranked as follows: 3,5 > 2,4 > 2,6 > 3,4 > 2,3 > 2,5. Attempts were made to correlate these activities with parameters such as pK_a of the phenol or with lipid solubility but with little success. By using radioactive dinitrophenols it was possible to measure the extent of penetration of isomeric dinitrophenols into yeast cells, and there was a parallelism between the extent of incorporation of the isomer into yeast cells and the uncoupling efficiency of that isomer in tests with cell-free systems. The results of an extensive series of investigations by Pinchot suggest that 2,4-dinitrophenol exerts its uncoupling action by preventing the association of a critical enzyme with a particulate molecular complex of the electron-transport sequence[105]. The possibility that 2,4-dinitrophenol and other phenols can act to inhibit directly various other enzymes has been advanced[619,620]. In this regard, attention is again called to the recent observation that nitrofurans can act as competitive inhibitors of the nitroductase enzymes capable of reducing *p*-nitrobenzoic acid.

V. THE ROLE OF NITRO AND NITROSO COMPOUNDS IN THE FORMATION OF METHEMOGLOBIN

In man, the major mechanism for transporting oxygen from the lungs to the tissues involves the reversible association of molecular oxygen with the protein hemoglobin. This material can be considered to be made up of two major components, the protein globin and the iron-containing porphyrin ring system termed heme. The normal reversible association of hemoglobin with oxygen occurs without a change in the ferrous oxidation state of the chelated iron atom. If the iron atom of the heme group is oxidized to the ferric state, the resulting methemoglobin is no longer able to combine with oxygen. If a significant fraction of the hemoglobin of the blood is transformed into methemoglobin then the oxygen-carrying capacity of the blood may be severely, even fatally, impaired. The physiological result resembles that observed with carbon monoxide poisoning because, in both cases, the tissues cannot be supplied with oxygen.

The introduction into the body of any substance which is capable of oxidizing hemoglobin either directly or following any metabolic transformation, may result in methemoglobinemia. A number of organic compounds cause the formation of methemoglobin, and this property is very much involved with the toxicity of many of these substances. The ingestion of nitric and nitrous acid esters and nitro and nitroso compounds is frequently observed to result in the formation of methemoglobin[621]. As might be expected, other chemicals such as quinones or chlorates can also act as oxidants of hemoglobin. Large doses of oxidized methylene blue (**109**) can result in the formation of methemoglobin in man. Interestingly, however. small doses of methylene blue are used as emergency therapy in cases of severe methemoglobenemia. This seemingly paradoxical situation arises because reduced methylene blue (**110**) can be formed

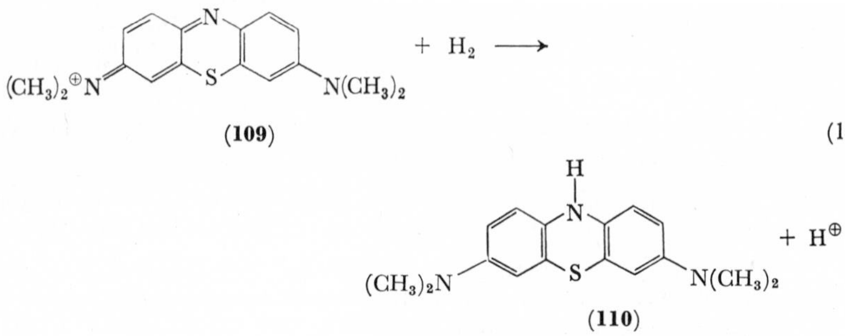

$(CH_3)_2{}^{\oplus}N$ ... $N(CH_3)_2$ + H_2 ⟶

(**109**) (18)

$(CH_3)_2N$ ... $N(CH_3)_2$ + $H^{\oplus}$

(**110**)

in the organism via a reduction reaction (such as 18) involving the biologically important coenzyme NADH (**2a**). Compound **110** then acts to reduce the ferric iron of methemoglobin to the normal ferrous state of hemoglobin. The resulting **109** may again be reduced and this cycle continued until the methemoglobin is effectively destroyed. On the other hand, if the capacity of the tissues to reduce **109** is greatly exceeded then the administered methylene blue may act as a toxic agent. The oxidation–reduction potentials of these substances are consistent with this behavior (Table 1).

It is possible that the pathway for reductive metabolism of, for example, an aromatic nitro compound involves as transient intermediates the nitroso and hydroxylamino derivatives (reaction 19).

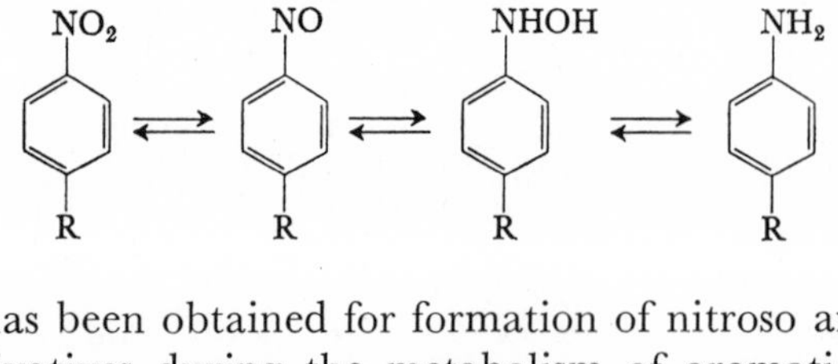

NO_2 (R) ⇌ NO (R) ⇌ NHOH (R) ⇌ NH_2 (R) (19)

Evidence has been obtained for formation of nitroso and hydroxylamino derivatives during the metabolism of aromatic nitro compounds by cell-free systems[622–625]. Phenylhydroxylamine is a potent producer of methemoglobin, both *in vitro* and *in vivo*[626–631]. The mechanism by which hydroxylamines act to produce methemoglobin is uncertain, although some interesting studies along those lines have been reported[632,633].

Aniline can also result in the formation of methemoglobin in animals. Significantly, when various anilines are administered to dogs or are filtered through excised lungs or livers of cats it is possible to detect nitrosobenzene[634–637]. Since nitrosobenzene can be detected following the administration of aniline *in vivo* the question which arises is whether it is produced with phenylhydroxylamine as a free intermediate as implied by reaction 19. The formation of phenylhydroxylamine during incubation of *N*-ethylaniline with a cell-free preparation from rabbit liver suggests that the hydroxylamine derivatives may indeed be normal, free intermediates[638]. Intravenous administration of *p*-chloroaniline to dogs results in the appearance of significant amounts of the administered dose in the form of the oxidized metabolite *p*-chloronitrosobenzene[639]. Similarly, the administration of 2-naphthylamine causes the appearance of the corresponding hydroxylamine[640]. These observations point up the apparently reversible nature of the metabolic transformations represented in reaction 19. In any event, it is clear that methemoglobin may be formed in higher organisms not only by the action of oxidants such as nitro and nitroso compounds, but also by the action of reduced compounds in the presence of oxygen. This again serves to emphasize the complex interrelationship between the various oxidative and reductive pathways of metabolism.

VI. ADDENDUM AND FINAL REMARKS

Since the completion of the main body of this chapter there have appeared several particularly pertinent articles dealing with biologically important nitro compounds. The isolation and characterization of miserotoxin, a naturally occurring alkyl nitro compound, was described in a recent communication[641]. Miserotoxin is a component of certain *Astragalus* species, particularly the so-called locoweeds or poison vetches. These weeds have been known for many years to cause both chronic and acute poisoning symptoms in livestock. It now appears that miserotoxin is the substance which causes the acute poisoning. Utilizing nuclear magnetic resonance and mass spectrometry, miserotoxin was identified as the β-D-glycoside of 3-nitro-1-propanol (**111**). This was confirmed by the

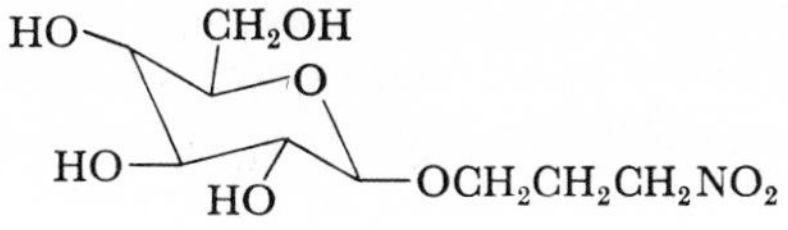

HO, CH_2OH, O, HO, HO, $OCH_2CH_2CH_2NO_2$

(**111**)

hydrolysis of miserotoxin to D-glucose and 3-nitro-1-propanol (identified by comparison with an authentic sample synthesized from 3-bromo-1-propanol). It was found that **111** was readily hydrolyzed in the rumen of livestock to D-glucose and 3-nitro-1-propanol, and that the administration to livestock of synthetic 3-nitro-1-propanol resulted in death with toxic symptoms identical to those caused by ingestion of a lethal amount of the timber milkvetch plant (*Astragalus miser*). This clearly establishes **111** as a major toxic component of this species because *A. miser* served as the biological source for the isolation of miserotoxin. It appears likely that the 3-nitro-1-propanol grouping of **111** is derived from the same source as the 3-nitropropionic acid (**22**) residue of the glycoside hiptagen. In view of the data available on the biosynthesis of 3-nitropropionic acid (see III. A.2.d) it would seem reasonable to conclude that the alcohol is derived from the acid. However, the authors report that they were unable to detect the acid in the same plant materials[641].

The proceedings of a conference on the pharmacological and chemotherapeutic properties of Niridazole and other antischistosomal compounds have recently been published. Niridazole, or 1-(5-nitro-2-thiazolyl)-2-imidazolidinone (**112**), is a potent drug in the treatment of trematode worm infections of man, a serious disease which is particularly prevalent in the tropics.

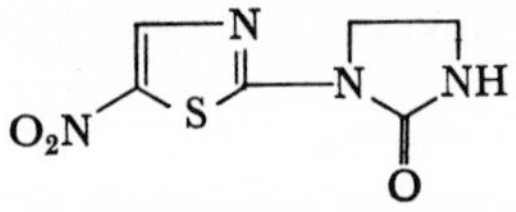

(**112**)

Among the papers presented is one dealing with the general topic of nitro heterocycles with antiparasitic effects[642]. The *in vivo* activities of a variety of nitrofurans, nitroimidazoles and nitrothiazoles were discussed, and the latter compounds (such as **112**) were particularly effective. The nitro group in the 5-position was regarded as essential for activity. Consistent with related studies described in the present review, it appears that a major route for the metabolism of **112** involves a NADPH-requiring reductase which transforms the nitro group to an amino group.[643,644] Other mechanisms for the detoxification and metabolism of drugs have been conveniently discussed[644].

The synthesis and structure-activity relationships of some substituted 2-nitroimidazole compounds have recently been described[645]. These materials may be considered to be derivatives of the natural antibiotic azomycin (**17**). The *in vivo* activity against *Trichomonas* *vaginalis* of a variety of alkyl-substituted 2-nitroimidazole structures such as **113** was measured. For the group where R_1 = methyl and R_3 = hydrogen, it was found that compounds with branched alkyl chains at R_2 were particularly effective chemotherapeutic agents, consistent with an earlier observation[646]. This effect may be due to the slower oxidative degradation which such branched chain compounds are expected to undergo *in vivo*, although this conclusion

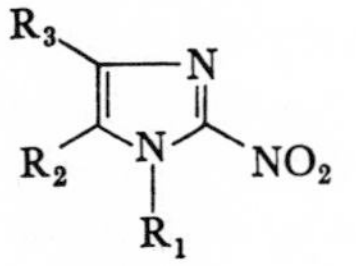

(113) 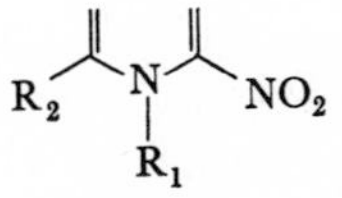 **(114)**

is subject to dispute. Consideration of numerous derivatives of the type of **113** led to the conclusion that the lower portion of the molecule (represented here as **114**) is responsible for the biological activity[645]. These authors also discuss the importance of the tautomerization (reaction 20) in affecting the biological activity.

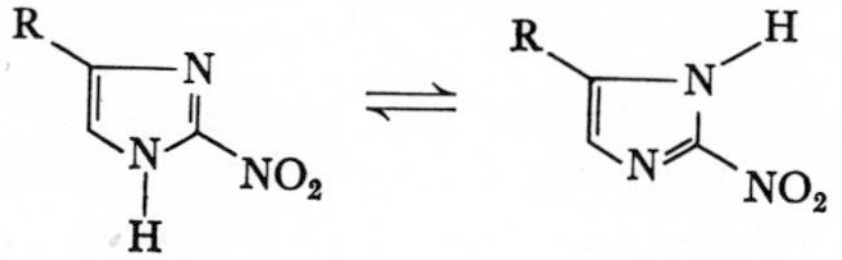

(20)

In this chapter we have examined numerous selected examples of nitro and nitroso compounds which are significant because of their biochemistry, pharmacology, or toxicology. There are probably no biological actions which can be uniquely ascribed to the nitro or nitroso groups. However, there are a number of recurring themes which hopefully are now apparent to the reader. Without a doubt the most important effect of nitro compounds on biological systems is that of the nitrophenols upon oxidative phosphorylation. Indeed many of the important medical uses of nitro-substituted aromatics can be considered to be examples of the selective uncoupling of a parasites' system of oxidative phosphorylation in preference to that of the host organism. It should be emphasized that nitrophenols and related compounds are toxic to humans as well as to lower organisms. For a time, nitrophenols were actually administered to humans as treatments for obesity, but this practice was soon halted because of the toxic side effects. Because the system of oxidative phosphorylation is a vital part of the energy-yielding metabolic reactions of all aerobic organisms, it should be apparent that any selective toxic effects are due, in part, to differences in solubility and penetration through membranes or related physicochemical phenomena which permit the penetration of the drug into the parasite in preference to the host.

Despite our generally inadequate knowledge of the metabolic transformations of compounds bearing the nitro and nitroso groups, another theme which is becoming apparent is the essential similarity of the biochemical transformations even in seemingly widely diverse living organisms. Such a conclusion follows readily, however, upon the assumption of a common progenitor of all existing organisms. Indeed the brilliant work of Margoliash and coworkers[647] provides direct evidence for this hypothesis. The oxidation–reduction reactions of the nitro and nitroso groups as well as the oxidative reactions of amines, etc., provide additional examples of common paths.

The organic chemist working with aromatic nitro compounds must be impressed with certain of the differences in their reactivity compared to the unsubstituted analogs. The biochemical reactivities are also influenced to varying extents by, for example, the electron-withdrawing properties of the nitro group. This is probably most apparent in the discussion of the cholinesterase inhibitors. In some of these systems it might seem possible to obtain fully as effective an insecticide by using for example, a trifluoromethyl substituent in place of a nitro substituent. However, it is nearly impossible at the present time to predict the effectiveness of a drug because the whole organism is so enormously complicated. As we have sought to emphasize, the chemical agent must be considered as an entirety—a complicated molecule interacting with a far more complicated biological receptor site. A major aim of biochemical pharmacology must be to define the nature of those receptor sites. Studies involving nitro- and nitroso-substituted compounds may be expected to be at the vanguard of these researches.

VII. REFERENCES

1. B. E. C. Banks in *The Chemistry of the Amino Group* (Ed. S. Patai), Interscience Publishers, New York, 1968, p. 499.
2. A. Mazur and B. Harrow, *Biochemistry: A Brief Course*, W. B. Saunders Co., Philadelphia, 1968.
3. H. R. Mahler and E. H. Cordes, *Biological Chemistry*, Harper and Row, New York, 1966; A. White, P. Handler, and E. L. Smith, *Principles of Biochemistry*, 4th ed, McGraw-Hill, New York, 1968.
4. L. E. Mortenson, R. C. Valentine, and J. E. Carnahan, *Biochem. Biophys. Res. Commun.*, **7**, 448 (1962); L. E. Mortenson, *Surv. Progr. Chem.*, **4**, 127 (1968).

5. K. Togawa and D. I. Arnon, *Biochim. Biophys. Acta*, **153**, 602 (1968).
6. L. E. Mortenson, *Proc. Natl. Acad. Sci.*, **52**, 272 (1964).
7. L. E. Mortenson, *Biochim. Biophys. Acta*, **127**, 18 (1966).
8. L. E. Mortenson, B. A. Morris, and D. Y. Jeng, *Biochim. Biophys. Acta*, **141**, 516 (1967).
9. I. R. Kennedy, J. A. Morris, and L. E. Mortenson, *Biochim. Biophys. Acta*, **153**, 777 (1968).
10. M. S. Naik and D. J. Nicholas, *Biochim. Biophys. Acta*, **118**, 195 (1966).
11. Y. Nagoi, R. F. Elleway, and D. J. Nicholas, *Biochim. Biophys. Acta*, **153**, 766 (1968).
12. E. Hackenthal, *Biochem. Pharmacol.*, **14**, 1313 (1965).
13. M. S. Naik and D. J. Nicholas, *Biochim. Biophys. Acta*, **113**, 490 (1966).
14. K. N. Subranamian, G. Padmanaban, and P. S. Sarma, *Arch. Biochem. Biophys.*, **124**, 535 (1968).
15. B. C. Radcliffe and D. J. Nicholas, *Biochim. Biophys. Acta*, **153**, 545 (1968).
16. J. S. Wessels, *Biochim. Biophys. Acta*, **109**, 357 (1965).
17. M. Hertogs and J. S. Wessels, *Biochim. Biophys. Acta*, **109**, 610 (1965).
18. F. F. del Campo, J. M. Ramirez, A. Paneque, and M. Losada, *Biochem. Biophys. Res. Commun.*, **22**, 547 (1966); A. Paneque, P. J. Apavicio, L. Cattalina, and M. Losada, *Biochim. Biophys. Acta*, **162**, 149 (1968).
19. P. Massini and G. Voorn, *Photochem. Photobiol.*, **6**, 851 (1967).
20. J. D. Kemp and D. E. Atkinson, *J. Bacteriol.*, **92**, 628 (1966).
21. J. A. Cole and J. W. Wimpenny, *Biochim. Biophys. Acta*, **162**, 39 (1968).
22. J. Ahmed and I. Morris, *Biochem. Biophys. Acta*, **162**, 32 (1968).
23. A. Addouki, F. D. Cahill, and J. F. Santos, *J. Biol. Chem.*, **243**, 2337 (1968).
24. F. Alvarado, *Biochem. Biophys. Acta*, **112**, 292 (1966).
25. F. Alvarado, *Biochim. Biophys. Acta*, **109**, 478 (1965).
26. F. Alvarado, *Comp. Biochem. Physiol.*, **20**, 461 (1967).
27. F. Aull, *Comp. Biochem. Physiol.*, **17**, 867 (1966).
28. A. Azzi and G. F. Azzone, *Biochim. Biophys. Acta*, **113**, 445 (1966).
29. R. D. Baker and D. B. Copp, *Experientia*, **21**, 510 (1965).
30. F. J. Bergersen, *Biochim. Biophys. Acta*, **130**, 304 (1966).
31. I. Betel and H. M. Klouwen, *Biochim. Biophys. Acta*, **131**, 453 (1967).
32. E. Blade, C. Blat, and L. Harel, *Biochim. Biophys. Acta*, **156**, 157 (1968).
33. J. J. Blum, *J. Gen. Physiol.*, **49**, 1125 (1966).
34. N. S. Bricker and S. Klahr, *J. Gen. Physiol.*, **49**, 483 (1966).
35. G. P. Brierley, W. E. Jacobus, and G. R. Hunter, *J. Biol. Chem.*, **242**, 2192 (1967).
36. F. J. Brinley, Jr., *J. Neurophysiol.*, **30**, 1531 (1967).
37. J. R. Bonk and D. S. Parsons, *Nature*, **208**, 785 (1965).
38. J. Bryla, Z. Kaniuga, and B. Frachkowiak, *Biochim. Biophys. Acta*, **143**, 285 (1967).
39. W. Cammer and R. W. Estabrook, *Arch. Biochem.*, **122**, 721 (1967).
40. E. Carafoli, *J. Gen. Physiol.*, **50**, 1849 (1967).
41. R. Cereijo-Sautalo, *Can. J. Biochem.*, **46**, 55 (1961).
42. E. L. Chambers and A. H. Whiteley, *J. Gen. Physiol.*, **68**, 289 (1966).
43. R. J. Chertok, W. H. Hulet, and B. Epstein, *Am. J. Physiol.*, **211**, 1379 (1966).
44. R. A. Chez, R. R. Palmer, and S. G. Schultz, *J. Gen. Physiol.*, **50**, 2357 (1967).
45. R. S. Cockrell, E. J. Harris, and B. C. Pressman, *Nature*, **215**, 1487 (1967).
46. B. Crocken and E. L. Tatum, *Biochim. Biophys. Acta*, **135**, 100 (1967).
47. J. T. Cummins, J. A. Strand, and B. E. Vaughn, *Biochim. Biophys. Acta*, **126**, 330 (1966).
48. P. F. Curran and M. Cereijido, *J. Gen. Physiol.*, **48**, 1011 (1965).

49. E. J. Davis and D. M. Gibson, *Biochem. Biophys. Res. Commun.*, **26**, 815 (1967).
50. A. Y. Divekar, N. R. Vaidya, and B. M. Braganca, *Biochim. Biophys. Acta*, **135**, 927 (1967).
51. G. R. Drapeau, T. I. Matula, and R. A. MacLeod, *J. Bacteriol.*, **92**, 63 (1966).
52. N. N. Durham and J. R. Martin, *Biochim Biophys. Acta*, **115**, 260 (1966).
53. M. Dydnska and E. J. Harris, *J. Physiol.* (*London*), **182**, 92 (1966).
54. C. J. Edmonds and J. Marriott, *J. Physiol.* (*London*), **194**, 479 (1968).
55. C. J. Edmonds and J. Marriott, *J. Physiol.* (*London*), **194**, 457 (1968).
56. E. Eidelberg, J. Fishman, and M. L. Hams, *J. Physiol.* (*London*), **191**, 47 (1967).
57. A. S. Fairhust and D. J. Jenden, *J. Cellular Comp. Physiol.*, **67**, 233 (1966).
58. M. Fong and H. Rasmussen, *Biochim. Biophys. Acta*, **153**, 88 (1968).
59. R. G. Faust, J. W. Hollifield, and M. G. Leadbetter, *Nature*, **215**, 1297 (1967).
60. M. E. Feldheim, H. W. Augstin, and E. Hofmann, *Biochem. Z.*, **344**, 238 (1966).
61. C. D. Fitch and R. P. Shields, *J. Biol. Chem.*, **241**, 3611 (1966).
62. S. Fukui and R. M. Hochster, *Can. J. Biochem.*, **43**, 1129 (1965).
63. N. Glick, E. Gillespie, and P. G. Scholefield, *Can. J. Biochem.*, **45**, 1401 (1967).
64. G. H. Gallagher and J. D. Judah, *Biochem. Pharmacol.*, **16**, 883 (1967).
65. R. D. Green and J. W. Miller, *J. Pharmacol. Exp. Therap.*, **152**, 439 (1966).
66. G. S. Groot and S. G. Van den Bergh, *Biochim. Biophys. Acta*, **153**, 22 (1968).
67. M. T. Hakala, *Biochim. Biophys. Acta*, **102**, 210 (1965).
68. R. W. Hardy and E. Knight, *Biochim. Biophys. Acta*, **122**, 520 (1966).
69. H. G. Hempling, *Biochim. Biophys. Acta*, **112**, 503 (1966).
70. I. Hassinen and M. Hallman, *Biochem. Pharmacol.*, **16**, 2155 (1967).
71. C. Hidalgo and M. Canessa-Fischer, *J. Cellular Comp. Physiol.*, **68**, 185 (1966).
72. V. Hopfer, A. L. Lehninger, and T. E. Thompson, *Proc. Nat. Acad. Sci.*, **59**, 484 (1968).
73. K. C. Huang, *Life Sci.* (Oxford), **4**, 1201 (1965).
74. K. C. Huang and W. R. Rout, *Amer. J. Physiol.*, **212**, 799 (1967).
75. L. J. Ignarro and R. E. Shideman, *J. Pharmacol. Exp. Therap.*, **159**, 59 (1968).
76. J. A. Jacquez and J. H. Sherman, *Biochem. Biophys. Acta*, **109**, 128 (1965).
77. C. L. Johnson, C. M. Mauritzen, and W. C. Starbuck, *Biochemistry*, **6**, 1121 (1967)
78. H. R. Kaback and E. R. Stadtman, *J. Biol. Chem.*, **243**, 1390 (1968).
79. B. Kiely and A. Martonosi, *J. Biol. Chem.*, **243**, 2273 (1968).
80. S. Kitahara, *Am. J. Physiol.*, **213**, 819 (1967).
81. R. L. Klein and A. P. Breland, *Comp. Biochem. Physiol.*, **17**, 39 (1966).
82. A. Kleinzeller, J. Kolinska, and I. Benes, *Biochem. J.*, **104**, 843 (1967).
83. M. Klingenberg, *Biochem. Z.*, **343**, 479 (1965).
84. P. G. Kohn, H. Newey, and D. H. Smyth, *Nature*, **215**, 1395 (1967).
85. A. Kotyk and A. Kleinzeller, *Biochim. Biophys. Acta*, **135**, 106 (1967).
86. H. Kromphardt, *European J. Biochem.*, **3**, 377 (1968).
87. L. Leive, *J. Biol. Chem.*, **243**, 2373 (1968).
88. C. Levinson and H. G. Hempling, *Biochim. Biophys. Acta*, **135**, 306 (1967).
89. G. Levy, N. J. Angelino, and T. Matsuzawa, *J. Pharm. Sci.*, **56**, 681 (1967).
90. H. H. Loh, P. Volfin, and E. Kun, *Biochemistry*, **7**, 726 (1968).
91. R. H. Lyon, P. Rogers, and W. H. Hall, *J. Bacteriol.*, **94**, 92 (1967).
92. K. P. McConnell and G. J. Cho, *Am. J. Physiol.*, **213**, 150 (1967).
93. A. G. Malenkov, S. A. Bogatyreva, and V. P. Bozhkova, *Exp. Cell. Res.*, **48**, 307 (1967).
94. J. Mayshak, O. C. Yoder, and K. C. Beamer, *Arch. Biochem.*, **113**, 189 (1966).
95. S. A. Mendoza, J. S. Handler, and J. Orloff, *Am. J. Physiol.*, **213**, 1263 (1967).
96. I. A. Menon and J. H. Quastel, *Biochem. J.*, **99**, 766 (1966).

97. S. M. Mossberg, G. Ross, and B. Weingarten, *Nature*, **212**, 1588 (1966).
98. P. Nijs, *Biochim. Biophys. Acta*, **153**, 70 (1968).
99. S. Nishi and K. Koketsu, *Life Sci.* (Oxford), **6**, 2049 (1967).
100. K. Nordstrom, *Acta Chem. Scand.*, **20**, 474 (1966).
101. F. M. Parkins, J. W. Hollifield, and A. J. McCaslin, *Biochim. Biophys. Acta*, **126**, 513 (1966).
102. F. G. Peron, J. L. McCarthy, and F. Guerra, *Biochim. Biophys. Acta*, **117**, 450 (1966).
103. O. H. Petersen and J. H. Poulsen, *Acta Physiol. Scand.*, **71**, 194 (1967).
104. J. Piatigorsky and A. H. Whiteley, *Biochim. Biophys. Acta*, **108**, 404 (1965).
105. G. B. Pinchot, *J. Biol. Chem.*, **242**, 4577 (1967).
106. M. J. Pine, *Biochem. Pharmacol.*, **17**, 75 (1968).
107. A. Politoff, S. J. Socolar, and W. R. Loewenstein, *Biochim. Biophys. Acta*, **135**, 791 (1967).
108. M. Pollay, *Am. J. Physiol.*, **210**, 275 (1966).
109. E. Quagliariello and F. Palmieri, *European J. Biochem.*, **4**, 20 (1968).
110. F. Reusser, *J. Bacteriol.*, **94**, 1040 (1967).
111. J. C. Riemersma, *Biochim. Biophys. Acta*, **153**, 80 (1968).
112. T. R. Riggs, M. W. Pan, and H. W. Feng, *Biochim. Biophys. Acta*, **150**, 92 (1968).
113. W. D. Riley, R. J. Delange, and G. E. Bratvold, *J. Biol. Chem.*, **243**, 2209 (1968).
114. G. Rindi and U. Ventura, *Experientia*, **23**, 175 (1967).
115. K. Ring and E. Heinz, *Biochem. Z.*, **344**, 446 (1966).
116. J. L. Robinson and J. P. Felber, *Biochem. Z.*, **343**, 1 (1965).
117. H. Rosenberg and J. M. La Nauze, *Biochem. Biophys. Acta*, **141**, 79 (1967).
118. R. Roskoski and D. F. Steiner, *Biochem. Biophys. Acta*, **135**, 717 (1967).
119. C. R. Ross and A. Farah, *J. Pharmacol. Exp. Therap.*, **151**, 159 (1966).
120. C. R. Ross, N. I. Pessah, and A. Farah, *J. Pharmacol. Exp. Therap.*, **160**, 381 (1968).
121. R. C. Rufin, E. S. Henderson, and E. S. Owens, *Cancer Res.*, **27**, 553 (1967).
122. B. M. Sahagian, I. Harding-Barlow, and H. M. Perry, *J. Nutr.*, **93**, 291 (1967).
123. D. Schachter, S. Kowarski, and J. D. Finkelstein, *Am. J. Physiol.*, **211**, 1131 (1966).
124. S. Schenker and B. Combes, *Am. J. Physiol.*, **212**, 295 (1967).
125. O. T. Schonherr and G. W. BorstPauwels, *Biochim. Biophys. Acta*, **135**, 787 (1967).
126. S. Segal, L. Schwartzman, and A. Blair, *Biochim. Biophys. Acta*, **135**, 127 (1967).
127. I. Seidman and J. Cascarano, *Am. J. Physiol.*, **211**, 1165 (1966).
128. L. Shear, J. D. Harvey, and K. G. Barry, *J. Lab. Clin. Med.*, **67**, 181 (1966).
129. E. C. Slater, *Bull. Soc. Chim. Biol.*, **48**, 1151 (1966).
130. B. K. Stem and W. E. Jensen, *Nature*, **209**, 789 (1966).
131. C. P. Sung and R. M. Johnstone, *Can. J. Biochem.*, **43**, 1111 (1965).
132. C. W. Tabor and H. Tabor, *J. Biol. Chem.*, 3714 (1966).
133. R. R. Tercafs, *Comp. Biochem. Physiol.*, **17**, 937 (1966).
134. Y. Tochino and L. S. Schanker, *Biochem. Pharmacol.*, **14**, 1557 (1965).
135. J. J. VanBuskirk and W. R. Frisell, *Biochim. Biophys. Acta*, **143**, 292 (1967).
136. W. G. Van der Kloot, *Comp. Biochem. Physiol.*, **17**, 75 (1966).
137. G. A. Vigers and F. O. Ziegler, *Biochem. Biophys. Res. Commun.*, **30**, 83 (1968).
138. R. W. Von Koroff, *Nature*, **214**, 23 (1967).
139. E. C. Weinbach and J. Gorfus, *Biochem. J.*, **106**, 711 (1968).
140. L. W. Wheeldon and A. L. Lehninger, *Biochemistry*, **5**, 3533 (1966).
141. A. H. Whiteley and E. L. Chambers, *J. Cellular Comp. Physiol.*, **68**, 309 (1966).
142. J. S. Willis, *Am. J. Physiol.*, **214**, 923 (1968).
143. H. H. Winkler and T. H. Wilson, *J. Biol. Chem.*, **241**, 2200 (1966).
144. P. Wins and E. Schoffeniels, *Arch. Intern. Physiol.*, **73**, 160 (1965).

145. P. Wins and E. Schoffeniels, *Biochim. Biophys. Acta*, **120**, 341 (1966).
146. K. Yabu, *Biochim. Biophys. Acta*, **135**, 181 (1967).
147. J. Yashphe, R. F. Rosenberger, and M. Shilo, *Biochim. Biophys. Acta*, **146**, 560 (1967).
148. B. Grabe, *J. Theoret. Biol.*, **7**, 112 (1964).
149. M. I. Aleem, *Biochim. Biophys. Acta*, **128**, 1 (1966).
150. S. M. Arfin, *Biochim. Biophys. Acta*, **136**, 233 (1967).
151. D. I. Arnon, H. Y. Toujimoto, and B. O. McSwain, *Nature*, **214**, 562 (1967).
152. A. Atsmon and R. P. Davis, *Biochim. Biophys. Acta*, **131**, 221 (1967).
153. H. Babod, R. Ben-Zvi, and A. Bdolah, *European J. Biochem.*, **1**, 96 (1967).
154. P. F. Baker and R. Presley, *J. Physiol.*, **186**, 47P (1966).
155. W. H. Bannister, *J. Physiol.*, **186**, 89 (1966).
156. B. J. Barnhart and C. T. Greegg, *Virology*, **32**, 687 (1967).
157. W. Bartley and E. R. Tustanoff, *Biochem. J.*, **99**, 599 (1966).
158. J. C. Batterton and C. Van Baalen, *Can. J. Microbiol.*, **14**, 341 (1968).
159. L. Beani, C. Brauchi, and F. Ledda, *Brit. J. Pharmacol.*, **27**, 299 (1966).
160. D. S. Beattie, R. E. Basford, and S. B. Koritz, *J. Biol. Chem.*, **242**, 3366 (1967).
161. F. Ben-Hamida and D. Schlessinger, *Biochim. Biophys. Acta*, **119**, 183 (1966).
162. J. E. Benbough, *J. Gen. Microbiol.*, **47**, 325 (1967).
163. C. Bergman and M. Joyeux, *Compt. Rend. Soc. Biol.* (Paris), **160**, 2039 (1966).
164. J. Bielawski and A. L. Lehninger, *J. Biol. Chem.*, **241**, 4316 (1966).
165. K. J. Blackburn, P. C. French, and R. J. Merrills, *Life Sci.* (Oxford), **6**, 1653 (1967).
166. P. V. Blair and F. A. Sollars, *Life Sci.* (Oxford), **6**, 2233 (1967).
167. L. Bongers, *J. Bacteriol.*, **93**, 1615 (1967).
168. G. W. Borst Pauwels, *J. Cellular Physiol.*, **69**, 241 (1967).
169. H. R. Bose and A. S. Levine, *Life Sci.* (Oxford), **5**, 403 (1966).
170. J. Broekhuysen, G. Deltour, and M. Ghislain, *Biochem. Pharmacol.*, **16**, 2077 (1967).
171. J. Buchanan and D. F. Topley, *Endocrinology*, **79**, 81 (1966).
172. F. L. Bygrave and A. L. Lehninger, *J. Biol. Chem.*, **241**, 3894 (1966).
173. A. I. Caplan and J. W. Greenwalt, *J. Cell. Biol.*, **36**, 15 (1968).
174. B. M. Carruthers, *Can. J. Physiol. Pharmacol.*, **45**, 269 (1967).
175. R. Cereijo-Santalo, *Can. J. Biochem.*, **45**, 897 (1967).
176. W. Chefurka and T. Dumas, *Biochemistry*, **5**, 3904 (1966).
177. A. Cherayil, J. Kandera, and A. Lajtha, *J. Neurochem.*, **14**, 105 (1967).
178. R. Cleland, *Science*, **160**, 192 (1968).
179. R. E. Click and D. P. Hackett, *Biochim. Biophys. Acta*, **142**, 403 (1967).
180. R. A. Conyers, W. J. Hensley, and M. D. Montague, *Arch. Intern. Pharmacodyn.*, **171**, 179 (1968).
181. J. E. Cremer, *Biochem. J.*, **104**, 212, 223 (1967).
182. R. H. Dahl and J. N. Pratley, *J. Cellular Biol.*, **33**, 411 (1967).
183. E. A. Davis and E. J. Johnson, *Can. J. Microbiol.*, **13**, 873 (1967).
184. B. Diehn and G. Tolbin, *Arch. Biochem.*, **121**, 169 (1967).
185. S. M. Dietrich and R. H. Burris, *J. Bacteriol.*, **93**, 1467 (1967).
186. G. Fassina, *Life Sci.* (Oxford), **6**, 825 (1967).
187. F. Galdiero, *Biochim. Biophys. Acta*, **126**, 54 (1966).
188. J. L. Gamble and J. A. Mazur, *J. Biol. Chem.*, **242**, 67 (1967).
189. J. T. Gatzy, *Am. J. Physiol.*, **213**, 425 (1967).
190. A. R. Gear, C. S. Rossi, and B. Reynafarje, *J. Biol. Chem.*, **242**, 3403 (1967).
191. A. Gosh and S. N. Battacharyya, *Biochim. Biophys. Acta*, **136**, 19 (1967).
192. O. Gonda and J. H. Quastel, *Biochem. J.*, **100**, 89 (1966).
193. E. Gorin and E. Shafrir, *Biochim. Biophys. Acta*, **137**, 189 (1967).

194. J. B. Griffin and R. Penniall, *Arch. Biochem.*, **114**, 67 (1966).
195. F. Guerra, F. G. Peron, and J. L. McCarthy, *Biochim. Biophys. Acta*, **117**, 433 (1966).
196. R. J. Guillory and E. Racker, *Biochim. Biophys. Acta*, **153**, 490 (1968).
197. Y. Hatefi and T. Fakouh, *Arch. Biochem.*, **125**, 114 (1968).
198. W. Heinen, *Arch. Biochem.*, **120**, 101 (1967).
199. R. J. Hay and J. Paul, *J. Gen. Physiol.*, **50**, 1663 (1967).
200. F. J. Hird and M. A. Marginson, *Arch. Biochem.*, **115**, 247 (1966).
201. H. Huddart and D. W. Wood, *Comp. Biochem. Physiol.*, **18**, 681 (1966).
202. D. A. Hudson and R. J. Levin, *J. Physiol.*, **195**, 369 (1968).
203. A. J. Ingenito and D. D. Bonnycastle, *Can. Bull. Physiol.*, **45**, 723 (1967).
204. J. J. Janke, A. Fleckenstein, and P. Marmier, *Pfluegers Arch. Ges. Physiol.*, **287**, 9 (1966).
205. L. Jarett and D. M. Kipnis, *Nature*, **216**, 714 (1967).
206. C. L. Johnson, B. Safer, and A. Schwartz, *J. Biol. Chem.*, **241**, 4513 (1966).
207. D. P. Kelley and P. J. Syrett, *J. Gen. Microbiol.*, **43**, 109 (1966).
208. G. A. Kimmich and H. Rasmussen, *Biochim. Biophys. Acta*, **131**, 413 (1967).
209. S. M. Kirpekar and A. R. Wakade, *J. Physiol.*, **194**, 609 (1968).
210. L. Kovac, H. Bednoarova, and M. Greksak, *Biochim. Biophys. Acta*, **153**, 32 (1968).
211. L. Kovac and S. Kuzela, *Biochim. Biophys. Acta*, **127**, 355 (1966).
212. S. Kozawa, K. Naito, and A. Yasui, *Japan. J. Pharmacol.*, **17**, 308 (1967).
213. G. D. Kuehn and B. A. McFadden, *J. Bacteriol.*, **95**, 937 (1968).
214. P. C. Laris and P. E. Letchworth, *J. Cellular Comp. Physiol.*, **69**, 143 (1967).
215. R. C. Lawrence, *J. Gen. Microbiol.*, **44**, 393 (1966).
216. R. C. Lawrence, *J. Gen. Microbiol.*, **46**, 65 (1967).
217. J. O. Lawe and L. H. Strickland, *Biochem. J.*, **104**, 158 (1967).
218. H. Lees, *Biochim. Biophys. Acta*, **131**, 310 (1967).
219. F. V. McCann, *Comp. Biochem. Physiol.*, **20**, 399 (1967).
220. L. H. Mantel, *Comp. Biochem. Physiol.*, **20**, 743 (1967).
221. D. Massaro, *Nature*, **215**, 646 (1967).
222. H. Meisner, *Cancer Res.*, **27**, 2077 (1967).
223. F. H. Milazzo and J. W. Fitzgerald, *Can. J. Microbiol.*, **13**, 659 (1967).
224. P. Mitchell and J. Moyle, *Biochem. J.*, **104**, 588 (1967).
225. R. A. Mitchell, R. D. Hill, and P. D. Boyer, *J. Biol. Chem.*, **242**, 1793 (1967).
226. J. V. Moller, *J. Physiol.*, **192**, 519 (1967).
227. B. D. Nelson, B. Highman, and P. D. Altland, *Am. J. Physiol.*, **213**, 1414 (1967).
228. V. N. Nigam, *Arch. Biochem.*, **120**, 214 (1967).
229. V. N. Nigam, *Arch. Biochem.*, **120**, 232 (1967).
230. V. N. Nigam, *Biochem. J.*, **105**, 515 (1967).
231. A. W. Norman, *Biochim. Biophys. Acta*, **118**, 655 (1966).
232. J. J. O'Neill and T. E. Duffy, *Life Sci.* (Oxford), **5**, 1849 (1966).
233. H. Oyama and S. Minakami, *J. Biochem.* (Tokyo), **61**, 103 (1967).
234. K. Ozawa, H. Araki, and K. Leta, *J. Biochem.* (Tokyo), **61**, 411 (1967).
235. K. Ozawa, K. Seta, and H. Todeda, *J. Biochem.* (Tokyo), **59**, 501 (1966).
236. S. Papa, E. J. DeHaan, and A. Francavilla, *Biochim. Biophys. Acta*, **143**, 438 (1967).
237. F. Paradisi, *Experientia*, **23**, 752 (1967).
238. M. Pfaffman and W. Holland, *Am. J. Physiol.*, **211**, 400 (1966).
239. G. B. Pinchot and B. J. Salmon, *Arch. Biochem.*, **115**, 345 (1966).
240. H. Rochman, G. H. Lathe, and M. J. Levell, *Biochem. J.*, **102**, 48 (1967).
241. C. R. Rossi, L. Galzigna, and A. Alexandre, *J. Biol. Chem.*, **242**, 2102 (1967).
242. S. Sandell, H. Low, and A. Von der Decken, *Biochem. J.*, **104**, 575 (1967).

243. G. Sauermann, *Z. Krebsforsch.*, **69**, 44 (1967).
244. L. W. Scheibel, H. J. Soz, and E. Bueding, *J. Biol. Chem.*, **243**, 2229 (1968).
245. D. Schlessinger and F. Ben-Hanida, *Biochim. Biophys. Acta*, **119**, 171 (1966).
246. E. Shrago, W. Brech, and K. Templeton, *J. Biol. Chem.*, **242**, 4060 (1967).
247. G. Shyamala, W. J. Lossow, and I. L. Chaikoff, *Biochim. Biophys. Acta*, **116**, 543 (1966).
248. P. A. Siegenthaler, M. M. Belsky, and S. Goldstein, *J. Bacteriol.*, **93**, 1281 (1967).
249. O. Sovik, I. Oye, and M. Rosell-Perez, *Biochim. Biophys. Acta*, **124**, 26 (1966).
250. A. M. Snoswell, *Biochemistry*, **5**, 1660 (1966).
251. N. Sperelakis and D. Lehmkuhl, *Am. J. Physiol.*, **213**, 719 (1967).
252. S. Streichman and Y. Avi-Dor, *Biochem. J.*, **104**, 71 (1967).
253. H. Takeda, T. Sukeno, and S. Tsuiki, *Biochem. Biophys. Res. Commun.*, **29**, 90 (1967).
254. H. Tedeschi and H. W. Horn, *Biochem. Biophys. Res. Commun.*, **28**, 752 (1967).
255. R. H. Vallejos and A. O. Stoppani, *Biochem. Biophys. Acta*, **131**, 295 (1967).
256. K. VanDam, *Biochem. Biophys. Acta*, **131**, 407 (1967).
257. K. VanDam, *Biochim. Biophys. Acta*, **128**, 337 (1966).
258. J. Venulet and A. Desperak-Naciazek, *J. Pharm. Pharmacol.*, **18**, 38 (1966).
259. P. Walter, H. A. Lardy, and D. Johnson, *J. Biol. Chem.*, **242**, 5014 (1967).
260. J. B. Warshaw, K. W. Lam, and D. R. Sanadi, *Arch. Biochem.*, **115**, 312 (1966).
261. E. C. Weinbach and J. Garfus, *J. Biol. Chem.*, **241**, 3708 (1966).
262. H. F. Welle and E. C. Slater, *Biochim. Biophys. Acta*, **143**, 1 (1967).
263. M. W. Whitehouse, *Biochem. Pharmacol.*, **16**, 753 (1967).
264. D. F. Wilson and B. Chance, *Biochim. Biophys. Acta*, **131**, 421 (1967).
265. D. F. Wilson and R. D. Merz, *Arch. Biochem.*, **119**, 470 (1967).
266. P. Wins and E. Schoffeniels, *Biochim. Biophys. Acta*, **135**, 831 (1967).
267. J. Ehrlich, Q. R. Bartz, R. M. Smith, D. A. Joslyn, and P. B. Burkholder, *Science*, **106**, 417 (1947).
268. M. C. Rebstock, H. M. Crooks, J. Controulis, and Q. R. Barte, *J. Am. Chem. Soc.*, **71**, 2458 (1949).
269. L. C. Vining and D. W. Westlake, *Can. J. Microbiol.*, **10**, 705 (1964).
270. D. Gottlieb, H. E. Carter, P. W. Robbins, and R. W. Burg, *J. Bacteriol.*, **84**, 888 (1962).
271. R. McGrath, M. Suddiquellah, L. C. Vining, F. Sala, and D. W. Westlake, *Biochem. Biophys. Res. Commun.*, **29**, 576 (1967).
272. C. D. Stralton and M. C. Rebstock, *Arch. Biochem. Biophys.*, **103**, 159 (1963).
273. H. Margreiter and F. Gapp in *Die Antibiotica*, Vol. I, (Ed. R. Brunner and G. Macher), Hans Carl, Nurnberg, 1962, p. 182.
274. J. Ehrlich, D. Gottlieb, P. R. Burkholder, L. E. Anderson, and T. G. Pridham, *J. Bacteriol.*, **56**, 476 (1948).
275. P. D. Shaw and D. Gottlieb in *Biogenesis of Antibiotic Substances* (Ed. Z. Hostalek and Z. Vanek), Czechoslovak Academy of Sciences, Prague, 1965, pp. 261–269.
276. E. F. Gale and J. Folkes, *Biochem. J.*, **53**, 493 (1953).
277. P. S. Sypard, N. Strauss, and H. P. Treffers, *Biochem. Biophys. Res. Commun.*, **7**, 477 (1962).
278. H. H. Ramsey, *Biochem. Biophys. Res. Commun.*, **23**, 353 (1966).
279. L. E. Schrader, L. Beevers, and R. H. Hageman, *Biochem. Biophys. Res. Commun.*, **26**, 14 (1967).
280. F. Lingens and O. Oltmanns, *Biochim. Biophys. Acta*, **130**, 336 (1966).
281. F. Lingens, H. Eberhardt, and O. Oltmanns, *Biochim. Biophys. Acta*, **130**, 345 (1966).
282. E. E. Snell, Ed., *Chemical and Biological Aspects of Pyridoxal Catalysis*, Macmillan, New York, 1961.

283. A. J. Glazko, L. M. Wolf, and W. A. Dill, *Proc. Soc. Exp. Biol., Med.*, **72,** 602 (1949).
284. A. J. Glazko, L. M. Wolf, W. A. Dill, and A. C. Bratton, *J. Pharmacol. Exp. Therap.*, **46,** 445 (1949).
285. H. L. Ley, J. E. Smadel, and T. T. Crockett, *Proc. Soc. Exp. Biol. Med.*, **68,** 9 (1948).
286. A. J. Glazko, W. A. Dill, and L. M. Wolf, *J. Pharmacol. Exp. Therap.*, **104,** 452 (1952).
287. R. Q. Thompson, M. Sturtevant, O. D. Bird, and A. J. Glazko, *Endocrinology*, **55,** 665 (1959).
288. G. N. Smith and C. Worrel, *Arch. Biochem. Biophys.*, **28,** 232 (1950).
289. J. R. Fouts and B. B. Brodie, *J. Pharmacol. Exp. Therap.*, **119,** 197 (1957).
290. I. Niedzwiecka-Namyslowska, Z. Loegler, and W. Ardelt, *Bull. Acad. Polon. Sci., Sér. Sci. Biol.*, **14,** 751 (1966).
291. I. Niedzwiecka-Namyslowska, Ph.D. Dissertation, Institute of Biochemistry and Biophysics, Polish Academy of Science, Warsaw, 1966.
292. F. D. Sompolinsky and Z. Saura, *J. Gen. Microbiol.*, **50,** 55 (1968).
293. E. F. Gale and J. Folkes, *Biochem. J.*, **53,** 493 (1953).
294. M. W. Nirenberg and J. H. Matthaei, *Proc. Nat. Acad. Sci. U.S.*, **47,** 1588 (1961).
295. A. S. Weisberger, S. Wolfe, and S. Armentraut, *J. Exp. Med.*, **120,** 161 (1964).
296. C. Biswas, J. Hardy, and W. S. Beck, *J. Biol. Chem.*, **240,** 3631 (1965).
297. W. Epstein and S. G. Schultz, *J. Gen. Physiol.*, **49,** 469 (1966).
298. P. Fortmagel and E. Freese, *J. Bacteriol.*, **95,** 1431 (1968).
299. T. J. Franklin, *Biochem. J.*, **105,** 371 (1967).
300. T. Fukuyama and E. J. Ordal, *J. Bacteriol.*, **90,** 673 (1965).
301. R. K. Ghambeer and R. L. Blakley, *J. Biol. Chem.*, **241,** 4710 (1966).
302. D. Goldman, S. G. Schultz, and W. Epstein, *Biochim. Biophys. Acta*, **130,** 546 (1966).
303. J. T. Holden and N. M. Utech, *Biochim. Biophys. Acta*, **135,** 517 (1967).
304. A. Kaji and D. Nakada, *Biochim. Biophys. Acta*, **145,** 508 (1967).
305. H. P. Kleber and H. Aurich, *Naturwiss.*, **53,** 234 (1966).
306. R. H. Lyon, P. Rogers, and W. H. Hall, *J. Bacteriol.*, **94,** 92 (1967).
307. B. Mills and D. T. Dubin, *Mol. Pharmacol.*, **2,** 311 (1966).
308. I. Morris, *Arch. Mikrobiol.*, **54,** 169 (1966).
309. M. J. Pine, *Biochem. Pharmacol.*, **17,** 75 (1968).
310. K. A. Pittman, S. Lakshamanan, and M. P. Bryant, *J. Bacteriol.*, **93,** 1499 (1967).
311. H. Rosenberg and J. M. La Nauze, *Biochim. Biophys. Acta*, **141,** 79 (1967).
312. G. Schatz, *J. Biol. Chem.*, **243,** 2192 (1968).
313. H. L. Schwartz, A. C. Carter, and D. M. Kydd, *Endocrinology*, **80,** 65 (1967).
314. M. A. Berberich, P. Venetianer, and R. F. Goldberger, *J. Biol. Chem.*, **241,** 4426 (1966).
315. G. W. Strandberg and P. W. Wilson, *Can. J. Microbiol.*, **14,** 25 (1968).
316. E. T. Young and R. L. Sinsheimer, *J. Mol. Biol.*, **30,** 165 (1967).
317. L. Dalgarno and F. Gros, *Biochim. Biophys. Acta*, **157,** 52 (1968).
318. I. H. Goldberg and K. Mitsugi, *Biochemistry*, **6,** 372 (1967).
319. C. W. Naidle, J. M. Kornfeld, and S. G. Knight, *Arch. Mikrobiol.*, **53,** 41 (1966).
320. A. M. Kroon, *Biochim. Biophys. Acta*, **72,** 391 (1963).
321. E. F. Gale, *Brit. Med. Bull.*, **16,** 11 (1960).
322. R. Rendi and S. Ochoa, *J. Biol. Chem.*, **237,** 3711 (1962).
323. A. S. Weisberger and S. Wolfe, *J. Lab. Clin. Med.*, **62,** 1020 (1963).
324. Z. Kucan and F. Lippmann, *J. Biol. Chem.*, **239,** 516 (1964).
325. D. Vasquez, *Biochim. Biophys. Acta*, **114,** 277 (1966).
326. D. Vasquez, *Symp. Soc. Gen. Microbiol.*, **16,** 169 (1966).
327. D. Vasquez, *Life Sci.* (Oxford), **6,** 381 (1967).

328. D. Vasquez and R. E. Monro, *Biochim. Biophys. Acta*, **142**, 155 (1967).
329. E. F. Galerd and J. P. Foulkes, *Biochem. J.*, **53**, 493 (1953).
330. D. Vazquez, *Nature*, **203**, 257 (1964).
331. A. S. Weisberger and S. Wolfe, *Federation Proc.*, **23**, 976 (1964).
332. B. Urbas and E. Gustak, *Croat. Chem. Acta*, **30**, 73 (1958).
333. L. M. Long and H. D. Troutman, *J. Am. Chem. Soc.*, **73**, 542 (1951).
334. N. Buu-Hoi, N. Hoan, P. Jazquignon, and H. Khoi, *Compt. Rend.*, **230**, 662 (1950).
335. M. C. Rebstock and E. L. Pfeiffer, *J. Am. Chem. Soc.*, **74**, 3207 (1952).
336. L. M. Long and N. Jenesel, *J. Am. Chem. Soc.*, **72**, 4299 (1950).
337. A. P. Phillips, *J. Am. Chem. Soc.*, **74**, 6125 (1952).
338. A. P. Phillips, *J. Am. Chem. Soc.*, **75**, 3621 (1953).
339. A. J. Glazko, W. H. Edgerton, W. A. Dill, and W. R. Lenz, *Antibiot. Chemotherapy*, **2**, 234 (1952).
340. S. Nakamura, *Pharm. Bull.* (Tokyo), **3**, 379 (1955).
341. O. K. Behrens, J. W. Corse, J. P. Lynette, R. G. Jones, Q. F. Soper, F. R. Van Abele, and C. M. Whitehead, *J. Biol. Chem.*, **175**, 796 (1948).
342. M. Isono, *J. Agr. Chem. Soc. Japan*, **28**, 526, 566 (1954).
343. S. S. Walkenstein, N. Chumakow, and J. Leifter, *Antibiot. Chemotherapy*, **4**, 1245 (1954).
344. R. T. Williams, *Detoxification Mechanisms*, John Wiley and Sons, New York, 1959.
345. M. Pailer, L. Belochlav, and E. Semonitsch, *Monatsh. Chem.*, **87**, 249 (1956).
346. M. Pailer, *Prog. Chem. Org. Nat. Prod.*, **18**, 70 (1960).
347. S. M. Kupchan and R. M. Doskotch, *J. Med. Pharm. Chem.*, **5**, 657 (1962).
348. J. R. Mose, *Planta Med.*, **11**, 72 (1963).
349. C. L. Carter and W. J. McChesney, *Nature*, **164**, 575 (1949).
350. M. T. Bush, O. Touster, and J. E. Brockman, *J. Biol. Chem.*, **188**, 685 (1951).
351. S. Nakamura and C. Shimoda, *J. Agr. Chem. Soc. Japan*, **28**, 909 (1954).
352. H. Raistrick and A. Stosal, *Biochem. J.*, **68**, 647 (1958).
353. M. P. Morris, C. Pogan, and H. E. Warmke, *Science*, **119**, 322 (1954).
354. J. W. Hylin and H. Matsumoto, *Arch. Biochem. Biophys.*, **93**, 542 (1960).
355. P. D. Shaw and N. Wang, *J. Bacteriol.*, **88**, 1629 (1964).
356. A. J. Brich, B. J. McLaughlin, H. Smith, and J. Winter, *Chem. Ind.* (London), **26**, 840 (1960).
357. J. H. Birkinshaw and A. M. Dryland, *Biochem. J.*, **93**, 478 (1964).
358. S. Gatenbeck and B. Forsgren, *Acta Chem. Scand.*, **18**, 1750 (1964).
359. P. D. Shaw and J. A. McCloskey, *Biochemistry*, **6**, 2247 (1967).
360. P. D. Shaw, *Biochemistry*, **6**, 2253 (1967).
361. K. Maede, *J. Antibiotics* (Tokyo), **A6**, 137 (1963).
362. H. Thrum and H. Bocker in *Biogenesis of Antibiotic Substances*, (Ed. Z. Vanek and Z. Hostalek), Czechoslovak Academy of Sciences, Prague, 1965, p. 233.
363. O. R. Gottlieb and M. T. Magelhaes, *J. Org. Chem.*, **24**, 2070 (1959).
364. O. R. Gottlieb, I. S. de Souza, and M. T. Magelhaes, *Tetrahedron*, **18**, 1137 (1962).
365. S. Hermann, *Naturwiss.*, **47**, 162 (1960).
366. K. Nishida, A. Kobayashi, and T. Nagahama, *Bull. Agr. Chem. Soc. Japan*, **19**, 77 (1955).
367. N. V. Riggs, *Chem. Ind.* (London), 926 (1956).
368. G. M. Bonser, *Brit. Med. J.*, **2**, 655 (1967).
369. G. L. Laqueur, O. Mickelsen, M. G. Whiting, and L. T. Kurland, *J. Nat. Cancer Inst.*, **31**, 919 (1963).
370. G. L. Laqueur, *Arch. Pathol. Anat. Physiol.*, **340**, 151 (1965).

371. G. L. Laqueur and H. Matsumoto, *J. Nat. Cancer Inst.*, **37**, 217 (1966).
372. G. L. Laqueur, quoted in Ref. 368.
373. H. Matsumoto and H. Higa, *Biochem. J.*, **98**, 20c (1966).
374. R. C. Shank and P. N. Magee, *Biochem. J.*, **105**, 521 (1967).
375. R. Schoental and P. N. Magee, *Brit. J. Cancer*, **16**, 92 (1962).
376. M. Horisberger and H. Matsumoto, *J. Labeled Compds.*, **4**, 164 (1968).
377. R. D. O'Brien, *Toxic Phosphorus Esters*, Academic Press, New York, 1960.
378. J. C. Webb, *Enzyme and Metabolic Inhibitors*, Vol. I, Academic Press, New York, 1963.
379. S. Friess and H. Baldridge, *J. Am. Chem. Soc.*, **78**, 2482 (1956).
380. V. A. Yakovlev, *Mekhanizm i Kinetika Fermentativnogo Kataliza*, Academy of Sciences USSR, Moscow, 1964, pp. 16–34.
381. E. F. Jansen, M. D. Nutting, and A. K. Balls, *J. Biol. Chem.*, **179**, 201 (1949).
382. B. S. Hartley and B. A. Kilby, *Biochem. J.*, **56**, 288 (1954).
383. Cf. M. L. Bender and F. J. Kezdy, *Ann. Rev. Biochem.*, **34**, 49 (1965).
384. W. N. Aldridge, *Biochem. J.*, **46**, 451 (1950).
385. W. N. Aldridge and A. N. Davison, *Biochem. J.*, **51**, 62 (1952).
386. W. N. Aldridge and A. N. Davison, *Biochem. J.*, **55**, 763 (1953).
387. Cf. F. Sanger, *Proc. Chem. Soc.*, 76 (1963).
388. I. B. Wilson, *J. Biol. Chem.*, **190**, 111 (1951).
389. I. B. Wilson, *J. Biol. Chem.*, **199**, 113 (1955).
390. I. B. Wilson and S. Ginsburg, *Biochim. Biophys. Acta*, **18**, 168 (1955).
391. I. B. Wilson, S. Ginsburg, and E. K. Meislich, *J. Am. Chem. Soc.*, **77**, 4286 (1955).
392. A. F. Childs, D. R. Davies, A. L. Green, and J. P. Rutland, *Brit. J. Pharmacol.*, **10**, 462 (1955).
393. F. Hobbiger, D. G. O'Sullivan, and P. V. Sodler, *Nature*, **182**, 1492 (1958).
394. F. Hobbiger, M. Pitman, and P. V. Sodler, *Biochem. J.*, **75**, 363 (1960).
395. S. M. Markov, N. A. Loshadkin, M. A. Cristova, and I. L. Knunyants, *Mekhanizm i Kinetika Fermentativnogo Kataliza*, Academy of Sciences USSR, Moscow, 1964, pp. 35–47.
396. T. R. Fukuto and R. L. Metcalf, *J. Am. Chem. Soc.*, **81**, 372 (1959).
397. W. M. Diggle and J. C. Gage, *Biochem. J.*, **49**, 491 (1951).
398. W. M. Diggle and J. C. Gage, *Nature*, **168**, 998 (1951).
399. R. O'Brien, *Nature*, **183**, 121 (1959).
400. A. N. Davison, *Biochem. J.*, **61**, 203 (1955).
401. K. P. DuBois, J. Doull, P. R. Salerno, and J. M. Coon, *J. Pharmacol. Exp. Therap.*, **95**, 79 (1949).
402. R. L. Metcalf and R. B. March, *J. Econ. Entomol.*, **42**, 721 (1949).
403. R. L. Metcalf and R. B. March, *Ann. Entmol. Soc. Am.*, **46**, 63 (1953).
404. M. K. Ahmed, J. E. Casida, and R. E. Nichols, *J. Agr. Food Chem.*, **6**, 740 (1958).
405. J. F. Gardocki and L. W. Hasleton, *J. Am. Pharm. Assoc.*, **40**, 491 (1951).
406. J. E. Pankaskie, F. C. Fountaine, and P. A. Dahm, *J. Econ. Entomol.*, **45**, 51 (1952).
407. J. W. Cook, *J. Agr. Food Chem.*, **5**, 859 (1957).
408. R. O'Brien and E. Y. Spencer, *J. Agr. Food Chem.*, **3**, 56 (1955).
409. J. M. Vandenbelt, C. Henrich, and S. G. Vandenberg, *Anal. Chem.*, **26**, 726 (1954).
410. F. W. Plapp and J. E. Casida, *J. Econ. Entomol.*, **51**, 800 (1958).
411. R. A. Benkeser and H. R. Krysiak, *J. Am. Chem. Soc.*, **75**, 2421 (1953).
412. H. H. Hodgson and R. Smith, *J. Chem. Soc.*, **1939**, 263.
413. W. Rusiecki and P. Kubikowski, *Toksykologia Wspolczesna*, Panstwowy Zaklad Wydawnictw Lekarskich, Warsaw, 1964, p. 385.

414. Z. Eckstein, J. Plenkiewiez, and S. Byrdy, *Bull. Acad. Polon. Sci., Sér. Sci. Chim.*, **8**, 623 (1960).
415. D. Plaut and Z. Eckstein, *Bull. Acad. Polon. Sci., Sér. Sci. Chim.*, **12**, 43 (1963).
416. Z. Eckstein and W. Sobotka, Conference on Scientific Problems of Plant Protection, Budapest, 1960 (unpublished).
417. A. N. Bates, D. M. Spencer, and R. L. Wain, *Ann. Appl. Biol.*, **50**, 21 (1962).
418. E. Schraustatter, W. Meister, and R. Gonnert, *Z. Naturforsch.*, **166**, 95 (1961).
419. Z. M. Baeq, J. Cheymol, M. J. Dallemagne, R. Hazard, J. LeBarre, J. J. Reuse, and M. Welsh, *Pharmacodynamie Biochemique*, Masson and Co., Paris, 1961, p. 634.
420. G. L. McNew and N. K. Sunderholm, *Phytopathology*, **39**, 721 (1949).
421. Z. Eckstein, E. Grochowski, and T. Urbański, *Bull. Acad. Polon. Sci., Sér. Sci. Chim.*, **7**, 289 (1959).
422. Z. Eckstein, E. Grochowski, R. Kowalik, and T. Urbański, *Bull. Acad. Polon. Sci. Sér. Sci. Chim.*, **11**, 687 (1963).
423. Z. Eckstein, *Osterr. Chemiker Ztg.*, **66**, 111 (1965).
424. A. Grob, Swiss Pat. 351,956 (1961); *Chem. Abstr.*, **55**, 24569 (1961).
425. R. E. Miller, U.S. Pat. 2,717,254 (1955); *Chem. Abstr.*, 8738 (1956).
426. M. Pianka, *J. Sci. Food Agr*, **14**, 48 (1963).
427. Z. Eckstein, Z. Ejmocki, W. Sobotka, and T. Urbański, Polish Pat. 43,915 (1960); *Chem. Abstr.* **60**, 4061 (1964).
428. Z. Eckstein, Z. Ejmocki, and I. Gwiazdecka, *Przemys Chem.*, **39**, 616 (1960).
429. A. Burger, M. L. Stein, and J. B. Clements, *J. Org. Chem.*, **22**, 143 (1957).
430. S. Byrdy, Z. Eckstein, and J. Plenkiewiez, *Bull. Acad. Polon. Sci., Sér. Sci. Chim.*, **9**, 627 (1961).
431. S. Byrdy, Z. Eckstein, R. Kowalid, and J. Plenkiewiez in *Nitro Compounds* (Ed. T. Urbański), Pergamon Press, London, 1964, p. 509.
432. A. C. Huitric, R. Pratt, Y. Okano, and W. D. Kumber, *Antibiot. Chemotherapy*, **6**, 294 (1956).
433. N. G. Clark, A. F. Hams, and B. E. Leggetter, *Nature*, **200**, 171 (1963).
434. G. Truffaut and I. Pastac, French Pat. 751,855 (1933).
435. A. S. Crafts, *Science*, **101**, 417 (1945).
436. A. H. Kirby, *World Rev. Pest Control*, **5** (1), 30 (1966).
437. M. Pianka, *J. Sci. Food Agr.*, **17**, 47 (1966).
438. M. Pianka and J. D. Edwards, *J. Sci. Food Agr.*, **18**, 355 (1967).
439. M. Pianka, Brit. Pat. 999,876 (1965); *Chem. Abstr.* **63**, 11436 (1965).
440. M. Pianka and K. M. Browne, *J. Sci. Food Agr.*, **18**, 447 (1967).
441. R. C. Brian, *Chem. Ind.* (*London*), **1955**, 1965.
442. M. Pianka and J. D. Edwards, *J. Sci. Food Agr.*, **19**, 60 (1968).
443. A. H. Kirby, E. L. Frick, L. D. Hunter, and R. P. Tew in *Nitro Compounds* (Ed. T. Urbański), Pergamon Press, London, 1964, p. 483.
444. E. W. Simon and G. E. Blackman, *J. Exp. Botany*, **4**, 235 (1953).
445. M. E. Karahl and G. H. A. Clowes, *J. Cellular Comp. Physiol.*, **11**, 21 (1938).
446. F. Burke and M. W. Whitehouse, *Biochem. Pharmacol.*, **16**, 209 (1967).
447. W. A. Sexton, *Chemische Konstitution and Biologische Wirkung*, Verlag Chemie, Weinheim, 1958, p. 205.
448. J. S. Pizey and A. Bates, *J. Sci. Food Agr.*, **12**, 542 (1961).
449. W. Sobotka, Z. Eckstein, and T. Urbański, *Roczniki Chem.*, **32**, 963 (1958).
450. K. H. Jacobsen, H. Wheelwright, J. Clem, and R. N. Shannon, *A.M.A. Arch. Ind. Health*, **12**, 617 (1955).
451. J. M. Barnes and P. N. Magee, *Brit. J. Ind. Med.*, **11**, 167 (1954).

452. R. M. Watrous, *Brit. J. Ind. Med.*, **4,** 111 (1947).
453. F. Weigley, *Brit. J. Ind. Med.*, **5,** 26 (1948).
454. C. E. Lewis, *J. Occupational Med.*, **6,** 91 (1964).
455. D. F. Heath and P. N. Magee, *J. Occupational Med.*, **19,** 276 (1962).
456. P. N. Magee and J. M. Barnes, *Advan. Cancer Res.*, **10,** 164 (1967).
457. P. N. Magee and J. M. Barnes, *Brit. J. Cancer,* **10,** 114 (1956).
458. H. Druckrey, R. Preussmann, G. Blum, S. Ivankovic, and J. Afkham, *Naturwiss.*, **50,** 100 (1963).
459. H. Druckrey, R. Preussmann, J. Afkham, and G. Blum, *Naturwiss.*, **49,** 451 (1962).
460. R. Schoental, *Nature,* **188,** 420 (1960).
461. D. Schmahl and C. Thomas, *Arzneimittel-Forsch.*, **12,** 585 (1962).
462. H. Druckrey and R. Preussmann, *Nature,* **195,** 111 (1962).
463. F. Arndt, B. Eistert, and W. Walter, *Naturwiss.*, **50,** 379 (1963).
464. R. Schoental and P. N. Magee, *Brit. J. Cancer,* **16,** 92 (1962).
465. H. Druckrey, D. Steinhoff, R. Preussmann, and S. Ivankovic, *Naturwiss.*, **50,** 735 (1963).
466. C. Thomas and D. Schmahl, *Z. Krebsforsch.*, **66,** 125 (1964).
467. A. Graffi and F. Hoffmann, *Acta Biol. Med. Ger.*, **16,** K1 (1966).
468. W. Lijinsky, K. Y. Lee, L. Tomatis, and W. H. Butler, *Naturwiss.*, **54,** 518 (1967).
469. B. A. Kihlman, *Action of Chemicals on Dividing Cells*, Prentice-Hall, Englewood Cliffs, N.J., 1966.
470. T. von Kreybig and W. Schmidt, *Arzneimittel-Forsch.*, **17,** 1093 (1967).
471. R. Schoental, *Nature,* **209,** 726 (1966).
472. H. Druckrey, R. Preussmann, G. Blum, S. Ivankovic, and J. Afkham, *Naturwiss.*, **50,** 100 (1963).
473. L. Pasternak, *Acta Biol. Med. Ger.*, **10,** 436 (1963).
474. A. H. Dutton and D. F. Heath, *Nature,* **178,** 644 (1956).
475. D. F. Heath and A. H. Dutton, *Biochem. J.*, **70,** 619 (1958).
476. D. F. Heath, *Nature,* **192,** 170 (1961).
477. I. Mizrahi and P. Emmelot, *Nature,* **193,** 1158 (1962).
478. I. Mizrahi and P. Emmelot, *Cancer Res.*, **22,** 339 (1962).
479. D. F. Heath, *Biochem. J.*, **85,** 72 (1962).
480. P. N. Magee and T. Hultin, *Biochem. J.*, **83,** 106 (1962).
481. P. N. Magee and E. Farber, *Biochem. J.*, **83,** 114 (1962).
482. K. Y. Lee, W. Lijinsky, and P. N. Magee, *J. Nat. Cancer Inst.*, **32,** 65 (1964).
483. K. Y. Lee and K. Spencer, *J. Nat. Cancer Inst.*, **33,** 957 (1964).
484. V. M. Craddock and P. N. Magee, *Biochem. J.*, **89,** 32 (1963).
485. P. N. Magee and K. Y. Lee, *Biochem. J.*, **91,** 35 (1964).
486. K. Y. Lee and W. Lijkinsky, *J. Nat. Cancer Inst.*, **37,** 401 (1966).
487. V. M. Craddock, *Biochem. J.*, **106,** 921 (1968).
488. V. M. Craddock and P. N. Magee, *Biochem. J.*, **100,** 721 (1966).
489. J. Rau and F. Lingens, *Naturwiss.*, **54,** 517 (1967).
490. J. G. Meyer-Bertenrath and V. Dege, *Z. Naturforsch.*, **22b,** 169 (1967).
491. B. Singer, H. Fraenkel-Conrat, J. Greenberg, and A. M. Michelson, *Science,* **160,** 1236 (1968).
492. B. A. Kihlman, *Exp. Cell Res.*, **20,** 657 (1960).
493. B. A. Kihlman, *Radiation Botany,* **1,** 43 (1961).
494. L. Pasternak, *Naturwiss.*, **49,** 381 (1962).
495. H. V. Malling, *Mutation Res.*, **3,** 537 (1966).
496. K. Miura, *Progr. Med. Chem.*, **5,** 320 (1967).

497. M. C. Dodd and W. B. Stillman, *J. Pharmacol. Exp. Therap.*, **82,** 11 (1944).
498. G. Berkelhammer and G. Asato, *Science*, **162,** 1146 (1968).
499. K. Skajius in *Nitro Compounds* (Ed. T. Urbański), Pergamon Press, London, 1964, pp. 475–482; see also: K. Skajius, K. Rubsenstein, and B. Ifversen, *Acta Chem. Scand.*, **14,** 1054 (1960).
500. M. C. Dodd, L. C. Cramer, and W. C. Ward, *J. Am. Pharm. Assoc.*, **39,** 313 (1950).
501. R. C. Bender and H. E. Paul, *J. Biol. Chem.*, **191,** 217 (1951).
502. H. E. Paul and R. C. Bender, *J. Pharmacol. Exp. Therap.*, **98,** 153 (1950).
503. A. H. Beckett and A. E. Robinson, *J. Pharm. Pharmacol.*, **8,** 1072 (1956).
504. A. H. Beckett and A. E. Robinson, *J. Med. Pharm. Chem.*, **1,** 135 (1959).
505. A. H. Beckett and A. E. Robinson, *J. Med. Pharm. Chem.*, **2,** 155 (1959).
506. J. J. Gavin, F. F. Ebetino, R. Freedman, and W. E. Waterbury, *Arch. Biochem. Biophys.*, **113,** 399 (1966).
507. J. Kranz and W. Evans, *J. Pharmacol.*, **85,** 324 (1945).
508. R. F. Raffauf, *J. Am. Chem. Soc.*, **72,** 753 (1950).
509. I. Yall and M. Green, *Proc. Exptl. Biol. Med.*, **79,** 306 (1952).
510. K. Hirano, S. Yoshina, K. Okamura, and I. Suzuka, *Bull. Chem. Soc. Japan,* **40,** 2229 (1967).
511. K. Fukui, A. Imamura, and C. Nagata, *Bull. Chem. Soc. Japan*, **33,** 122 (1960).
512. F. Yoneda and Y. Nitta, *Chem. Pharm. Bull.* (*Tokyo*), **12,** 1264 (1964).
513. M. T. Umar and M. Mitchard, *Biochem. Pharmacol.*, **17,** 2057 (1968).
514. D. Ernaelsteen and R. Williams, *Gastroenterology*, **41,** 590 (1961).
515. M. West and H. J. Zimmerman, *J. Am. Med. Assoc.*, **162,** 637 (1956).
516. K. J. Murphy and M. D. Innis, *J. Am. Med. Assoc.*, **204,** 104 (1968).
517. S. Jokela, *Gastroenterology*, **53,** 306 (1967).
518. K. J. Murphy, *Med. J. Australia*, **2,** 207 (1966).
519. E. J. Sotter, *J. Urol.*, **96,** 86 (1966).
520. P. A. Sollaccio, C. A. Ribando, and W. J. Grace, *Ann. Intern. Med.*, **65,** 1284 (1966).
521. C. H. Walton, *Can. Med. Assoc. J.*, **94,** 40 (1966).
522. W. G. Strauss and L. M. Griffin, *J. Am. Med. Assoc.*, **199,** 765 (1967).
523. R. H. Frankenfeld, *Ann. Intern. Med.*, **66,** 1055 (1967).
524. G. A. Pankey, *Med. Clin. N. Am.*, **51,** 925 (1967).
525. T. M. Nicklaus and A. B. Snyder, *Arch. Intern. Med.*, **121,** 151 (1968).
526. C. J. DeMasi, *Arch. Intern. Med.*, **120,** 631 (1967).
527. P. A. Moody, *Genetics of Man*, W. W. Norton and Co., New York, 1967, p. 212.
528. F. Olivarius, *Ugeskrift Laeger*, **118,** 783 (1956).
529. L. W. Longhridge, *Lancet*, **2,** 1133 (1962).
530. A. Palmlov and G. Tunevall, *Svenska Lakartidn.*, **53,** 2864 (1956).
531. W. J. Martin, K. B. Corbin, and D. C. Utz, *Proc. Mayo Clin.*, **37,** 288 (1962).
532. T. Urbański, B. Serafinowa, S. Malinowski, S. Slopek, I. Kamienska, J. Venulet, and K. Jakemowska, *Gruzlica*, **20,** 157 (1952).
533. T. Urbański *et al.*, *Gruzlica*, **22,** 681 (1954).
534. T. Urbański *et al.*, *Gruzlica*, **26,** 889 (1958).
535. T. Urbański, *Nature,* **166,** 267 (1950).
536. T. Urbański, S. Hornung, S. Slopek, and J. Venulet, *Nature,* **170,** 753 (1952).
537. T. Urbański, *Nature,* **168,** 562 (1951).
538. F. Leonard, F. A. Barkley, E. V. Brown, F. E. Anderson, and D. M. Green, *Antibiot. Chemotherapy*, **6,** 261 (1956).
539 W. Logemann, L. Almirante, and I. de Carneri, *Farmaco,* **13,** 139 (1957).

540. I. de Carneri, *Z. Tropenmed. Parasitol.*, **9**, 32 (1958).
541. I. de Carneri, G. Coppi, L. Almirante, and W. Logemann, *Antibiot. Chemotherapy*, **10**, 626 (1960).
542. C. Cosar and L. Julou, *Ann. Inst. Pasteur* (Paris), **96**, 238 (1959).
543. Y. Chin, Y. Wu, B. Serafin, T. Urbański, and J. Venulet, *Nature*, **186**, 170 (1960); see also: Y. Chin, Y. Wu, B. Serafin, T. Urbański, J. Venulet, and K. Jakimowska, *Bull. Acad. Polon. Sci., Sér. Sci. Chim.*, **8**, 109 (1960).
544. J. Venulet and M. Wutkiewicz, *Acta Physiol. Poland*, **18**, 50 (1967).
545. R. Gonnert and E. Schraufstetter, *Arzneimittel Forsch.*, **10**, 881 (1960).
546. R. Strufe and R. Gonnert, *Arzneimittel Forsch.*, **10**, 886 (1960).
547. G. Hecht and C. Gloxhuber, *Arzneimittel Forsch.*, **10**, 889 (1960).
548. H. Uehleke, *Progress in Drug Research* (Ed. E. Jucker), Birkhauser, Basle, 1965, p. 218.
549. R. Gonnert, J. Johannis, E. Schraufstetter, and R. Strufe, *Med. Chem.* (Verlag Chemie, Weinheim), **7**, 540, 1963.
550. T. Urbański *et al.* in *Nitro Compounds* (Ed. T. Urbański), Pergamon Press, London, 1964, pp. 463–468.
551. K. Jakimowski, M. Wutkiewicz, and J. Venulet, *Acta Physiol. Polon.*, **15**, 701 (1964).
552. M. Wutkiewicz and J. Venulet, *Acta Physiol. Polon.*, **16**, 885 (1965).
553. T. Urbański, C. Radzikowski, Z. Ledochowski, and W. Czarnocki, *Nature*, **178**, 1351 (1956).
554. T. Urbański, D. Gurne, S. Slopek, H. Mordarska, and M. Mordarski, *Nature*, **187**, 426 (1960).
555. Z. Eckstein, P. Gluzinski, E. Grochowski, M. Mordarski, and T. Urbański, *Bull. Acad. Polon. Sci., Sér. Sci. Chim.*, **10**, 331 (1962).
556. J. B. Chylinska, E. Grochowski, M. Mordarski, and T. Urbański, *Acta Union Int. Contre Cancer*, **20**, 118 (1964).
557. C. Radzikowski, Z. Ledochowski, A. Ledochowski, T. Nazurewicz, H. Wisniewski, and S. Schwan, *Patol. Polska*, **9**, 339 (1958).
558. C. Radzikowski, Z. Ledochowski, A. Ledochowski, M. Ruprecht, and M. Hrabowska, *Patol. Polska*, **13**, 39 (1962).
559. Z. Ledochowski, A. Ledochowski, and C. Radzikowski, *Acta Union Int. Contre Cancer*, **20**, 122 (1964).
560. C. Radzikowski, A. Ledochowski, M. Hrabowska, B. Stefanska, B. Horowska, J. Konopa, E. Morawska, and M. Urbanska, *Arch. Immunol. Therap. Exp.*, **15**, 126 (1967).
561. Z. Ledochowski, M. Serozynska, and C. Radzikowski, *Nowotwory*, **14**, 317 (1964); *Chem. Abstr.*, **62**, 16837*h* (1965).
562. C. Radzikowski, M. Urbanska, T. Michalik, M. Mysliwski, and M. Hrabowska, *Arch. Immunol. Therap. Exp.*, **15**, 148 (1967).
563. C. Radzikowski, Unpublished data.
564. H. Krysicka-Doczkal, H. Szafranowa, and J. Venulet, Unpublished data.
565. S. L. Brenner, F. H. Crick, and A. Orgel, *J. Mol. Biol.*, **3**, 121 (1961).
566. L. S. Lerman, *J. Mol. Biol.*, **3**, 18 (1961).
567. L. S. Lerman, *Proc. Natl. Acad. Sci. U.S.*, **49**, 94 (1963).
568. F. Leonard and F. E. Anderson, *J. Am. Chem. Soc.*, **77**, 4425 (1955).
569. W. Kessler, M. I. Blake, and C. E. Miller, *J. Am. Pharm. Assoc.*, **45**, 570 (1957).
570. F. T. Galysk, G. H. Bergman, and C. E. Miller, *J. Am. Pharm. Assoc.*, **47**, 141 (1958).
571. R. Preussman, *Arzneimittel Forsch.*, **8**, 638 (1958).

572. W. C. Ross, *Biological Alkylating Agents*, Butterworth and Co., London, 1962.
573. M. M. Rath and J. C. Krantz, *J. Pharmacol. Exp. Therap.*, **76**, 33 (1942).
574. L. A. Crandall, C. D. Leake, S. A. Leevenhart, and C. W. Meuhlberger, *J. Pharmacol. Exp. Therap.*, **37**, 283 (1929).
575. J. C. Krantz, G. G. Lu, F. K. Bell, and H. F. Cascorbi, *Biochem. Pharmacol.*, **11**, 1095 (1962).
576. C. R. Marshall, *J. Pharmacol. Exp. Therap.*, **83**, 106 (1945).
577. F. W. Oberst and F. H. Snyder, *J. Pharmacol. Exp. Therap.*, **93**, 444 (1948).
578. L. A. Heppel and R. J. Hilmore, *J. Biol. Chem.*, **183**, 129 (1950).
579. F. E. Hunter and L. Ford, *J. Pharmacol. Exp. Therap.*, **113**, 186 (1955).
580. P. Needleman and J. C. Krantz, *Biochem. Pharmacol.*, **14**, 1225 (1965).
581. P. Needleman and F. E. Hunter, *Mol. Pharmacol.*, **1**, 77 (1965).
582. S. Bitny-Szlachto, *Acta Polon. Pharm.*, **17**, 384 (1960).
583. S. Bitny-Szlachto, S. J. Kosinski, and M. Niedzielska, *Acta Polon. Pharm.*, **20**, 371 (1963).
584. S. Bitny-Szlachto and A. Kaminski, *Roczniki Wojsk, Inst. Hig. Epidemiol.*, **2**, 199 (1962).
585. G. L. Ellman, *Arch. Biochem. Biophys.*, **82**, 70 (1959).
586. E. V. Jensen, *Science*, **130**, 1319 (1959).
587. W. L. Sutton in *Industrial Hygiene and Toxicology*, Vol. 2 (Ed. F. A. Patty), John Wiley and Sons, New York, 1962, p. 2069.
588. J. Treon and F. R. Dutra, *Arch. Ind. Hyg. Occup. Med.*, **5**, 52 (1952).
589. J. H. Weatherby, *Arch. Ind. Health*, **11**, 102 (1955).
590. E. W. Scott, *J. Ind. Hyg. Toxicol.*, **24**, 226 (1942).
591. E. W. Scott, *J. Ind. Hyg. Toxicol.*, **25**, 20 (1943).
592. K. Takagi and I. Takayanagi, *Arch. Intern. Pharmacodyn.*, **155**, 373 (1965).
593. M. Georgescu and N. Carp, *Biochem. Pharmacol.*, Suppl., **12**, 160 (1963).
594. E. Skog, *Acta Dermatovener* (Stockholm), **46**, 386 (1966).
595. C. L. Meneghini and F. Rantuccio, *Giorn. Ital. Dermatol.*, **106**, 457 (1965).
596. C. L. Meneghini, F. Rantuccio, and G. Cozza, *Dermatologica*, **132**, 425 (1966).
597. R. M. Parkhouse and R. W. Dutton, *Immunochem.*, **4**, 431 (1967).
598. J. Zikan and O. Kotynek, *Biopolymers*, **6**, 681 (1968).
599. T. J. Gill, D. S. Papermaster, and H. W. Kunz, *J. Biol. Chem.*, **243**, 287 (1968).
600. A. H. Good, P. S. Traylor, and S. J. Singer, *Biochemistry*, **6**, 873 (1967).
601. J. P. Lamelin, W. E. Paul, and B. Benacerraf, *J. Immunol.*, **100**, 1058 (1968).
602. J. R. Little and H. N. Eisen, *Biochemistry*, **7**, 711 (1968).
603. J. R. Little and H. N. Eisen, *Biochemistry*, **6**, 3119 (1967).
604. V. Nussenzweig and B. Benacerraf, *J. Exp. Med.*, **126**, 727 (1967).
605. C. W. Parker, S. M. Gott, and M. C. Johnson, *Biochemistry*, **5**, 2314 (1966).
606. S. Schlossman and H. Levine, *J. Immunol.*, **98**, 211 (1967).
607. H. Metzger, L. Wofsy, and S. J. Singer, *Arch. Biochem. Biophys.*, **103**, 206 (1963).
608. F. Sanger, *Cold Spring Harbor Symp. Quant. Biol.*, **14**, 153 (1949).
609. E. A. Kabat, *Structural Concepts in Immunology and Immunochemistry*, Holt, Rinehart and Winston, New York, 1968, Chapter 6.
610. D. Robinson, J. N. Smith, and R. T. Williams, *Biochem. J.*, **50**, 228 (1951).
611. D. V. Parke, *Biochem. J.*, **62**, 339 (1956).
612. I. D. Storey, *Biochem. J.*, **95**, 209 (1965).
613. G. A. Tomlinson and S. J. Yaffe, *Biochem. J.*, **99**, 507 (1966).
614. M. Jaffe, *Z. Physiol. Chem.*, **2**, 47 (1878).
615. H. G. Bray, W. V. Thorpe, and P. B. Wood, *Biochem. J.*, **44**, 39 (1949).

616. J. J. Betts, S. P. Jones, and W. V. Thorpe, *Biochem. J.*, **61,** 611 (1955).
617. D. Robinson, J. N. Smith, and R. T. Williams, *Biochem. J.*, **50,** 221 (1951).
618. J. F. Burke and M. W. Whitehouse, *Biochem. Pharmacol.*, **16,** 209 (1967).
619. R. T. Wedding and M. K. Black, *Plant Physiol.*, **38,** 157 (1962).
620. R. T. Wedding, C. Hansch, and T. R. Fukuto, *Arch. Biochem. Biophys.*, **121,** 9 (1967).
621. H. Uehleke, *Progr. Drug. Res.*, **8,** 229 (1965).
622. H. Uehleke, *Biochem. Pharmacol.*, Suppl., **12,** 159 (1963).
623. H. Uehleke, *Naturwiss.*, **50,** 355 (1963).
624. H. Uehleke, *Arch. Exp. Pathol. Pharmakol.*, **247,** 412 (1964).
625. W. L. Lipschitz, *Z. Physiol., Chem.*, **109,** 189 (1920).
626. B. Issekutz, *Arch. Exp. Pathol. Pharmakol.*, **193,** 551, 567, 569 (1939).
627. F. Jung, *Arch. Exp. Pathol. Pharmakol.*, **195,** 208 (1940).
628. J. Haan, M. Kiese, and A. Warner, *Arch. Exp. Pathol. Pharmakol.*, **235,** 365 (1954).
629. W. L. Lipschitz, *Arch. Exp. Pathol. Pharmakol.*, **205,** 305 (1948).
630. M. Kiese and M. Soetbeer, *Arch. Exp. Pathol. Pharmakol.*, **207,** 426 (1949).
631. M. Kiese, K. H. Plattig, and C. Schneider, *Arch. Exp. Pathol. Pharmakol.*, **231,** 170 (1957).
632. B. E. Wahler, G. Schoffa, and H. G. Thom, *Arch. Exp. Pathol. Pharmakol.*, **236,** 20 (1959).
633. N. Holzer and M. Kiese, *Arch. Exp. Pathol. Pharmakol.*, **238,** 546 (1960).
634. M. Kiese, *Arch. Exp. Pathol. Pharmakol.*, **235,** 360 (1959).
635. M. Kiese, *Naturwiss.*, **46,** 384 (1959).
636. N. Holzer and M. Kiese, *Arch. Exp. Pathol. Pharmakol.*, **238,** 546 (1960).
637. M. Kiese and H. Uehleke, *Arch. Exp. Pathol. Pharmakol.*, **242,** 117 (1961).
638. M. Kiese, E. Rauscher, and G. Renner, *Biochem. Pharmacol.*, Suppl., **12,** 159 (1963).
639. M. Kiese, *Arch. Exp. Pathol. Pharmakol.*, **245,** 484 (1963).
640. R. Heringlake, M. Kiese, G. Renner, and W. Wenz, *Arch. Exp. Pathol. Pharmakol.*, **239,** 370 (1960).
641. F. R. Stermitz, F. A. Norris, and M. C. Williams, *J. Am. Chem. Soc.*, **91,** 4599 (1969).
642. P. Schmidt, K. Erchenberger, A. O. Ilvespaa, and M. Wilhelm, *Ann. N.Y. Acad. Sci.*, **160** (Art. 2), 530 (1969).
643. J. W. Faigle and H. Keberle, *Ann. N.Y. Acad. Sci.*, **160** (Art. 2), 544 (1969).
644. J. R. Gillette, *Ann. N.Y. Acad. Sci.*, **160** (Art. 2), 558 (1969).
645. G. C. Lancini, E. Lazzari, V. Arioli, and P. Bellani, *J. Med. Chem.*, **12,** 775 (1969).
646. K. Butler, H. L. Howes, J. E. Lynch, and D. K. Pirie, *J. Med. Chem.*, **10,** 891 (1967).
647. C. Nolan and E. Margoliash, *Ann. Rev. Biochem.*, **37,** 727 (1968).

CHAPTER **5**

The synthesis and reactions of trinitromethyl compounds

LLOYD A. KAPLAN*

Chemistry Research Department,
U.S. Naval Ordnance Laboratory,
Silver Spring, Maryland

* The author wishes to express his appreciation to the U.S. Naval Ordnance Laboratory for granting him the time and the facilities used in the preparation of this manuscript.

I. INTRODUCTION

Though trinitromethane has been known for over 100 years[1], no systematic general study of the chemistry of trinitromethyl compounds, with the possible exception of Hantzsch's[2] studies which were limited in that they were concerned with an investigation of the salts of trinitromethane and the utility of the silver salt in metathetical reactions with alkyl halides, had been initiated until the work of Schimmelschmidt[3] during World War II. Shortly after the war, a program initiated by the U.S. Department of the Navy to investigate this area of chemistry as a possible source of high-energy materials was responsible in part for the rapid growth of interest in the chemistry of trinitromethyl compounds. Similar interests undoubtedly existed in the Soviet Union, for commencing with the mid-1950's, as results of these investigations became declassified, reports began to appear quite regularly in the open literature both in the United States and the Soviet Union. At present, the open literature also contains the collected papers of two symposia[4,5] dealing for the most part with the chemistry of trinitromethyl compounds. Finally, mention should also be made of several recent reviews[6–9] and one text[10] covering certain aspects of the chemistry of trinitromethyl compounds.

II. GENERAL CHARACTERISTICS OF THE TRINITROMETHYL GROUP

It is possible to conceive of two broad divisions of the chemistry of trinitromethyl compounds which differ considerably. The first, is that of the tetrahedrally hybridized trinitromethyl group, simply exemplified by substituted trinitromethanes such as the 1,1,1-trinitroalkanes. This is to be contrasted with the trigonally hybridized trinitromethyl group present in the trinitromethide ion. It should be pointed out that the chemistry of substituted 1,1-dinitromethide ions, though not specifically considered in this section, generally parallels that of the trinitromethide ion. The alkyl and halodinitromethides are generally better nucleophiles than the trinitromethide ion whereas cyano and 2-Y-vinyldinitromethide ions (Y = CO_2CH_3, SO_2CH_3, NO_2, etc.), in which the charge on the carbanion is more delocalized, are considerably poorer nucleophiles.

A. Tetrahedrally Hybridized Trinitromethyl Group

The placing of three electron-withdrawing nitro groups on the same carbon atom creates a functional group that is extremely acid strengthening via an inductive electron withdrawal. Thus, 4,4,4-trinitrobutyric acid, $pK_a = 3.64$[11], is about 0.5 pK_a units stronger than 4,4,4-trifluorobutyric acid, $pK_a = 4.15$[12]. A quantitative measure of this inductive electron withdrawal was obtained by Hine and Bailey[13] who reported for trinitromethyl a value of $\sigma^* = 4.54$. With the possible exception of fluorodinitromethyl, $\sigma^* = 4.38$[14], this appears to be the largest σ^* value reported for an uncharged substituent.

One of the consequences of this strong inductive electron withdrawal is that hydrogen atoms α to a trinitromethyl function in a 1,1,1-trinitroalkyl group are quite susceptible to being removed as protons. When coupled with the fact that a nitro group is a good leaving group, this supplies an excellent driving force for the decomposition of 1,1,1-trinitroalkyl groups in alkaline media by means of an E2-type elimination of the elements of nitrous acid.

A second reactive site is a nitro substituent of the trinitromethyl group. Due to steric crowding, non-bonded repulsions of the nitro-oxygen atoms, loss of a nitro group would lower the free energy of the system. A second effect[15] to be considered is that due to the presence of multiple nitro substituents the electronegativity of the carbon atom of the trinitromethyl function is increased as compared to a nitromethyl function. We may then say that on the average a nitro substituent of a trinitromethyl group is more electropositive than in a nitromethyl group and therefore it should be open to nucleophilic attack. The driving force for this reaction would be the loss of non-bonded oxygen repulsions as well as the creation of a resonance stabilized 1,1-dinitromethide ion (equation 1). The

$$N\colon \curvearrowright O_2N\text{—}C(NO_2)_2R \longrightarrow RC(NO_2)_2^- \qquad (1)$$

chemistry of the tetrahedrally hybridized trinitromethyl group is replete with reactions of this type involving both intermolecular and intramolecular nucleophilic displacements, with the latter category often yielding some rather deep-seated rearrangements. One quite often finds a given nucleophile behaving simultaneously as a base toward the hydrogen atoms α to the trinitromethyl group and a nucleophile toward a nitro substituent of the trinitromethyl group. Thus, the products obtained from a given reaction can be a complex mixture. Subtle changes in the down-chain molecular structure, solvent, or attacking reagent can often cause wide, if not seemingly random, variation in the product composition.

B. Trigonally Hybridized Trinitromethyl Group

Assuming a completely planar conformation in solution, one would predict an almost complete lack of nucleophilic character for trinitromethide ion since the p-electron pair on carbon should be rather well delocalized by the π system of the three nitro groups. However, the acidities (Table 1) of mono-, di-, and trinitromethanes exhibit a 'saturation effect'; the effect of a second and third nitro substituent upon the acidity of nitromethane is not additive. By comparison, the stepwise substitution of hydrogen by cyano groups produces a regular increase in the acidity of cyanomethane. Thus, linear cyano groups operate efficiently as p-electron delocalizers in cyanocarbanions as contrasted with an increasingly damped resonance interaction with increasing substitution of nitro groups in nitro carbanions. A second observation is that replacing a nitro group in trinitromethane with a cyano group increases the acidity by a factor of more than one million. The pK_a of cyanodinitromethane is -6.2[16]. A cyano group, $\sigma^* = 1.30$, is a somewhat poorer electron-withdrawing substituent via an inductive interaction than a nitro group, $\sigma^* = 1.40$. Therefore, the increased acidity of cyanodinitromethane relative to trinitromethane must be due to differences in carbanion stability rather than C–H bond strengths in the undissociated methanes. The obvious conclusion is that trinitromethide ion is *not* completely planar in solution. The substitution of the linear cyano group for a nitro group reduces the non-bonded oxygen repulsions of the nitro groups which would exist in a planar trinitromethide ion, and permits cyanodinitromethide ion

TABLE 1. Acidities of substituted methanes[15].

	pK_a		pK_a
CH_4	40		
CH_3NO_2	11	CH_3CN	25
$CH_2(NO_2)_2$	4	$CH_2(CN)_2$	12
$CH(NO_2)_3$	0	$CH(CN)_3$	-5^a

[a] R. H. Boyd, *J. Am. Chem. Soc.*, **83**, 4288 (1961).

to have a planar conformation, or certainly more nearly so, in solution.

Results of solid state conformation determinations support the above conclusion. Dickens[17] has determined the structure of trinitromethide ion in hydrazinium trinitromethide and finds that the crystal is composed of two crystallographically independent trinitromethide ions arising from different hydrogen bonding environments in the crystal lattice. Though crystallographically different, both trinitromethide ions have quite similar conformations with respect to the orientation of the O–N–O planes of the three nitro groups about the carbon atom. Thus, the atoms N–C(–N)–N lie in a plane with the O–N–O planes of the nitro groups making dihedral angles with the N–C(–N)–N plane of 7, 8, and 41° in one anion and 4, 5, and 74° in the other. In the crystal, trinitromethide ion appears to be an asymmetric propeller having one blade considerably twisted and the other two blades twisted only very slightly out of the plane of the hub. Supporting the hypothesis that non-bonded oxygen repulsions are responsible for the lack of coplanarity of trinitromethide ion, it was found[17] that the N–C–N angle between the more nearly planar nitro groups has opened to 124 and 127°, with the greater spread belonging to the anion in which the non-planar nitro group is 74° out of the N–C(–N)–N plane.

By contrast, Grigor'eva and coworkers[18] report that cyanodinitromethide ion in rubidium cyanodinitromethide is completely planar in the crystal and suggest that the p-electron pair on the carbanion is delocalized by the π systems of both the nitro and cyano groups. Thus, we would expect that the trinitromethide ion would have considerable nucleophilic character and indeed behave more like a substituted 1,1-dinitromethide ion in which the substituent, an orthogonal nitro group, can only reduce the p-electron density on the carbanion by an inductive electron withdrawal. However, the cyanodinitromethide ion should be an extremely poor nucleophile. This is corroborated by the observation[19] that under conditions where trinitromethide ion adds to methyl acrylate at a specific rate which is $\frac{1}{30}$ that of alkyldinitromethide ions, cyanodinitromethide ion is totally unreactive.

III. SYNTHETIC APPROACHES TO TRINITROMETHYL COMPOUNDS

We may divide the synthetic approaches to trinitromethyl compounds into two categories. The first, and undoubtedly more important since it has considerably greater synthetic utility and versatility, makes use of the trinitromethide ion as a nucleophile in reactions with saturated and unsaturated substrates. Thus, we find that trinitromethide ion readily adds to a variety of α,β-unsaturated systems of the general structure $CH_2{=}CHY$, where Y is a conjugating electron-withdrawing substituent, to yield the corresponding 3-Y-1,1,1-trinitropropyl derivatives. Additions to a transient $>N^+{=}CH_2$ double bond such as is generated in the Mannich reaction[20] and to the carbonyl function also occur quite readily, although additions to the latter are quite limited. It is possible to effect 1,2 additions to unconjugated olefin systems by utilizing the mercury salt of trinitromethane as the addend. Though considerably more restricted in its application, nucleophilic displacements of halogen from a saturated carbon atom by trinitromethide ion can also be used to introduce the trinitromethyl function.

The second technique for introducing the trinitromethyl function might well be called the 'hammer and tongs' procedure since it involves the stepwise construction of the trinitromethyl function. It is at present of little value because of the lack of suitable procedures for converting the rather readily obtained substituted 1,1-dinitromethide ions to the corresponding 1,1,1-trinitromethyl compounds. This area of trinitromethyl chemistry, the nitration of substituted 1,1-dinitromethide ions, undoubtedly merits further investigation.

A. Trinitromethide Ion as a Nucleophile

I. Addition to α,β-unsaturated systems

The addition of carbon acids to α,β-unsaturated systems is generally termed the Michael reaction. The variation in the structure of the donor as well as the acceptor molecules which undergo reaction is extremely large[21]. However, it is required that the acceptor molecule have a double or a triple bond conjugated with an electron-withdrawing substituent such as COR, SO_2R, NO_2, CN, etc. The reaction may be depicted by equations 2–4, in which HAn is the conjugate acid of the carbanion addend, Y is a $-T$ substituent in the Ingold notation[22] and HA is a proton donor, such as the solvent, present in the reaction mixture. The reaction is said to be

$$\text{HAn} \rightleftharpoons \text{H}^+ + \text{An}^- \tag{2}$$

$$\text{An}^- + \text{CH}_2{=}\text{CHY} \rightleftharpoons \text{AnCH}_2\bar{\text{C}}\text{HY} \tag{3}$$

$$\text{AnCH}_2\bar{\text{C}}\text{HY} + \text{HA} \rightleftharpoons \text{AnCH}_2\text{CH}_2\text{Y} + \text{A}^- \tag{4}$$

subject to catalysis by bases[21]. However, kinetic studies of the reactions of barbituric acid with β-nitrostyrene[23] and alkylmalonic and acetoacetic esters with acrylonitrile[24] showed that the function of the base catalyst is to generate a sufficient concentration of the carbanion addend (equation 2) in those cases where its conjugate acid is essentially undissociated in the reaction solvent. The rate-determining step in both of these systems is the addition of the carbanion to the α,β-unsaturated acceptor molecules.

a. Mechanism of trinitromethane additions. Unlike most other carbon acids, trinitromethane is an exceedingly strong acid, $pK_a = 0.1$[16]. Therefore, it should not require base catalysis to generate a sufficient concentration of its conjugate base to add readily to α,β-unsaturated systems. Indeed, the converse is true. Additions of trinitromethane are subject to acid catalysis. Hine and Kaplan[25] investigated the kinetics and mechanism of the addition (and the reverse reaction) of trinitromethane to β-nitrostyrene in methanol and found that the forward reaction is subject to general acid catalysis. The reverse reaction, proceeding by an E1cB mechanism, is subject to general base catalysis. The reaction sequence is summarized by equations 5 and 6. This system is completely reversible

$$C(NO_2)_3^- + C_6H_5CH{=}CHNO_2 \underset{k_{-1}}{\overset{k_1}{\rightleftharpoons}} C_6H_5CH(C(NO_2)_3)\bar{C}HNO_2 \tag{5}$$

$$C_6H_5CH(C(NO_2)_3)\bar{C}HNO_2 + BH^+ \underset{k_{-2}}{\overset{k_2}{\rightleftharpoons}} C_6H_5CH(C(NO_2)_3)CH_2NO_2 + B \tag{6}$$

with either protonation of the intermediate carbanion $C_6H_5CH[C(NO_2)_3]\bar{C}HNO_2$ or deprotonation of the adduct 1,1,1,3-tetranitro-2-phenylpropane (equation 6) being rate determining. When the decomposition was studied in methyloxonium chloride solutions, a change in the rate-controlling step was observed. Under these more strongly acidic conditions, the rate, $k_2[C_6H_5CH[C(NO_2)_3]\bar{C}HNO_2][BH^+]$, $BH^+ = MeOH_2^+$, is so rapid that the rate of decomposition of the intermediate carbanion, $k_{-1}[C_6H_5CH[C(NO_2)_3]\bar{C}HNO_2]$ becomes rate controlling.

In contrast to the relative straightforwardness of the β-nitrostyrene system, the nature of the products obtained from the addition of trinitromethane to methyl acrylate is quite sensitive to the acidity of the reaction medium. It was observed that the yield of the normal Michael adduct, methyl 4,4,4-trinitrobutyrate (**1**), falls off markedly from 90% at pH 1–2 to 55% at pH 4–5[26]. That this was not due to a pH-dependent reversal of the adduct **1** to trinitromethane and methyl acrylate was shown by the inability to recover unreacted trinitromethane from the reaction mixture. Instead, a co-product, dimethyl 4,4-dinitro-2-hydroxypimelate (**2**), was isolated in 27% yield together with 55% of the normal Michael adduct **1** (equation 7). Substituting methyl vinyl ketone for methyl acrylate afforded the structurally analogous co-product, 5,5-dinitro-3-hydroxy-2,8-nonanedione. Under more alkaline conditions, a

$$C(NO_2)_3^- + CH_2{=}CHCO_2Me \xrightarrow[30\%\ MeOH]{HOAc}$$

$$\underset{(\mathbf{1})}{C(NO_2)_3CH_2CH_2CO_2Me} + \underset{(\mathbf{2})}{MeO_2CCH_2CH_2C(NO_2)_2CH_2CHOHCO_2Me} \quad (7)$$

second co-product, the potassium salt of methyl 4,4-dinitro-2-butenoate (**3**), was isolated together with **1** and **2**. Subsequent investigation of the trinitromethane–methyl acrylate system[27] showed that **2** and **3** were not primary reaction products and that in the synthetic sequence **2** was formed via a second Michael addition of the primary rearrangement product, methyl 4,4-dinitro-

$$C(NO_2)_3^- + CH_2{=}CHCO_2Me \xrightarrow{OAc^-} \mathbf{1} + \underset{(\mathbf{4})}{^-C(NO_2)_2CH_2CHOHCO_2Me} \quad (8)$$

$$\mathbf{4} + CH_2{=}CHCO_2Me \longrightarrow \mathbf{2} \quad (9)$$

$$\mathbf{1} \xrightarrow{base} \underset{(\mathbf{3})}{^-C(NO_2)_2CH{=}CHCO_2Me} \quad (10)$$

2-hydroxybutyrate (**4**), to methyl acrylate (equations 8 and 9). The olefin **3** was shown to be a product evolving from the elimination of nitrous acid from methyl 4,4,4-trinitrobutyrate (**1**) under alkaline conditions (equation 10). Rearrangement products analogous to the α-hydroxy ester **4** have been isolated with other acrylic augends[27,28].

The mechanisms and pH dependency of the reactions taking place in the trinitromethane–methyl acrylate system were elucidated by Kaplan and Glover[29] who studied the kinetics of the reactions in both acid and near-neutral media. The stoichiometry of the primary reactions occurring in this system are summarized by equations 11–13, where HA is water, hydronium ion and in near-neutral media, acetic acid from the buffer system used. Reactions 12 and 13 were *not* reversible under the conditions used in these kinetic runs.

The kinetic results obtained are expressed by the specific rate equations 14–16, where k_T, k_M, and k_D are, respectively, the observed

$$C(NO_2)_3^- + CH_2{=}CHCO_2Me \underset{k_{-1}}{\overset{k_1}{\rightleftharpoons}} C(NO_2)_3CH_2\bar{C}HCO_2Me \quad (\mathbf{5}) \tag{11}$$

$$\mathbf{5} + HA \xrightarrow{k_{HA}} C(NO_2)_3CH_2CH_2CO_2Me \; (\mathbf{1}) + A^- \tag{12}$$

$$\mathbf{5} \xrightarrow{k_2} \mathbf{4} + NO_2^- \tag{13}$$

specific rates for the disappearance of trinitromethide ion, the formation of the normal Michael adduct **1**, and the formation of the α-hydroxy ester **4** at a given acidity. These workers[29] found that in 50% dioxane both k_M and k_T increased on increasing the acidity of

$$k_T = \frac{k_1(\sum k_{HA}[HA] + k_2)}{k_{-1} + \sum k_{HA}[HA] + k_2} \tag{14}$$

$$k_M = \frac{k_1 \sum k_{HA}[HA]}{k_{-1} + \sum k_{HA}[HA] + k_2} \tag{15}$$

$$k_D = \frac{k_1 k_2}{k_{-1} + \sum k_{HA}[HA] + k_2} \tag{16}$$

the reaction medium. However, at high acidities where $\sum k_{HA}[HA] \gg k_{-1} \approx k_2$, k_M and k_T approached a limiting value of k_1, the specific rate of addition of trinitromethide ion to methyl acrylate (equation 11). The mechanism for the formation of the normal Michael adduct **1** (equations 11 and 12) in the methyl acrylate system is identical with the mechanism suggested for the formation of 1,1,1,3-tetranitro-2-phenylpropane (equations 5 and 6) in the β-nitrostyrene system.

In an aqueous reaction medium, both Kaplan and Glover[29] and Novikov and coworkers[30] observed that the specific rate of disappearance of trinitromethide ion, k_T, was constant over a wide acidity range. It was inferred from this observation[30] that the formation of the normal Michael adduct **1** was also an uncatalyzed reaction in this solvent system. However, this was shown to be incorrect[29] by dissecting k_T into its components k_M and k_D. The change in pH dependence of k_T on changing the reaction medium from 50% dioxane to water was rationalized[29] by assuming that at intermediate acidities in 50% dioxane $k_{-1} \approx \sum k_{HA}[HA] \approx k_2$, but on going to water, $\sum k_{HA}[HA] \gg k_{-1} \ll k_2$. For the latter condition,

k_T (equation 14) becomes equal to k_1, and equation 15 reduces to equation 17. Therefore, the specific rate of formation of **1** would still

$$k_M = \frac{k_1 \sum k_{HA}[HA]}{k_{-1} + \sum k_{HA}[HA]} \tag{17}$$

be subject to acid catalysis. Sufficient kinetic data were not obtained in the aqueous system to confirm the change in the relative magnitudes of k_{-1}, k_2, and $\sum k_{HA}[HA]$ on going from 50% dioxane to water.

The mode of formation of the α-hydroxy ester **4** from the intermediate carbanion **5** was shown to be first order in the carbanion **5** (equation 16). However, the presence of a water concentration term as in $k_D \propto k_1 k_2[H_2O]$ could not be ruled out. The process for the formation of **4** competed for the carbanion **5** with the reaction leading to the normal adduct **1**. Since equation 18 describes this system, the reaction path governed by equations 11 and 12 is

$$\frac{k_M}{k_D} = \frac{\sum k_{HA}[HA]}{k_2} \tag{18}$$

favored over the path governed by equations 11 and 13 in more strongly acidic media. Tracer experiments utilizing O^{18}-enriched solvent introduced only about 6% of the isotope enrichment into the α-hydroxy ester **4** if the solvent is assumed to be directly involved in attacking the α-carbon atom to yield the hydroxyl group. It was concluded that the formation of **4** occurred by either of two first-order processes both of which involved intramolecular nucleophilic attack of the α-carbon atom in **5** upon the oxygen atom of a nitro group of the trinitromethyl function to yield the transition state **6**. Collapse of **6** with solvent participation, predominantly with attack at nitrogen, yields **4**. An alternate route involves collapse to

$C(NO_2)_2$—CH_2—$\underline{C}HCO_2Me$, $^+N{=}\ddot{O}$, $^-:\ddot{O}$ **(5)** ⟶ $C(NO_2)_2$—CH_2—$CHCO_2Me$, $\ddot{N}$—$\ddot{O}$, $^-:\ddot{O}$, OH_2, OH_2 **(6)** (19)

$^-C(NO_2)_2CH_2CH(ONO)CO_2Me \xrightarrow{H_2O}$ **4** + HNO_2

(7)

the nitrite ester **7** which is rapidly hydrolyzed under the reaction conditions to **4** (equation 19). No distinction could be made between the two reaction paths from the experimental evidence obtained.

The reaction path for the retrogradation of the trinitromethide ion addition product to α,β-unsaturated systems is quite sensitive to the down-chain structure of the molecule. Neglecting at this time, the possibility of the base-catalyzed elimination of the elements of nitrous acid (equation 20) as a competing reaction (see section IV.B) and considering only the fate of the carbanion **8** generated

$$C(NO_2)_3CH_2CH_2Y \xrightarrow{OH^-} [C(NO_2)_2{=}CHCH_2Y] \xrightarrow{-H^+} {}^-C(NO_2)_2CH{=}CHY \quad (20)$$

by α-proton abstraction (equation 21), there are at least two modes of decomposition possible for **8**. The first of these paths (equation 22) is the reversal to trinitromethide ion and the α,β-unsaturated

$$C(NO_2)_3CH_2CH_2Y + B \longrightarrow \underset{(\mathbf{8})}{C(NO_2)_3CH_2\bar{C}HY} + BH^+ \quad (21)$$

$$\mathbf{8} \longrightarrow C(NO_2)_3^- + CH_2{=}CHY \quad (22)$$

$$\mathbf{8} \longrightarrow \underset{(\mathbf{9})}{{}^-C(NO_2)_2CH_2CHOHY} + NO_2^- \quad (23)$$

augend. The second path (equation 23) is the conversion of **8** to an α-hydroxydinitro carbanion **9** which is structurally analogous to the α-hydroxy derivative **4** whose formation from **8** in the addition reaction is favored in near-neutral media[27–29]. In these retrograde Michael reactions, the mode of decomposition appears to depend upon whether either or both of the hydrogen atoms α to the trinitromethyl group have been replaced by an alkyl or an aryl substituent.

It was shown[25] that 1,1,1,3-tetranitro-2-phenylpropane (β-phenyl substituent) reverses quantitatively to trinitromethide ion. Nikolaeva and coworkers[31] similarly observed that trinitromethyl ketones

$$C(NO_2)_3CRR^1CH_2COR^2 \xrightarrow{OH^-} C(NO_2)_3^- + RR^1C{=}CHCOR^2 \quad (24)$$

(**10**), $R = H$; $R^1 = Me$; $R^2 = C_6H_5$
(**11**), $R = R^1 = R^2 = Me$
(**12**), $R = H$; $R^1 = C_6H_5$; $R^2 = Me$

substituted in the β position yielded only trinitromethide ion (equation 24) when allowed to retrograde in strongly alkaline media. Analogous results were obtained by Novikov and coworkers[32] who investigated the reaction of 1,1,1,3-tetranitro-2-alkylpropanes in strongly alkaline or near-neutral media. In every instance, only reversal of the adducts **13** to trinitromethide ion and the nitro olefin **14** was observed (equation 25).

$$\underset{(\mathbf{13})}{C(NO_2)_3CHRCH_2NO_2} + B \longrightarrow C(NO_2)_3^- + \underset{(\mathbf{14})}{RCH{=}CHNO_2} + BH^+ \quad (25)$$

R = Me, Et, *n*-Pr

B = OH^-, OMe^-, OAc^-, C_5H_5N

By contrast with the β-mono- or β-disubstituted derivatives which undergo a normal retrograde Michael reaction in alkaline media, it was noted that trinitromethyl ketones having the general structure

$$\underset{(\mathbf{15})}{C(NO_2)_3CH_2CHRCOR^1} \longrightarrow \underset{(\mathbf{16})}{^-C(NO_2)_2CH_2CHROHCOR^1} \quad (26)$$

R = H; R^1 = Me, CH_2OAc, C_6H_5, $CH{=}CHMe_2$

R = Me; R^1 = C_6H_5

15 yield the α-hydroxy derivative **16** when allowed to retrograde in strongly alkaline or near-neutral (sodium acetate or potassium nitrite) solutions[31]. These workers incorrectly interpreted the result in potassium nitrite solution as due to reaction with nitrite ion rather than hydroxide. Since solutions of potassium nitrite generate pH's greater than 8, the reaction (equation 26) is undoubtedly effected by the hydroxide ion present in the reaction medium (see equations 21–23).

The reaction of the α-substituted analogs of **13** with alkali took a surprisingly different course than the reaction of ketones **15**. The tetranitro compounds **17** afforded a rearranged tetranitro derivative **18** instead of α-hydroxy derivatives similar to **16**[33–35]. The formation of **18** was viewed as occurring by an intramolecular nucleophilic displacement by the first formed α-carbanion **19** upon the nitrogen of a nitro group in the trinitromethyl function[35] (equation 27). A similar intramolecular displacement on the nitrogen of a nitro group in a trinitromethyl function by a carbanion has been proposed for one of the steps in the conversion of 2,2,2-trinitroethyl chloride to 1,1,2,2-tetranitroethane in the presence of nitrite and hydroxide ions[36].

$$\underset{(\mathbf{17})}{C(NO_2)_3CH_2CHRNO_2} \rightleftharpoons \underset{(\mathbf{19})}{C(NO_2)_3CH_2\bar{C}RNO_2} \rightleftharpoons \underset{(\mathbf{20})}{C(NO_2)_3CH_2CR{=}NO_2H} \quad (27)$$

$$\mathbf{19} \longrightarrow \underset{(\mathbf{18})}{^-C(NO_2)_2CH_2CR(NO_2)_2}$$

R = H, Me, Et

Novikov and coworkers[34] observed that the *aci*-form **20** isomerized to **18** more rapidly than the true nitro derivative **17**. From this they concluded that the rearrangement of **17** to **18** proceeds with the prior formation of the *aci*-form **20**. This seems unlikely since the base-catalyzed isomerization of the true form **17** to the *aci*-form **20** requires the intermediacy of the α-carbanion **19**. It is undoubtedly the α-carbanion that undergoes isomerization in these reactions as well as in the transformation of **15** to **16**. A rationalization for the relative reactivities observed is that α-proton abstraction is rate determining in the isomerization of **17** to **18** just as it is in the normal retrograde Michael reaction of trinitromethane adducts[25]. However, since proton transfers from oxygen and nitrogen are generally much faster than those from carbon[25], the rearrangement of the α-carbanion **19** becomes rate determining in the transformation of **20** to **18**. No satisfactory explanation has been given for the change in the site of nucleophilic attack of the α-carbanion from oxygen (equation 26) to nitrogen (equation 27) on changing the down-chain substituent from acyl to nitro.

The sensitivity of the reaction course of the α-carbanions to the presence of a β-substituent deserves comment. In general, it appears[25,31–35] that if R is alkyl or aryl, the adduct **21** will retrograde to trinitromethide ion (equation 28). The directive influence of the

$$\underset{\mathbf{(21)}}{C(NO_2)_3CHRCH_2Y} \xrightarrow{\text{base}} C(NO_2)_3^- + RCH{=}CHY \qquad (28)$$

$$R = \text{alkyl or aryl}$$

β-substituent probably evolves from the availability of a more energetically favorable reaction path (equation 26 or 27) for the α-carbanions which yields the α-hydroxy or rearranged tetranitro derivatives. If this transformation occurs by a cyclic transition state such as **6** (equation 19) or the equivalent four-membered ring structure for attack on nitrogen, then the presence of a β-substituent would tend to increase the energy of the transition state for this conversion due to non-bonded interactions of the β-substituent with the nitro groups in the γ position. Displacement of the resonance stabilized trinitromethide ion, the alternate path for α-carbanion reaction could then become the more energetically favored reaction path.

b. Reactivity of α,β-unsaturated systems. Information as to the relative reactivity of activated vinyl compounds with trinitromethide ion is scant, and generally reactivity rules are based upon product yields.

In the one quantitative study available, Novikov[37] found the reactivity ordering toward trinitromethide ion to be CN < CO_2R < CO_2H < $CONH_2$ for the unsaturated system $CH_2{=}CHY$. Substitution in the α position had a rate-retarding effect. By comparison, the rate-enhancing effect of the substituent Y for the addition of alkoxide ions was found to be $CONH_2$ < CO_2R < CN < SO_2R < COR[38]. The different reactivity ordering obtained may be due to the fact that the rates of trinitromethide ion addition were carried out in acidic media in which the equilibrium shown in equation 29

$$CH_2{=}CHY + H_3O^+ \rightleftharpoons \underset{(\mathbf{22})}{CH_2{=}CHYH^+} \quad (29)$$

may have made a significant contribution for Y = $CONH_2$ and COOH. Since addition of trinitromethide ion to the protonated form **22** would occur considerably more rapidly than to the unprotonated form, this may account for the inverted ordering observed with trinitromethide ion[37]. In place of more complete data, a reactivity sequence according to the magnitude of σ_{para} for the substituent Y is probably as good as any. There seems to be little doubt that α-alkyl or aryl substitution is rate retarding. The presence of a conjugatively electron-withdrawing substituent in the β position such as in maleic or fumaric acid derivatives renders the double bond inactive to addition by trinitromethide ion.

c. Trinitromethide ion adducts and their reactions. A wide variety of acrylic augends have been utilized in Michael additions with trinitromethide ion. The reader is referred to reviews[4,5,7] of the subject for a survey of the various adducts that have been prepared. It should be noted that a diversity of chemistry can be performed on the down-chain structure of these trinitromethyl adducts without affecting the trinitromethyl function. As examples are the transformation of trinitrobutyric acid to 3,3,3-trinitropropyl isocyanate (**24**) via the acid chloride (**23**) (equation 30)[39]. The isocyanate **24** under-

$$C(NO_2)_3CH_2CH_2CO_2H \xrightarrow{SOCl_2} \underset{(\mathbf{23})}{C(NO_2)_3CH_2CH_2COCl} \xrightarrow{NaN_3}$$

$$C(NO_2)_3CH_2CH_2CON_3 \xrightarrow{heat} \underset{(\mathbf{24})}{C(NO_2)_3CH_2CH_2NCO} \quad (30)$$

goes typical reactions such as amine, urea, and urethane formation[39]. Transesterification of methyl 4,4,4-trinitrobutyrate has been accomplished even with such electronegatively substituted alcohols as 2,2,2-trinitroethanol by utilizing fuming sulfuric acid as a catalyst[40].

The reaction of **23** with sodium peroxide under conventional conditions is reported to yield bis(4,4,4-trinitrobutyryl) peroxide[41]. It is also possible to selectively reduce the carbonyl function in the presence of the trinitromethyl group (equation 31)[42].

$$C(NO_2)_3CH_2CH_2COCH_2OH \xrightarrow{NaBH_4} C(NO_2)_3CH_2CH_2CHOHCH_2OH \quad (31)$$

2. Addition to carbonyl compounds

a. Scope and mechanism of the reaction. The addition of polynitroalkanes to a variety of aldehydes has been amply reviewed in the literature[6,7]. Although trinitromethane affords good yields of the formaldehyde addition product 2,2,2-trinitroethanol (**25**)[43], unlike 1,1-dinitroalkanes it does not yield isolable adducts with other aldehydes or ketones. Synthetic attempts to force the reaction by using strained carbonyl compounds such as 2,2,4,4-tetramethylcyclobutanedione-1,3 did not produce either a mono- or bis(trinitromethyl)carbinol[44].

Though stable addition products of trinitromethane to aldehydes other than formaldehyde were not isolable, the formation of 2-alkyl-2,2,2-trinitroethanols (**26**) was shown to occur in solution

$$HC(NO_2)_3 + RCHO \rightleftharpoons \underset{(\mathbf{26})}{C(NO_2)_3CRHOH} \quad (32)$$

(equation 32). Rondestvedt and coworkers[45] observed that in dioxane, the equilibrium (R = *n*-Pr) lay far in the direction of the carbinol **26** and was attained relatively slowly. The hydroxyl band of the alcohol **26** was located at 3.05 μ in the infrared spectrum. These workers[45] reported that carbinol formation was also detected spectroscopically in carbon tetrachloride with *n*-butyraldehyde as the substrate.

A more quantitative study of the trinitromethane–carbonyl compound equilibrium was carried out by Hall[46] who determined values of the equilibrium constant for reaction 33 in aqueous acid.

$$H_2O + C(NO_2)_3\text{—}Y\text{—}OH \rightleftharpoons C(NO_2)_3^- + Y(OH)_2 + H^+ \quad (33)$$

$$Y = \text{—}CH_2\text{—}, \text{—}CHMe\text{—}, (CH_2)_3C<, \text{—}CMe_2\text{—}$$

The extent of dissociation of the trinitromethylcarbinols was found to increase in the order $CH_2 < CHMe < (CH_2)_3C < CMe_2$. The values of the equilibrium constants obtained are reproduced in Table 2. Unfortunately, values of the equilibrium constant were obtained at only one temperature so that the enthalpies and

TABLE 2. Dissociation constants for trinitromethylcarbinols, $C(NO_2)_3$–Y–OH.

Y	K, M^{-1}
CH_2	7.80×10^{-7}
CHMe	2.80×10^{-4}
$(CH_2)_3C$	6.5×10^{-3}
CMe_2	[a]

[a] No detectable amount of carbinol was produced.

entropies of the reaction could not be calculated. However, it would appear from the data in Table 2 that the equilibrium position is extremely sensitive to the size of the alkyl group attached to the carbinol carbon. This is probably due to the non-bonded interactions between the trinitromethyl group and the alkyl substituent. The fact that a ketone such as cyclobutanone, which should have a more favorable enthalpy of reaction because of release of I-strain[47], has a less favorable carbinol equilibrium than acetaldehyde, strongly suggests that the steric factor mentioned above governs the position of the carbinol equilibrium.

The instability of trinitromethyl carbinols as compared to other substituted dinitromethyl carbinols is attributed to the increased stability of the trinitromethide ion relative to alkyl and halodinitromethide ions[48]. This hypothesis is supported by the observation that the equilibrium constant for the dissociation of 2-cyano-2,2-dinitroethanol is about seven powers of ten larger than for 2,2,2-trinitroethanol[48]. Similar differences observed in the acidities of cyanodinitromethane, pK = −6.2[48], and trinitromethane, p$K \approx$ O[48] are also attributable to the relative stabilities of trinitromethide and cyanodinitromethide ions.

Though studies of the mechanism of formation of 2,2,2-trinitroethanol have not been carried out, we may extrapolate from data[49] obtained from a study of the kinetics of the addition of 1,1-dinitroethane to formaldehyde. A reasonable mechanism for the formation of 2,2,2-trinitroethanol would then involve a rate-determining addition of trinitromethide ion to formaldehyde followed by a rapid protonation of the resulting 2,2,2-trinitroethoxide ion.

b. Reactions of 2,2,2-trinitroethanol. The chemistry of 2,2,2-trinitroethanol (**25**) is at variance with that of other alcohols. The attachment of the inductively electron-withdrawing trinitromethyl group

($\sigma^* = 4.54$) to the carbinol function effectively cancels the oxygen basicity of the hydroxyl group. In fact, the alcohol becomes reasonably acidic as evidenced by the observation[46] that aqueous solutions of **25** exhibit the spectrum of its progenitor trinitromethide ion. At pH's greater than 6, the equilibrium lies well in the direction of trinitromethide ion and formaldehyde.

The facile dissociation of **25** in weakly acidic or alkaline media precludes the preparation of 2,2,2-trinitroethoxide ion and it has not even been possible to utilize synthetically its transitory existence as an intermediate in the reversal of **25** to trinitromethide ion and formaldehyde. Thus, preparation of 2,2,2-trinitroethoxy derivatives via nucleophilic displacement reactions can not be achieved.

The alcohol **25** can be esterified by reaction with neat acid chlorides[50] although a more suitable procedure involves the use of aluminum chloride or other metal halide catalysts which have also been found to give superior results with trihaloethanols[40,51]. An alternate procedure, attractive since it circumvents the preparation of the acid chloride, involves the direct esterification by using polyphosphoric acid as the reaction medium[52,53]. With sterically hindered or electronegatively substituted carboxylic acids, either metal halide catalysis or polyphosphoric acid are the only suitable esterification procedures. The only reported esterification of an acid chloride by **25** using base catalysis is the formation of bis(2,2,2-trinitroethyl) carbonate from phosgene in the presence of pyridine or pyridine *N*-oxide[54]. This compound had previously been prepared by utilizing the aluminum chloride catalysis route[55].

Though the preparation of 2,2,2-trinitroethyl alkyl ethers has not been accomplished, the synthesis of 2,2,2-trinitroethyl acetals and formals has been carried out in good yield. Shipp and Hill[56] have reported the preparation of bis(2,2,2-trinitroethyl) formal (**27**) from the reaction of **25** and formaldehyde in concentrated sulfuric acid. Formal formation probably occurs by attack of protonated formaldehyde, CH_2OH^+, upon unprotonated **25** to afford the hemiacetal which dehydrates to the alkoxycarbonium ion $C(NO_2)_3CH_2OCH_2^{\oplus}$. Subsequent attack of the alkoxycarbonium ion upon a second alcohol molecule followed by transfer of a proton to the reaction medium yields the formal **27**. This procedure was successfully utilized with other electronegatively substituted alcohols[56]. Its success probably lies in the fact that these alcohols are incompletely protonated in concentrated sulfuric acid.

Mixed acetals of the general structure $CH_3CH(OR)OCH_2C(NO_2)_3$ have been prepared by the addition of **25** to alkyl vinyl ethers in the presence of catalytic amounts of hydrogen chloride[57] (equation 34). The preparation of bis(2,2,2-trinitroethyl) acetals by the sulfuric

$$C(NO_2)_3CH_2OH + ROCH{=}CH_2 \xrightarrow{HCl} C(NO_2)_3CH_2OCH(OR)CH_3 \quad (34)$$

acid method[56] is not possible due to the rapidity with which aldehydes with α-hydrogens undergo the acid-catalyzed aldol condensation.

The synthesis of ortho esters of **25** with the aid of metal halide catalysis has been accomplished. Hill[58] has reported that tetrakis-(2,2,2-trinitroethyl)orthocarbonate, tris(2,2,2-trinitroethyl)orthoformate, and tris(2,2,2-trinitroethyl)orthobenzoate are obtained from the reaction of **25** with carbon tetrachloride, chloroform, and benzotrichloride, respectively, in the presence of anhydrous ferric chloride. He has suggested[59] that the reaction involves a nucleophilic attack of the alcohol upon the ion pair $Cl_3C^{\oplus}Fe_2Cl_7^{\ominus}$ in the primary step to yield the complex $C(NO_2)_3CH_2OCCl_3{\cdot}Fe_2Cl_6$ which undergoes further substitution of chlorine by the alcohol until the ortho ester is produced. Similar intermediates could be proposed for the formation of the orthoformate and orthobenzoate from chloroform and benzotrichloride.

The replacement of the hydroxyl group of **25** by chlorine has also been effected. However, most of the usual synthetic procedures for this conversion are of either limited or no use. Thus, the reaction of **25** with phosphorus pentachloride gives predominantly tris(2,2,2-trinitroethyl) phosphate[36]. With neat thionyl chloride, the alcohol **25** affords bis(2,2,2-trinitroethyl) sulfite (**28**). Using anhydrous ferric chloride as a catalyst, the corresponding chlorosulfite ester **29**

$$\underset{(\mathbf{28})}{[C(NO_2)_3CH_2O]_2SO} \qquad \underset{(\mathbf{29})}{C(NO_2)_3CH_2OSOCl}$$

is produced. The successful conversion of **25** to 2,2,2-trinitro-1-chloroethane is accomplished by treating the alcohol with either thionyl chloride or sulfuryl chloride in the presence of catalytic amounts of pyridine, quinoline, or piperidinium chloride. These workers[36] found that the esters **28** and **29** are smoothly converted to 2,2,2-trinitro-1-chloroethane by thionyl chloride and a catalytic amount of pyridine. Attempts to prepare 2,2,2-trinitro-1-bromoethane by similar procedures were unsuccessful[36]. However, the preparation of 2,2,2-trinitroethyl-1-fluoroethane from **25** and sulfur tetrafluoride has been recently described[60].

A second and rather unique procedure evolves from the conversion of ketals to alkyl halides by reaction with phosphorus pentachloride[61]. Thus, the reaction of ethyl 2,2,2-trinitroethyl acetal (*vide supra*) with phosphorus pentachloride in benzene affords relatively good yields of 2,2,2-trinitro-1-chloroethane[62].

3. The Mannich reaction

The utilization of trinitromethane as the active hydrogen component in the Mannich reaction provides a synthetically valuable route to trinitroethyl amines and their derivatives. Studies of the mechanism of the Mannich reaction[20,63] suggest that the rate-controlling step may involve the addition of the anion of the active hydrogen compound, in this case trinitromethide ion, to a cationic intermediate such as **30**. The intermediate **30** is derived from the prior condensation of the amine and aldehyde components in the reaction mixture (equations 35–37).

$$CH_2O + R_2NH \rightleftharpoons R_2NCH_2OH \qquad (35)$$

$$R_2NHCH_2OH \rightleftharpoons \underset{(\mathbf{30})}{R_2\overset{+}{N}{=}CH_2} \qquad (36)$$

$$R_2\overset{+}{N}H{=}CH_2 + C(NO_2)_3^- \rightleftharpoons R_2NCH_2C(NO_2)_3 \qquad (37)$$

An alternate procedure makes use of **25** as a source of the active hydrogen component, trinitromethane, as well as formaldehyde. Since the equilibrium position of the reaction forming **25** lies well in the direction of its precursors, trinitromethane and formaldehyde at pH 5 or greater[46], the addition of the amine substrate to buffered 2,2,2-trinitroethanol solutions has been found to be a suitable procedure for the preparation of Mannich bases. Frankel and Klager[64] took advantage of this procedure to prepare a series of mono- and bis-Mannich bases from polynitroalkylamines (equations 38 and 39). These workers were also able to obtain the bis-Mannich base $CH_2[NHCH_2C(NO_2)_3]_2$ from methylenediamine.

$$RCH_2NH_2 + C(NO_2)_3CH_2OH \longrightarrow \underset{(\mathbf{31})}{RCH_2NHCH_2C(NO_2)_3} \qquad (38)$$

$$R = CH_3C(NO_2)_2CH_2,\ C(NO_2)_3CH_2$$

$$R(CH_2CH_2NH_2)_2 + C(NO_2)_3CH_2OH \longrightarrow \underset{(\mathbf{32})}{R[CH_2CH_2NHCH_2C(NO_2)_3]_2} \qquad (39)$$

$$R = C(NO_2)_2,\ NNO_2$$

The synthesis of 2,2,2-trinitroethylamine (**31**) has not been accomplished. Instead of **31**, Schenck[65] reported that the reaction of **25** even with only 1 equivalent of ammonia yields bis(2,2,2-trinitroethyl)amine (**32**) rather than the monoamine **31**. An interesting route to **32** uses hexamethylenetetramine as the ammonia source[66]. The substitution of urea for ammonia in the reaction with **25** gives rise to bis(2,2,2-trinitroethyl)urea[65].

Using trinitromethane as the active hydrogen component together with formaldehyde and amino alcohols, Feuer and Swarts[67] were able to prepare 2,2,2-trinitroethylaminocarbinols (**33**) (equation 40). Feuer[68,69] also observed that although some amides would not yield

$$HC(NO_2)_3 + CH_2O + RNH_2 \longrightarrow \underset{(\mathbf{33})}{RNHCH_2C(NO_2)_3} \quad (40)$$

$$R = (CH_2)_2OH,\ (CH_2)_3OH,\ CMe(CH_2OH)_2$$

the Mannich base when treated with trinitromethane and formaldehyde, the corresponding *N*-methylol derivatives and their benzoates reacted smoothly with trinitromethane to produce the expected Mannich base.

Though Mannich bases could be isolated from the reactions described, the workers in this area of trinitromethyl chemistry have reported that most of the adducts derived from amine bases were not particularly stable. This is probably due to the presence of a facile path for the reversal of the Mannich equilibrium involving the unshared p pair on nitrogen (equation 41). The methylene imonium

$$R\ddot{N}H{-}CH_2{-}C(NO_2)_3 \longrightarrow R\overset{\oplus}{N}H{=}CH_2 + C(NO_2)_3^{\ominus} \quad (41)$$

ion probably degrades to the amine and formaldehyde. The driving force for this reaction is supplied by the expulsion of the resonance-stabilized trinitromethide ion. This hypothesis is supported by the fact that nitration of the Mannich bases to the corresponding *N*-nitramines enhances the stability of these adducts[67]. This would be expected, since delocalization of the p pair on the amine nitrogen onto the nitro group of the nitramine reduces the electron density on the amine nitrogen. Furthermore, bis(2,2,2-trinitroethyl)urea in which the p pairs on the amine nitrogen are delocalized by the carbonyl function, exhibits better stability than the Mannich bases derived from amines.

4. The three-body reaction

Though trinitromethane yields only isolable carbonyl compound adducts with formaldehyde, the addition of an alcohol, mercaptan, or amide to a mixture of trinitromethane and an aldehyde affords the three-body product **34** (equation 42). Rondestvedt and coworkers[45] carried out a thorough investigation of the kinetics and

$$\text{RCHO} + \text{HC(NO}_2)_3 + \text{HYR}' \longrightarrow \underset{(\mathbf{34})}{\text{RCH(YR}')\text{C(NO}_2)_3} \qquad (42)$$

$$\text{Y} = \text{O, S, CONH}$$

mechanism of this reaction using the trinitromethane–*n*-butyraldehyde–ethanol system as a model. The mechanism presented in equations 43–45 for the formation of **34** (R = *n*-Pr, R′ = Et, Y = O) in dioxane solutions was consistent with their results.

$$n\text{-PrCHO} + \text{EtOH} \xrightleftharpoons{\text{HC(NO}_2)_3} \underset{(\mathbf{35})}{n\text{-PrCH(OH)OEt}} \qquad (43)$$

$$\mathbf{35} + \text{HC(NO}_2)_3 \rightleftharpoons \underset{(\mathbf{36})}{n\text{-Pr}\overset{\oplus}{\text{C}}\text{HOEt}\cdot\text{C(NO}_2)_3^{\ominus}} + \text{H}_2\text{O} \qquad (44)$$

$$\mathbf{36} \longrightarrow \underset{(\mathbf{37})}{n\text{-PrCH(OEt)C(NO}_2)_3} \qquad (45)$$

The sequence involves formation of the hemiacetal **35** catalyzed by undissociated trinitromethane. Reaction of **35** with undissociated trinitromethane affords the alkoxycarbonium ion **36** as an ion pair with trinitromethide ion. Collapse of the ion pair **36** yields the trinitromethide alkylate **37**. Though **37** equilibrated with excess ethanol fairly rapidly to form *n*-butyraldehyde diethyl acetal (**38**) and trinitromethane, **38** was not formed in the three-body reaction in the absence of excess ethanol. This was explained by assuming that collapse of **36** was more rapid than the diffusion of ethanol from the body of the dilute solution into the solvent cage about **36**. This mechanism was also consistent with the observation that the trinitromethane–aldehyde–mercaptan system affords high yields of trinitromethyl thioethers **34** (Y = S), but equilibration of dithioacetals with trinitromethane does not afford the same product. Thus, any dithioacetal formed constitutes a reaction dead end. By contrast, it is interesting to note that the acetal **38** equilibrates with trinitromethane to produce the ether **37**.

These workers[45] were also able to show that the aldehyde–trinitromethane equilibrium was not involved in the formation of the ether **37** (equation 46). The incorporation of this equilibrium into the reaction sequence for the formation of **37** did not fit the observed kinetics. Furthermore, the formation of the alkoxycarbonium ion,

$$n\text{-PrCHO} + \text{HC(NO}_2)_3 \rightleftharpoons \underset{(\mathbf{39})}{n\text{-PrCH(OH)C(NO}_2)_3} \qquad (46)$$

$n\text{-Pr}\overset{\oplus}{\text{C}}\text{HC(NO}_2)_3$ (**40**), from **39** at these acidities is unattractive when one considers the feeble basicity of trinitromethylcarbinols[56] as well as the destabilizing effect of the trinitromethyl group upon the cation **40.**

In the absence of ethanol or trinitromethane, it was observed[45] that the ether **37** in dioxane solution slowly dissociated to the extent of 2 to 5 mole % before the reaction attained equilibrium. However, instead of the products being *n*-butyraldehyde, ethanol, and trinitromethane, **37** appeared to dissociate according to equation 47. Support for this reaction path comes from the work of Shechter and

$$\underset{(\mathbf{37})}{n\text{-PrCH(OEt)C(NO}_2)_3} \longrightarrow \text{HC(NO}_2)_3 + \text{CH}_3\text{CH}_2\text{CH=CHOEt} \quad (47)$$

Cates[70] who observed that trinitromethane readily adds to vinyl alkyl ethers to afford trinitromethyl ethers which are structurally analogous to **37.**

5. Nucleophilic displacements at saturated carbon

Prior to a study of the mechanism of the alkylation of trinitromethide ion by Hammond and coworkers[71], little preparative use had been made of trinitromethide ion as a nucleophile in displacement reactions at saturated carbon. Hantzsch[2] had obtained 1,1,1-trinitroethane (**41**) by reacting silver trinitromethide with methyl iodide in ether. However, attempts to alkylate the silver salt with other alkyl halides or to substitute the potassium salt for the silver salt of trinitromethane in the preparation of **41** were unsuccessful. A later report[72] showed that potassium trinitromethide could be alkylated with methyl iodide if the reaction were carried out in acetone.

The alkylation of silver trinitromethide with a variety of mono-, di-, and tribenzylic iodides was effected under reaction conditions similar to those used by Hantzsch[2]. These workers[73] obtained mono-, bis-, and tris(trinitroethyl)benzenes together with considerable quantities of unstable red oils which they assumed were *O*-alkylation products. The formation of *O*-alkylates in these reactions would not be unexpected since trinitromethide ion would be classified as an ambident ion[74] and could therefore alkylate at either carbon or oxygen. The apparently larger amount of *O*-alkylate formed with the benzyl derivatives as compared to methyl iodide is consistent with the postulate that alkylation at the most electronegative atom of an ambident ion generally increases with increasing S_N1 character of the reaction[74].

The alkylation of silver trinitromethide with methyl iodide in acetonitrile solution exhibited overall third-order kinetics[71]. Thus, the participation of silver ion as an electrophile as well as trinitromethide ion as a nucleophile is required in the rate-determining or prior steps. These workers[71] observed that the reaction was 4000 times faster in acetone than in acetonitrile. They rationalized this observation by assuming that the electrophilicity of silver ion is reduced in acetonitrile due to coordination of silver ion with the p pairs on the nitrogen atom of the solvent. This hypothesis is supported by the good solubility of silver salts in acetonitrile as compared to acetone.

Attempts to extend their kinetic studies to isopropyl iodide did not yield integral-order kinetics. Synthetic examination of this system showed that *O*-alkylate was probably being formed as no *C*-alkylate could be isolated. The reaction was not following the same path as the alkylation of methyl iodide. This observation also fits well with the proposal[74] that alkylation at the more electronegative atom of an ambident ion should increase as the S_N1 character of the reaction increases.

A rather extensive synthetic investigation of the reactivity of halide substrates and the effect of structure of the halide substrate upon the reaction course was carried out by these workers[71]. They observed that the *C*-alkylation reaction to yield 1,1,1-trinitroalkanes (**42**) occurred in reasonably good yield, 28 to 65 %, with primary alkyl iodides using acetonitrile as the reaction solvent (equation 48).

$$AgC(NO_2)_3 + RI \longrightarrow \underset{(\mathbf{42})}{RC(NO_2)_3} + AgI \qquad (48)$$

$$R = Me, Et, n\text{-}Bu, n\text{-}C_6H_{13}, n\text{-}C_8H_{17}, CH_2{=}CHCH_2$$

However, together with the *C*-alkylate **42** varying amounts of the alkyl nitrate $RONO_2$ (**43**) were formed as well. When R = n-C_8H_{17}, a 71 % yield of the ester **43** was obtained together with a 28 % yield of the *C*-alkylate **42**.

Two routes suggested for the formation of the nitrate ester **43** are reaction of silver nitrate, formed by the decomposition of silver trinitromethide (equation 49), with the alkyl iodide and *O*-alkylation

$$2C(NO_2)_3^- \longrightarrow 2NO_3^- + 2NO + 2CO_2 + N_2 \qquad (49)$$

to yield an alkyl nitronate **44**, which would subsequently decompose by a multistep path to the nitrate ester **43** (equation 50). The primary route to **43** probably involves the *O*-alkylation sequence.

A variety of other halide substrates such as α-halo acids and

$$\begin{matrix} RO \\ \quad\diagdown^{\oplus} \\ \quad\; N{=}C(NO_2)_2 \\ \quad\diagup \\ ^{\ominus}O \end{matrix} \longrightarrow RONO_2 \qquad (50)$$

(**44**) (**43**)

esters, α-halo ketones, α-halo acetals, and acetylenic halides were used as alkylating agents. None of these substrates yielded the *C*-alkylate in spite of the fact that all of these substrates afforded good yields of the by-product silver halide. The products were generally complex mixtures which appeared to contain a nitrate ester as one component. In one instance, it was possible to obtain a *C*-alkylate in low yield from glycidyl iodide by using methyl acetate instead of acetonitrile for the reaction medium.

The possibility of one-electron-transfer reactions, as observed previously in the *C*-alkylation of the anion of 2-nitropropane[75], to yield radical ion intermediates has not been ruled out in the *C*-alkylation of trinitromethide ion. It would appear that this is another area of trinitromethyl chemistry that merits further investigation.

6. Reactions of mercury trinitromethide

An extensive study of the reaction of mercury trinitromethide (**45**) with various olefinic and active hydrogen substrates was carried out by Novikov and coworkers[76]. Infrared examination of the mercury salt **45** indicated that it has a covalent structure in the solid state. However, in aqueous or alcoholic solutions, it dissociates according to equations 51 and 52. The values of K_{51} and K_{52} in water are 1.47×10^{-2} *M* and 6.90×10^{-5} *M* at 20°, respectively. No evidence for the existence of the tautomeric form of the mercury

$$\underset{(\mathbf{45})}{Hg[C(NO_2)_3]_2} \rightleftharpoons \underset{(\mathbf{46})}{HgC(NO_2)_3^{\oplus}} + C(NO_2)_3^{\ominus} \qquad (51)$$

$$\mathbf{46} \rightleftharpoons Hg^{2+} + C(NO_2)_3^{\ominus} \qquad (52)$$

salt **45** in which oxygen is bonded to mercury was found either in the solid state or in solution.

From synthetic studies[77], they observed that the mercury salt **45** yielded substitution products with aromatic substrates (equations 53 and 54). Both *ortho* and *para* substitution products **47** were isolated and with the exception of benzene, the other substrates could be mercuriated in either alcohol or aprotic solvents.

When the benzene ring is substituted with electron-withdrawing

$$C_6H_5R + \mathbf{45} \longrightarrow \underset{(\mathbf{47})}{RC_6H_4HgC(NO_2)_3} + HC(NO_2)_3 \qquad (53)$$
$$R = H, Me, OMe, NMe_2$$

$$\text{C}_4\text{H}_4\text{Y} + \mathbf{45} \longrightarrow \text{C}_4\text{H}_3\text{Y}\text{—HgC(NO}_2)_3 + HC(NO_2)_3 \qquad (54)$$
$$Y = O, S$$

substituents, e.g. *m*-dinitrobenzene and *o*-nitroanisole, a 1:1 complex of the mercury salt **45** with the aromatic substrate is obtained[78]. Although these complexes are quite stable, treatment with strong alkali regenerates the aromatic nitro compound together with trinitromethide ion (equation 55). The inability of symmetrically trisubstituted nitro aromatics such as 1,3,5-trinitrobenzene and 3,5-dinitroanisole to form a complex with **45**, suggests that the adducts are charge-transfer complexes with the site of bonding being

$$O_2NC_6H_5{\cdot}Hg[C(NO_2)_3]_2 \xrightarrow{OH^\ominus} C_6H_5NO_2 + HgO + C(NO_2)_3^\ominus \qquad (55)$$

meta to the nitro substituents. These positions would have the highest electron density in the aromatic ring.

With aniline[77], the product, *N*-(trinitromethylmercuri)aniline, arises from a replacement of the amine hydrogen. This observation led these workers[79] to investigate the reaction of **45** with other 'active' hydrogen compounds. The reaction followed a path similar to the reaction with aniline. Some of the results are summarized in equation 56.

$$CH_2XY + Hg[C(NO_2)_3]_2 \longrightarrow CHXYHgC(NO_2)_3 + HC(NO_2)_3 \qquad (56)$$
$$X = CO_2Et, MeCO, MeCO, NO_2$$
$$Y = CO_2Et, CO_2Et, MeCO, CO_2Et$$

When olefinic substrates were used, the elements $HgC(NO_2)_3^+$ and $C(NO_2)_3^-$ added to the double bond to form 1,1,1-trinitro-3-trinitromethylmercurialkanes **48** (equation 57)[80]. Isobutylene did not yield an addition product.

$$RCH{=}CHR' + Hg[C(NO_2)_3]_2 \longrightarrow \underset{(\mathbf{48})}{RCH[C(NO_2)_3]CH[HgC(NO_2)_3]R'} \qquad (57)$$
$$R = H, Me, C_6H_5;\ R' = H$$
$$R + R' = (CH_2)_3, (CH_2)_4$$

With a large excess of the olefinic substrate, the initially formed adduct **48** acts as a source of both electrophilic and nucleophilic components to produce the bis adduct **49** (equation 58). This is not a disproportionation reaction as it did not take place in the absence of an excess of olefin. The same bis adducts were obtained

$$\mathrm{RCH[C(NO_2)_3]CH[HgC(NO_2)_3]R' + RCH{=}CHR' \longrightarrow [RCH[C(NO_2)_3]CHR']_2Hg} \quad (58)$$

$$\mathrm{R = R' = H} \qquad \mathbf{(49)}$$

from the addition of trinitromethylmercuribenzene to olefins[80]. However, this reaction probably proceeded via the phenylmercuri-trinitroalkane **50** which then disproportionated to the bis adduct since diphenyl mercury was also formed in the reaction (equation 59).

$$\mathrm{C_6H_5HgC(NO_2)_3 + CH_2{=}CH_2 \longrightarrow \underset{(\mathbf{50})}{[C_6H_5HgCH_2CH_2C(NO_2)_3]} \longrightarrow (C_6H_5)_2Hg + \underset{(\mathbf{49})}{[C(NO_2)_3CH_2CH_2]_2Hg}} \quad (59)$$

Structurally similar addition products were obtained when unsaturated alcohols or esters were utilized as the olefinic substrates[81]. Bis adducts of the type **49** were obtained when an excess of the unsaturated substrate was used.

The monoolefin adducts **48** were found to react smoothly and in high yield with halogens, hydrogen halides, and alkali halide salts to form trinitroalkylmercury halides **51** (equation 60). Although cleavage of the carbon–mercury bond was not reported for the

$$\mathrm{(NO_2)_3CCH_2CH_2HgC(NO_2)_3 + ZX \longrightarrow \underset{(\mathbf{51})}{(NO_2)_3CCH_2CH_2HgX} + ZC(NO_2)_3} \quad (60)$$

$$\mathrm{ZX = HCl, Br_2, KI}$$

trinitroalkyl mercury derivatives, the conversion of dinitroalkyl-mercury chlorides **52** to dinitroalkyl bromides **53** was accomplished by refluxing with bromine and catalytic amounts of benzoyl peroxide in carbon tetrachloride solution (equation 61)[82]. The lack, at present, of a suitable method for cleavage of the carbon–mercury

$$\mathrm{\underset{(\mathbf{52})}{HC(NO_2)_2CHMeCH_2HgCl} + Br_2 \xrightarrow{Bz_2O_2} \underset{(\mathbf{53})}{HC(NO_2)_2CHMeCH_2Br} + HgClBr} \quad (61)$$

bond in the trinitroalkyl mercury derivatives somewhat reduces the synthetic value of this reaction as a preparative tool in trinitromethyl chemistry.

Information as to the mechanism of formation of the adducts **48** was also obtained from these studies. In aprotic solvents such as nitromethane, these workers[83] suggest the reaction sequence given in equations 62–63 for the addition of mercury trinitromethide to olefins. The electrophilic reagent **46** is formed by the dissociation of mercury trinitromethide in polar solvents[76].

$$CH_3CH{=}CH_2 + \underset{(\mathbf{46})}{HgC(NO_2)_3^{\oplus}} \longrightarrow \underset{(\mathbf{54})}{CH_3\overset{\oplus}{C}HCH_2HgC(NO_2)_3} \quad (62)$$

$$\mathbf{54} + C(NO_2)_3^{\ominus} \longrightarrow \underset{(\mathbf{48}),}{CH_3CH[C(NO_2)_3]CH_2HgC(NO_2)_3} \quad (63)$$

In aqueous or alcoholic media, a different mechanism was suggested[83] for this addition reaction (equations 64–66). In these

$$Hg[C(NO_2)_3]_2 + ROH \rightleftharpoons HgOR[C(NO_2)_3] + HC(NO_2)_3 \quad (64)$$

$$CH_3CH{=}CH_2 + HgOR[C(NO_2)_3] \rightleftharpoons \underset{(\mathbf{55})}{CH_3CH(OR)CH_2HgC(NO_2)_3} \quad (65)$$

$$CH_3CH{=}CH_2 + Hg[C(NO_2)_3]_2 \longrightarrow \underset{(\mathbf{48}),\ R\,=\,Me,\ R'\,=\,H}{CH_3CH[C(NO_2)_3]CH_2HgC(NO_2)_3} \quad (66)$$

media, hydrolysis of mercury trinitromethide (equation 64), which has been shown to give a strongly acidic reaction in an aqueous solution[2], would produce the hydroxy (R = H) or alkoxy mercury trinitromethide. They suggest that in acid media an equilibrium exists between **55** and its precursors. However, the adduct **48** (R = Me, R′ = H) is quite stable in acidic media[81]. Therefore, the formation of **48** (R = Me, R′ = H) drains the system of **55**.

As evidence in support of this hypothesis, they report[83] that phenylcyclopropane yields the γ-methoxy derivative **56** (equation 67), and vinyl ethyl ether forms trinitromethylmercuriacetaldehyde (**57**) (equation 68) rather than adducts which are structurally similar to **48** when the reaction is carried out in methanol and water,

$$C_6H_5\underset{\diagdown\ \diagup \atop CH_2}{CH{-}CH_2} + Hg[C(NO_2)_3]_2 \xrightarrow{MeOH} \underset{(\mathbf{56})}{C_6H_5CH(OMe)CH_2CH_2HgC(NO_2)_3} \quad (67)$$

$$CH_2{=}CHOEt + Hg[C(NO_2)_3]_2 \xrightarrow{H_2O} [(NO_2)_3CHgCH_2CH(OEt)OH] \longrightarrow \underset{(\mathbf{57})}{(NO_2)_3CHgCH_2CHO} \quad (68)$$

respectively. In each of these cases the reaction with alkoxy- or hydroxymercury trinitromethide affords a product (**56** or **57**) which cannot readily reverse to its precursors under the reaction conditions.

Though this is a reasonable rationale for the mechanism of addition to olefins, it seems more probable in the light of other studies[84] that the reversal of **55** in acid media is not to its precursors (equation

65) but involves the loss of ROH from the protonated adduct **58** to form a cationic intermediate such as **59**. The ion **59** can return by adding alkoxide ion or go on to the stable adduct by adding trinitromethide ion.

$$\underset{\underset{\oplus}{\text{HOR}}}{CH_3CHCH_2HgC(NO_2)_3}\ (\mathbf{58}) \rightleftharpoons \underset{(\mathbf{59})}{CH_3CH\overset{HgC(NO_2)_3}{\overset{\oplus}{\text{—}}}CH_2} \xrightarrow{\ominus(NO_2)_3C} CH_3CH[C(NO_2)_3]CH_2HgC(NO_2)_3 \quad (69)$$

(**48**), R = Me, R′ = H

B. The "Hammer and Tongs" Technique

The synthesis of trinitromethyl compounds by this technique involves the conversion of a nitromethyl derivative to the corresponding dinitromethyl derivative which is, in turn, nitrated to the trinitromethyl compound. Although there are good synthetic procedures for converting the nitromethyl function to the dinitromethyl function, this synthetic route to trinitromethyl compounds is not often used because of the lack of a good, general procedure for the conversion of the dinitromethyl function to the corresponding trinitromethyl group.

Starting with a nitroalkane, this is converted to the α-chloro derivative[85]. The α-chloronitroalkane **60** is transformed into the 1,1-dinitroalkane **61** under ter Meer conditions (equation 70). The

$$RCH_2NO_2 \longrightarrow \underset{(\mathbf{60})}{RCHClNO_2} \longrightarrow \underset{(\mathbf{61})}{RC(NO_2)_2^-} \quad (70)$$

reader is referred to a brief review[7] of the synthetic possibilities of this reaction. An alternate procedure for the conversion of nitroalkanes to *gem*-dinitroalkanes is an oxidative nitration technique utilized by Kaplan and Shechter[86]. In this procedure, the anion of a nitroalkane, when allowed to react with a mixture of silver nitrate and alkali nitrite in an aqueous alkaline solution, is converted to the corresponding *gem*-dinitroalkane (equation 71). The yields of *gem*-dinitroalkane are generally quite good and this procedure has the

$$\underset{(\mathbf{62})}{RCH{=}NO_2^-} \xrightarrow{Ag^+,\ NO_2^-} \mathbf{61} \quad (71)$$

advantage of not requiring the preparation of the α-chloronitroalkane which is the starting point for the ter Meer reaction.

The nitration of 1,1-dinitroalkanes to the corresponding trinitromethyl compounds has been described by Plummer[87]. This procedure utilizes tetranitromethane in alkaline media as the nitrating agent. In this way, a number of substituted 1,1,1-trinitroalkanes **63** were prepared in modest yield (equation 72).

$$RC(NO_2)_2^- + C(NO_2)_4 \xrightarrow{OH^-} \underset{\mathbf{(63)}}{RC(NO_2)_3} \qquad (72)$$

$$R = C_6H_5CH_2CH_2,\ Me_2CH,\ Me_2CHCH_2,\ Me_3C,\ Et,\ n\text{-Pr},\ n\text{-Bu},\ n\text{-Am}$$

Aside from this nitration procedure which appears to be suitable for the preparation of unsubstituted 1,1,1-trinitroalkanes, there are only a few selected reports of the nitration of 1,1-dinitroalkanes and other intermediates to 1,1,1-trinitromethyl compounds. Novikov and coworkers[88–90] observed that selected arylaldoximes, arylnitrolic acids, and arylnitromethanes could be converted to 1,1,1-trinitroalkanes with dinitrogen tetroxide (equation 73). The yields of the trinitromethyl derivative are quite good, however, there are

$$YC_6H_4X + N_2O_4 \longrightarrow YC_6H_4C(NO_2)_3 \qquad (73)$$

$$X = CH{=}NO_2^-,\ C(NO_2){=}NOH,\ CH{=}NOH$$

$$Y = p\text{-Cl},\ p\text{-NO}_2,\ m\text{-NO}_2$$

no reports of this procedure producing trinitromethyl derivatives with other aromatic or aliphatic substrates.

Other examples involving the nitration of dinitroalkanes or other intermediates are the conversion of 1,1,2,2-tetranitroethane to hexanitroethane[88], cyanoacetic acid to trinitroacetonitrile[89], and acetylene[90] or ketene[91] to tetranitromethane.

IV. CHARACTERIZATION OF THE TRINITROMETHYL GROUP

It has been observed that trinitroalkanes have a low-intensity absorption band at about 280 mμ in the ultraviolet spectrum. There appears to be little, if any, interaction between the nitro groups, and in hexane solution 1,1,1-trinitroalkyl groups have a molar extinction of 98 $\pm$ 7[92,93] at 280 mμ.

Of greater utility, is the conversion of the trinitromethyl group to the corresponding substituted 1,1-dinitromethide ion (see section IV.A) which has an intense absorption maximum between 350 and 400 mμ. The correlation of the position of the absorption maximum of substituted 1,1-dinitromethide ions with the σ^* parameter of the substituent has been accomplished[94]. The quantitative reduction of the trinitromethyl group to the corresponding 1,1-dinitromethide ion with alkaline hydrogen peroxide is the basis of a method for the assay of trinitromethyl compounds developed by Glover[95].

A quite useful degradative procedure for the trinitromethyl group has been reported by Kamlet and coworkers[96]. It involves conversion of a 1,1-dinitromethide ion to the carboxylic acid by refluxing with aqueous acid. Thus, the trinitromethyl compound is first converted to the 1,1-dinitromethyl derivative (see section IV.A) which in turn is transformed into the carboxylic acid (equation 74).

$$RC(NO_2)_3 \longrightarrow RC(NO_2)_2^- \longrightarrow RCOOH \qquad (74)$$

V. REACTIONS OF THE TRINITROMETHYL GROUP

As stated previously (see section I.A. 1), trinitromethyl compounds are susceptible to attack by nucleophiles at the nitro group and by bases at the hydrogens α to the trinitromethyl group. The first of these reactions forms a substituted 1,1-dinitromethide ion (equation 1) whereas the second produces a 1,1-dinitroethylenic intermediate which, depending upon the nature of the down-chain substituents can yield a variety of products (e.g. see equation 20).

A. Nucleophilic Displacements on the Nitro Group

This reaction is exemplified by the reduction of trinitromethyl compounds to dinitromethide ions by hydroperoxide ion[95,97] or iodide ion[98]. Studies of the rate and mechanism of the reduction of trinitromethyl compounds have only been carried out with tetranitromethane. Lv'ova and coworkers[99] investigated the kinetics of the reaction of tetranitromethane with both iodide and nitrite ions in 70% ethanol. They observed that the reaction was first order in tetranitromethane and first order in the nucleophile. With excess nitrite ion, quantitative yields of trinitromethane were obtained. However, with equimolar nitrite ion and tetranitromethane, only a 50% conversion of tetranitromethane to trinitromethane was realized. This result could be explained by assuming that the attack of nitrite ion occurred on the nitro group to form trinitromethide ion and dinitrogen tetroxide (equation 75). Reaction of dinitrogen tetroxide with water produces both nitrous and nitric acids (equation 76). The equilibrium 77, shifted far in the direction of nitrous acid and nitrate ion, consumes a second mole of nitrite ion per mole of trinitromethide ion produced in the reaction. Because of the reduced conversion to trinitromethide ion under equimolar conditions, these workers[99] described the over-all reaction by equation 78. Reaction of nitrogen trioxide with water produces 2 moles of nitrous acid. Thus, equation 78 is essentially the summation of equations 75–77.

$$C(NO_2)_4 + NO_2^- \longrightarrow C(NO_2)_3^- + N_2O_4 \quad (75)$$

$$N_2O_4 + H_2O \longrightarrow HNO_2 + HNO_3 \quad (76)$$

$$HNO_3 + NO_2^- \rightleftharpoons NO_3^- + HNO_2 \quad (77)$$

$$C(NO_2)_4 + 2KNO_2 \longrightarrow C(NO_2)_3^-K^+ + KNO_3 + N_2O_3 \quad (78)$$

The reaction of iodide ion with tetranitromethane was described by equations 79–80[99]. It would seem more reasonable that the rate-determining step involves the formation of trinitromethane and

$$C(NO_2)_4 + I^- \xrightarrow{\text{slow}} C(NO_2)_3^- + \tfrac{1}{2}I_2 + NO_2 \quad (79)$$

$$NO_2 + I^- \xrightarrow{\text{fast}} NO_2^- + \tfrac{1}{2}I_2 \quad (80)$$

nitryl iodide (equation 81). Rapid reaction of nitryl iodide with excess iodide ion (equation 82) would form the observed products, nitrite ion and iodine. They also noted that the specific rate of reaction of tetranitromethane with nitrite ion, $6.8 \times 10^{-3}\ M^{-1}$

$$C(NO_2)_4 + I^- \xrightarrow{\text{slow}} C(NO_2)_3^- + NO_2I \quad (81)$$

$$NO_2I + I^- \xrightarrow{\text{fast}} NO_2^- + I_2 \quad (82)$$

sec^{-1}, is about 225 times slower than its reaction with iodide ion. This is the relative reactivity ordering expected when one considers the relative nucleophilicities of nitrite and iodide ions[100].

The results of Glover's studies[101] of the kinetics of the reaction of nitrite ion with tetranitromethane led to the same mechanism for the formation of trinitromethide ion. He also confirmed the suggestion[99] that 1 mole of nitrate ion was produced per mole of trinitromethide ion formed. He observed that hydroxide ion reacted with tetranitromethane to form both trinitromethide ion and carbonate. The reaction producing trinitromethide ion gave rise to nitrate ion rather than nitrite ion. This reaction was therefore assumed to involve nucleophilic attack by hydroxide ion on the nitrogen atom of the nitro group (equation 83).

$$HO^{\delta-} \cdots \overset{\overset{O^-}{|}}{\underset{\underset{O}{\|}}{N^+}} \cdots {}^{\delta-}C(NO_2)_3 \longrightarrow HONO_2 + C(NO_2)_3^- \quad (83)$$

The carbonate-forming reaction, which yields 4 moles of nitrite per mole of carbonate, could involve a rate-determining attack on carbon to form the transitory intermediate trinitromethanol (equation 84). Successive eliminations of nitrous acid and hydration would transform the intermediate trinitromethanol into carbonate and nitrite ions.

$$OH^- + C(NO_2)_4 \longrightarrow C(NO_2)_3OH + NO_2^- \quad (84)$$

Another contribution to the mechanism of the reaction of tetranitromethane with nucleophiles was made by Hoffsommer[102] who studied the reaction of tetranitromethane with hydroperoxide and alkyl hydroperoxide ions. For these nucleophiles, the specific rate of reaction was about one thousand times faster than the specific rate of reaction of tetranitromethane with iodide ion[99]. The enhanced nucleophilicity of hydroperoxide ions has been attributed to the 'alpha effect'[103]. From the kinetic and analytical results, the stoi-

$$C(NO_2)_4 + ROO^- + H_2O \longrightarrow C(NO_2)_3^- + O_2 + ROH + NO_2^- + H^+ \quad (85)$$
$$R = \text{alkyl or H}$$

chiometry of the reaction is given by equation 85. When the reaction was carried out in an O^{18}-enriched aqueous system with *t*-butyl hydroperoxide ion as the nucleophile, the only oxygen-containing product found to be enriched in the O^{18} isotope was nitrite ion. This observation, together with the fact that the reaction is first order in both tetranitromethane and *t*-butyl hydroperoxide ion suggested that it is the oxygen atom of a nitro group which is attacked by the nucleophile in the rate-determining step (equation 86). Isotopically enriched solvent attack at the nitrogen atom of **64** would produce

$$ROO^- \dashrightarrow O{=}\overset{\oplus}{N}(O^\ominus){-}C(NO_2)_3 \longrightarrow ROOO\ddot{N}(O^\ominus)\ (\mathbf{64}) + C(NO_2)_3^\ominus \quad (86)$$
$$R = t\text{-Bu}$$

the observed products, *t*-butyl alcohol and oxygen both unenriched and nitrite ion carrying the O^{18} label (equation 87).

$$RO{-}O{-}O{-}\ddot{N}(O^-)\ (\mathbf{64}) \leftarrow\!\!-\!- {}^{18}OH_2 \longrightarrow ROH + O_2 + HNO^{18}O \quad (87)$$

The conversion of tetranitromethane to trinitromethane has been accomplished with other nucleophilic reagents. Hydrazine[95], thiosulfate ion[104], sulfite and hydrogensulfite ions[104], and arsenite ion[104] are typical of those nucleophiles which react. Nucleophilic displacements upon a nitro group of 1,1,1-trinitroethane by a variety of nucleophiles to yield 1,1-dinitroethane have been reported[105]. As examples are 1-butanethiol and 2-nitropropyl anions. The co-products obtained from these reactions were the dimers di-*n*-butyl disulfide and 2,3-dimethyl-2,3-dinitrobutane. It was suggested that these reactions may be proceeding by a route which involves radical intermediates rather than an ionic displacement mechanism[105]. For synthetic purposes, the reduction of 1,1,1-trinitromethyl compounds to the corresponding 1,1-dinitromethyl derivative with iodide ion appears to be the best choice.

The reaction of the halotrinitromethanes with nucleophiles appears to be more complex[104]. From synthetic results, it is noted that fluorotrinitromethane affords good yields of fluorodinitromethane when hydroperoxide ion is used as the nucleophile[106]. However, chlorotrinitromethane yields a mixture of trinitromethane and chlorodinitromethane with hydroperoxide ion, and only trinitromethane is produced when the bromo derivative is allowed to react with hydroperoxide ion[104]. These results can be correlated with the electronegativity of the halogen atom bonded to the trinitromethyl group. Thus, the less electronegative the halogen atom, the more susceptible it should be to nucleophilic attack. The observed transition from nucleophilic displacement on the nitro group to nucleophilic displacement on the halogen atom by hydroperoxide ion would then be expected.

B. Nitrous Acid Elimination Reactions

The products derived from the reaction of 1,1,1-trinitromethyl compounds with bases can be considered as having arisen from a substituted 1,1-dinitroethylene intermediate. The reaction of 1,1,1-trinitroethane with methoxide[107], ethoxide[107], cyanide[108], amine

$$CH_3C(NO_2)_3 + B{:}^- \longrightarrow C(NO_2)_2{=}CH_2 + BH + NO_2^- \qquad (88)$$

$$C(NO_2)_2{=}CH_2 + B{:}^- \longrightarrow {}^-C(NO_2)_2CH_2B \qquad (89)$$

$$\mathbf{(65)}$$

$$B{:}^- = CN^-, OMe^-, OEt^-, {}^-CH(CO_2Et)_2, R_3N, (H_2N)_2C{=}NH, (CH_2)_5NH, Me_2NH, \text{ and } NH_3$$

bases[105,109], and the anion of diethyl malonate[105] affords products which are typical (equations 88–89). With uncharged amine bases as the reagent, the product is not the carbanion **65** but the zwitterionic species, $^{\ominus}C(NO_2)_2CH_2B^{\oplus}$ (**66**)[105,109].

When 1,1,1-trinitromethyl derivatives of the type $C(NO_2)_3$-CH_2CH_2Y, where Y is an electron-withdrawing substituent by a resonance effect, are treated with a variety of bases, the products obtained are 1-Y-3,3-dinitro-1-propenes **68**[27,35,110] rather than the β-substituted addition product analogous to **65** (equations 90 and 91). The driving force for the formation of **68** by α-proton abstraction

$$C(NO_2)_3CH_2CH_2Y + B:^- \longrightarrow \underset{\mathbf{(67)}}{C(NO_2)_2{=}CHCH_2Y} + BH + NO_2^- \quad (90)$$

$$\mathbf{67} + B:^- \longrightarrow \underset{\mathbf{(68)}}{^-C(NO_2)_2CH{=}CHY} + BH \quad (91)$$

$$Y = CO_2Me,\ CONH_2,\ CO_2H,\ SO_2Me,\ CN,\ NO_2$$

from **67** is supplied by a low-energy transition state which probably looks much like the planar, resonance-stabilized carbanion[110,111] product **68**. It should be noted that protonation of **68** affords $HC(NO_2)_2CH{=}CHY$ **69** rather than **67**[110]. The carbon acids **69** are from 2 to 2000 times stronger than trinitromethane[110]. Therefore, it would be expected that the 1,1-dinitroethylenes **67** would be still stronger acids and α-proton loss from **67** would be the favored reaction path when compared with nucleophilic addition to the double bond.

An investigation of the kinetics and mechanism of the formation of carbanions **68** from the corresponding 1,1,1-trinitromethyl derivatives has been carried out in aqueous media[112]. The reaction is first order in both 1,1,1-trinitromethyl substrate and hydroxide ion. Neither catalysis by buffer bases nor rate enhancement when the reaction was carried out in the presence of added nucleophilic reagents such as bromide or thiosulfate ions was observed. Therefore, the mechanism for the formation of **68** is best described by equations 92 and 93 where the rate-determining step is a concerted elimination of the elements of nitrous acid. Proton loss from **67** is extremely rapid and is probably catalyzed by any base present in the reaction mixture.

$$C(NO_2)_3CH_2CH_2Y + OH^- \xrightarrow{\text{slow}} \underset{\mathbf{(67)}}{C(NO_2)_2{=}CHCH_2Y} + H_2O + NO_2^- \quad (92)$$

$$C(NO_2)_2{=}CHCH_2Y + B \xrightarrow{\text{very fast}} \underset{\mathbf{(68)}}{^-C(NO_2)_2CH{=}CHY} + BH \quad (93)$$

When the kinetics of the elimination reaction were studied in the presence of thiosulfate ion, a very weak base and exceptionally good nucleophile, the first-formed product was not carbanion **68** but a species to which on the basis of its ultraviolet absorption maximum ($\sim$365 mμ) structure **70** was assigned (equation 94). The thiosulfate ester **70** represents a trapping product of the 1,1-dinitroethylene **67** and is structurally analogous to the addition products **65** isolated from the reaction of 1,1,1-trinitroethane with various bases (equation 89). The ester **70** was slowly converted to

$$C(NO_2)_2{=}CHCH_2Y + S_2O_3^{2-} \longrightarrow {}^{-}C(NO_2)_2\underset{\displaystyle OS_2O_2^{-}}{\underset{|}{C}}HCH_2Y \qquad (94)$$

(**67**) (**70**)

the carbanion **68** by a general base catalyzed process. The mechanism for this conversion (E1cB) probably involves a rate-determining proton removal to yield ion **71** (equation 95) which loses thiosulfate ion to form carbanion **68** (equation 96). The reversal of reaction

$${}^{-}C(NO_2)_2\underset{\displaystyle OS_2O_2^{-}}{\underset{|}{C}}HCH_2Y + B: \overset{\text{slow}}{\rightleftharpoons} {}^{-}C(NO_2)_2\underset{\displaystyle OS_2O_2^{-}}{\underset{|}{C}}H\bar{C}HY + BH \qquad (95)$$

(**71**)

$$\mathbf{71} \longrightarrow {}^{-}C(NO_2)_2CH{=}CHY + S_2O_3^{2-} \qquad (96)$$

(**68**)

B: = any buffer base present including solvent

94, displacement of thiosulfate ion by the electron pair on the dinitromethyl function in **70**, is probably not the preferred reaction path since this p pair is rather well delocalized by the nitro groups. A structurally similar 1,1-dinitroethylene trapping product has been reported[35] as one of the species formed in the reaction of 4,4,4-trinitrobutyramide with ammonia.

The intermediacy of 1,1-dinitroethylenes has been proposed for numerous reactions in polynitro aliphatic chemistry. As examples are the conversion of 2,2,2-trinitro-1-chloroethane to 1,1,2,2-tetranitroethane with nitrite ion[36], the formation of 2,2,4,4-tetranitrobutanol from 2,2-dinitroethanol[113] and the reactions of 2-bromo-2,2-dinitroethyl acetate with various anions[114,115].

Novikov and coworkers[116] reported the only preparation of a relatively stable 1,1-dinitroethylene derivative **72** by reacting dinitrogen tetroxide with β-nitrostyrene (equation 97). The olefin **72** reacted readily with alcohols and alkoxide ions to form the ethers of 2,2-dinitro-1-phenylethanol.

$$C_6H_5CH{=}CHNO_2 + N_2O_4 \longrightarrow C_6H_5CH{=}C(NO_2)_2 \qquad (97)$$

(**72**)

C. Miscellaneous Reactions

Altukhov and coworkers[117] observed that tetranitromethane underwent heterolytic addition to styrene to ultimately yield *N*-(α-phenyl-β-nitroethoxy)-3,3-dinitro-5-phenylisoxazolidine (**73**) (equation 98). They suggested that the reaction proceeds with the

$$C(NO_2)_4 + 2C_6H_5CH{=}CH_2 \longrightarrow \text{5-}C_6H_5\text{-3,3-}(NO_2)_2\text{-isoxazolidine-N-}OCH(C_6H_5)CH_2NO_2 \quad (98)$$

(**73**)

formation of the π-complex ion pair **74** which collapses to yield the *O*-alkylate **75** rather than a *C*-alkylate product of trinitromethide ion (equation 99). The isoxazolidene **73** is the product of the 1,3-

$$[C_6H_5\overset{NO_2}{=}CH_2]^{\oplus\ominus}C(NO_2)_3 \longrightarrow C_6H_5CHCH_2NO_2\text{(}ON(O^{\ominus}){=}C(NO_2)_2\text{)} \quad (99)$$

(**74**) (**75**)

dipolar addition of **75** to a second mole of styrene. This reaction was extended to a variety of olefinic substrates by these workers[118]. They also observed that the use of radical or ionic catalysts had essentially no effect upon the product yield. This suggested that the first step in the reaction was the formation of a charge-transfer complex of tetranitromethane with the olefin which then rearranges to the π-complex ion pair **74**[117].

Other examples of 1,3-dipolar cycloaddition reactions to olefins using a trinitromethane derivative have been reported by Tartakovskii and coworkers[119–121] who utilized the *O*-methyl ether of trinitromethane as the 1,3-dipole component. A wide variety of 5-substituted *N*-methoxy-3,3-dinitroisoxazolidines were prepared by this route. These workers[122] also observed that the *O*-methyl ether of trinitromethane decomposed at ambient temperatures to yield as the main product 2,2,2-trinitroethanol.

$$C(NO_2)_2{=}N(O^{\ominus})OCOCH_3 \xrightarrow{CH_3COCl} C(NO_2)_2ClN(OCOCH_3)_2 \longrightarrow$$

(**76**)

$$C(NO_2)Cl{=}NOCOCH_3 + CH_3CO_2NO_2 \quad (100)$$

(**77**)

The acylation of trinitromethide ion has also been reported[123,124]. However, instead of the *O*-acetyl derivative **76**, acetic chloroformonitrolic anhydride (**77**) was obtained. The suggested pathway[124] for the formation of the anhydride **77** is presented in equation 100.

VI. REFERENCES

1. O. Schischkoff, *Ann. Chem.*, **101**, 216 (1857).
2. A. Hantzsch and A. Rinckenberger, *Chem. Ber.*, **32**, 628, 637 (1899); A. Hantzsch and R. Caldwell, *Chem. Ber.*, **39**, 2474 (1906); **45**, 110 (1912); **54**, 2617 (1921).
3. Hunter Report, BIOS 1919/22 1G, July 3, 1946.
4. Nitro Paraffins (Ed. H. Feuer), *Tetrahedron*, **19**, Suppl. 1 (1963).
5. Nitro Compounds (Ed. T. Urbański), *Tetrahedron*, Suppl. 1 (1964).
6. G. A. Shevkhgeimer, N. F. Pyatakov, and S. S. Novikov, *Usp. Khim.*, **28**, 485 (1959).
7. P. Noble, Jr., F. G. Borgardt, and W. L. Reed, *Chem. Rev.*, **64**, 19 (1964).
8. S. L. Ioffe, V. A. Tartakovskii, and S. S. Novikov, *Usp. Khim.*, **35**, 19 (1966).
9. V. I. Erashko, S. A. Shevelev, and A. A. Fainzil'berg, *Usp. Khim.*, **35**, 719 (1966).
10. V. V. Perekalin, *Unsaturated Nitro Compounds*, Goskhimizadt, Leningrad, 1961.
11. M. I. Fauth, A. C. Richardson, and G. W. Nauflett, *Anal. Chem.*, **38**, 1947 (1966).
12. A. L. Henne and C. J. Fox, *J. Am. Chem. Soc.*, **73**, 2323 (1951).
13. J. Hine and W. C. Bailey, *J. Org. Chem.*, **26**, 2098 (1961).
14. L. A. Kaplan and H. B. Pickard, Unpublished results.
15. R. G. Pearson and R. L. Dillon, *J. Am. Chem. Soc.*, **75**, 2439 (1953).
16. T. N. Hall, *J. Org. Chem.*, **29**, 3587 (1964).
17. B. Dickens, *Chem. Comm.*, **1967**, 246.
18. N. V. Grigor'eva, N. V. Margolis, I. N. Shokhov, I. V. Tselinskii and V. V. Mel'nikova, *Zh. Strukt. Khim.*, **8**, 175 (1967).
19. L. A. Kaplan and H. B. Pickard, *Abstracts of Papers of 152nd National Meeting of the American Chemical Society*, Sept. 1966, Paper S9.
20. S. V. Lieberman and E. C. Wagner, *J. Org. Chem.*, **14**, 1001 (1949).
21. E. D. Bergmann, D. Ginsburg, and R. Pappo, *Organic Reactions*, Vol. X, John Wiley and Sons, New York, 1959, p. 179 *et seq.*
22. C. K. Ingold, *Structure and Mechanism in Organic Chemistry*, Cornell University Press, Ithaca, N.Y., 1953, pp. 308–408.
23. M. J. Kamlet and D. J. Glover, *J. Am. Chem. Soc.*, **78**, 4556 (1956).
24. U. Schmidt and H. Kubitzek, *Chem. Ber.*, **93**, 866 (1960).
25. J. Hine and L. A. Kaplan, *J. Am. Chem. Soc.*, **82**, 2915 (1960).
26. L. A. Kaplan and M. J. Kamlet, *J. Org. Chem.*, **27**, 780 (1962).
27. L. A. Kaplan, *J. Org. Chem.*, **29**, 2256 (1964).
28. M. J. Kamlet and L. A. Kaplan, *J. Org. Chem.*, **28**, 2128 (1963).
29. L. A. Kaplan and D. J. Glover, *J. Am. Chem. Soc.*, **88**, 84 (1966).
30. S. S. Novikov, L. A. Nikonova, and V. I. Slovetskii, *Izv. Akad. Nauk SSSR, Otd. Khim. Nauk*, **1965**, 395.
31. A. D. Nikoleva, L. K. Popov, and G. K. Kamai, *Zh. Organ. Khim.*, **2**, 1369 (1966).
32. S. S. Novikov, K. K. Babievskii, S. A. Shevelev, I. S. Ivanova, and A. A. Fainzil'berg, *Izv. Akad. Nauk SSSR, Otd. Khim. Nauk*, **1962**, 1853.
33. S. S. Novikov, A. A. Fainzil'berg, S. A. Shevelev, I. S. Korsakova, and K. K. Babievskii, *Dokl. Akad. Nauk SSSR*, **124**, 589 (1959).

34. S. S. Novikov, A. A. Fainzil'berg, S. A. Shevelev, I. S. Korsakova, and K. K. Babievskii, *Dokl. Akad. Nauk SSSR*, **132,** 846 (1960).
35. M. J. Kamlet, J. C. Dacons, and J. C. Hoffsommer, *J. Org. Chem.*, **26,** 4881 (1961).
36. F. G. Borgardt, A. K. Seeler, and P. Noble, Jr., *J. Org. Chem.*, **31,** 2806 (1961).
37. S. S. Novikov, L. A. Nikonova, V. I. Slovetskii, and I. S. Ivanova, *Izv. Akad. Nauk SSSR, Ser. Khim*, **1965**, 1066.
38. R. N. King, G. C. Tesow, and D. R. Moore, *J. Org. Chem.*, **32,** 1091 (1967).
39. M. H. Gold, M. B. Frankel, G. B. Linden, and K. Klager, *J. Org. Chem.*, **27,** 334 (1962).
40. M. E. Hill, *J. Am. Chem. Soc.*, **75,** 3020 (1953).
41. A. I. Shreibert, N. V. Elaskov, A. P. Khardin, and V. I. Ermarchenko, *Zh. Organ. Khim.*, **3,** 1755 (1967).
42. H. Shechter, D. E. Ley, and L. Zeldin, *J. Am. Chem. Soc.*, **74,** 3664 (1952).
43. H. Feuer and T. Kucera, *J. Am. Chem. Soc.*, **77,** 5740 (1955).
44. L. A. Kaplan, Unpublished results.
45. C. S. Rondestvedt, Jr., M. Stiles, and A. L. Krieger, *Tetrahedron*, **19,** Suppl. 1, 197 (1963).
46. T. N. Hall, *Tetrahedron*, **19,** Suppl. 1, 115 (1963).
47. H. C. Brown, R. S. Fletcher, and R. B. Johannesen, *J. Am. Chem. Soc.*, **73,** 212 (1951).
48. T. N. Hall, *J. Org. Chem.*, **29,** 3587 (1964).
49. T. N. Hall, *J. Org. Chem.*, **30,** 3157 (1965).
50. H. Feuer, H. B. Hass, and R. D. Lowrey, *J. Org. Chem.*, **25,** 2070 (1960).
51. M. E. Hill, *J. Am. Chem. Soc.*, **76,** 2329 (1954).
52. K. Klager and J. P. Kispersky, *J. Am. Chem. Soc.*, **77,** 5433 (1955).
53. L. W. Kissinger, M. Schwartz, and W. E. McQuistion, *J. Org. Chem.*, **26,** 5203 (1961).
54. T. N. Hall, *J. Org. Chem.*, **33,** 4557 (1968).
55. M. E. Hill, Unpublished results.
56. K. G. Shipp and M. E. Hill, *J. Org. Chem.*, **31,** 853 (1966).
57. S. S. Novikov and G. A. Shvekhgeimer, *Izv. Akad. Nauk SSSR, Otd. Khim Nauk*, **1960,** 307.
58. M. E. Hill, U.S. Pat. 3,306,939 (1967); *Chem. Abstr.*, **66,** 104,676 (1967).
59. M. E. Hill, *J. Org. Chem.*, **30,** 411 (1965).
60. K. Baum, *J. Org. Chem.*, **33,** 1293 (1968).
61. F. Straus and H. Blankenhorn, *Ann. Chem.*, **415,** 232 (1918).
62. S. S. Novikov and G. A. Shvekhgeimer, *Izv. Akad. Nauk SSSR, Otd. Khim Nauk*, **1960,** 2026.
63. E. R. Alexander and E. J. Underhill, *J. Am. Chem. Soc.*, **71,** 4014 (1949).
64. W. B. Frankel and K. Klager, *J. Chem. Eng. Data*, **7,** 412 (1962).
65. R. Schenck and G. A. Wetterholm, U.S. Pat. 2,731,460 (1956); *Chem. Abstr.*, **50,** 7125 (1956).
66. W. J. Murray and C. W. Sauer, U.S. Pat. 3,006,957 (1961); *Chem. Abstr.*, **56,** 2330 (1962).
67. H. Feuer and W. A. Swarts, *J. Org. Chem.*, **27,** 1455 (1962).
68. H. Feuer and U. E. Lynch-Hart, *J. Org. Chem.*, **26,** 391 (1961).
69. H. Feuer and U. E. Lynch-Hart, *J. Org. Chem.*, **26,** 587 (1961).
70. H. Shechter and H. L. Cates, *J. Org. Chem.*, **26,** 51 (1961).
71. G. S. Hammond, W. D. Emmons, C. O. Parker, B. M. Graybill, J. H. Waters, and M. F. Hawthorne, *Tetrahedron*, **19,** Suppl. 1, 177 (1963).
72. F. Holahan, T. Castorina, J. Autera, and S. Helf, *J. Am. Chem. Soc.*, **84,** 756 (1962).
73. W. S. Reich, G. G. Rose, and W. Wilson, *J. Chem. Soc.*, **1947,** 1234.

74. N. Kornblum, R. A. Smiley, R. K. Blackwood, and D. C. Iffland, *J. Am. Chem. Soc.*, **77**, 6269 (1955).
75. R. C. Kerber, G. Urry, and N. Kornblum, *J. Am. Chem. Soc.*, **87**, 4520 (1965).
76. V. I. Slovetskii, V. A. Tartakovskii, and S. S. Novikov, *Izv. Akad. Nauk SSSR, Otd. Khim. Nauk*, **1962**, 1400.
77. S. S. Novikov, T. I. Godovikova, and V. A. Tartakovskii, *Izv. Akad. Nauk SSSR, Otd. Khim. Nauk*, **1960**, 505.
78. S. S. Novikov, T. I. Godovikova, and V. A. Tartakovskii, *Izv. Akad. Nauk SSSR, Otd. Khim. Nauk*, **1960**, 863.
79. S. S. Novikov, T. I. Godovikova, and V. A. Tartakovskii, *Izv. Akad. Nauk SSSR, Otd. Khim. Nauk*, **1960**, 669.
80. V. A. Tartakovskii, S. S. Novikov, and T. I. Godovikova, *Izv. Akad. Nauk SSSR, Otd. Khim. Nauk*, **1961**, 1042.
81. S. S. Novikov, V. A. Tartakovskii, T. I. Godovikova, and B. G. Gribov, *Izv. Akad. Nauk SSSR, Otd. Khim. Nauk*, **1962**, 272.
82. V. A. Tartakovskii, B. G. Gribov, and S. S. Novikov, *Izv. Akad. Nauk SSSR, Otd. Khim. Nauk*, **1965**, 1074.
83. S. S. Novikov, V. A. Tartakovskii, T. I. Godovikova, and B. G. Gribov, *Izv. Akad. Nauk SSSR, Otd. Khim. Nauk*, **1962**, 276.
84. N. S. Zefirov, L. P. Prikazchikova, and Y. K. Yur'ev, *Dokl. Akad. Nauk SSSR*, **152**, 869 (1963).
85. D. R. Levering, *J. Org. Chem.*, **27**, 2930 (1962).
86. R. B. Kaplan and H. Shechter, *J. Am. Chem. Soc.*, **83**, 3535 (1961).
87. C. W. Plummer, U.S. Pat. 2,991,315 (1961); *Chem. Abstr.*, **67**, 53,650 (1967).
88. L. Hunter, *J. Chem. Soc.*, **123**, 543 (1923).
89. C. O. Parker, W. D. Emmons, A. S. Pangano, H. A. Rolewicz, and K. S. McCallum, *Tetrahedron*, **17**, 79 (1962).
90. K. J. Orton and P. V. McKie, *J. Chem. Soc.*, **117**, 283 (1920).
91. G. Darzens and M. Levy, *Compt. Rend.*, **229**, 1081 (1949).
92. V. I. Slovetskii, V. A. Shlyapochnikov, K. K. Babievskii, A. A. Fainzil'berg, and S. S. Novikov, *Izv. Akad. Nauk SSSR, Otd. Khim. Nauk*, **1960**, 1709.
93. V. I. Slovetskii, V. A. Shlyapochnikov, S. A. Shevelev, A. A. Fainzil'berg, and S. S. Novikov, *Izv. Akad. Nauk SSSR, Otd. Khim. Nauk*, **1961**, 330.
94. M. J. Kamlet and D. J. Glover, *J. Org. Chem.*, **27**, 537 (1962).
95. D. J. Glover, *Tetrahedron*, **19**, Suppl. 1, 219 (1963).
96. M. J. Kamlet, L. A. Kaplan, and J. C. Dacons, *J. Org. Chem.*, **26**, 4371 (1961).
97. D. J. Glover, *J. Org. Chem.*, **29**, 990 (1964).
98. D. J. Glover and M. J. Kamlet, *J. Org. Chem.*, **26**, 4731 (1961).
99. M. S. Lv'ova, V. I. Slovetskii, and A. A. Fainzil'berg, *Izv. Akad. Nauk SSSR, Ser. Khim. Nauk*, **1966**, 649.
100. J. E. Leffler and E. Grunwald, *Rates and Equilibria of Organic Reactions*, John Wiley and Sons, New York, 1963, pp. 251 *et seq.*
101. D. J. Glover, *J. Phys. Chem.*, **74**, 21 (1970).
102. W. F. Sager and J. C. Hoffsommer, *J. Phys. Chem.*, **73**, 4155 (1969).
103. J. O. Edwards and R. G. Pearson, *J. Am. Chem. Soc.*, **84**, 16 (1962).
104. L. A. Kaplan and H. B. Pickard, Unpublished results.
105. L. Zeldin and H. Shechter, *J. Am. Chem. Soc.*, **79**, 4708 (1957).
106. H. G. Adolph and M. J. Kamlet, *J. Org. Chem.*, **34**, 45 (1969).
107. J. Meisenheimer, *Chem. Ber.*, **36**, 434 (1903).
108. J. Meisenheimer and M. Schwarz, *Chem. Ber.*, **39**, 2546 (1906).

109. M. J. Kamlet and J. C. Dacons, *J. Org. Chem.*, **26**, 3005 (1961).
110. N. E. Burlinson, L. A. Kaplan, W. B. Moniz, and C. F. Poranski, Jr., *Abstracts of Papers, 4th Middle Atlantic Regional Meeting of the American Chemical Society*, Feb. 1969, p. 74.
111. J. R. Holden and C. Dickenson, *J. Am. Chem. Soc.*, **90**, 1975 (1968).
112. L. A. Kaplan, *Abstracts of Papers, 4th Middle Atlantic Regional Meeting of the American Chemical Society*, Feb. 1969, p. 90.
113. K. Klager, J. P. Kispersky, and E. Hamel, *J. Org. Chem.*, **26**, 4368 (1961).
114. M. B. Frankel, *J. Org. Chem.*, **23**, 813 (1958).
115. L. J. Winters and W. E. McEwen, *Tetrahedron*, **19**, Suppl. 1, 49 (1963).
116. S. S. Novikov, V. M. Belikov, and V. F. Dem'yanenko, *Izv. Akad. Nauk SSSR, Otd. Khim. Nauk*, **1960**, 1295.
117. K. V. Altukhov, V. A. Tartakovskii, V. V. Perekalin, and S. S. Novikov, *Izv. Akad. Nauk SSSR, Ser. Khim.*, **1967**, 197.
118. K. V. Altukhov and V. V. Perekalin, *Zh. Organ. Khim.*, **3**, 2003 (1967).
119. V. A. Tartakovskii, I. E. Chlenov, S. S. Smagin, and S. S. Novikov, *Izv. Akad. Nauk SSSR, Seriya Khim.*, 583 (1964).
120. V. A. Tartakovskii, I. E. Chlenov, and G. V. Lagodzinskaya, and S. S. Novikov, *Dokl. Akad. Nauk SSSR*, **161**, 136 (1964).
121. V. A. Tartakovskii, O. A. Duk'yanov, N. I. Shlykova, and S. S. Novikov, *Zh. Organ. Khim.*, **4**, 231 (1968).
122. V. A. Tartakovskii, S. I. Ioffe, O. P. Shitov, S. S. Smagin and I. E. Chlenov, *Izv. Akad. Nauk SSSR, Ser. Khim.*, **1967**, 1614.
123. S. A. Shevelev, V. I. Erashko, B. G. Sankov, and A. A. Fainzil'berg, *Izv. Akad. Nauk SSSR, Ser. Khim.*, **1967**, 1630.
124. S. A. Shevelev, V. I. Erashko, B. G. Sankov, and A. A. Fainzil'berg, *Izv. Akad. Nauk SSSR, Ser. Khim.*, **1968**, 382.

CHAPTER **6**

Polynitroaromatic addition compounds*

THOMAS N. HALL

U.S. Naval Ordnance Laboratory, White Oak
Silver Spring, Maryland

and

CHESTER F. PORANSKI, JR.

U.S. Naval Research Laboratory
Washington, D.C.

* The authors thank Dr. H. G. Adolph and Dr. W. B. Moniz for helpful discussions.

I. INTRODUCTION

Polynitroaromatic compounds react with bases to form brightly colored solids or solutions. Although a definite and simple stoichiometry is often indicated by elemental analysis of the solid or spectroscopic studies of the solution, the exact nature of the bonding in the product or complex has been a controversy since 1882[1a]. Recently, considerable interest has been shown in the reaction of polynitroaromatic compounds with bases because the products are often the type of addition compound which has been proposed by Bunnett and others[1b] as an intermediate in activated nucleophilic aromatic substitution reactions.

The application of modern techniques and theory has shown that the products of the interaction of bases with polynitroaromatics fall into one of the classes of charge-transfer complexes, as defined by Mulliken[2a]. The four types of interaction which have been identified are: (1) addition of one or more molecules of the base to the nitroaromatic ring; (2) abstraction of a proton from the nucleus or substituent of the nitro compound; (3) transfer of an electron from the base to the polynitroaromatic compound, resulting in the formation of radical anions; and (4) formation of a donor–acceptor complex between neutral molecules, as defined by Briegleb[2b]. This chapter will treat the first three types of interaction. Various aspects of all four types of interaction have been reviewed.[2b–2e]

II. ALKOXIDE AND HYDROXIDE ION EQUILIBRIA

A. Alkoxide Ion Equilibria

1. Product structure

a. Early chemical studies. In 1895 Lobry de Bruyn reported the isolation and analysis of a red solid obtained from the reaction of a methanolic solution of *sym*-trinitrobenzene with a molar equivalent of potassium hydroxide[3]. The formula suggested for this solid was $[C_6H_3(NO_2)_3KOCH_3]_2{\cdot}H_2O$. Victor Meyer proposed that the color produced by the reaction of excess alkali on *sym*-trinitrobenzoic acid was due to nuclear proton abstraction[4a], and that de Bruyn's compound had formula **1**[4b]. Lobry de Bruyn argued that proton

$$C_6H_2K(NO_2)_3 + CH_3OH + \tfrac{1}{2}H_2O$$

(1)

abstraction was unlikely because boiling xylene solutions of *sym*-trinitrobenzene and other polynitroaromatics with sodium failed to evolve hydrogen[5]. Hantzsch and Kissel suggested that *sym*-trinitrobenzene and other nitroaromatics reacted with the calculated amount of potassium methoxide by addition to a nitro group, forming such structures as **2**[6].

$$(NO_2)_2C_6H_3{\cdot}N(=O)(-OK)(-OCH_3)$$

(2)

In 1898 Jackson and Boos reported that the products of the reaction of a variety of sodium alkoxides with picryl chloride analyzed for the composition $C_6H_2(NO_2)_3OR{\cdot}NaOR$ (R = methyl, ethyl, propyl, isopentyl and benzyl)[7]. In 1900 Jackson and Gazzolo suggested that the product of the reaction of methyl picrate with sodium methoxide had structure **3** or **4**[8]. They reasoned as follows.

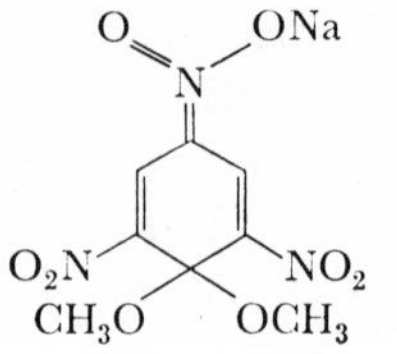

(**3**) 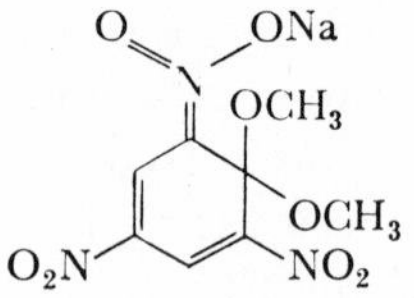 (**4**)

1. The two methyl groups were equivalent. They found that 'soaking' their compound in benzyl alcohol converted it to the corresponding dibenzyl derivative, while boiling the latter with methanol regenerated the original dimethyl derivative. Under comparable conditions they also established that methyl picrate was inert to benzyl alcohol, and benzyl picrate inert to methanol. Thus they argued that analogs of structure **2** were ruled out because such structures would permit substitution of only one methyl group by a benzyl group.

2. The intense color of their compound could be accounted for by the quinoid structure of **3** or **4** but not by analogs of **2** since a pale yellow color was reported for **5**.

3. Their compound was decomposed immediately by hydrochloric acid to methyl picrate, while acidification of **5** gave a stable acid.

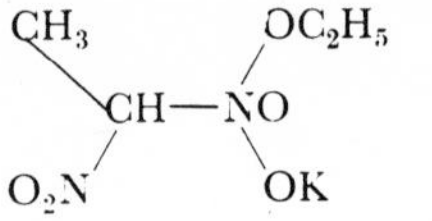

(**5**)

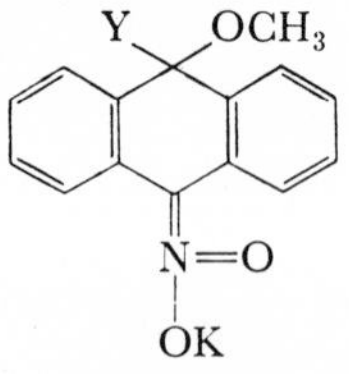

(**6a**), Y = H
(**6b**), Y = OCH_3

In 1902 Meisenheimer reported the isolation of potassium methoxide adducts of 9-nitroanthracene and 9-nitro-10-methoxyanthracene, and proposed structures **6a** and **6b** for these products[9]. Meisenheimer considered the structure of the dialkoxy derivatives reported by Jackson[7] to be analogous to **6**, and, apparently unaware of Jackson's earlier paper in 1900[8], proposed structure **7** for the dimethoxy derivative[9]. He ruled out structure **8** and the Hantzsch formulation **9** by isolating a greater than 50% yield of ethyl picrate by the acidification of either the potassium ethoxide adduct of methyl picrate or the potassium methoxide adduct of ethyl picrate. He reasoned that **8** and **9** should have given only methyl picrate. He

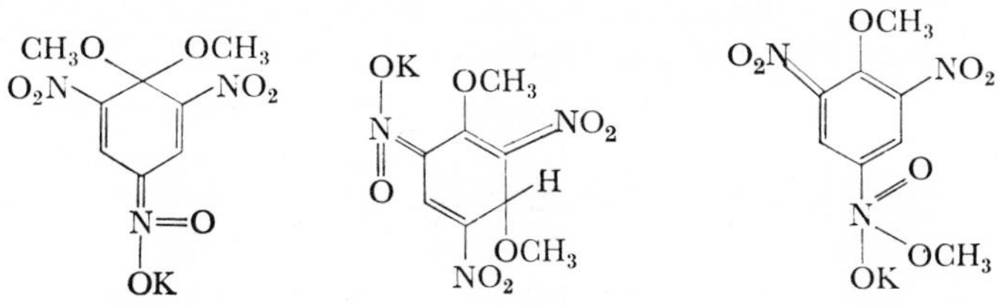

(7) (8) (9)

also found that acidification of the potassium ethoxide adduct of isobutyl picrate or the potassium isobutoxide adduct of ethyl picrate gave a mixture of ethyl and isobutyl picrates, with the latter predominating.

In 1903 Jackson and Earle[10], realizing that Meisenheimer had not made a complete product analysis, showed that acidification of the methyl picrate–sodium isopentoxide adduct gave nearly equal proportions of methyl and isopentyl picrates. They also showed that the product was not an equimolar mixture of the two possible symmetrical alkoxy adducts.

Thus it is clear that Jackson and Meisenheimer, apparently independently, arrived at essentially the same structural assignments for the alkoxide complexes of alkyl picrates, and that it would be fitting to designate these complexes as Jackson–Meisenheimer compounds rather than Meisenheimer compounds as is usually done.

b. X-Ray diffraction. Recently the results of three crystal structure determinations of picryl ether–metal alkoxide complexes confirm the Jackson–Meisenheimer structural assignment[11,12], with the exception, of course, that the metal appears as a cation and the nitroaromatic moiety carries a formal negative charge.

Since the most reliable parameters were found for the ethyl picrate–potassium ethoxide complex ($r = 0.064$)[11], the parameters for this complex will be presented here. The C-1 carbon of the ring is tetrahedral, the C-1–C-2, and C-1–C-6 bonds being 1.514 Å and the C-2–C-1–C-6 angle being 107.8°. The C-2–C-3 and C-5–C-6 bonds are shortened from the normal aromatic length to 1.347 Å, accommodating the hybridization change at C-1. The ring and the substituent nitro groups are essentially coplanar; the nitro groups are not extensively rotated with respect to the plane of the ring as they are in the parent ether[13]. The C–N bond at C-4 is shorter than the other two equivalent C–N bonds: 1.390 Å *vs.* 1.449 Å. The N–O distance for the nitro group bonded to C-4 is 1.246 Å *vs.* 1.226 Å for the other N–O distances. Using the structural parameters for this complex and the parent ether[13] Destro *et al.*[11], calculated that complexing ethyl picrate with ethoxide ion decreases the electronic charge of the ring from 5.64 to 4.33 π-electrons (cf. ref 14).

c. Nuclear magnetic resonance. In 1964 Crampton and Gold reported the chemical shifts for the methyl and aromatic protons of methyl picrate and its potassium methoxide complex in dimethyl sulfoxide solution: methyl picrate, 4.07 ppm (3) and 9.07 ppm (2), respectively; methyl picrate–potassium methoxide complex, 3.03 ppm (6) and 8.64 ppm (2), respectively[15]. Chemical shifts are expressed in parts per million of the applied field using tetramethylsilane as an internal reference and the numbers in parentheses are the relative intensities. These workers note that the changes in chemical shifts and relative intensities produced by complex formation are compatible with an assignment of **10** as the structure of the

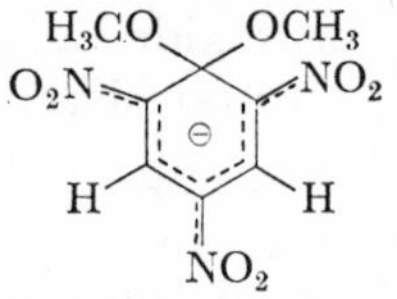

H_3CO OCH_3
O_2N NO_2
$\ominus$
H H
NO_2

(**10**)

complex, i.e., the Jackson–Meisenheimer formulation with the nitroaromatic carrying a formal negative charge. The negative charge increases the screening of the protons present in the anionic species, resulting in a shift of the resonances to higher fields. The absence of spin–spin coupling in the spectrum of the complex provides evidence for the presence of an equivalent pair of ring protons and an equivalent pair of methoxy groups. Crampton and Gold point out that a donor–acceptor complex between methoxide ion and methyl picrate with rapid exchange between the methoxide ion and the methoxy group could also rationalize the observed data. They ruled this possibility out by noting that solutions containing both methyl picrate and the methyl picrate–methoxide complex showed separate resonances for the two species.

Crampton and Gold also compared the spectrum of *sym*-trinitrobenzene with that of its monopotassium methoxide complex in dimethyl sulfoxide[15]. The single resonance of *sym*-trinitrobenzene was at 9.21 ppm. In the complex a doublet at 8.42 ppm (2), a broad peak at 6.14 ppm (1), and a sharp singlet at 3.10 ppm (3) were observed. The spectrum of the complex is consistent with structure **11**. The resonance at 6.14 ppm, in the methynyl proton region, is clearly due to a proton bonded to a carbon which has been made tetrahedral by methoxide addition. These authors attribute the broadness of the signal to unresolved spin–spin coupling to the two other ring protons. The resonance from these protons occurs at 8.42 ppm and is a doublet due to spin–spin coupling with the methynyl proton. The resonance at 3.10 ppm is due to the protons

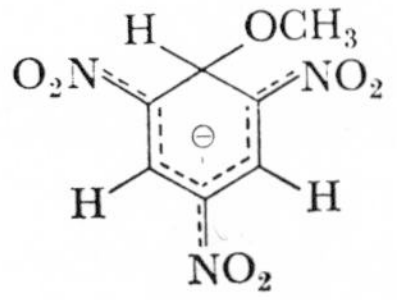

H OCH$_3$
O_2N NO_2
⊖
H H
NO_2

(**11**)

of the methoxy group. Crampton and Gold pointed out that the observed spectrum is not compatible with that expected from a species formed by abstraction of a ring proton or with that of a complex in which the equivalence of the ring protons is preserved.

Following Crampton and Gold's study there appeared many papers on the nuclear magnetic resonance of alkoxide interactions with polynitroaromatics[16–27]. Much of the data was repetitious. Two points are worth making concerning these papers. (1) One should not assume that the *initial* addition of alkoxide to a solution of a polynitroaromatic will necessarily generate the same species as is precipitated from such a solution. For example, Servis has shown that the addition of 1 equivalent of sodium methoxide to methyl picrate in dimethyl sulfoxide generates species **12** initially; after 15

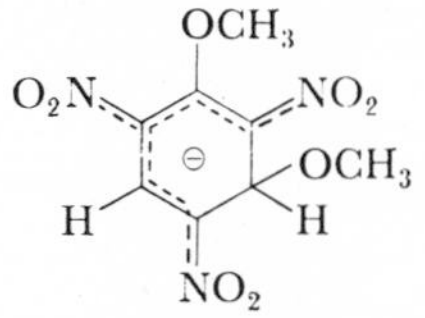

OCH$_3$
O_2N NO_2
⊖ OCH$_3$
H H
NO_2

(**12**)

minutes standing the initial spectrum has changed to that of **10**, the thermodynamically more stable species[17,23], and the anion of the species which has been isolated from solution. Addition of more than 1 equivalent of sodium methoxide to this 'aged' solution generates species **13**. Earlier, Foster and Fyfe had reported these spectral changes as the direct conversion of methyl picrate to **10** and then to **13**[16]. (2) Picramide and its *N*-alkyl and *N*-aryl derivatives are

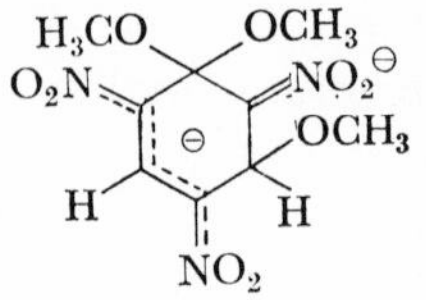

(**13**)

attacked by alkoxide at the 3 and/or 5 positions[21,23]; attack at the 1 position has never been observed. Ionization of the NH proton usually competes with this addition and in some cases the ionized 3-adduct is formed in excess base, *e.g.*, **14**[21,23]. Table 1 gives the chemical shifts

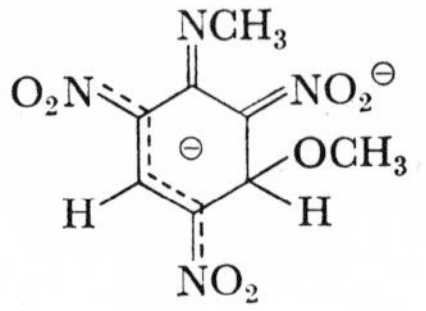

(**14**)

of the ring protons for the 3-methoxide adduct of a variety of picryl derivatives. The chemical shift of the methynyl proton, H_β, is nearly independent of structure. In this connection, the adduct of 9-nitroanthracene, **6a**, shows the methynyl proton resonance at

TABLE 1. Chemical shifts of the 3-methoxide adduct of some picryl derivatives.

R in 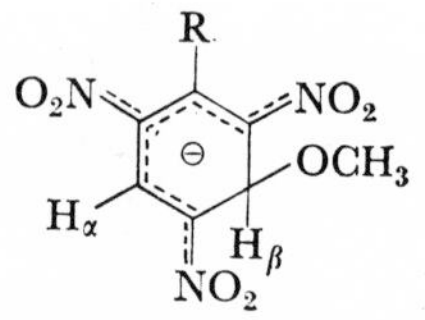	Average δ for ring protons[a]		References
	H_α	H_β	
H	8.45 (8.42)	6.16 (6.14)	15–17 (21)
OCH_3	8.42 (8.48)	6.13 (6.20)	16, 17, 25 (21)
NH_2	8.43 (8.61)	6.14 (6.14)	17, 23 (21)
$NHCH_3$	8.50 (8.48)	6.14 (6.16)	17, 23 (21)
$N(CH_3)_2$	8.49 (8.46)	6.18 (6.17)	23 (21)
NH^-	8.67	6.06	23
$N(CH_3)^-$	8.70 (8.64)	6.16 (6.10)	23 (21)
$N(C_6H_5)^-$	8.71 (8.68)	6.17 (6.18)	23 (21)

[a] In parts per million (ppm) from internal $Si(CH_3)_4$ reference. Solvent is $(CH_3)_2SO$ except for the values in parentheses for which the solvent is 50–50 mole % $(CH_3)_2SO$–CH_3OH.

Table 2. Chemical shifts of some dimethoxy- and ethylenedioxycyclohexadienides.

R_1	R_2[a]	Average δ[b] 3-H	R_1	R_2	References
NO_2	H	8.67 (8.51)	—	8.67 (8.51)	14–17, 25 (14)
CN	H	8.74	—	8.29	14
H	H	8.70 (8.55)	5.08 (5.30)	7.25 (6.83)	14, 20, 22 (24)
C_4H_4[c]		9.33 (9.06)	—	—	25, 26a (26b)

[a] In the formulas ... or

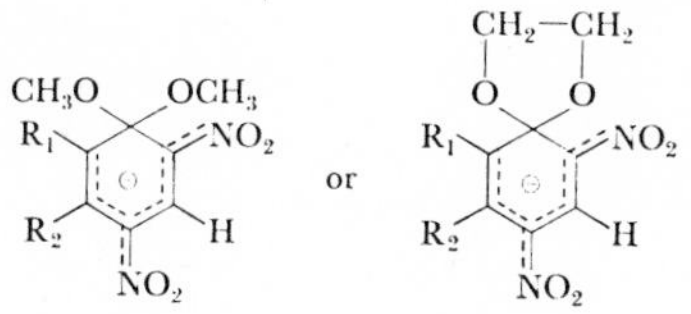

[b] Chemical shift in parts per million (ppm) from internal $Si(CH_3)_4$ reference in $(CH_3)_2SO$. Value in parentheses is for the spiro ether. [c] 1-Methoxide adduct of 1-methoxy-2,4-dinitronaphthalene.

4.96 ppm[25]. Table 2 gives the chemical shifts of the ring protons of some nitroaromatic complexes in which methoxide attack has occurred at a carbon bearing a methoxy group, or which have been formed by cyclization of a hydroxyethyl ether.

d. Infrared spectroscopy. For a variety of picryl ethers, 1-alkoxy-2,4-dinitrobenzenes and 1-alkoxy-2,4-dinitronaphthalenes the infrared absorption spectrum of the alkoxide adduct is consistent with that expected for addition of methoxide to the 1-carbon, and is distinctly different from that of the parent ether.[22,26a,28,29] This difference in infrared spectra has been used as evidence that these adducts are not donor–acceptor complexes[22,26a,28]; spectra of the latter have been shown to resemble closely the sum of the donor and acceptor spectra[30]. Several new, strong bands appear in the ketal region of the adduct spectra; these bands are absent in the spectra of the parent ethers.[19,22,28,29] Both the symmetrical and asymmetrical N–O stretching frequencies of the picryl ether adducts appear at about 50 cm^{-1} less than the corresponding bands of the parent ether[28,29].

e. Electron spin resonance. Solutions of *m*-dinitrobenzene in potassium *t*-butoxide–*t*-butyl alcohol[31a], potassium *t*-butoxide–*t*-butyl alcohol–dimethyl sulfoxide[31b], and basic acetonitrile[32] have been shown to contain the radical anion of *m*-dinitrobenzene. The spectrum observed for a 0.01 *M* solution of *m*-dinitrobenzene in dimethyl sulfoxide–*t*-butyl alcohol (80:20), containing 0.005 *M* potassium *t*-butoxide, agrees well with the theoretical spectrum calculated for the radical anion of *m*-dinitrobenzene[31b]. For this radical anion, the following values of the hyperfine coupling constant (a measure of the interaction between the nucleus and the unpaired electron) were reported: $a_N = a_{4,6-H} = 4.28$ gauss, $a_{2-H} = 3.10$ gauss, $a_{5-H} = 1.05$ gauss[31b]. The radical anion of *m*-dinitrobenzene can also be generated by reaction with a variety of carbanion donors[31b].

In contrast, electron transfer to *sym*-trinitrobenzene is not important in potassium *t*-butoxide–*t*-butyl alcohol–dimethyl sulfoxide[31b]. In potassium *t*-butoxide–*t*-butyl alcohol, 2,4-dinitrotoluene, and a variety of *o*- and *p*-nitroalkylbenzene derivatives were found to form radical anions[31b]. 2,4-Dinitrotoluene apparently forms radical anions by electron transfer between ionized and unionized 2,4-dinitrotoluene molecules.

f. Isotopic exchange. Crampton and Gold have shown that the colored species generated reversibly by the reaction of sodium methoxide and *m*-dinitrobenzene in tritiated methanol–dimethyl sulfoxide is unreactive in the process of exchanging the 2-hydrogen of the dinitrobenzene[33]. They found that the maximum exchange rate was considerably less than the rate of color formation and that the dinitrobenzene recovered from a tritiated solution containing sufficient sodium methoxide to produce the maximum extinction coefficient still contained tritium after quenching with protic acid.

The hydrogen-exchange reactivity of *sym*-trinitrobenzene is considerably less than that of *m*-dinitrobenzene. The rate of exchange of the protons of *sym*-trinitrobenzene is extremely sensitive to the conditions used. This compound fails to undergo hydrogen exchange in 8 *M* sodium hydroxide in D_2O[34], and in pyridine–D_2O[35,38], but shows considerable deuterium enrichment in 0.02 *M* sodium hydroxide in C_2H_5OD[36] and in 0.01 *M* sodium deuteroxide in dimethylformamide–D_2O(90:10)[37]. In the latter solvent the trinitrobenzene was 93.8% deuterated after a 24-hour reaction period at room temperature, as compared to the equilibrium value of 96.9%[37]. Buncel and Symons suggest that a dipolar solvent such as dimethylformamide increases the nucleophilicity of OD^- and the exchange rate[37]. In contrast, *sym*-trinitrotoluene dissolved in 0.1 *M* sodium deuteroxide in dimethylformamide-D_2O(90:10) exchanges its methyl protons, but not its ring protons at room temperature.[39]

g. Acidity function correlations. The ionization ratios for proton abstraction from an aromatic hydrocarbon, ArH, should correlate with the acidity function H_-[40]. In equation 1, K_{ArH} is the ionization constant for ArH. Some workers have used the symbol H_M in place

$$H_- \equiv \mathrm{p}K_{\mathrm{ArH}} + \log\,([\mathrm{Ar}^-]/[\mathrm{ArH}]) \tag{1}$$

of H_- (see section II.A 2b). The ionization ratios for addition of methoxide ion should be correlated with the acidity function J_-[41,42].

$$J_- \equiv \mathrm{p}K + \log\,([\mathrm{ArH{\cdot}OCH_3}^-]/[\mathrm{ArH}]) \tag{2}$$

In equation 2, K is the equilibrium constant for the reaction shown

$$\mathrm{ArH} + \mathrm{CH_3OH} \rightleftharpoons \mathrm{ArH{\cdot}OCH_3}^- + \mathrm{H}^+ \tag{3}$$

in equation 3, and the ionization ratio refers to the equilibrium shown in equation 4.

$$\mathrm{ArH} + \mathrm{CH_3O}^- \rightleftharpoons \mathrm{ArH{\cdot}OCH_3}^- + m\mathrm{CH_3OH} \tag{4}$$

2,4-Dinitroaniline, for which the primary reaction with methoxide ion is ionization of the amino proton[21], has been used to set up the H_- acidity scale[43,44]. Spectrophotometric data indicate that another reaction may be important above 2 M methoxide concentration. Examples of acidity function correlations for nitroaromatic–alkoxide ion equilibria may be found in Rochester's work[45]. Consecutive equilibria for reactions of methoxide ion with picrate ion and with the 1-methoxide adduct of methyl picrate follow J_- rather than H_-.

2. Equilibrium spectrophotometric measurements

a. Inherent problems. Many papers have appeared in the last 15 years concerning the electronic spectra of basic solutions of polynitroaromatic compounds. Spectral envelopes have often been used to 'identify' the type or types of species present in such solutions. For example, the similarity of the spectral envelopes of *sym*-trinitrobenzene and methyl picrate in methanolic sodium methoxide has been used as evidence that the trinitrobenzene forms the 2-methoxide adduct[46a]. Spectrophotometric measurements of the absorptivities of alkoxide solutions of polynitroaromatic compounds have been used to establish the thermodynamic and pseudo-thermodynamic quantities for the reactions involved. Unfortunately, the very nature of the reactivity of polynitroaromatic compounds often poses problems which make these quantities semiquantitative at best and their interpretation equivocal. Some of the problems are as follows.

(1) Decomposition of the reactant and/or the product. It is well known that a substituent, such as a halogen, of a polynitroaromatic

compound can undergo facile nucleophilic displacement. Unfortunately, a rather general reaction of a polynitroaromatic compound is nucleophilic displacement of a nitro group by alkoxide ion[46–49], and by hydroxide ion[50–52]. Many of these reactions are light catalyzed[46c,50b–d,52]. *N,N*-Dimethylpicramide reacts with methoxide ion to form methyl picrate[46b] and with hydroxide ion to form picrate ion[50a]. Needless to say, ethers of di- and trinitrophenols undergo hydrolysis in the presence of hydroxide ion[54] and transetherification in the presence of alkoxide or phenoxide ion[55–58].

(2) Competing equilibria. Changes of the shape of the spectral envelope with time or base concentration are hard to detect because many distinctly different species have similar absorption characteristics. Such changes may well escape detection if only a narrow band is monitored.

(3) Low concentration of base. The concentration of methoxide ion must be rather low in order to study the equilibria of the more reactive nitroaromatic compounds, such as methyl picrate. Unless buffer solutions are used, determinations at these low concentrations of methoxide ion are clearly subject to systematic errors.

(4) High concentration of base. Equilibria involving the less reactive dinitroaromatics, such as dinitroanisole, must be studied at such high base concentrations that acidity functions are required. Thus, the exact interpretation of the equilibrium constant depends on the appropriateness of the acidity function chosen and the slope of the ionization ratio–acidity function plot.

(5) Sequential equilibria. The equilibrium constant for a polynitroaromatic compound with alkoxide will be incorrect if sequential equilibria escape detection. Indirect calculations of extinction coefficients are particularly susceptible to the effects of sequential equilibria since reaction with a second methoxide ion produces a strong hypsochromic shift in λ_{max} (see Table 3).

The direct determination of the extinction coefficient for the product of an alkoxide–polynitroaromatic reaction is usually impossible because degradation and/or sequential equilibria very often occur at the base concentration needed for the complete conversion of the nitroaromatic to the product. The usual expedient is the indirect calculation of the extinction coefficient by an extrapolation procedure based on the Ketelaar equation[59] or the Benesi–Hildebrand equation[60]. Using the latter equation, one can determine the extinction coefficient for the 1:1 complex, C, formed by a polynitroaromatic, ArH, and alkoxide ion, RO^-, by plotting the reciprocal of the apparent molar extinction coefficient of C *vs*. the reciprocal of the alkoxide ion concentration. A linear plot should be obtained if ArH and RO^- do not absorb at the wavelength in question and if the formal concentration of ArH is much less than that of the alkoxide ion. The intercept on such a plot at $1/[RO^-] = 0$ is the reciprocal of the true molar extinction coefficient of C, ε, and the slope is $1/K\varepsilon$, where K is the equilibrium constant for the reaction in concentration units. Many authors consider that the stoichiometry of the reaction has been established as 1:1 if a linear Benesi–Hildebrand plot has been obtained. More complex stoichiometry can be established by the method of continuous variation developed by Job[61]. The occurrence of an invariant isosbestic point[62,63] is good evidence that competing or sequential equilibria are absent. Finally it should be mentioned that for some reactions it is advantageous to determine the stoichiometry kinetically and the equilibrium constant as the ratio of the forward and reverse rate constants.

b. pK values for polynitroaromatic compounds in alkoxide solution. Table 3 gives the pK values and the electronic spectral data reported for polynitroaromatic compounds in methanolic methoxide solution. Methanol is the solvent most frequently used for quantitative spectrophotometric study of alkoxide–polynitroaromatic equilibria. Taking the data of Table 3 at face value one notes:

(1) the pK of a picryl derivative is always less than that of the corresponding 2,4- or 2,6-dinitrophenyl derivative;

(2) the pK's for 1-substituted 2,4,6-trinitrobenzenes increase in the order (picryl)NH $\ll$ $OCH_3 \leq (C_6H_5)NH < NH_2 < H < N(CH_3)_2 = CH_3$;

(3) the pK's of 1-substituted 2,4-dinitrobenzenes increase in the order $NHN{=}CR_1R_2 < (C_6H_5)NH < NH_2 < OCH_3$;

(4) 2,6-dinitroanisole is slightly more acidic than 2,4-dinitroanisole;

(5) the methoxide adducts of methyl picrate and 1-methoxy-2,4-dinitronaphthalene have about the same stability, implying that the stabilization of the adduct by a nitro group and by a fused ring is about the same;

(6) the product of the primary equilibrium usually absorbs at a longer wavelength than the products generated by further reaction of the primary product.

However, it is dangerous to draw definite conclusions from the data of Table 3 because one does not know for sure what products are

TABLE 3. Summary of the p*K*'s, acidity function correlations, and electronic spectral data for some di- and trinitroaromatics in methanol[a].

Nitroaromatic[b]	p*K* (°C)	Acidity function	λ_{max}, mμ(log ε) for product[c]	Isosbestic pt., mμ(log ε)	Ref
Amines					
Pi_2NH	4.21 (20)		420 (4.42)		64b
PiNHPh	13.60 (20)		435	390 (4.07)	65a
$PiNH_2$	15.34 (20)		410 (4.35)	335	65a
$PiNH_2$	15.34 (20)		400 (4.47)	335	46b
$PiN(CH_3)_2$	16.07 (25)				46b
DiNHPh	17.16 (20)	H_M	510 (4.14) 405	300 (3.20) 380 (4.10)	65b
$DiNH_2(1)$	18.15 (25)	H_-	515 *383*	299 (3.55)	45
$DiNH_2(1)$	18.35 (20)	H_M	510 (4.00) 390		65b
$DiNH_2(2)$	21.06 (25)	J_-	326 (4.34)	365 (4.12)	45
Ethers					
$PiOCH_3(1)$	13.03 (25)		410 (4.38) 486 (4.21) 250 (3.94)		53
$PiOCH_3(1)$	13.58 (25)[d]				66
$PiOCH_3(2)$	19.81 (25)	J_-	480	431 (4.10)	45
$PiOCH_3(3)$	19.80 (25)	H_-	299		45
1-Methoxy-2,4-dinitro-naphthalene	14.56 (25)		495 (4.47)		26a
4-Cyano-2,6-dinitroanisole	16.53 (43)[e]		531.2 353.4		67
2,6-Dinitro-anisole(1)	19.01 (20)	H_M	595 (4.40) 350 (3.47) 300 (3.90)		68
2,6-Dinitro-anisole(2)	20.78 (20)	H_M	305 (4.26)		68
$DiOCH_3(1)$	20.24 (25)	J_-	*500* (4.33) 345	316 (3.83)	45
$DiOCH_3(1)$	19.62 (20)	H_M	500 (4.41) 340 (4.12)	315 (3.80)	68
$DiOCH_3(1)$	20.48 (25)[f]		495		69
$DiOCH_3(1)$	21.22 (25)[d]				69
$DiOCH_3(2)$	21.66 (25)	J_-	302 (4.30)	329 (3.96)	45
$DiOCH_3(2)$	22.24 (20)	H_M	305	330 (3.98)	68
$PiO^-(1)$	20.25 (25)	J_-	470	394 (4.00)	45
$PiO^-(2)$	17.69 (25)		394 (4.40)	408 343 292	45

TABLE 3—*Continued*

Nitroaromatic[b]	pK (°C)	Acidity function	λ_{max}, mμ(log ε) for product[c]	Isosbestic pt., mμ(log ε)	Ref
Misc					
PiH	15.71 (20)		500 (4.26) *425**		65a
PiH	15.73 (28)		500* *425* (4.49)		46a
DiNHN= $CH(CH_2)_2CH_3$	15.92 (20)		500* *450**	415	65a
$PiCH_3$	16.07 (20)		*520* (4.09) 420*		65a
sym- Trinitroxylene	17.20 (20)		610		65a

[a] All pK values were determined from equilibrium spectrophotometric measurements unless otherwise indicated. The equilibria of the nitroaromatic SH with methanol are either for deprotonation of the ring or substituent of SH or for the formation of an addition compound, i.e., $SH + CH_3OH \rightleftharpoons S^- + CH_3OH_2^+$ or $SH + CH_3OH \rightleftharpoons (SH{\cdot}CH_3O)^- + H^+$. The pK values for these equilibria were calculated from the equilibrium constants for the reaction of SH with CH_3O^- and the autoprotolysis constant for methanol, $1.2 \times 10^{-17}(20°)$, used by Schaal[64]. [b] The following abbreviations are used: Pi = 2,4,6-trinitrophenyl, Di = 2,4-dinitrophenyl, Ph = phenyl. The numbers in parentheses following the compound name refer to successive equilibria with methanol. For example, the primary equilibrium of a nitroaromatic with methanol to form the product C_1 is designated by (1), the equilibrium of C_1 with methanol to form C_2 by (2), and the equilibrium of C_2 with methanol to form C_3 by (3). [c] For the spectral region 300–350 mμ to 550–600 mμ. An italicized λ_{max} value indicates the strongest peak of those reported. An asterisk after the λ_{max} value designates that the value was estimated from a plot. [d] Calculated from spectrophotometric measurements of the forward and reverse rate constants. [e] Nmr determination. [f] 0.2 *M* sodium methoxide.

formed *under the spectrophotometric conditions*. The ethers probably do form the 1-methoxide adducts since nuclear magnetic resonance studies have shown that the product isolated from the reaction of methoxide ion with methyl picrate and with 4-cyano-2,6-dinitroanisole in methanol is the 1-adduct[67]. Nuclear magnetic resonance studies have also shown that the 1-methoxide adduct is the product isolated from the reaction of 2,4-dinitroanisole and methoxide ion in methanol–dioxane, and from the reaction of methoxide ion with 1-methoxy-2,4-dinitronaphthalene in methanol–benzene[22,26a]. The importance of the solvent and aging time in determining the products formed by polynitrophenyl ethers is emphasized by Servis who reported that in dimethyl sulfoxide the 3-methoxide adduct of methyl picrate is formed initially, but rearranges to the more stable 1-adduct[23]. Those equilibria which follow the H_- acidity function can be assumed to involve proton transfer. Thus, p$K(1)$ for 2,4-dinitroaniline can be assumed to refer to the ionization of the NH proton. In the case of the dinitroanisoles, p$K(1)$ should be for methoxide addition and the associated equilibria should follow J_-. It is interesting that both J_- and H_M have been used to describe the primary equilibrium of 2,4-dinitroanisole.

The p$K(1)$ for picramide may refer to the ionization of the NH proton, the addition of methoxide ion, or possibly a combination of both reactions, since these two modes of interaction occur simultaneously in 50–50 mole % methanol–dimethyl sulfoxide containing methoxide ion[21]. In the latter solvent, *N,N*-dimethylpicramide has been shown to form a very stable 3-methoxide adduct[21]. By inference, then, the more acidic *N*-phenylpicramide can be assumed to undergo ionization of the NH proton as its initial reaction with methoxide ion.

c. Thermodynamic functions for alkoxide–polynitroaromatic equilibria. Table 4 gives values of $\Delta F°$, ΔH, and $\Delta S°$ reported for alkoxide–polynitroaromatic equilibria. The enthalpy change is small for all the equilibria listed, the absolute values all being less than 6 kcal/mole. All $\Delta S°$ values are positive except that for reaction iv. An important factor which makes the entropy change favorable for these reactions is the net desolvation which accompanies these reactions. The alkoxide ion, whose negative charge is essentially confined to the oxygen, requires much more specifically oriented solvent molecules than the product, whose negative charge is well dispersed. For the reactions given in Table 4 it is clear that proportionality between ΔH and $\Delta S°$ does not exist, even if the ethers and non-ethers are considered separately. These reactions are not unusual in this respect, however, because, as Leffler has pointed out, reactions which form or destroy carbanions usually fail to exhibit such proportionality[75].

For the methyl ethers one observes the following. $\Delta S°$ is nearly zero (although positive)—all values are in the range 4 ± 3 eu (except reaction iv). The stability of the products is, therefore, almost completely determined by ΔH. The ΔH (and $\Delta H^{\ddagger}$) values for reactions ii, iii, and v agree well with those calculated by Miller[76] on the basis of appropriate bond dissociation energies, electron affinities, and solvation energies. For these reactions the exothermicity is ordered according to the resonance stabilization expected for the addition products.

For the non-ethers the stability of the product is favored by the entropy change but not by the enthalpy change. The $\Delta S°$ values

TABLE 4. Thermodynamic and pseudothermodynamic functions for alkoxide–polynitroaromatic equilibria.

Reaction number	Reactants[a]	$\Delta F°$, kcal/mole	ΔH, kcal/mole	$\Delta H^‡$, kcal/mole[b]	$\Delta S°$, eu	Ref
i	$PiOCH_3 + CH_3O^-$ (H_2O)	−6.14	−5.1	12.4	4	70
ii	$PiOCH_3 + CH_3O^-$ (slow)	−4.56	−2.8[c]	9.5	6.7	66[d]
iii	1-Methoxy-2,4-dinitronaphthalene + CH_3O^-	−3.21	−2.7	13.2	1.0	26a
iv	$PiOCH_3 + C_2H_5O^-$ (fast)	(−1.1)[e]	−3.3	9.8	−7.5[f]	72a
v	$DiOCH_3$ + 0.2 *M* CH_3O^-	+4.86	5.6	16.8	2.7	69
vi	$PiNH_2 + C_2H_5O^-$	−4.84	2.0	11.1	23	73
vii	$PiH + C_2H_5O^-$	−4.61	0.3	11.1	16.5	72b
viii	$PiCH_3 + C_2H_5O^-$ (slow)	−4.50	3.6	13.0	27	73
ix	$PiCH_3 + C_2H_5O^-$ (fast)	(2.36)[e]	4.2	11.8	22[f]	72c
x	$PiCH_3$ + 3-$CH_3C_6H_4O^-$ (C_2H_5OH)	+0.1	6.5	15.7	21	74

[a] Solvent is protonated alkoxide unless given in parentheses. Abbreviations used: Pi = 2,4,6-trinitrophenyl, Di = 2,4-dinitrophenyl. [b] For formation of the product. Values of $\Delta S^‡$ (formation), and $\Delta H^‡$ and $\Delta S^‡$ for decomposition of the product are given in the references cited. [c] Direct calorimetric determination gives −7.15 kcal/mole[71]. [d] See also ref 53. [e] Calculated as the ratio of the forward and reverse rate constants. [f] Based on an extrapolation of low-temperature data.

fall into the range 22 ± 5 eu and are significantly more positive than the values found for the ethers.

Why should the reaction of methyl picrate and methoxide ion, reaction ii, be more exothermic by 3.1 kcal/mole than the reaction of *sym*-trinitrobenzene with ethoxide ion, reaction vii? An alkoxide addition product is very probably formed in each case, in the 1 position of methyl picrate and in the 2 position of *sym*-trinitrobenzene. A possible explanation for this exothermicity difference is the electronegativity effect which Hine has suggested to explain the increase in C–F bond strength when the carbon hybridization is changed from sp^2 to the less electronegative sp^3 state[77]. Applied to reactions ii and vii, this rationale would say that conversion of an aromatic carbon bonded to oxygen to a tetrahedral carbon bonded to oxygen, reaction ii, is more exothermic than a similar conversion involving a carbon bonded to hydrogen, reaction vii, because oxygen is more electronegative than hydrogen[78].

Why does $\Delta S°$ favor reaction vii much more than reaction ii? A possible explanation is that the non-bonded interactions between the alkoxy carbon and the oxygens of the nitro groups limit the rotational freedom of the adduct formed by reaction ii. Such interactions may well be significant because the X-ray diffraction structure determination of the ethoxide adduct of ethyl picrate shows that the ring and substituent nitro groups are coplanar[11].

3. Kinetic studies

The rates of reaction of alkoxide ion with polynitroaromatic compounds have been measured spectrophotometrically in order to establish the stoichiometry and activation parameters. Table 4 gives the values for enthalpy of activation, i.e., $\Delta H^{\ddagger}_{\text{formation}}$, for ten reactions. Values of $\Delta S^{\ddagger}_{\text{formation}}$, $\Delta H^{\ddagger}_{\text{decomposition}}$, and $\Delta S^{\ddagger}_{\text{decomposition}}$ are available in the references cited. There is no simple relation between ΔH and $\Delta H^{\ddagger}_{\text{formation}}$ for the reactions of the ethers with methoxide ion or for the reactions of the non-ethers with ethoxide ion. It might be pointed out that if ΔH and $\Delta H^{\ddagger}$ are to be correlated for reactions ii, iii, and v, ΔH for ii should be more exothermic.

sym-Trinitrotoluene and methyl picrate each undergo two reactions with ethoxide ion: a rapid reaction (ix and iv) which can be followed at low temperatures, and a slow reaction (viii and ii) which is normally associated with these compounds. In the case of *sym*-trinitrotoluene, the fast reaction, ix, requires high trinitrotoluene concentrations and produces a brown solution, while the slow reaction, viii, occurs at normal spectrophotometric concentrations (10^{-4} to 10^{-5} M) and produces the characteristic purple solutions. Caldin, *et al.* have discussed the nature of the fast and slow reactions and have proposed that reaction iv, and possibly ix, yield a donor–acceptor complex or an ion–dipole complex[72c,79]. The recent work by Servis[23] suggests, however, that reaction iv yields the 3-ethoxide adduct of methyl picrate.

Rates of decomposition of the products of several of the reactions listed in Table 4 have been measured in acid solution. The decomposition rates follow the Brønsted catalysis relation for the products formed by reactions iv ($\alpha = 0.56$), vii ($\alpha = 0.67$), viii ($\alpha = 0.40$), and x ($\alpha = 0.44$, 0.84 for phenols). The decomposition rate for the product of reaction viii, considering ethanol as an acid, follows the Brønsted relation. This is not true for the product of reaction vii.

The four α values are significantly less than 1, indicating that the decomposition involves a rate determining proton transfer.[72a,b,73,74] The rates of uncatalyzed decomposition of a variety of alkoxide adducts of several alkyl ethers of picric acid have been measured in water[80]; the activation energies were 17–19 kcal/mole and the log A values were 9–10. The rates of decomposition of the ethoxide adduct of ethyl picrate have been determined for pressurized aqueous dimethyl sulfoxide solutions[81]; for the uncatalyzed reaction the activation parameters obtained are consistent with a bimolecular mechanism ($\Delta V^* = -5.6$ cm^3/mole, large negative ΔS^*); for the acid-catalyzed reaction the activation parameters obtained ($\Delta V^* =$ 18 cm^3/mole, slightly positive ΔS^*) are consistent with the A1 mechanism.

The nature of the slow reaction of *sym*-trinitrotoluene with alkoxides and other bases is controversial. At the outset, two points should be made: (1) alkoxide solutions of trinitrotoluene are not stable at room temperature; (2) the major species generated by the action of alkoxide on trinitrotoluene has not been identified by an unequivocal method (such as nuclear magnetic resonance spectroscopy, for example). Although Servis failed to observe the nuclear magnetic resonance spectrum of trinitrotoluenide in dimethyl sulfoxide–sodium methoxide solutions of trinitrotoluene[23], Sitzmann and Kaplan have shown that the carbanion is at least one of the species formed by the action of methoxide on alcoholic solutions of trinitrotoluene[82]. They reacted 0.22 M trinitrotoluene and 0.22 M sodium methoxide in 2:1 tetrahydrofuran–methanol-*d* (CH_3OD) for 30 seconds at 0°, quenched the reaction with DCl in D_2O and separated the unreacted trinitrotoluene from the products. A nuclear magnetic resonance spectrum of the recovered trinitrotoluene showed an average of one methyl proton replaced by deuterium, indicating that the methyl hydrogens are rapidly exchanged in this particular solvent system. Analysis by mass spectroscopy showed peaks for mono-, di-, and trideuterated trinitrotoluene, with the dideuterate predominating. Moniz, *et al.*[83] have followed the changes in the nuclear magnetic resonance spectrum of trinitrotoluene in CD_3OD solution containing piperidine–piperidine·HCl. They observed that the methyl peak, originally sharp, began to broaden and lose intensity immediately after the addition of the piperidine–piperidine·HCl, and after 17 hours had practically disappeared. No changes in the peak due to the ring protons occurred. Trinitrotoluene has also been reported to undergo exchange with deuterium in D_2O–pyridine[35], and in basic dimethylformamide–D_2O[39].

Caldin and Long have concluded that the reaction of ethoxide ion with *sym*-trinitrotoluene and *sym*-trinitrobenzene produces different types of products because the decomposition rate of the trinitrotoluene product fits the Brønsted relation for its decomposition in other acids, while that of the trinitrobenzene product does not fit the Brønsted relation derived for its acid decomposition in the same solvent[73]. They inferred that the trinitrotoluene product is a deprotonated form of trinitrotoluene, since they thought that the trinitrobenzene product was an ethoxide adduct. The well-known condensation of trinitrotoluene with aldehydes and the formation of 2,2′,4,4′,6,6′-hexanitrostilbene by the reaction of trinitrotoluene with sodium hypochlorite in tetrahydrofuran–methanol[84] are best explained by assuming that the trinitrotoluene is first converted to trinitrotoluenide. Of course, the exchange experiments and proposed mechanisms do not prove that this anion is the only, or even the major species, generated by the action of alkoxide ion on trinitrotoluene.

Caldin has studied the effect of a wide variation in temperature ($>100°$) on the rates of proton transfer to the 'anion of trinitrotoluene,' using acetic acid and monochloroacetic acid[85], and hydrofluoric acid[86,87]. A non-linear Arrhenius plot was obtained for the hydrofluoric acid rates[87], in contrast to the linear plots obtained for the carboxylic acid rates[85]. Caldin appears to prefer quantum mechanical tunneling as an explanation for the non-linear Arrhenius plot[87,88], although other rationales are certainly possible[88].

B. Hydroxide Ion Equilibria

Table 5 summarizes the pK values and electronic spectral data reported for di- and trinitroaromatics in water. Changing the solvent to methanol raises the pK by 1.5 to 5 units (cf. Table 3). The species generated in these solutions have been identified in only a few cases. Most authors appear to assume that the reactions in water solutions are analogous to those in methanolic solutions. It should be mentioned that the equilibria studied in aqueous solutions containing appreciable concentrations of ethylenediamine may involve addition of the amine rather than hydroxide ion. It is interesting that for *N,N*-dimethylpicramide in aqueous sodium hydroxide, the rate equation derived from spectral measurements required the 1:2 stoichiometry

$$N,N\text{-dimethylpicramide} + 2OH^- \rightleftharpoons \text{complex} \quad (5)$$

shown in equation 5[50d]. No evidence for 1:1 stoichiometry could be found in water, but this was clearly established in methanol[46b].

TABLE 5. p*K*'s and electronic spectral data for some di- and trinitroaromatics in water[a].

Nitroaromatic compound	p*K*	λ_{max}, mμ[b,c]	$\lambda_{isosbestic}$, mμ[c]
Amines			
PiNHPh	10.20	440	300*
$PiNH_2(1)$	12.25	420	
	12.88[d]	410*	
$PiNH_2(2)$	17.55	420, 480 infl	440
3,6-Dinitrocarbazole	13.05	490	370
DiNHPh	14.65	*430**, 495 plat	405
$DiNH_2$	15.80	545	375*, 440*
Methylated benzenes			
$PiCH_3(1)$	14.45	515	
$PiCH_3(2)$	17.55	*470*, 530*	495*, 540*
1,3-Dimethyl-2,4,6-trinitrobenzene	16.05	*410*, 570	300
$DiCH_3$	17.12	*410*, 660	
1,3-Dimethyl-2,4-dinitrobenzene	< 19		
Misc			
PiOH(1)	−0.327[e]	357, 195[f]	222, 307[f]
	+0.46[g]	353*	307
PiOH(2)	15.10	380	360
PiH(1)	14.40	515	480
	14.0 ± 0.3[h,i]	445, 485[j]	
PiH(2)	17.55	455, *520*	475, 545
DiH	16.80	550	

[a] Unless otherwise indicated, all p*K* values were determined from equilibrium spectrophotometric measurements and all data are those reported by Schaal for ethylenediamine–water solutions at 20° [89]. The equilibria of the nitroaromatic SH with water are either for deprotonation of the ring or substituent of SH or for the formation of an addition compound, i.e., $SH + H_2O \rightleftharpoons S^- + H_3O^+$ or $SH + H_2O \rightleftharpoons (SH{\cdot}OH)^- + H^+$. The numbers in parentheses following the compound name refer to successive equilibria with water. For example, the primary equilibrium of a nitroaromatic with water to form the product C_1 is designated by (1) and the equilibrium of C_1 with water to form C_2 by (2). The following abbreviations are used: Pi = 2,4,6-trinitrophenyl, Di = 2,4-dinitrophenyl, Ph = phenyl, infl = inflection, plat = plateau midpoint. [b] An italicized value indicates the strongest absorption of those reported. [c] An asterisk indicates that the value was estimated from a plot. [d] For aqueous NaOH at 25° [50d]. [e] Conductometric value for 25°. [f] Reference 90. [g] Reference 91. This reference cites previous literature values for p*K*(1) at 25°, ranging from 0.22 to 0.82. [h] For aqueous NaOH at 25° [50a]. This reference also cites previous literature p*K* values, ranging from 13.6 to 14.2, and a polarographic value of 11.86. [i] Reference 92. [j] Reference 93.

The wide variation in the pK(1) values reported for picric acid is due to the difficulty in interpreting the spectral changes unambiguously and to the effect of neglected association equilibria on conductivity[90]. The small difference between pK(1) and pK(2) for *sym*-trinitrobenzene partially accounts for the wide variation in the values reported for pK(1). Gold and Rochester note that pK(2) for *sym*-trinitrobenzene may be for deprotonation of the hydroxide ion adduct formed in the primary equilibrium[50a].

C. Kinetic Evidence for Alkoxide and Hydroxide Addition Intermediates in Aromatic Nucleophilic Substitution

Several kinetic investigations of the nucleophilic substitution of the halogen in nitro-activated phenyl halides by hydroxide, methoxide, and phenoxide ions indicate that addition compounds are intermediates. Gaboriaud and Schaal, who studied the rates of hydrolysis of picryl chloride by continuous flow techniques detected a reversibly formed intermediate, presumably the 1-hydroxide addition product of picryl chloride[94]. The equilibrium constant for the formation of the intermediate was reported to be 1.12 (21.5°), and the ratio of the rate constants for its decomposition to picryl chloride and picric acid was reported as 15.5:1 (21.5°). Earlier, Farmer, who studied the formation of methyl picrate and $KO_2NC_6H_2(NO_2)_2(OCH_3)_2{\cdot}CH_3OH$ from picryl chloride and potassium methoxide, proposed that the 1-methoxide addition product of picryl chloride was an intermediate[95a]. He analyzed the product composition *vs.* time and showed that picryl chloride was more reactive with methoxide than methyl picrate. Farmer also proposed that the reaction of picryl chloride with phenoxide ion proceeds via the 1-phenoxide addition intermediate[95b]. The reactivities of some 4-nitrohalobenzenes and 2,4-dinitrohalobenzenes with azide, methoxide, and thiomethoxide ions have been studied by Miller[96]. The individual reactivities and overall reactivity patterns were interpreted as evidence for the addition type of intermediate required by the two-stage mechanism of activated nucleophilic substitution. Murto has discussed the suitability of the two-stage mechanism for substitution of the halogens of picryl chloride and fluoride by hydroxide ion[97].

An aromatic carbon bonded to a nitro or thiomethyl group is capable of forming methoxide addition products. Schaal and Peure have shown that the rate of decomposition of *p*-dinitrobenzene in methanolic methoxide solution is proportional to H_M, and have concluded that an addition intermediate is formed according to equation 6[98]. An analogous mechanism was proposed for the reaction of *o*-dinitrobenzene and methoxide ion although here the decomposition rate was found to be proportional to $[H_M + \log (CH_3OH)]$[99].

$$p\text{-}C_6H_4(NO_2)_2 + CH_3O^- \underset{\text{rapid}}{\rightleftharpoons} [CH_3O(NO_2)C_6H_4{=}N^+(O^-)_2] \xrightarrow{\text{slow}} p\text{-}CH_3OC_6H_4NO_2 + NO_2^- \qquad (6)$$

Murto has observed a red color during the reaction of 1,2,4,6-tetranitrobenzene with hydroxide ion and has suggested a two-stage mechanism for the reaction[97]. Gitis has reported that the reaction of 2,4-dinitrothioanisole with potassium methoxide makes the otherwise inert thiomethyl group susceptible to nucleophilic attack by dimethylamine, and has proposed the formation of a 1-methoxide addition intermediate[100].

III. COMPOUND FORMATION BY AMMONIA AND ALIPHATIC AMINES

A. Introduction

In 1905 Kraus and Franklin reported that the intensely colored liquid ammonia solutions of *sym*-trinitrobenzene, *sym*-trinitrotoluene, and 2,4-dinitrotoluene possessed salt-like conductivities[101]. The conductivity of solutions of *m*-dinitrobenzene in liquid ammonia[102], of *sym*-trinitrobenzene in pyridine and ethanolic diethylamine[35], and of 2,6-dinitrotoluene in liquid ammonia[104] was found to increase with time. For 2,6-dinitrotoluene the rate of increase in conductivity was found to follow a first-order rate equation[104]. The conductivity of the blue solutions of *m*-dinitrobenzene in liquid ammonia was ascribed to the electron-transfer process shown in equation 7[102]. Lewis and Seaborg, in attempting to explain why the color intensities of *m*-dinitrobenzene–amine solutions did not parallel the amine basicity, proposed that the intense color of ammonia solutions is

$$m\text{-}C_6H_4(NO_2)_2 + 2NH_3 \longrightarrow [m\text{-}C_6H_4(NO_2)_2]^{2-} + 2NH_3^+ \qquad (7)$$

due to addition of ammonia to the 2-carbon of the dinitrobenzene and that the adduct was stabilized by hydrogen bonds from the ammonia hydrogens to the oxygens of the adjacent nitro groups[105]; this stabilization would be less in the case of dimethylamine, for example. Canbäck has made qualitative estimates of the intensities of the colors produced at −72° by the action of aliphatic amines on 1,3-dinitrobenzene, 2,4-dinitrotoluene, 2,6-dinitrotoluene, 1,3-dimethyl-4,6-dinitrobenzene, and dinitromesitylene[106]. He thought that the colors produced by the primary and secondary amines were due to addition products, such as structure **15**, and that tertiary

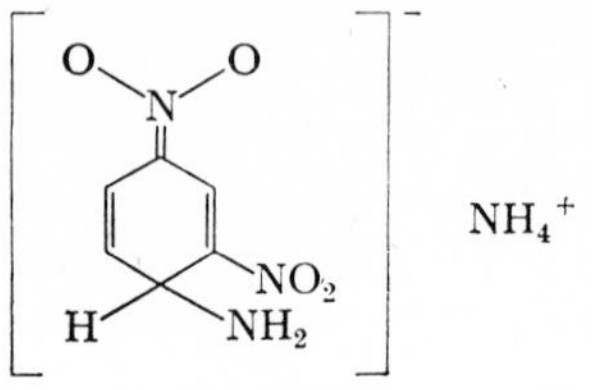

(**15**)

amines, such as triethylamine and *N*-ethylpiperidine, failed to produce colored solutions because the nitrogen of the tertiary amines is much less nucleophilic than the nitrogen of the substituted amide ions derived from primary and secondary amines. Wheland and coworkers found that electrolysis of *m*-dinitrobenzene in liquid ammonia produced hydrogen at the cathode and nitrogen at the anode, in addition to reduction products of *m*-dinitrobenzene; the formation of **15** was suggested as a rationale[107]. Miller and Wynne-Jones[103] have interpreted the interaction of *sym*-trinitrobenzene with diethylamine in ethanol in terms of Mulliken's 'inner and outer complexes[108],' suggesting that the inner complex dissociated in polar solvents into two radical ions. A weak electron-spin resonance spectrum was obtained from a *sym*-trinitrobenzene–diethylamine solution[103] (*vide infra*).

B. Nuclear Magnetic Resonance Studies

The results of nuclear magnetic resonance studies of the interactions of polynitroaromatic compounds with aliphatic amines in dimethyl sulfoxide can be summarized as follows.

(1) Ammonia, and either primary or secondary aliphatic amines react with *sym*-trinitrobenzene to form an addition compound whose structure is typified by **16**[20,23,109–111]. Table 6 gives the average

(16), R = H or alkyl

chemical shifts reported for the ring protons of the adducts. The chemical shifts of H_β in the amine adducts are about 0.5 ppm upfield from the shifts of the corresponding protons in the alkoxide adducts of trinitrobenzene derivatives (see Table 1). Since oxygen is more electronegative than nitrogen, this difference in chemical shift is not unexpected[112].

(2) *sym*-Trinitrobenzene shows no detectable interaction with tertiary aliphatic amines in dimethyl sulfoxide solution[113]. However, Strauss and Johanson have reported the isolation and characterization of a stable zwitterionic adduct from the reaction of *sym*-trinitrobenzene, triethylamine, and acrylonitrile[114a]. However, this adduct results from the addition of acrylonitrile, and not the amine, to the ring.

(3) Although the product of the action of triethylamine on methyl picrate was originally postulated to be the zwitterion **17**[23], new

TABLE 6. Chemical shifts for the ring protons of the aliphatic amine adducts of *sym*-trinitrobenzene.

R in	Average chemical shift[a] $H\alpha$	$H\beta$	References
NH_2	8.32	5.52	109
$NHCH_3$	8.44	5.69	109, 110
$N(CH_3)_2$	8.50	5.61	109, 110
$N(C_2H_5)_2$	8.42	5.63	23, 109–111
$NHCH_2CH_2OH$	8.43	5.70	110
N(CH₂—CH₂)₂CH₂ (piperidino)	8.47	5.56	109, 110

[a] In parts per million (ppm) from $(CH_3)_4Si$. Dimethyl sulfoxide used for solvent.

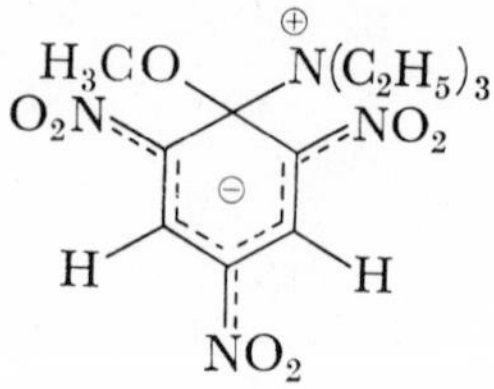

(17)

information based on conductivity measurements, nuclear magnetic resonance, and ultraviolet spectra has shown the product to be the tetralkylammonium picrate[114b]; similarly, the reaction between diethylamine and methyl picrate results in the formation of methyldiethylammonium picrate[114b].

(4) Diethylamine adds to the 3-carbon of ethyl picrate to form **18**[20], while dimethylamine adds to the 1-carbon of methyl picrate to form **19**[23]. In this connection, Servis has noted that adduct **20**, formed by the reaction of methoxide ion with *N,N*-dimethylpicramide, and adduct **19** do not show interconvertability[23].

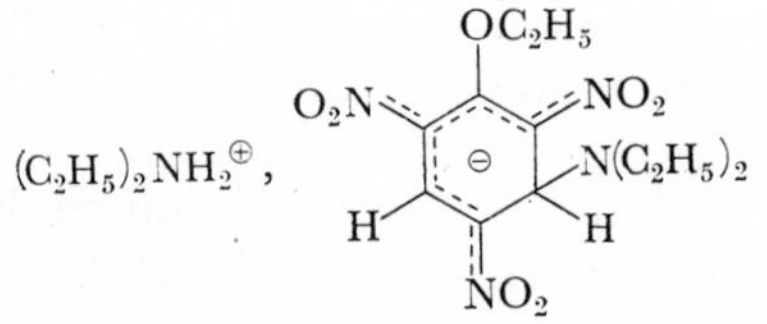

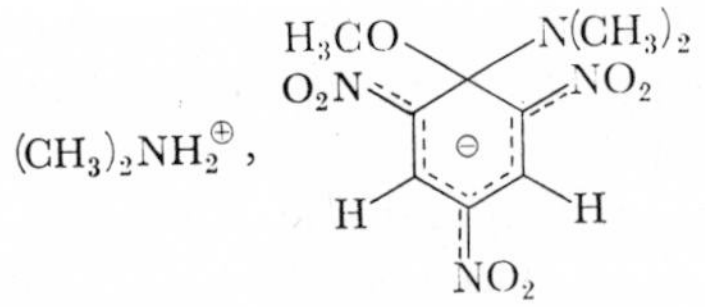

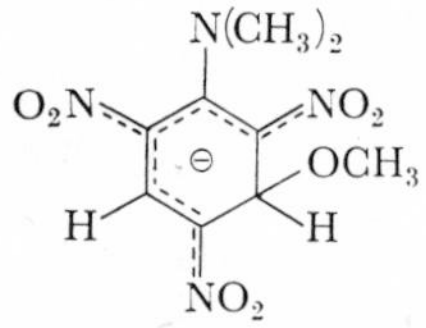

(18)

(19)

(20)

C. Thermodynamic Functions

Although nuclear magnetic resonance studies of the interaction of trinitrobenzene derivatives with amines in dimethyl sulfoxide indicate that the overall stoichiometry should be 1 : 2, spectrophotometric determinations of the equilibrium constants for these reactions in other solvents show that the reaction of one polynitroaromatic molecule may require more than two amine molecules. For example, a 1:4 stoichiometry was reported for the trinitrobenzene–dimethylamine reaction in dioxane[115]; the data are also consistent with the coexistence of equilibrated 1:1, 1:2, and 1:3 complexes[116]. The latter interpretation accommodates the data for the reaction of trinitrobenzene with methylamine, dimethylamine, ethylamine, or diethylamine in chloroform or dioxane[117]. The van't Hoff factor has been reported to be two for *sym*-trinitrobenzene and 2,4,6-trinitro-*t*-butylbenzene and three for methyl picrate in ethanolamine[118].

Briegleb and coworkers have made a detailed spectrophotometric and conductometric analysis of the sequential equilibria involving *sym*-trinitrobenzene and piperidine dissolved in acetonitrile[119] and cyclohexane[120]. These workers have concluded that the expected anion, **21**, is formed by the addition of piperidine to *sym*-trinitro-

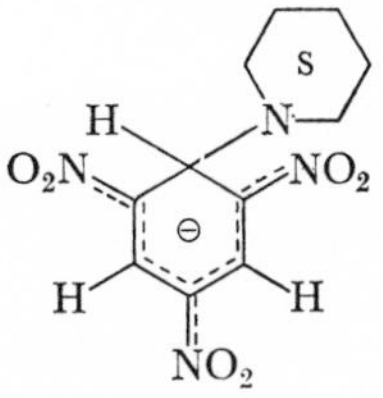

S
H N
O_2N NO_2
⊖
H H
NO_2

(**21**)

benzene. In acetonitrile, the concentration of piperidine determines the extent to which the other product, piperidinium ion, is free, ion-paired to **21**, or associated with a neutral piperidine molecule. In cyclohexane, the anion **21** is ion-paired with protonated polymers of piperidine; if the concentration of piperidine is low, the trinitrobenzene forms only a donor–acceptor complex with piperidine.

The equilibria 8a–8e describe the interactions between *sym*-trinitrobenzene and piperidine, represented by T and PH respectively, in acetonitrile[119]. In equations 8a and 8b, $(TP^- PH_2^+)$

$$T + 2PH \rightleftharpoons (TP^- PH_2^+) \qquad K = 10.4\ M^{-2} \tag{8a}$$

$$(TP^- PH_2^+) \rightleftharpoons TP^- + PH_2^+ \qquad K = 5.5 \times 10^{-3}\ M \tag{8b}$$

$$TP^- + PH_2^+ + PH \rightleftharpoons TP^- + P_2H_3^+ \qquad K = 23.5\ M^{-1} \tag{8c}$$

$$T + 2PH \rightleftharpoons TP^- + PH_2^+ \qquad K = 5.65 \times 10^{-2}\ M^{-1} \tag{8d}$$

$$\Delta H = -12.8 \text{ kcal/mole};\ \Delta S = -49 \text{ eu}$$

$$T + 3PH \rightleftharpoons TP^- + P_2H_3^+ \qquad K = 1.33\ M^{-2} \tag{8e}$$

$$\Delta H = -18.9 \text{ kcal/mole};\ \Delta S = -64 \text{ eu}$$

represents an ion-pair. The equilibrium constant, K, for 20° is given for each reaction. For reactions 8d and 8e, the negative entropy changes can be accounted for by the decrease in total number of molecules and by the increase in solvation required by the ions. The reactions of alkoxides with polynitroaromatics, on the other hand, exhibit positive entropy changes. (See section II.A. 2c).

Equations 9a–9c describe the interactions between *sym*-trinitrobenzene and piperidine in cyclohexane[120]. In contrast to Miller and Wynne-Jones[103], Liptay and Tamberg report that solutions of *sym*-

Donor–acceptor complex

$$\mathrm{T} + \mathrm{PH} \rightleftharpoons \mathrm{T}\cdots\mathrm{PH} \qquad K = 1.04\ M^{-1} \qquad \Delta H = -2.7\ \mathrm{kcal/mole} \tag{9a}$$

Ion-pair formation

$$\mathrm{T} + 3\mathrm{PH} \rightleftharpoons (\mathrm{TP}^{-}\ \mathrm{P_2H_3}^{+}) \qquad K = 0.21\ M^{-3} \qquad \Delta H = -15\ \mathrm{kcal/mole} \tag{9b}$$

Ion-pair formation

$$\mathrm{T} + 4\mathrm{PH} \rightleftharpoons (\mathrm{TP}^{-}\ \mathrm{P_3H_4}^{+}) \qquad K = 0.31\ M^{-4} \qquad \Delta H = -20\ \mathrm{kcal/mole} \tag{9c}$$

trinitrobenzene and aliphatic amines show no electron spin resonance spectrum initially, and only a weak spectrum after aging[120].

D. Electronic Spectra

Table 7 gives the characteristic absorption maxima for solutions of *sym*-trinitrobenzene in four solvents containing ammonia or an aliphatic amine. Except for the chloroform solution of trinitrobenzene and diethylamine, the colored solutions exhibit two maxima, one in the 430–470 mμ range, and the other, weaker in intensity, in the 505–575-mμ range. By way of comparison, *sym*-trinitrobenzene in acetonitrile containing potassium hydroxide shows maxima at 431 mμ and 500 mμ[119], and ethyl picrate in methanolic potassium methoxide has maxima at 417 mμ and 486 mμ[121]. Many of the spectral envelopes of the ethanol and chloroform solutions change drastically after an aging period of only 10 minutes[121].

For dimethyl sulfoxide solutions, we know from nuclear magnetic resonance studies that the colored products formed are of type **16**[109]. In contrast to the intensely colored dimethyl sulfoxide solution of methyl picrate and triethylamine[114], a dimethyl sulfoxide solution of *sym*-trinitrobenzene and triethylamine is virtually colorless[109],

TABLE 7. Visible absorption maxima for solutions of *sym*-trinitrobenzene containing ammonia or an aliphatic amine.

	λ_{max}, mμ			
Type of amine	$(CH_3)_2SO$[109]	CH_3CN[119]	$CHCl_3$[a]	C_2H_5OH[a]
NH_3	454, 542	448, 541	455, 545	432, 507
RNH_2	452, 538 ($R = CH_3$)	446, 532 ($R = Ph\text{-}CH_2$)	458, 540 ($R = C_2H_5$)	443, 525 ($R = CH_3$)
R_2NH	450, 528 ($R = CH_3$)	—	478 ($R = C_2H_5$)	444, 515 ($R = CH_3$)
Piperidine	448, 525	444, 521	—	—
R_3N	negl ($R = CH_3$)[b]	455, 513 ($R = n$-Bu)	469, 573 ($R = C_2H_5$)	430, 505 ($R = CH_3$)
		467, 565 ($R = C_2H_5$)	None ($R = CH_3$)[c]	

[a] Estimated from plots given in ref 121. [b] Negligible absorption above 400 mμ. [c] No maxima above 380 mμ.

indicating little or no addition of the amine to trinitrobenzene. The nuclear magnetic resonance spectrum of a dimethyl sulfoxide solution of *sym*-trinitrobenzene and trimethyl- or triethylamine shows that the amine causes no change in the position, intensity or broadness of the proton resonance of the trinitrobenzene[109]; this observation is thus consistent with the spectrophotometric measurements. The reactivity of tertiary amines with *sym*-trinitrobenzene in the other solvents is clearly not consistent with the results for dimethyl sulfoxide solutions. Crampton and Gold have noted that traces of impurities in the trimethylamine, possibly primary or secondary amines, increase the color intensity of trinitrobenzene–trimethylamine–dimethyl sulfoxide solutions[109]. Finally, any discussion of the effect of solvent on the absorption maxima given in Table 7 should be deferred until companion nuclear magnetic resonance spectra are available for product identification.

E. Mechanism of Formation

Kinetic evidence for the two-stage mechanism of activated nucleophilic aromatic substitution by amines has elucidated the mechanism of formation of the stable amine adducts of nitroaromatic compounds. The effect of base concentration on the second-order rate coefficient, k_2, for the nucleophilic attack by primary and secondary amines on 2,4-dinitrophenyl ethers and 2,4-dinitrofluorobenzene indicates that the zwitterion, **22**, is an intermediate[122–124].

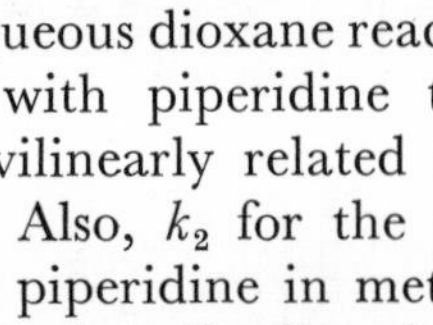

X, $\overset{+}{N}HR_1R_2$, H, NO_2, H, H, NO_2

(**22**), X = OCH_3, OC_6H_5, or F

For example, k_2 for the aqueous dioxane reaction of methyl or phenyl 2,4-dinitrophenyl ether with piperidine to form *N*-(2,4-dinitrophenyl)piperidine is curvilinearly related to the concentration of sodium hydroxide[122a,123]. Also, k_2 for the reaction of methyl 2,4-dinitrophenyl ether with piperidine in methanol is nearly linearly dependent on the concentration of sodium hydroxide[122c]. The phenyl ether is subject to general base catalysis by piperidine[123]. Finally, k_2 for the reaction of 2,4-dinitrofluorobenzene with the butylamines in benzene increases with amine concentration; for *t*-butylamine the increase is nearly linear, while for the other butylamines the increase is much less than linear[124]. In contrast, values of k_2 for the reaction of 2,4-dinitrofluorobenzene with *n*-butylamine in methanol and with aniline in methanol, *t*-butyl alcohol, or aqueous dioxane are at most mildly augmented by the addition of base[122b].

The final product of these activated nucleophilic substitution reactions, an *N*-substituted 2,4-dinitroaniline, results from the solvent- or base-catalyzed deprotonation of the zwitterion **22** either in a concerted process, or in a two-step process, i.e., a rapid deprotonation followed by ejection of methoxide, phenoxide, or fluoride ion in a slow step. Thus, these results suggest that the stable primary and secondary amine adducts are formed by deprotonation of zwitterionic intermediates analogous to **22**.

IV. COMPOUND FORMATION IN ALKALINE KETONE SOLUTIONS

A. Introduction

In 1886 Janovsky reported that an intense purple color is produced when aqueous alkali is added to an acetone solution of *m*-dinitrobenzene[125]. Since then, colored alkaline acetone solutions of a wide variety of aromatic compounds having two[126–130] or three[129–131] nitro groups *meta* to each other have been examined from the viewpoint of the qualitative detection[128b,129] and quantitative determination[127,128b,130,131] of these compounds. The acetone 'complexes' are formed at much lower base concentrations than the alkoxide addition compounds[128a]. Thus the presence of acetone makes it easier to establish characteristic absorption curves for the dinitroaromatics because the degradation of the latter is slower. Under normal Janovsky conditions, *o*- and *p*-dinitrobenzene do not give strong colors[128c], although Gitis has reported the isolation of solid complexes of potassium acetonate with each of these isomers[134]. Aromatic compounds closely related to *m*-dinitrobenzene, such as 3,3'-dinitroazoxybenzene, also give colors under Janovsky conditions[128a].

Color is also developed under Janovsky conditions if the acetone is replaced by a ketone with an α-hydrogen atom or by an aldehyde[128a,135–137,138a]. Zimmermann has developed a colorimetric determination of 17-ketosteroids which involves the reaction of the steroid with excess *m*-dinitrobenzene in ethanolic potassium hydroxide[139]. Canbäck has described a spectrophotometric method for differentiating active cardiac glucosides from the inactive forms by measuring the rate of change of the optical density for solutions of the glucosides in aqueous alcoholic sodium hydroxide containing *m*-dinitrobenzene[138b].

B. Chemical and Spectrophotometric Evidence for Structure

Canbäck suggested that the product of the Janovsky reaction had structure **23**[141]; his evidence was that the absorption spectrum of the

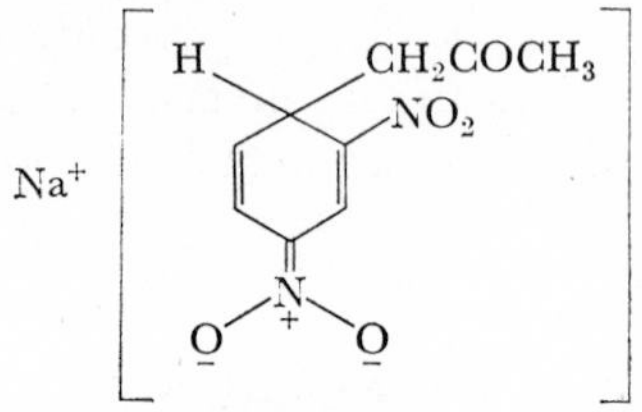

(**23**) 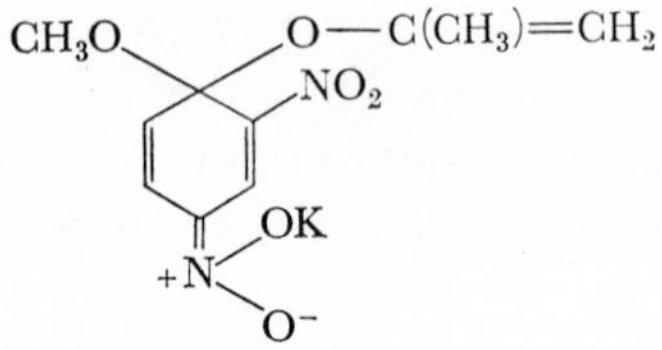 (**24**)

Janovsky product and of an ethanolic ethoxide solution of *sym*-trinitrobenzene were similar. He felt that **23** was a typographic formalism however, and that instead of a true carbon–carbon bond between the acetonate group and the ring carbon, only ion–dipole interaction existed. Gitis proposed that the Janovsky product of 2,4-dinitroanisole is **24**[134], the product of addition of the enolate of acetone to the 1-carbon of the anisole, or alternatively a mixture of both types of addition products, **23** and **24**[133]. Ryzhova and coworkers suggested that the 1:2 complexes of *m*-dinitrobenzene and potassium acetonate are donor–acceptor complexes[136]. Kimura provided the first convincing evidence that structure **23** is correct by showing that the Janovsky products of phenyl picrate and picryl chloride gave **25a** and **25b**, respectively, on oxidation with acidic

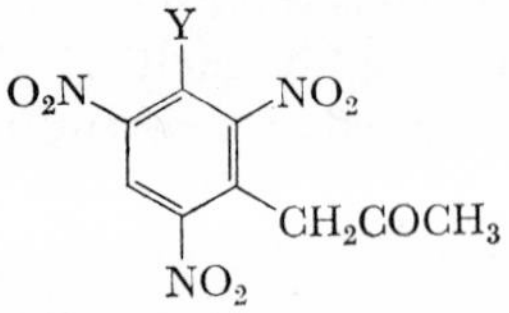

(**25a**), Y = OC_6H_5
(**25b**), Y = Cl

hydrogen peroxide[132]. Confirmatory chemical evidence was later presented by Severin who showed that the Janovsky product of *sym*-trinitrobenzene was reduced by sodium borohydride to **26** which was converted to **27** by the action of bromine[142]. A similar reaction sequence was carried out with 2,4-dinitrophenol[143].

If the Janovsky complex has Canbäck's structure **23** its formation should be reversed by acidification. Reversibility has indeed been demonstrated by isolating essentially pure *m*-dinitrobenzene and *sym*-trinitrobenzene on acidification of alkaline acetone solutions of

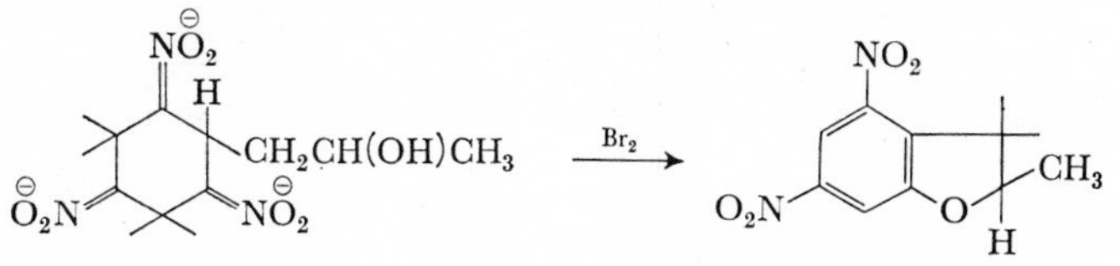

(**26**) (**27**)

these compounds[142,144]. Kimura isolated the classical Janovsky complex of *m*-dinitrobenzene and acetone as the potassium salt[145]; the analysis of the purple solid agreed with that required by the Canbäck structure.

In contrast to the large molar excess of carbonyl compound over *m*-dinitrobenzene used in the Janovsky reaction, the Zimmermann reaction uses a molar ratio of *m*-dinitrobenzene to carbonyl compound of at least one[144]. In the reaction of *m*-dinitrobenzene and acetone, the solution is bluish purple under Janovsky conditions (absorption maximum at 570 mμ), while the solution is red under Zimmermann conditions (absorption maximum at 497 mμ)[144]. Ishidate and Sakaguchi suggested in 1950 that the color generated by the Zimmermann reaction was due to **29**, formed by oxidation of the Janovsky complex **28**, by excess *m*-dinitrobenzene[146]. This

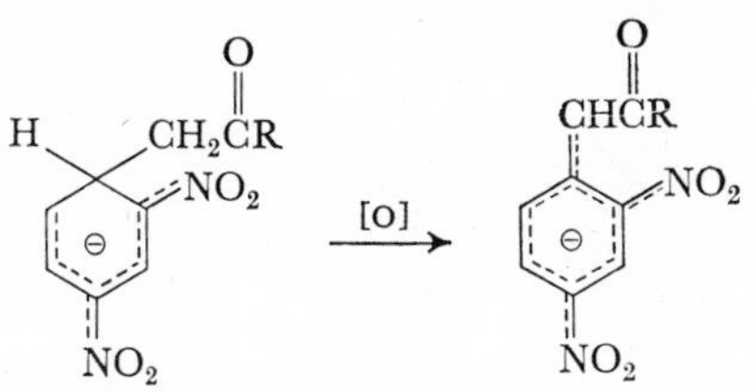

(**28**) (**29**)

suggested structure was supported by the isolation of 3,3′-dinitroazobenzene from the Zimmermann reaction of 17-ketosteroids[140], and by the isolation of *m*-nitroaniline and 2,4-dinitrobenzyl phenyl ketone from the Zimmermann reaction of acetophenone[137]. Kimura has reported the isolation of the potassium salt of **29** (R = methyl) from the Zimmermann reaction of acetone and *m*-dinitrobenzene[145].

As confirmation of **29** as the product of the Zimmermann reaction, a basic solution of **30a** or **30b** had essentially the same absorption spectrum as that generated by the Zimmermann reaction of acetophenone or acetone[137,144].

(**30a**), R = C_6H_5
(**30b**), R = CH_3

C. Electronic Absorption Spectra

Table 8 gives a comparison of the absorption maxima for the Janovsky complexes of some derivatives of *m*-dinitrobenzene and *sym*-trinitrobenzene. For all of the compounds listed in this table the stronger absorption occurs at the shorter wavelength. Spectral data for other derivatives of *m*-dinitrobenzene are given by Pollitt and Saunders[128a] and Newlands and Wild[128b]. The long-wavelength absorption of the Janovsky complex is red shifted (by 20 to 161 mμ)

TABLE 8. Visible absorption maxima for some di- and trinitrobenzene derivatives under Janovsky conditions.

Substituent, Y	λ_{max}, mμ		
	2,4,6-trinitro-Y-benzene [a]	2,4-dinitro-Y-benzene [b]	3,5-dinitro-Y-benzene [b]
H	465, 560	576, 692	576, 692
OCH_3	445, 500–520	576	590
OC_2H_5	445, 500–520	580	—
COOR	460, 555 (C_2H_5)	556, 685 (CH_3)	406, 561, 625 (CH_3)
Cl	450, 520–530	548, 666	563
CH_3	460, 500–530	580, 665	580
NR_2	425? (C_2H_5)	572 (CH_3)	614 (CH_3)

[a] From ref 132 [b] From ref 128a

with respect to that of the corresponding methoxide adduct[128a]. The position of the absorption maximum of the classical Janovsky complex (*m*-dinitrobenzene and alkaline acetone) is strongly dependent on solvent[145], while substituting another ketone for acetone has little effect on λ_{max}[144]. On the other hand, values of λ_{max} for Zimmermann products are nearly independent of solvent[145] and strongly dependent on the ketone[144].

D. Nuclear Magnetic Resonance Spectroscopy

Foster and Fyfe confirmed Canbäck's structural assignment for the Janovsky complex by nuclear magnetic resonance[147]. Referring to complex **31** they reported the following parameters for a solution

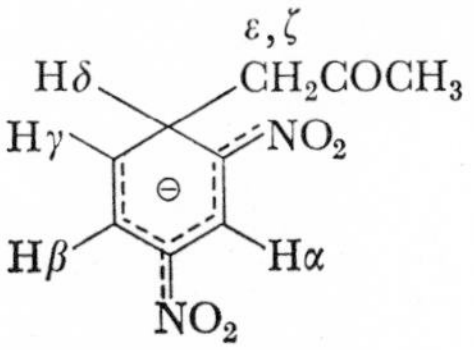

(**31**)

of *m*-dinitrobenzene and sodium methoxide in 1:1 v/v dimethyl sulfoxide–acetone: $\delta(H_\alpha)$ 8.32 ppm, $\delta(H_\beta)$ 6.61 ppm, $\delta(H_\gamma)$ 5.38 ppm, $\delta(H_\delta)$ 4.17 ppm, $J(H_\alpha\text{–}H_\beta)$ 1.9 Hz, $J(H_\beta\text{–}H_\gamma)$ 10.2 Hz, $J(H_\gamma\text{–}H_\delta)$ 5.0 Hz, $J(H_\delta\text{–}H_\varepsilon) \simeq 5.0$ Hz, and $J(H_\delta\text{–}H_\zeta) \simeq 10.0$ Hz, where δ is the chemical shift and J is the coupling constant. No evidence for structural isomers of **31** was found. Nuclear magnetic resonance has also shown that picryl derivatives react with acetone in dimethyl sulfoxide containing triethylamine to form 3-acetonyl adducts[25]. *N,N*-Dimethylpicramide forms the 3,5-diacetonyl adduct after 2 days if excess triethylamine is used[148]. Table 9 gives the nuclear magnetic resonance parameters of several ketone adducts of trinitrobenzene.

Strauss and Schran have identified by spectroscopic methods the red solid precipitated from the reaction of *sym*-trinitrobenzene, acetone, and diethylamine as the bicyclic product **32**[149]. For compound **32** dissolved in acetone-d_6, the following chemical shift values were obtained: for H_α, a singlet at 8.52 ppm; for H_β, a poorly resolved doublet (multiplet?) at 4.53 ppm, and for H_γ, a triplet at 5.72 ppm. The relative intensities of the three resonances were 1:2:1. This solid was also found to have λ_{max} in ethanol at 510 mμ (log ε 4.48) and the broad infrared absorption at 1425–1520 cm^{-1} expected for the asymmetric N–O stretching frequency in a system containing a

TABLE 9. NMR spectral data for some ketonic adducts of *sym*-trinitrobenzene[a].

Adduct structure: 1-$C(H_\gamma)(R)(R')$-1-H_β-2,4,6-trinitrocyclohexadienide anion (ring hydrogens H_α)

Reactant ketone	R	R′	Chemical shifts[b] Hα	Chemical shifts[b] Hβ	Coupling constant[c] $J_{H_\beta-H_\gamma}$
$CH_3C(=O)CH_3$[d]	H	$C(=O)CH_3$	8.35	5.08	5.5 (t)
$CH_3CH_2C(=O)CH_2CH_3$	CH_3	$C(=O)CH_2CH_3$	8.45	5.31	3 (d)
$CH_3C(=O)CH_2CH_3$	H	$C(=O)CH_2CH_3$ (20%)	8.35	5.04	6 (t)
	CH_3	$C(=O)CH_3$ (80%)	8.46	5.35	3 (d)
$CH_3C(=O)CH(CH_3)_2$	H	$C(=O)CH(CH_3)_2$	8.34	5.05	7 (t)
$CH_3C(=O)CH_2CH_2C(=O)CH_3$	H	$C(=O)CH_2CH_2C(=O)CH_3$ (25%)	8.35	5.07	5 (t)
	$CH_3C(=O)$	$CH_2C(=O)CH_3$ (75%)	8.45	5.35	3 (d)

[a] Taken from ref 149. Base is $(C_2H_5)_3N$. Solvent is dimethyl sulfoxide unless otherwise indicated. [b] In parts per million (ppm) from $Si(CH_3)_4$. [c] In hertz (Hz); d = doublet, t = triplet. [d] In acetone-d_6.

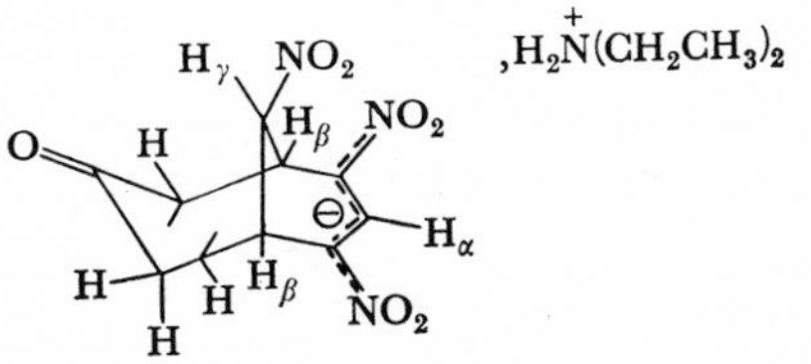

$,H_2\overset{+}{N}(CH_2CH_3)_2$

(32)

negative charge delocalized over two nitro groups. The chemical shifts for the compound analogous to **32** which formed by the reaction of *sym*-trinitrobenzene, dibenzyl ketone and triethylamine in dimethyl-d_6 sulfoxide are H_α, 8.6 ppm, H_β, 4.4 and 4.7 ppm, and H_γ, 6.0 ppm[150]. The fact that *sym*-trinitrobenzene and acetone give the bicyclic compound **32** in the presence of diethylamine, but only the non-cyclic addition product **33** in the presence of triethylamine,

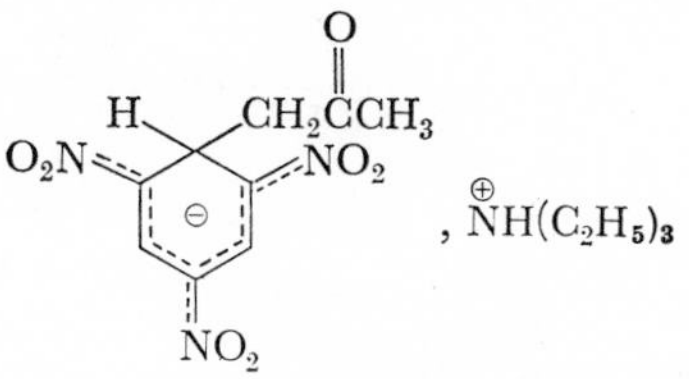

$, \overset{\oplus}{N}H(C_2H_5)_3$

(33)

has been rationalized by Strauss and Schran by assuming that the reaction with acetone and diethylamine proceeds through an enamine intermediate.

V. COMPOUND FORMATION BY OTHER NUCLEOPHILIC REAGENTS

A. Sulfite Ion

1. Introduction

Muraour reported in 1924 that *sym*-trinitrobenzene and *sym*-trinitrotoluene reacted with aqueous sodium sulfite to form colored solutions from which the nitroaromatic could be regenerated[151]. Čůta and Beránek showed spectrophotometrically that *sym*-trinitrobenzene formed a 1:1 complex with aqueous sulfite and suggested that the complex was addition compound **34** (Y = H), formed according to equation 10[152]. They felt that the changes which occurred in the spectral envelope as the sulfite concentration was increased were due to the addition of a second or third sulfite ion. In support of the existence of multiple sulfite addition products is Henry's isolation of a solid of composition *sym*-trinitrobenzene·$(Na_2SO_3)_2$[153].

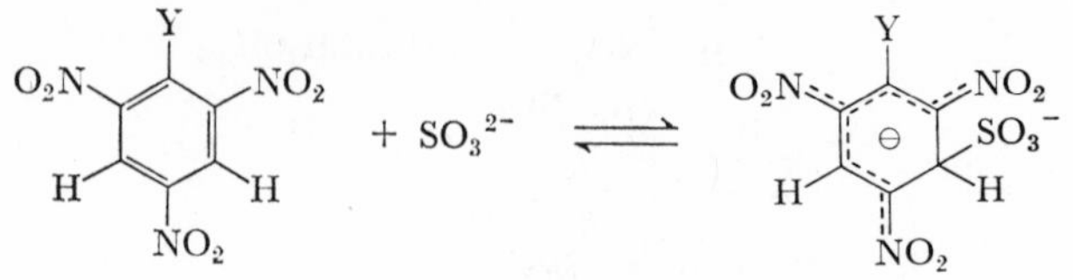

$$+ \; SO_3^{2-} \rightleftharpoons \qquad (10)$$

(**34**)

2. Structural assignment of the products

Crampton has confirmed by nuclear magnetic resonance spectroscopy that in water solutions picryl derivatives react with sulfite ion according to equations 10 or 11, depending on sulfite ion concentration, to form addition products **34** and **35**, respectively[154]. In

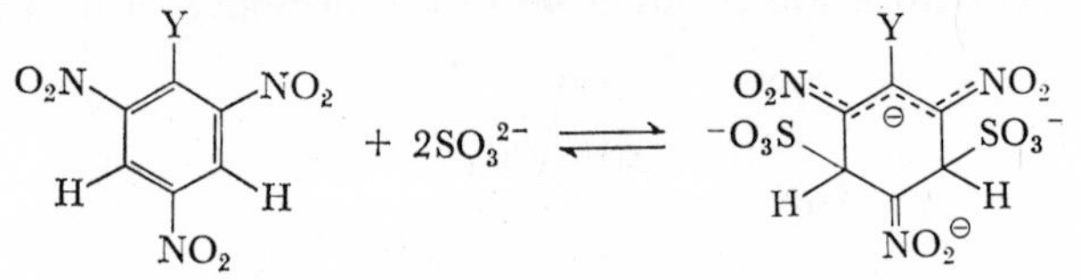

$$+ \; 2SO_3^{2-} \rightleftharpoons \qquad (11)$$

(**35**)

$Y = H, OCH_3, NH_2, NHCH_3, NHC_6H_5$ and $N(CH_3)_2$

contrast to the methoxide reactions, no nuclear magnetic resonance evidence was found for addition of sulfite to the 1-carbon of methyl picrate or for exchange of amino protons of picramide derivatives. However, conversion of picramide and its *N*-methyl and *N*-phenyl derivatives to the mono adduct **34** did result in a shift to lower field of the amino proton resonance; in the case of picramide two resonances of equal intensity were observed at chemical shifts of 10.10 and 10.57 ppm. Crampton suggested that these shifts to lower fields and the magnetic non-equivalence of the amino protons of picramide were due to hydrogen bonding of the amino protons as shown in structure **36**. The chemical shifts for the mono- and disulfite adducts of the picryl derivatives studied by Crampton are given in Table 10.

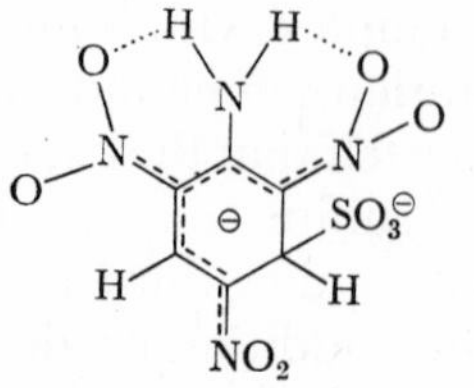

(**36**)

TABLE 10. Structural assignments, spectral data, and equilibrium constants for sulfite complexes of *sym*-trinitrobenzene and its derivatives[a].

Reactant (2,4,6-trinitro-1-Y-benzene) Y =	Product structure[b]	λ_{max}, mμ (log ε) for product	Equilibrium constant (ionic strength)[c,d]	Chemical shifts, ppm, for product		
				H_α	H_β	CH_3
H	**34**	462 (4.39)	$K_1 = 2.5 \times 10^2$ (ind)	8.30	6.00	
H	*e*	462 (4.37)	$K_1 = 2.67 \times 10^2$ (0.144)[e]			
H	**35**	490 (4.22)	$K_2 = 2.25 \times 10^3$ (0.3)	8.60[f]	6.05	
CH_3	*e*	465 (4.17)	$K_1 = 5.6$ (0.144)[e]			
CHO	*e*	458 (4.21)	$K_1 = 2.15 \times 10^3$ (0.144)[e]			
OCH_3	**34**	446 (4.18)	$K_1 = 2.10 \times 10^2$ (ind)	8.35	6.05	3.85
OCH_3	**35**	430 (4.08)	$K_2 = 1.9 \times 10^5$ (0.3)		6.02	4.08
NH_2	**34**	426 (4.43)	$K_1 = 1.01 \times 10^4$ (ind)	8.38	6.10	
NH_2	**35**	421 (4.45)	$K_2 = 1.9 \times 10^5$ (0.3)		6.17	
$NHCH_3$	**34**	418 (4.38)	$K_1 = 5.4 \times 10^4$ (ind)	8.30	6.15	3.10
$NHCH_3$	**35**	402 (4.32)	$K_2 = 9.7 \times 10^7$ (0.3)		6.07, 6.20	3.13
$N(CH_3)_2$	**34**	420 (4.30)	$K_1 = 5.4 \times 10^4$ (ind)	8.35	6.15	3.03
$N(CH_3)_2$	**35**	417 (4.30)	$K_2 = 3.3 \times 10^9$ (0.3)		6.26	3.07

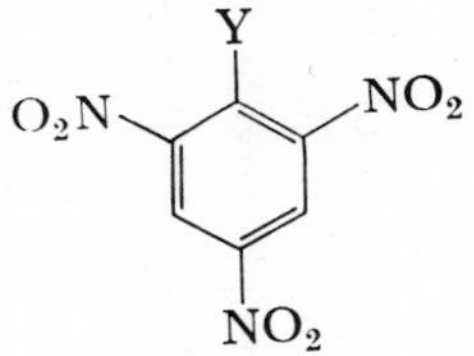

[a] All assignments and data taken from ref 154 except as noted otherwise. Water is the solvent for all determinations except the chemical shifts for which water–dimethyl sulfoxide (30:70, v/v) was used.

[b]

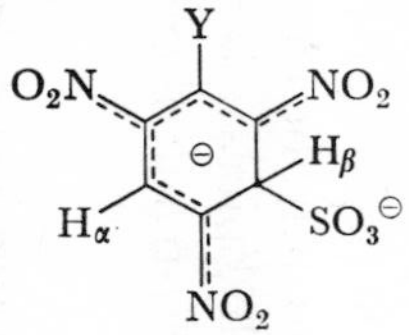

(**34**) 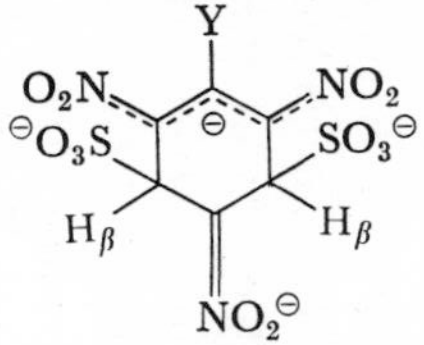 (**35**)

[c] K_1 and K_2 are the equilibrium constants for equations 10 and 11, respectively. [d] At 20° except for those values footnoted by *e* which are at 25°. "ind" means independent of ionic strength. [e] Taken from ref 155. K_1 refers to a 1:1 equilibrium typified by equation 10; however, structure **34** is not necessarily the structure of the product. [f] For Y = H.

3. Stability constants

Equilibrium constants for reaction 10 (K_1) and 11 (K_2) in aqueous solution have been determined spectrophotometrically by Crampton[154] and Norris[155] (see Table 10). For *sym*-trinitrobenzene the two K_1 values agree well with each other and with a value of K_1 determined by the partition method (2.25×10^2 M^{-1}, 25°)[155], but differ markedly from Čůta and Beránek's value of K_1 (5.12×10^2 M^{-1}, 25°, $\mu = 0$)[152]. This discrepancy apparently cannot be reconciled on the basis of ionic strength differences[155]. For reaction 10 (Y = H) Norris has calculated $\Delta H° = -4.0$ kcal/mole and $\Delta S° = -2.26$ eu from the temperature dependence of K_1. For ethoxide addition, ΔH is also small and $\Delta S°$ is significantly more positive. (See Table 4.) The value of $\Delta S°$ for sulfite addition might be expected to be more negative than that for ethoxide addition because the sulfite addition product has a negative charge on the sulfite group (in addition to the dispersed negative charge on the ring) which requires rather specific solvent orientation.

The carbon basicity of the sulfite ion toward picryl derivatives is greater than that of hydroxide ion. For example, the ratio of the formation constants for monosulfite and monohydroxide adducts of *sym*-trinitrobenzene is about 100[50a,154,155]; this ratio for picramide is *ca.* 300[50d,154,155]. Concerning substituent effects on adduct stability it is interesting that in water the 3-sulfite adduct of methyl picrate has approximately the same stability as the monosulfite adduct of *sym*-trinitrobenzene, while in methanol the 1-methoxide adduct of methyl picrate is much more stable than the monomethoxide adduct of *sym*-trinitrobenzene (see Table 3).

B. Thioethoxide and Thiophenoxide Ions

1. Nuclear magnetic resonance studies of the products

Crampton has investigated the products of the reaction of *sym*-trinitrobenzene, picramide, *N*-methylpicramide, and 2,4-dinitroaniline with sodium thioethoxide and sodium thiophenoxide in methanol–dimethyl sulfoxide (15:85, v/v)[156]. His results and interpretations can be summarized as follows.

(*a*) Addition of sodium thioethoxide to a solution of one of the trinitroaromatics produces the addition product **37.** The chemical shifts of the ring and amino protons of the trinitroaromatic reactant

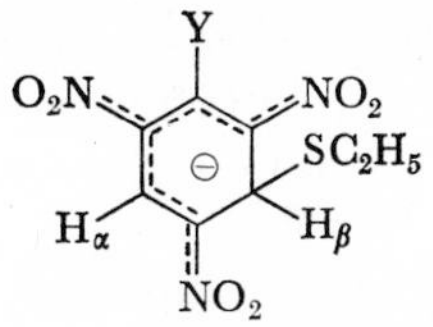

(**37**), Y = H, NH_2 or $NHCH_3$

and of **37** are given in Table 11. If the mole ratio of sodium thioethoxide to *N*-methylpicramide is two, only one resonance for the ring protons of the product is observed at 6.00 ppm, implying addition of thioethoxide to the 3- and 5-carbons of the ring.

(*b*) The formation of the 3-adduct, **37**, from picramide and *N*-methylpicramide is accompanied by a downfield shift of the amino proton resonance. Also, the amino protons of the picramide adducts are magnetically non-equivalent. These effects are similar to those produced by the action of sulfite ion on these compounds and the rationales invoked by Crampton are identical[154] (*vide supra*).

(*c*) For equilibrated solutions of *sym*-trinitrobenzene and its monothioethoxide adduct, the line width of the resonance of the ring protons of both species increases with the percentage of methanol in the methanol–dimethyl sulfoxide solvent. Rate constants for the decomposition of the complex to *sym*-trinitrobenzene and thioethoxide ion were calculated from these line widths, and were found to vary from 1.7 sec^{-1} for 7.5 vol % methanol to 20 sec^{-1} for 57.5 vol % methanol.

TABLE 11. Chemical shifts of the thioethoxide adducts of some picryl derivatives[a].

Reactant Y	Reactant chemical shifts[b]		Product chemical shifts[b,c]		
	Ring H	Amino H	H_α	H_β	Amino H
H	9.20	—	8.32	5.75	—
NH_2	9.14	9.10	8.47	5.73	9.90 10.65
$NHCH_3$	8.95	9.25	8.40	6.00	10.80

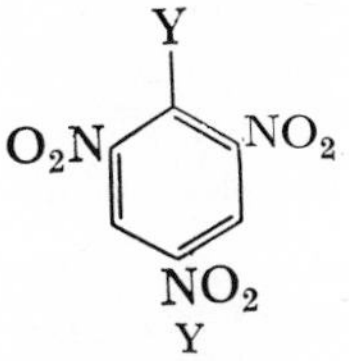

[a] From ref 156. Solvent is CH_3OH–$(CH_3)_2SO$ (15:85, v/v). [b] In parts per million (ppm) from $(CH_3)_4Si$. [c] For the 3-thioethoxide adduct, with H_α and H_β as in structure **37**.

(*d*) Sodium thioethoxide causes deprotonation of the amino group of 2,4-dinitroaniline, but not of picramide or *N*-methylpicramide. In contrast the amino groups of the latter two compounds are extensively deprotonated by methoxide in methanol–dimethyl sulfoxide[21,23]. These results are not unexpected, for the carbon basicity, relative to hydrogen basicity, of an oxygen base is known to be much less than that of the corresponding thio base[157].

(*e*) Solutions of *sym*-trinitrobenzene or picramide containing sodium thiophenoxide have a single aromatic proton resonance. As the formal concentration of sodium thiophenoxide is increased the resonance is shifted upfield until the limiting values, 7.45 and 7.10 ppm, respectively, are reached near a molar ratio of nitroaromatic to phenoxide ion of one. Apparently the resonances for the ring protons of the nitroaromatic and its thiophenoxide addition product are combined due to a rate of equilibration which is fast on the nuclear magnetic resonance time scale.

2. Equilibrium spectrophotometric measurements

Values of the formation constants, K_f, for the 1:1 complexes of *sym*-trinitrobenzene and each of the four bases, methoxide, phenoxide, thioethoxide, and thiophenoxide have been determined in methanol by the Benesi-Hildebrand method from equilibrium spectrophotometric measurements[156]. The values of pK_f along with λ_{max} and ε for the complexes, and pK_i for the conjugate acids of the bases in methanol are given in Table 12. The presence of dimethyl sulfoxide markedly increases the value of K_f. For example, log [K_f(40 vol % dimethyl sulfoxide)/K_f(methanol)] is about 3 for methoxide and 1.4 for thiophenoxide[156]. It is well known of course that the addition of dimethyl sulfoxide to methanol increases the effective basicity of the solvent[161].

TABLE 12. Electronic spectral data and formation constants for 1:1 complexes of *sym*-trinitrobenzene with sulfur and oxygen bases[a].

Base (B^-)	λ_{max}, mμ (log ε)	log K_f	pK_i	log ($K_f \cdot K_i$)	log [$K_{HB}^{CH_3B}$][157]
CH_3O^-	422 (4.48)	1.18	16.7[158]	−15.5	2.0
$C_6H_5O^-$	—	< −2.7	14.1[159]	−16.8	1.2
(Alkyl)S^-	460 (4.46)C_2H_5	3.54	15[b]	−11.5	10.3 (CH_3)
$C_6H_5S^-$	464 (4.45)	0.29	10.9[160]	−10.6	9.9

[a] Taken from ref 156; for methanol at 20°. [b] Estimated by Crampton[156] from the pK_i for water.

It is informative to compare the carbon basicity of the four bases in these complexes with that in other carbon compounds. According to Hine[157], the R basicity of the base B^-, with respect to its proton basicity, is defined as the equilibrium constant K_{HB}^{RB} for equation 12. Thus, the trinitrocyclohexadienyl basicity of B^-, with respect

$$ROH + HB \rightleftharpoons RB + H_2O \qquad (12)$$

to its proton basicity, is the equilibrium constant for equation 13.

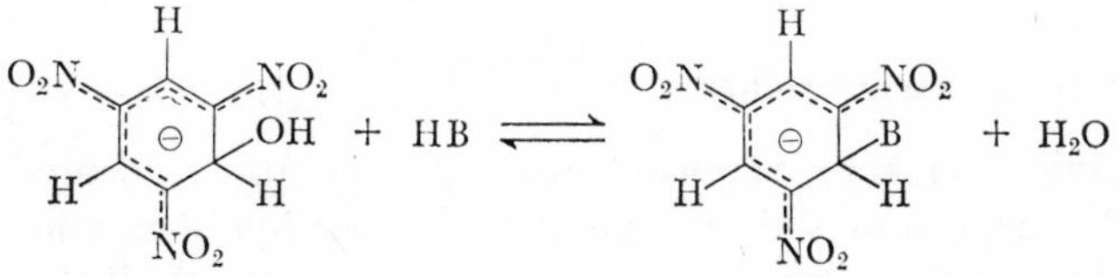

$$\text{1-H-3,5-(NO}_2)_2\text{-4-NO}_2\text{-6-OH-6-H cyclohexadienyl anion} + HB \rightleftharpoons \text{corresponding 6-B-6-H cyclohexadienyl anion} + H_2O \qquad (13)$$

Now the equilibrium constant for equation 13 is equal to $K_f \cdot K_i$ divided by the equilibrium constant for equation 14. Although the

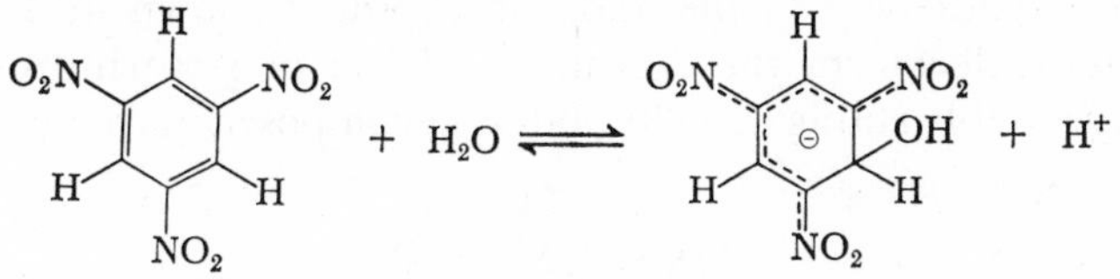

$$\text{1,3,5-trinitrobenzene} + H_2O \rightleftharpoons \text{1,3,5-trinitro-6-hydroxy-6-H cyclohexadienyl anion} + H^+ \qquad (14)$$

latter has not been evaluated in methanol, the difference in the logarithms of the equilibrium constants for equation 13 for any two bases is clearly the same as the difference in log $(K_f \cdot K_i)$ for the two bases. From Table 12 we see that log $(K_f \cdot K_i)$ for B^- = thioethoxide is 4 units greater (more basic) than for B^- = methoxide, and that log $(K_f \cdot K_i)$ for B^- = thiophenoxide is at least 6.2 units greater than for B^- = phenoxide. For comparison, log $(K_{HB}^{C_6H_5B})$ is 4.6 units greater for B^- = thiomethoxide than for B^- = methoxide[157]. The enhancement of carbon basicity produced by thio substitution is more pronounced if the carbon atom is classically aliphatic. For example, log $(K_{HB}^{CH_3B})$ for B^- = thiomethoxide or thiophenoxide is about 8 units larger than for B^- = methoxide or phenoxide.

C. Cyanide Ion

I. Chemical studies

Hepp, in 1882[162], and Hantzsch and Kissel, in 1899[6], reported the isolation of a red-violet salt from the reaction of potassium cyanide and *sym*-trinitrobenzene. Foster, however, was not able to reproduce these results[163]. The addition of potassium cyanide to *m*-dinitrobenzene in aqueous methanol or aqueous ethanol yields a purple solution from which 2-nitro-6-methoxy- (or 6-ethoxy-) benzonitrile can be isolated[164]. The methoxy compound can in fact be prepared by this method in 20 % yield[165]. The dinitrobenzenes, in the presence of sodium cyanide, have been observed to fluoresce at room temperature; they and *sym*-trinitrobenzene have been observed to phosphoresce in the presence of sodium cyanide at liquid air temperature[166].

2. Structural assignments

Although Vickery reported that the complex formed by *m*-dinitrobenzene and sodium cyanide in dimethylformamide has no detectable nuclear magnetic resonance spectrum[167], Buncel[168] and Norris[169a] have shown by nuclear magnetic resonance that cyanide ion converts *sym*-trinitrobenzene, *sym*-trinitrotoluene, and *sym*-trinitrobenzaldehyde to the addition products shown in Table 13. In deuterochloroform the 3-cyanide adduct of *sym*-trinitrotoluene, **38**, is reasonably stable at $-30°$, but is decomposed at room tempera-

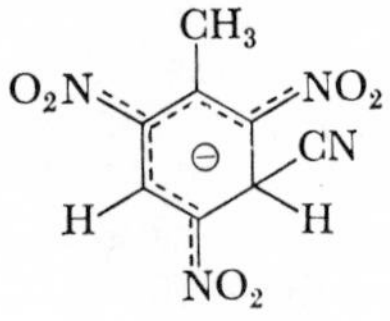

(**38**)

ture after several hours to as yet unidentified products[168]. In contrast, in dimethyl sulfoxide the 3-methoxide adduct of methyl picrate is formed rapidly, but is gradually replaced by the 1-methoxide adduct, the more thermodynamically stable form[17,23]. It is significant that cyanide ion is the only nucleophile which has been shown to convert *sym*-trinitrotoluene to a product identifiable by nuclear magnetic resonance.

3. Absorption spectra

Vickery has measured the time dependence of λ_{max} values for dimethyl sulfoxide solutions of sodium cyanide containing nitrobenzene, *m*-dinitrobenzene, and *sym*-trinitrobenzene[167]. The nitrobenzene solution has a transient green color, while the *m*-dinitrobenzene solution changes from blue to red. Vickery claims to have isolated a deep red solid complex of *m*-dinitrobenzene and sodium cyanide from a dimethylformamide solution, and gives λ_{max} and ε

TABLE 13. Chemical shifts for the cyanide addition products of some *sym*-trinitrobenzene derivatives[a].

Shifts in parent compound			Shifts in complex		
			Ring		
Ring	Other H	Complex	H_α	H_β	Other H
9.36	—	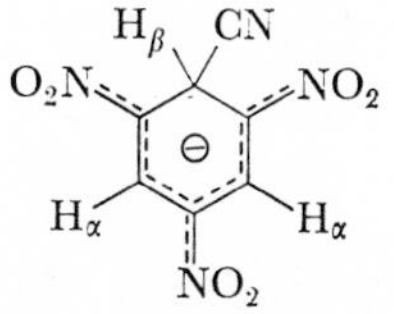 [structure: 1-H_β-1-CN-2,4,6-trinitro-3,5-H_α cyclohexadienide anion]	8.42	5.48	—
9.37	10.73 (CHO)	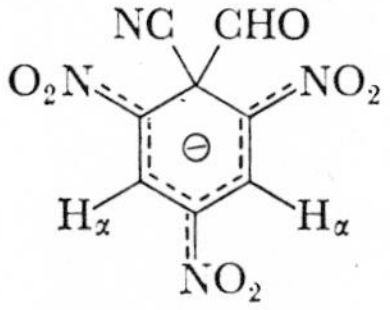 [structure: 1-NC-1-CHO-2,4,6-trinitro-3,5-H_α cyclohexadienide anion]	8.55	—	10.14 (CHO)
9.04	2.82 (CH_3)	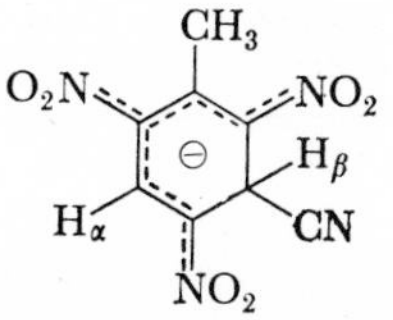 [structure: 1-CH_3-3-H_β-3-CN-2,4,6-trinitro-5-H_α cyclohexadienide anion]	8.65	5.76	2.74 (CH_3)

[a] In parts per million (ppm) from $(CH_3)_4Si$. Solvent is $CDCl_3$ at $-30°$. Taken from ref 168 and 169a. Source of the cyanide ion is tetraphenylarsonium cyanide.

for the solid dissolved in methanol. The initial visible spectrum of *sym*-trinitrobenzene–cyanide ion solutions has been attributed to structure **39**[152,163,169b], and a 1:1 stoichiometry has been established for the interaction from the spectrophotometric measurements[152].

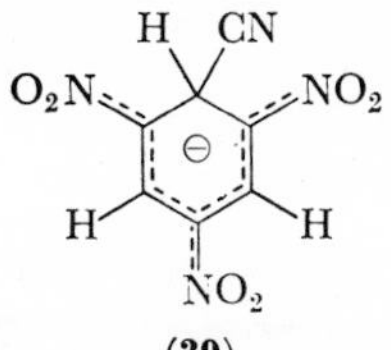

[structure: H, CN on sp^3 carbon; O_2N, NO_2, H, H, NO_2 on anionic ring]

(**39**)

Norris has measured the infrared and visible absorption spectra at about $-30°$ of deuterochloroform solutions of *sym*-trinitrobenzene containing about a molar equivalent of tetraphenylarsonium cyanide[169a]. Since the formal concentrations of trinitrobenzene and cyanide were essentially the same as those used for the nuclear magnetic resonance measurements (Table 13), we know at least the major species, i.e. **39**, responsible for the infrared and visible absorption spectra. For these solutions Norris reported that the following absorptions are associated with **39**: infrared (in cm^{-1})—1495 (m), 1410–1400 (w), 1235 (s), 1190 (s), and 1050 (m); visible—448 and 561 mμ. From the extinction coefficient at 561 mμ for the complex in chloroform at 25.3°, $2.25 \times 10^4\ M^{-1}\ cm^{-1}$ [169b], the extinction coefficient for the absorption at 448 mμ was calculated to be $4.05 \times 10^4\ M^{-1}\ cm^{-1}$ under the same conditions.

D. Salts of Mononitroalkanes

Fyfe has studied the reactions of salts of mononitroalkanes with di- and trinitroaromatics in dimethyl sulfoxide[170]. His results are summarized below.

Salts of mononitroalkanes react with *sym*-trinitrobenzene and its derivatives in dimethyl sulfoxide to produce red solutions; the salt

TABLE 14. Spectral data for complexes of *sym*-trinitrobenzene with salts of mononitroalkanes[a].

Adduct structure (1-R-1-H_β-2,4,6-trinitrocyclohexadienide, ring protons H_α) R	λ_{max}, mμ (log ε)[b]	Chemical shifts[c]		Coupling constant, Hz	
		H_α	H_β	$J_{\alpha,\beta}$	$J_{\beta,\gamma}$
$NO_2\overset{\gamma}{C}H_2$—	456 (4.34), 568 (4.04)[d]	8.41	5.35	0.0	7.5
$NO_2\overset{\gamma}{C}HCH_3$	452 (4.38), 552 (4.08)	8.47[e]	5.67	0.5	3.2
$NO_2\overset{\gamma}{C}H(C_2H_5)$	452 (4.34), 552 (4.04)	8.46	5.63	1.0	3.5
$NO_2C(CH_3)_2$	450 (4.36), 545 (4.06)	8.51	5.90	1.4	—

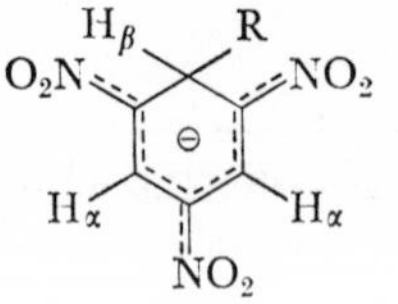

[a] Data from ref 170. Solvent $(CH_3)_2SO$. Cation = $(C_2H_5)_3NH^+$ except where noted otherwise. [b] Adduct was isolated and redissolved. [c] In parts per million (ppm) from internal $(CH_3)_4Si$. [d] Cation = $K^{\oplus}$. [e] Additional splitting due to non-equivalent H_α.

of nitromethane reacts with *m*-dinitrobenzene in dimethyl sulfoxide–nitromethane (60:40) to produce a pinkish purple solution. Electronic spectral data for the products, as reported by Fyfe, are given in Tables 14 and 15. The spectral envelopes of the products of the reactions of *m*-dinitrobenzene with salts of nitromethane and acetone

TABLE 15. Spectral data for nitromethide adducts of some polynitroaromatics[a].

Adduct structure	λ_{max}, mμ	Chemical shifts[b]			Coupling constants, Hz
		H_α	H_β	H_γ	
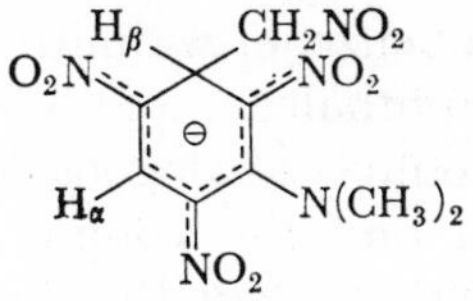 H_β, CH_2NO_2, O_2N, NO_2, ⊖, H_α, $N(CH_3)_2$, NO_2	490	8.4	5.77[c]	—	$J_{\alpha,\beta} = 1$
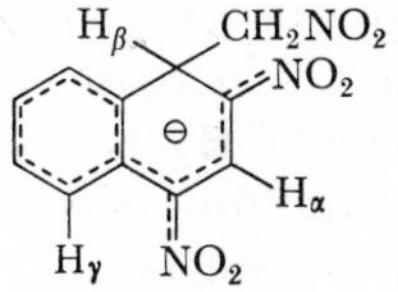 H_β, CH_2NO_2, NO_2, ⊖, H_α, H_γ, NO_2	538	6.78	5.13	8.67	
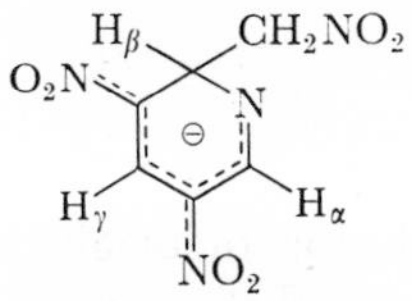 H_β, CH_2NO_2, O_2N, N, ⊖, H_γ, H_α, NO_2	515[d] 570 sh	8.34	5.85[e]	8.52	$J_{\alpha,\gamma} = 2$
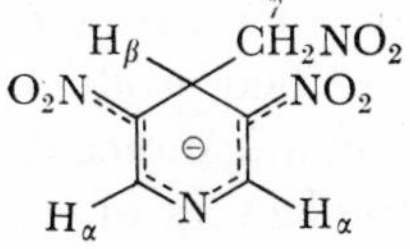 H_β, γ, CH_2NO_2, O_2N, NO_2, ⊖, H_α, N, H_α	515[d] 570 sh	8.22	5.30	—	$J_{\beta,\gamma} = 3.5$
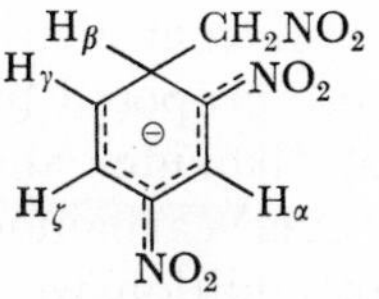 H_β, CH_2NO_2, H_γ, NO_2, ⊖, H_ζ, H_α, NO_2	562[f]	8.30[f]	4.18	5.38[g]	$J_{\alpha,\zeta} = 1.9$ $J_{\gamma,\zeta} = 10.2$ $J_{\beta,\gamma} = 5.0$

[a] Taken from ref 170. Solvent is CH_3NO_2; base is $(C_2H_5)_3N$. [b] In parts per million (ppm) from internal $(CH_3)_4Si$. [c] H_β has four lines (each split into a doublet by H_α); ring asymmetry causes nonequivalent methylene hydrogens; $J = 5$ and 8 Hz. [d] Due to a mixture of the 2- and 4-adducts. [e] Four lines, $J = 5$ and 7 Hz. [f] Solvent is $(CH_3)_2SO$–CH_3NO_2 (60:40); base is $NaOCH_3$. [g] $H_\zeta = 6.59$ ppm.

are quite similar; the λ_{max} values for the products of the reactions of *sym*-trinitrobenzene with salts of nitromethane and ethyl malonate are similar[128a,170], but are red shifted from λ_{max} for the 1-methoxide adduct of methyl picrate. The color generated by the reaction of *m*-dinitrobenzene and a mononitroalkane in basic aqueous methanol is the basis of a colorimetric determination of mononitroalkanes[171]. The species responsible for the color is apparently the nitroalkane adduct of *m*-dinitrobenzene, since Fyfe has prepared the nitroalkane adduct of *sym*-trinitrobenzene by adding the nitroalkane to a solution of the methoxide adduct of *sym*-trinitrobenzene.

Fyfe has shown by nuclear magnetic resonance spectroscopy that salts of mononitroalkanes react with *sym*-trinitrobenzene, *m*-dinitrobenzene, *N*,*N*-dimethylpicramide, 2,4-dinitronaphthalene, and 3,5-dinitropyridine to form carbanion addition products analogous to the Janovsky complexes. The chemical shifts and the coupling constants for the products, given in Tables 14 and 15, show that the products cannot be formed by attack of the nitrogen or oxygen of the nitroalkane. From Table 14 one can see that an increase in branching in R increases $J_{\alpha\beta}$ and shifts the resonance of H_β downfield. The chemical shifts and coupling constants for the products of addition of $O_2NCH_2^{\ominus}$ and $CH_3COCH_2^{\ominus}$ to *m*-dinitrobenzene are quite similar[147,170].

E. Salts of Ethyl Malonate and Ethyl Acetoacetate

Jackson and Gazzolo, in 1900, reported the isolation of brightly colored solids from the reaction of trinitroaromatics and the sodium salt of ethyl malonate or ethyl acetoacetate, using molar ratios of aromatic to sodium salt of 1 : 3[8]. Analyses of these unpurified products were consistent with the composition of one trinitroaromatic molecule (methyl picrate or *sym*-trinitrobenzene) and three molecules of the sodium salt. Jackson and Earle suggested that these products were formed by addition of the enolate of the ester to the nitroaromatic ring, forming addition compounds analogous to those proposed by Jackson and Gazzolo for the reaction products of alkoxides and nitroaromatics[10]. Enolate addition was preferred over carbanion addition because the products of the latter type of addition were thought to be stable to hydrolysis.

Pollitt and Saunders have reported electronic spectral data for *m*-dinitrobenzene and its negatively substituted derivatives in dimethylformamide solutions containing sodium hydroxide and ethyl malonate or ethyl methylmalonate[128a] (see Table 16). Although

TABLE 16. Electronic spectral data for basic solutions of *m*-dinitrobenzene derivatives and malonate esters[a].

Derivative (1-Y-3,5-dinitrobenzene, O_2N, NO_2) Y	λ_{max}, mμ[b]	
	Ethyl malonate	Ethyl methylmalonate
H	365, *570*	360, *563*
Cl	363, *556*	365, *552*
COO^-	568	561
$CONH_2$	398, *552*	397, *547*
$COOCH_3$	*406*, 542	402, *528*
CN	404, 536[c]	404, *525*
NO_2	*460*, 568	*454*, 538

[a] Taken from ref 128a. Solvent is dimethylformamide. [b] The italicized value refers to the stronger of the two peaks. [c] Peaks are of the same intensity.

the spectra were considered intermediate between Meisenheimer and Janovsky type spectra, it should be pointed out that no structural assignments for the products can be made from these spectral data alone.

Recently, Baudet isolated a red solid from the reaction of equimolar quantities of 2,4-dinitrofluorobenzene, ethyl malonate, and triethylamine[172]. The solid reacted with water and alcohols to form ethyl 2,4-dinitrophenylmalonate, showed no electron spin resonance spectrum in dimethylformamide, and had the following significant infrared absorptions: 1680 cm^{-1}, characteristic of cyclohexadienes, and 1725 and 1745 cm^{-1} carbonyl absorptions; electronic absorption occurred at 397 and 510 mμ in dimethylformamide and at 370 and 520 mμ in dimethyl sulfoxide. Although Baudet failed to obtain elemental analyses for the solid and did not measure its nuclear magnetic resonance spectrum, he appears justified in concluding that the solid is **40**, the intermediate required by the two-stage

F $CH(CO_2C_2H_5)_2$, NO_2, H, H, H, NO_2, $,(C_2H_5)_3NH^+$

(**40**)

mechanism of activated nucleophilic substitution. Enolate addition cannot be definitely ruled out by Baudet's study, but it seems unlikely because a rearrangement would then be required to form ethyl 2,4-dinitrophenylmalonate.

F. Azide Ion

Caveng and Zollinger have studied the reaction of azide ion with methyl picrate and picryl azide by nuclear magnetic resonance spectroscopy and visible spectrophotometry[173]. At temperatures

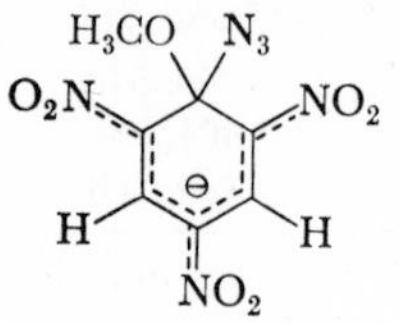

H_3CO N_3 O_2N NO_2 ⊖ H H NO_2

(**41**)

between −30° and −45° the nuclear magnetic resonance measurements showed that azide ion adds to the 1-carbon of methyl picrate to form **41**; however, the solutions are unstable, turning brown and evolving nitrogen gas at 0°. Table 17 gives the chemical shifts for

TABLE 17. Chemical shifts for the ring and methyl protons of methyl picrate and its 1-azide addition compound in four solvents[a].

		Chemical shift for ring protons[b]		Chemical shift for CH_3 group[b]	
Solvent	Temp, °C	Methyl picrate	Addition compd	Methyl picrate	Addition compd
CD_3CN	−40 to −46	8.85	8.59	4.05	3.05
$(CD_3)_2CO$	−40	9.01	8.66	4.16	3.03
$CH_3CON(CH_3)_2$	−30	9.39	8.78	—	—
$CH_3CON(CH_3)_2$–CH_3CN (4:1)	−40	9.18	8.63	—	—

[a] Taken from ref 173. Molarity of the methyl picrate is 0.2 except for the solvent pair which was 0.16 *M* in methyl picrate. Mole ratio of methyl picrate to tetraethylammonium azide was 2 except for $(CD_3)_2CO$ for which the ratio was *ca.* 3. [b] In parts per million (ppm) from internal $(CH_3)_4Si$.

41 and unreacted methyl picrate under conditions of slow exchange in four solvents. The temperature dependence of the difference between the chemical shifts of methyl picrate and its azide addition product was measured and used to calculate activation parameters. The absorption spectrum of **41** in dimethylacetamide has two maxima in the visible, 419 and 506 mμ (log ε 4.30 and 4.10, respectively).

Solutions of picryl azide and tetraethylammonium azide in a solvent composed of dimethylacetamide, acetonitrile, and acetone in a 2:2:1 ratio behave qualitatively like the methyl picrate–azide solutions in that they are stable below 0° (λ_{max} 365–370 and 410–415 mμ), but above this temperature turn brown and evolve nitrogen. However, for mole ratios of azide ion to picryl azide between 0 and 2, these solutions show only one sharp aromatic proton resonance, varying in position at −50° from 9.07 to 9.00 ppm. Since the chemical shift of the aromatic protons of the addition compound would be expected to be in the range of 8.6 to 8.8 ppm, it appears that only a very small fraction of the picryl azide can be converted to the complex.

Picryl chloride gives a fleeting color with azide ion, but the addition compound, if formed, is too unstable for nuclear magnetic resonance measurements. In this connection it should be mentioned that reaction of *p*-nitrofluorobenzene with the dimethylamine present in dimethylformamide led to the erroneous conclusion that azide ion forms a σ-complex with *p*-nitrofluorobenzene in dimethylformamide[174].

G. Bicarbonate Ion

Heating *sym*-trinitrobenzene in 76% aqueous methanol in the presence of sodium or potassium bicarbonate or sodium carbonate has been found by Izzo to give a 60–80% yield of 3,5-dinitroanisole[175]. Unchanged *sym*-trinitrobenzene, and no dinitroanisole, was obtained if these bases were replaced by sodium acetate, dibasic sodium phosphate, sodium iodide, ammonium carbonate, carbon dioxide or sufficient sodium hydroxide to give an initial pH of 11. As expected, complete degradation occurred if 0.6 *N* sodium hydroxide was used. Izzo suggested that the peculiar specificity of this reaction is due to the formation and decomposition of **42**, a bicarbonate addition product of *sym*-trinitrobenzene, as shown in equation 15. This mechanism is similar to that proposed by Bunnett and Rauhut for the von Richter reaction[176].

$$1,3,5\text{-}(NO_2)_3C_6H_3 + \bar{O}\text{—}C(=O)\text{—}OH \longrightarrow \textbf{(42)} \xrightarrow{-OH^-} \xrightarrow{CH_3OH} \longrightarrow 3,5\text{-}(NO_2)_2C_6H_3OCH_3 + HNO_2^{\cdot} + CO_2 \qquad (15)$$

VI. REFERENCES

1. (a) P. Hepp, *Ann. Chem.*, **215**, 345 (1882).
   (b) J. F. Bunnett, *Quart. Rev.* (London), **12**, 1 (1958).
2. (a) R. S. Mulliken, *J. Am. Chem. Soc.*, **74**, 811 (1952); *J. Phys. Chem.*, **56**, 801 (1952).
   (b) G. Briegleb, *Elektronen–Donator–Acceptor–Komplexe*, Springer Verlag, Berlin, 1961.
   (c) E. Buncel, A. R. Norris, and K. E. Russell, *Quart. Rev.*, **22**, 123 (1968).
   (d) P. Buck, *Angew. Chem. Internat. Ed.*, **8**, 120 (1969).
   (e) M. R. Crampton in *Advances in Physical Organic Chemistry*, Vol. 7 (Ed. V. Gold), Academic Press, London, 1969.
3. C. A. Lobry de Bruyn and F. H. van Leent, *Rec. Trav. Chim.*, **14**, 150 (1895).
4. (a) V. Meyer, *Chem. Ber.*, **27**, 3153 (1894).
   (b) V. Meyer, *Chem. Ber.*, **29**, 848 (1896).
5. C. A. Lobry de Bruyn, *Rec. Trav. Chim.*, **14**, 89 (1895).
6. A. Hantzsch and H. Kissel, *Chem. Ber.*, **32**, 3137 (1899).
7. C. L. Jackson and W. F. Boos, *Am. Chem. J.*, **20**, 444 (1898).
8. C. L. Jackson and F. H. Gazzolo, *Am. Chem. J.*, **23**, 376 (1900).
9. J. Meisenheimer, *Ann. Chem.*, **323**, 205 (1902).
10. C. L. Jackson and R. B. Earle, *Am. Chem. J.*, **29**, 89 (1903).
11. R. Destro, C. M. Gramaccioli, and M. Simonetta, *Acta Cryst.*, **24B**, 1369 (1968).
12. H. Ueda, N. Sakabe, J. Tanaka, and A. Furasaki, *Bull. Chem. Soc. Japan*, **41**, 2866 (1968).
13. C. M. Gramaccioli, R. Destro, and M. Simonetta, *Acta Cryst.*, **24B**, 129 (1968).
14. P. Caveng, P. B. Fischer, E. Heilbronner, A. L. Miller, and H. Zollinger, *Helv. Chim. Acta*, **50**, 848 (1967).
15. M. R. Crampton and V. Gold, *J. Chem. Soc.*, **1964**, 4293.
16. R. Foster and C. A. Fyfe, *Tetrahedron*, **21**, 3363 (1965).

17. K. L. Servis, *J. Am. Chem. Soc.*, **87**, 5495 (1965).
18. M. R. Crampton and V. Gold, *Chem. Commun.*, **1965**, 256.
19. R. Foster, C. A. Fyfe, and J. W. Morris, *Rec. Trav. Chim.*, **84**, 516 (1965).
20. R. Foster and C. A. Fyfe, *Rev. Pure Appl. Chem.*, **16**, 61 (1966).
21. M. R. Crampton and V. Gold, *J. Chem. Soc., B*, **1966**, 893.
22. W. E. Byrne, E. J. Fendler, J. H. Fendler, and C. E. Griffin, *J. Org. Chem.*, **32**, 2506 (1967).
23. K. L. Servis, *J. Am. Chem. Soc.*, **89**, 1508 (1967).
24. C. E. Griffin, E. J. Fendler, W. E. Byrne, and J. H. Fendler, *Tetrahedron Lett.*, **1967**, 4473.
25. R. Foster, C. A. Fyfe, P. H. Emslie, and M. I. Foreman, *Tetrahedron*, **23**, 227 (1967).
26. (a) J. H. Fendler, E. J. Fendler, W. E. Byrne, and C. E. Griffin, *J. Org. Chem.*, **33**, 977 (1968).
    (b) J. H. Fendler, E. J. Fendler, W. E. Byrne, and C. E. Griffin, *J. Org. Chem.*, **33**, 4141 (1968).
27. C. A. Fyfe, *Tetrahedron Lett.*, **1968**, 659.
28. L. K. Dyall, *J. Chem. Soc.*, **1960**, 5160.
29. R. Foster and D. L. Hammick, *J. Chem. Soc.*, **1954**, 2153.
30. R. A. Friedel, *J. Phys. Chem.*, **62**, 1341 (1958).
31. (a) G. A. Russell and E. G. Janzen, *J. Am. Chem. Soc.*, **84**, 4153 (1962).
    (b) G. A. Russell, E. G. Janzen, and E. T. Strom, *J. Am. Chem. Soc.*, **86**, 1807 (1964).
32. (a) D. H. Geske and A. H. Maki, *J. Am. Chem. Soc.*, **82**, 2671 (1960).
    (b) A. H. Maki and D. H. Geske, *J. Chem. Phys.*, **33**, 825 (1960).
33. M. R. Crampton and V. Gold, *J. Chem. Soc., B*, **1966**, 498.
34. J. A. A. Ketelaar, A. Bier, and H. T. Vlaar, *Rec. Trav. Chim.*, **73**, 37 (1954).
35. R. E. Miller and W. F. K. Wynne-Jones, *J. Chem. Soc.*, **1959**, 2375.
36. M. S. Kharasch, W. G. Brown, and J. McNab, *J. Org. Chem.*, **2**, 36 (1937).
37. E. Buncel and E. A. Symons, *Can. J. Chem.*, **44**, 771 (1966).
38. R. E. Miller and W. F. K. Wynne-Jones, *J. Chem. Soc.*, **1961**, 4886.
39. E. Buncel, K. E. Russell, and J. Wood, *Chem. Commun.*, **1968**, 252.
40. M. A. Paul and F. A. Long, *Chem. Rev.*, **57**, 1 (1957).
41. C. H. Rochester, *Trans. Faraday Soc.*, **59**, 2820 (1963).
42. V. Gold and B. W. V. Hawes, *J. Chem. Soc.*, **1951**, 2102.
43. R. A. M. O'Ferrall and J. H. Ridd, *J. Chem. Soc.*, **1963**, 5030.
44. R. Schaal and G. Lambert, *J. Chim. Phys.*, **59**, 1164 (1962).
45. C. H. Rochester, *J. Chem. Soc.*, **1965**, 2404.
46. (a) V. Gold and C. H. Rochester, *J. Chem. Soc.*, **1964**, 1692.
    (b) V. Gold and C. H. Rochester, *J. Chem. Soc.*, **1964**, 1697.
    (c) V. Gold and C. H. Rochester, *J. Chem. Soc.*, **1964**, 1704.
47. S. S. Gitis and I. G. L'vovich, *J. Gen. Chem. USSR*, **34**, 2262 (1964).
48. A. F. Holleman and F. E. van Haeften, *Rec. Trav. Chim.*, **40**, 68 (1921).
49. S. F. Acree, *J. Am. Chem. Soc.*, **37**, 1909 (1915).
50. (a) V. Gold and C. H. Rochester, *J. Chem. Soc.*, **1964**, 1710.
    (b) V. Gold and C. H. Rochester, *J. Chem. Soc.*, **1964**, 1717.
    (c) V. Gold and C. H. Rochester, *J. Chem. Soc.*, **1964**, 1722.
    (d) V. Gold and C. H. Rochester, *J. Chem. Soc.*, **1964**, 1727.
51. E. Tommila and J. Murto, *Acta Chem. Scand.*, **16**, 53 (1962).
52. V. Gold and C. H. Rochester, *Proc. Chem. Soc.*, **1960**, 403.
53. V. Gold and C. H. Rochester, *J. Chem. Soc.*, **1964**, 1687.
54. J. Murto and E. Tommila, *Acta Chem. Scand.*, **16**, 63 (1962).
55. J. Murto, *Ann. Acad. Sci. Fennicae Ser. A II*, **117**, 1 (1962).

56. Y. Ogata and M. Okano, *J. Am. Chem. Soc.*, **71**, 3211 (1949).
57. S. S. Gitis and A. V. Ivanov, *J. Org. Chem. USSR*, **2**, 107 (1966).
58. S. S. Gitis, A. J. Glaz, and A. Sh. Glaz, *J. Org. Chem. USSR*, **3**, 1577 (1967).
59. J. A. A. Ketelaar, C. van de Stolpe, A. Goudsmit, and W. Dzcubas, *Rec. Trav. Chim.*, **71**, 1104 (1952).
60. H. A. Benesi and J. H. Hildebrand, *J. Am. Chem. Soc.*, **71**, 2703 (1949).
61. P. Job, *C. R. Acad. Sci. Paris*, **180**, 928 (1925); *Ann. Chim.* (Paris), [10] **9**, 113 (1928).
62. H. L. Schlaefer and O. Kling, *Angew. Chem.*, **68**, 667 (1956).
63. H. H. Jaffé and M. Orchin, *Theory and Applications of Ultraviolet Spectroscopy*, John Wiley and Sons, New York, 1962, p. 562.
64. (a) G. Lambert and R. Schaal, *C. R. Acad. Sci. Paris*, **255**, 1939 (1962).
    (b) G. Lambert and R. Schaal, *J. Chim. Phys.*, **59**, 1170 (1962).
65. (a) R. Schaal and G. Lambert, *J. Chim. Phys.*, **59**, 1151 (1962).
    (b) R. Schaal and G. Lambert, *J. Chim. Phys.*, **59**, 1164 (1962).
66. T. Abe, T. Kumai, and H. Arai, *Bull. Chem. Soc. Japan*, **38**, 1526 (1965).
67. J. E. Dickeson, L. K. Dyall, and V. A. Pickles, *Aust. J. Chem.*, **21**, 1267 (1968).
68. F. Terrier, P. Pastour, and R. Schaal, *C. R. Acad. Sci. Paris*, **260**, 5783 (1965).
69. C. F. Bernasconi, *J. Am. Chem. Soc.*, **90**, 4982 (1968).
70. J. Murto and E. Kohvakka, *Suomen Kemistilehti*, **39B**, 128 (1966).
71. J. H. Fendler, *Chem. Ind.* (London), **1965**, 764.
72. (a) J. B. Ainscough and E. F. Caldin, *J. Chem. Soc.*, **1956**, 2528.
    (b) J. B. Ainscough and E. F. Caldin, *J. Chem. Soc.*, **1956**, 2540.
    (c) J. B. Ainscough and E. F. Caldin, *J. Chem. Soc.*, **1956**, 2546.
73. E. F. Caldin and G. Long, *Proc. Roy. Soc. Ser. A*, **228**, 263 (1955).
74. J. A. Blake, M. J. B. Evans, and K. E. Russell, *Can. J. Chem.*, **44**, 119 (1966).
75. J. E. Leffler, *J. Org. Chem.*, **20**, 1202 (1955).
76. J. Miller, *J. Am. Chem. Soc.*, **85**, 1628 (1963).
77. J. Hine, R. Wiesboeck, and O. B. Ramsay, *J. Am. Chem. Soc.*, **83**, 1222 (1961).
78. L. Pauling, *The Nature of the Chemical Bond*, 3rd ed., Cornell University Press, Ithaca, N.Y., 1960, p. 90.
79. E. F. Caldin, *J. Chem. Soc.*, **1959**, 3345.
80. J. Murto and J. Vainionpää, *Suomen Kemistilehti*, **39B**, 133 (1966).
81. J. Murto and A. Viitala, *Suomen Kemistilehti*, **39B**, 138 (1966).
82. M. E. Sitzmann and L. A. Kaplan, (Naval Ordnance Laboratory, Silver Spring, Md.) Private Communication, 1969.
83. W. B. Moniz, T. N. Hall, and C. F. Poranski, Jr., Unpublished Results.
84. K. G. Shipp and L. A. Kaplan, *J. Org. Chem.*, **31**, 857 (1966).
85. J. B. Ainscough and E. F. Caldin, *J. Chem. Soc.*, **1960**, 2407.
86. E. F. Caldin and R. A. Jackson, *J. Chem. Soc.*, **1960**, 2413.
87. E. F. Caldin and M. Kasparian, *Discussions Faraday Soc.*, **39**, 25 (1965).
88. E. F. Caldin, *Chem. Rev.*, **69**, 135 (1969).
89. R. Schaal, *J. Chim. Phys.*, **52**, 784, 796 (1955).
90. D. J. G. Ives and P. G. N. Moseley, *J. Chem. Soc., B*, **1966**, 757.
91. M. M. Davis and M. Paabo, *J. Res. Nat. Bur. Stand. Sect. A*, **67**, 241 (1963).
92. T. Abe, *Bull. Chem. Soc. Japan*, **33**, 41 (1960).
93. F. Čůta and J. Písecký, *Chem. Listy*, **51**, 433 (1957); *Collection Czech. Chem. Commun.*, **23**, 628 (1958).
94. R. Gaboriaud and R. Schaal, *C. R. Acad. Sci. Paris, Ser. C*, **265**, 1376 (1967).
95. (a) R. C. Farmer, *J. Chem. Soc.*, **1959**, 3425.
    (b) R. C. Farmer, *J. Chem. Soc.*, **1959**, 3430.
96. K. C. Ho, J. Miller, and K. W. Wong, *J. Chem. Soc., B*, **1966**, 310.

97. J. Murto, *Acta Chem. Scand.*, **20**, 310 (1966).
98. R. Schaal and F. Peure, *Bull. Soc. Chim. Fr.*, **1963**, 2638; *C. R. Acad. Sci. Paris*, **256**, 4020 (1963).
99. R. Schaal and J. C. Latour, *Bull. Soc. Chim. Fr.*, **1964**, 2177.
100. S. S. Gitis, *J. Org. Chem. USSR*, **1**, 903 (1965).
101. E. C. Franklin and C. A. Kraus, *J. Am. Chem. Soc.*, **27**, 191 (1905).
102. M. J. Field, W. E. Garner, and C. C. Smith, *J. Chem. Soc.*, **127**, 1227 (1925).
103. R. E. Miller and W. F. K. Wynne-Jones, *Nature*, **186**, 149 (1960).
104. W. E. Garner and H. F. Gilbe, *J. Chem. Soc.*, **1928**, 2889.
105. G. N. Lewis and G. T. Seaborg, *J. Am. Chem. Soc.*, **62**, 2122 (1940).
106. T. Canbäck, *Acta Chem. Scand.*, **3**, 946 (1949).
107. J. D. Farr, C. C. Bard, and G. W. Wheland, *J. Am. Chem. Soc.*, **71**, 2013 (1949).
108. R. S. Mulliken, *J. Am. Chem. Soc.*, **72**, 600 (1950).
109. M. R. Crampton and V. Gold, *J. Chem. Soc., B*, **1967**, 23.
110. R. Foster and C. A. Fyfe, *Tetrahedron*, **22**, 1831 (1966).
111. R. Foster and C. A. Fyfe, *Tetrahedron*, **23**, 528 (1967).
112. J. W. Emsley, J. Feeney, and L. H. Sutcliffe, *High Resolution Nuclear Magnetic Resonance Spectroscopy*, Vol. 2, Pergamon Press, Oxford, England, 1966, Chapter 10.
113. M. R. Crampton and V. Gold, *Chem. Commun.*, **1965**, 549.
114. (a) M. J. Strauss and R. G. Johanson, *Chem. Ind.* (London), **1969**, 242.
(b) K. L. Servis, Private Communication, 1969.
115. R. Foster, D. L. Hammick, and A. A. Wardley, *J. Chem. Soc.*, **1953**, 3817.
116. M. M. Labes and S. D. Ross, *J. Org. Chem.*, **21**, 1049 (1956).
117. R. Foster, *J. Chem. Soc.*, **1959**, 3508.
118. V. Baliah and V. Ramakrishnan, *Rec. Trav. Chim.*, **78**, 783 (1959).
119. G. Briegleb, W. Liptay, and M. Cantner, *Z. Phys. Chem.* (Frankfurt am Main), **26**, 55 (1960).
120. W. Liptay and N. Tamberg, *Z. Elektrochem.*, **66**, 59 (1962).
121. R. Foster and R. K. Mackie, *Tetrahedron*, **16**, 119 (1961).
122. (a) J. F. Bunnett and R. H. Garst, *J. Am. Chem. Soc.*, **87**, 3879 (1965).
(b) J. F. Bunnett and R. H. Garst, *J. Am. Chem. Soc.*, **87**, 3875 (1965).
(c) J. F. Bunnett and R. H. Garst, *J. Org. Chem.*, **33**, 2320 (1968).
123. J. F. Bunnett and C. Bernasconi, *J. Am. Chem. Soc.*, **87**, 5209 (1965).
124. F. Pietra and D. Vitali, *J. Chem. Soc., B*, **1968**, 1200.
125. J. V. Janovsky and L. Erb, *Chem. Ber.*, **19**, 2155 (1886); J. V. Janovsky, *Chem. Ber.*, **24**, 971 (1891).
126. S. S. Gitis, G. M. Oksengendler, and A. Ya. Kaminskii, *J. Gen. Chem. USSR*, **29**, 2948 (1959); S. S. Gitis and A. Ya. Kaminskii, *J. Gen. Chem. USSR*, **30**, 3771 (1960).
127. T. Abe, *Bull. Chem. Soc. Japan*, **32**, 887 (1959).
128. (a) R. J. Pollitt and B. C. Saunders, *J. Chem. Soc.*, **1965**, 4615.
(b) M. J. Newlands and F. Wild, *J. Chem. Soc.*, **1956**, 3686.
(c) E. Sawicki and T. W. Stanley, *Anal. Chim. Acta*, **23**, 551 (1960).
129. F. L. English, *Anal. Chem.*, **20**, 745 (1948); S. A. H. Amas and H. J. Yallop, *Analyst* (London), **91**, 336 (1966).
130. A. Ya. Kaminskii and S. S. Gitis, *J. Org. Chem. USSR*, **2**, 1780 (1966).
131. T. Abe, *Bull. Chem. Soc. Japan*, **32**, 775 (1959); **34**, 1776 (1961).
132. M. Kimura, *Pharm. Bull.* (Tokyo), **3**, 75 (1955).
133. S. S. Gitis and A. Ya. Kaminskii, *J. Gen. Chem. USSR*, **33**, 3226 (1963).
134. S. S. Gitis, *J. Gen. Chem. USSR*, **27**, 1956 (1957).
135. B. von Bittó, *Ann. Chem.*, **269**, 377 (1892).

136. G. L. Ryzhova, T. A. Rubtsova, and N. A. Vasil'eva, *J. Gen. Chem. USSR*, **36,** 2022 (1966).
137. T. J. King and C. E. Newall, *J. Chem. Soc.*, **1962**, 367.
138. (a) T. Canbäck, *Svensk Farm. Tidskr.*, **54,** 1 (1950).
(b) T. Canbäck, *Svensk Farm. Tidskr.*, **54,** 225 (1950).
139. W. Zimmermann, *Z. Physiol. Chem.*, **233,** 257 (1935); **245,** 47 (1937).
140. O. Neunhoeffer, K. Thewalt, and W. Zimmermann, *Z. Physiol. Chem.*, **323,** 116 (1961).
141. T. Canbäck, *Farm. Revy*, **48,** 153 (1949).
142. T. Severin and R. Schmitz, *Angew. Chem. Internat. Ed.*, **2,** 266 (1963).
143. T. Severin and H. Temme, *Chem. Ber.*, **98,** 1159 (1965).
144. R. Foster and R. K. Mackie, *Tetrahedron*, **18,** 1131 (1962).
145. M. Kimura, M. Kawata, and M. Nakadate, *Chem. Ind.* (London), **1965,** 2065.
146. M. Ishidate and T. Sakaguchi, *J. Pharm. Soc. Japan*, **70,** 444 (1950).
147. C. A. Fyfe and R. Foster, *Chem. Commun.*, **1967**, 1219.
148. R. Foster and C. A. Fyfe, *J. Chem. Soc., B*, **1966,** 53.
149. M. J. Strauss and H. Schran, *J. Am. Chem. Soc.*, **91,** 3974 (1969).
150. R. Foster, M. I. Foreman, and M. J. Strauss, *Tetrahedron Lett.*, **1968,** 4949.
151. H. Muraour, *Bull. Soc. Chim. Fr.*, **35,** 367 (1924).
152. F. Čůta and E. Beránek, *Collection Czech. Chem. Commun.*, **23,** 1501 (1958).
153. R. A. Henry, *J. Org. Chem.*, **27,** 2637 (1962).
154. M. R. Crampton, *J. Chem. Soc., B*, **1967,** 1341.
155. A. R. Norris, *Can. J. Chem.*, **45,** 175 (1967).
156. M. R. Crampton, *J. Chem. Soc., B*, **1968,** 1208.
157. J. Hine and R. D. Weimer, Jr., *J. Am. Chem. Soc.*, **87,** 3387 (1965).
158. R. P. Bell, *The Proton in Chemistry*, Cornell University Press, Ithaca, N.Y., 1959, p. 37.
159. B. D. England and D. House, *J. Chem. Soc.*, **1962,** 4421.
160. B. W. Clare, D. Cook, E. C. F. Ko, Y. C. Mac, and A. J. Parker, *J. Am. Chem. Soc.*, **88,** 1911 (1966).
161. R. Stewart, J. P. O'Donnell, D. J. Cram, and B. Rickborn, *Tetrahedron*, **18,** 917 (1962).
162. P. Hepp, *Ann. Chem.*, **215,** 360 (1882).
163. R. Foster, *Nature*, **176,** 746 (1955).
164. C. A. Lobry de Bruyn, *Rec. Trav. Chim.*, **2,** 205 (1883).
165. A. Russell and W. G. Tebbens in *Org. Syn.*, Coll. Vol. III, (Ed. E. C. Horning) John Wiley and Sons, New York, 1955, p. 293.
166. R. Foster and D. L. Hammick, *Nature*, **171,** 40 (1953).
167. B. Vickery, *Chem. Ind.* (London), **1967,** 1523.
168. E. Buncel, A. R. Norris, and W. Proudlock, *Can. J. Chem.*, **46,** 2759 (1968).
169. (a) A. R. Norris, *J. Org. Chem.*, **34,** 1486 (1969).
(b) A. R. Norris, *Can. J. Chem.*, **45,** 2703 (1967).
170. C. A. Fyfe, *Can. J. Chem.*, **46,** 3047 (1968).
171. M. R. F. Ashworth and E. Gramsch, *Mikrochim. Acta*, **1967,** 358.
172. P. Baudet, *Helv. Chim. Acta*, **49,** 545 (1966).
173. P. Caveng and H. Zollinger, *Helv. Chim. Acta*, **50,** 861 (1967).
174. R. Boulton, J. Miller, and A. J. Parker, *Chem. Ind.* (London), **1960,** 1026; **1963,** 492.
175. P. T. Izzo, *J. Org. Chem.*, **24,** 2026 (1959).
176. J. F. Bunnett and M. M. Rauhut, *J. Org. Chem.*, **21,** 944 (1956).

Author Index

Numbers in parentheses are reference numbers and show that an author's work is referred to although his name is not mentioned in the text. Numbers in italics indicate the pages on which the full references appear.

C

H

I

J

K

L

O

P

S

U

Subject Index

C

D

Q

R